Automotive Electricity

Wiley Automotive Series

Automotive Electricity

John Remling
Board of Cooperative Educational Services
Valhalla, New York

John Wiley & Sons
New York   Chichester
Brisbane   Toronto   Singapore

Cover painting by Edward A. Butler

The author would like to thank the following for the chapter opening illlustrations:

Chapter 4—courtesy of ESB, Inc.
Chapters 5, 8, and 11—courtesy of American Motors Corporation.
Chapters 6, 7, 12, and 13—courtesy of Ford Motor Company
Chapter 9—courtesy of Chrysler Corporation
Chapters 10 and 15—courtesy of Chevrolet Division, G.M.
Chapter 14—courtesy of Champion Spark Plug Co.

Library of Congress Cataloging in Publication Data:

Remling, John, 1928–
  Automotive electricity.

  (Wiley automotive series)
  Includes index.
  1. Automobiles—Electric equipment—Maintenance and
repair. 1. Title. II. Series.
TL272.R442   1987   629.2′548   86-11042
ISBN 0-471-80508-4 (pbk.)

Printed in the United States of America

10 9 8 7 6 5 4 3 2 1

Preface

Concerns for our environment and the need for energy conservation have brought about revolutionary changes in automotive design and construction. Those changes are most obvious to those in the automotive service industry. A service technician is required to perform increasingly varied diagnostic and repair procedures. And, because of their interdependency, a technician must also have a broader understanding of the operation of the various automotive systems.

This text provides the means by which the reader can gain the necessary knowledge and develop the necessary diagnostic and repair skills to become proficient in servicing the electrical systems of domestic and imported cars. Intended for use in automotive courses in post-secondary and technical education, it has been written so that it can easily be used in secondary vocational programs and by working technicians who wish to increase their knowledge and upgrade their skills.

Each chapter contains units of instruction that combine, in a developmental sequence, principles of operation, problem diagnosis, and repair procedures. Emphasis is placed on the information and knowledge required to develop salable skills. The basic principles of operation, the function of components, and the relationship of components are presented just prior to the diagnostic and repair procedures where that knowledge is needed. A progression of jobs provides the means by which the knowledge gained and the skills developed can be measured and evaluated.

At the end of each chapter is a self-test. The test items are similar in form and content to the items used in the ASE (Automotive Service Excellence) certification tests for Electrical Systems given by the National Institute for Automotive Service Excellence.

I am very grateful for the comments and suggestions of the many individuals who reviewed this text, and extend special thanks to the members of the Wiley staff, without whose help this book would not have been possible.

John Remling

To the Reader

The field of automotive repair service offers many opportunities for interesting, gainful employment. If you have the desire and the ability to be an auto mechanic, education and practical training will help you become a good one.

You may think that all it takes to be a good mechanic is the ability to remove and install a few parts—that all the skills needed are in your hands. It is true that a mechanic must have some highly developed hand skills. But hand skills are useless unless you know when and where to apply them. How will you know which part to replace or adjust? How will you know what to do to keep that part from failing again? As a mechanic you must check automotive systems to find the causes of many kinds of problems. But that is not enough. You must check related systems and parts to be sure that the problems will not occur again. Such checking requires a skill called *diagnosis*. Diagnosis is the basis for all repair, and it requires knowledge.

As a mechanic you must have a working knowledge of all the systems that make up an automobile. You must know the ways in which those systems work and relate to one another. Without that knowledge you cannot make an accurate diagnosis. Without an accurate diagnosis you may do unnecessary work, and you may replace parts that do not need replacing. As a mechanic you will use your knowledge to diagnose problems and to determine needed repairs. You will then use your hand skills to make the repairs.

As a mechanic you will need knowledge, diagnostic skills, and repair skills. Specifically, you will need

Knowledge of:

1 The function and operation of automotive systems and their parts
2 The names of parts
3 Automotive theory
4 Measurement and related mathematics and science
5 Hand tools and other equipment
6 Shop practices and safety

Diagnostic skills, such as:

1 Recognizing malfunctions
2 Isolating sources of trouble
3 Using test equipment
4 Interpreting test results
5 Analyzing failure
6 Evaluating completed repairs

Repair skills, such as:

1 Lubricating and adjusting parts
2 Repairing, overhauling, or replacing parts

This book can help you acquire some of this knowledge and some of those skills.

John Remling

Contents

*For the remaining chapters, we list in a developmental sequence, jobs whose learning objective is either knowledge (K), diagnostic skills (D), or repair skills (R).

Introduction

The electrical system is considered by many to be the most complicated and confusing of all the automotive systems. Although it is complicated, the electrical system is not at all confusing to a technician who has a knowledge of basic electrical fundamentals. With an understanding of some basic principles, you can determine the cause of problems in various electrical circuits and perform the required repairs.

This textbook, which is one of several in a series dealing with a broad range of automotive service areas, covers the automotive electrical system. It has been written to help you gain basic knowledge and to develop basic diagnostic and repair skills.

The goal of this book is to enable you to:

1 Service automotive electrical systems to meet industry standards.
2 Pass the ASE certification test for Electrical Systems given by the National Institute for Automotive Service Excellence (NIASE).

In studying this text, you will find many words with which you are familiar. But some of those words will have meanings that are different from those you already know. You will also find some words with which you are not familiar. Although many of those words are descriptive, what they describe is not always clear.

Like any other technical field, automotive service has a language of its own. A technician may use the language of a layperson, especially when explaining to a car owner the need for a certain repair. However, that same technician must understand and use the language of the trade when talking to a parts dealer or another technician and when reading a service manual.

In this book you will find the language of the trade easy to learn. When a new technical term is first used, it is printed in *italic* type. Also, a definition or explanation of the term is provided. To help yourself add the new term to your vocabulary, you should study the illustrations related to the text. Studying the illustrations will help you in three ways.

1 It will help you better understand the definition or explanation.
2 It will help you recognize various automotive parts when you see them on a car.
3 It will help you understand the function or operation of the parts and their relationship to one another.

Throughout this book you will be advised to consult the service manual for the car on which you are working. Not all car makers use the same components and circuits in their electrical systems. Many variations are used, even among cars built by the same manufacturer. Different components and different circuits require different service procedures. Service manuals provide you with drawings and photos that show where and how to perform certain service procedures on the particular make and model of car on which you are working. At times, a special tool is required, and its use is shown and explained. A special sequence of steps sometimes is given that will save you time. Precautions are given so that you may perform the procedures without damaging parts or injuring yourself.

Even if you know the correct procedure for any given job, you still need the car maker's service manual. It provides specifications. Specifications are measurements that must be taken before, during, and after assembling the components of automotive systems. Only by working to specifications can you be sure that the job is well done.

In most chapters of this book, you will find pages that present jobs for you to perform. Some of the jobs will require you to identify parts or the function of parts. Such jobs are really tests to help you determine whether you have learned the information presented in the text material that precedes them. They require you to furnish information within a certain time. Do not attempt to perform a job until you are sure you understand the material that precedes it.

Some jobs require that you perform diagnostic and repair operations on a component or in a circuit.

Those jobs are tests also. They test your skills in performing service procedures to the specifications of the car manufacturer. They require you to complete the procedure within a specified time. Those jobs should be started only after you have practiced them and have gained the necessary skills.

Each job is a checkpoint that you must pass to achieve your goals. It is placed at a point where you can measure the knowledge, diagnostic skills, or repair skills you have gained. Since the material that follows each job is built on the preceding material, you should use the job to determine whether or not you are ready to advance. If you do not achieve a satisfactory performance on a particular job, you know that you have not gained sufficient knowledge or skills to move on. You should review the material preceding the job and, when necessary, practice the procedures that revealed your weakness.

At the end of each chapter, you will find a self-test. The test is provided so that you can see how well you have learned the material presented in the chapter. All the self-test questions and incomplete statements are of the multiple-choice type. They are all similar in form and content to the questions used in the ASE certification tests given by NIASE. When taking the self-test, read each incomplete statement or question very carefully. Read each of the answers or completion choices and choose the one that best answers the question or best completes the statement. After responding to all the test items, check your responses against the answer key located at the back of the book. If you have chosen an incorrect response, review the appropriate material. The answer key also indicates where that material can be found.

Automotive Electricity

Chapter 1 Automotive Electrical System Service —An Overview

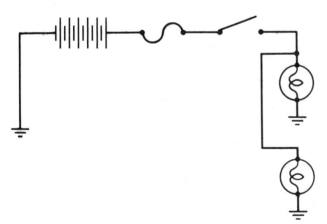

The electrical system is the most extensive system in an automobile. It consists of several subsystems with dozens of different circuits, and contains hundreds of various devices connected by thousands of feet of wire. Considered in its entirety, the electrical system appears complicated and confusing. However, when each circuit is viewed separately, it is seen to be rather simple. With an understanding of some basic principles of electricity, you can form a mental picture of each circuit.

This chapter will introduce you to the automotive electrical system, will acquaint you with some of the subsystems and circuits, and will explain what is required to gain knowledge of electrical fundamentals. It also will provide an overview of the diagnostic and repair skills that you will acquire from the proper use of this text.

This chapter also explains the need for service manuals and other reference materials, and explains why auto mechanics supply their own hand tools. Three Jobs are provided as your objectives in this chapter. Those Jobs are as follows:

A
Locate specifications in manufacturers' manuals
B
Locate specifications in a comprehensive service manual
C
Identify basic hand tools

AUTOMOTIVE ELECTRICAL SYSTEM SERVICE

The automotive electrical system consists of many different circuits. In most instances, service in those circuits must be based on diagnosis, and that diagnosis is based on knowledge. This text provides a means by which you can gain that knowledge and develop basic diagnostic and repair skills.

Basic Electricity

Electricity often is considered a mysterious form of energy because it is invisible. However, electricity follows certain natural laws. With a knowledge of those laws, you will be able to form a mental picture of what cannot be seen. This chapter presents some basic electrical principles to help you understand those laws and to use them to control electricity and to direct it to perform work.

The word *electricity* is derived from the word *electron*. An electron is a part of an atom, and an electrical current is the flow of electrons. For electrons to flow, a path or *circuit* must be provided. (See Figure 1.1.) That circuit must be made up of ma-

terials that will allow the flow of electrons. Those materials are called *conductors*. Materials that resist the flow of electrons are called *insulators*. In this chapter, you will learn about electron flow and why some materials are good conductors and why others are good insulators.

The amount of current that flows in a circuit can be measured in a unit of measurement called an *ampere*. (See Figure 1.2.) The amount of current that flows in a circuit is dependent on the pressure ap-

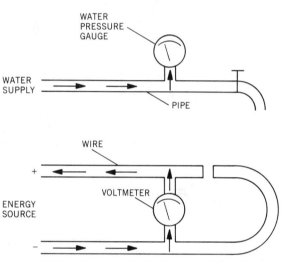

Figure 1.3 The voltage, or electrical pressure, in a circuit can be measured in much the same manner as pressure in a water pipe. You will develop skills in the use of a voltmeter and learn to interpret your readings by performing various jobs.

Figure 1.1 For current to flow, a path or circuit must be provided. In this simple circuit, current flows from the battery through a wire to the bulb. After flowing through the filament of the bulb, the current flows back to the battery through the return wire.

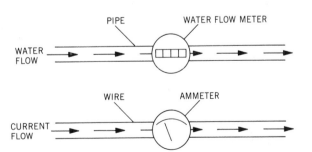

Figure 1.2 Just as the amount of water that flows through a pipe can be measured, the current that flows through a conductor can be measured. In this text, you will use an ammeter to measure current flow.

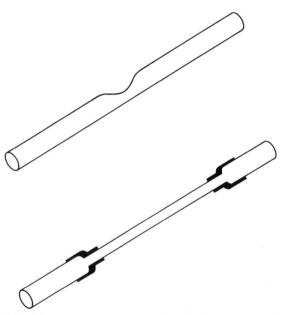

Figure 1.4 In a water system, a damaged or undersized pipe can restrict the flow of water. In an electrical circuit, a damaged or undersized wire offers resistance to the flow of current.

plied to the circuit and the resistance in the circuit. Pressure is measured in *volts* (Figure 1.3) and resistance is measured in *ohms* (Figure 1.4). Much of your work as an automotive electrician will require that you measure current flow, electrical pressure, and resistance. In this chapter, you will gain a working knowledge of those factors and their relationships. (See Figure 1.5.)

An automobile may have hundreds of circuits, but they all can be classified as *series circuits, parallel circuits,* or a combination of both series and parallel circuits. (See Figures 1.6, 1.7, and 1.8.) You will learn about the advantages and disadvantages of those circuits, and how those circuits are wired. You will also learn of the various components used in a circuit, and their functions in a circuit. (See Figure 1.9.)

Since an automobile contains so many circuits, tracing the wires in any one circuit would be difficult without a *wiring diagram.* A wiring diagram is similar to a road map in that you can use it to follow the route that current takes to flow from one part of a circuit to another. By studying the material in this chapter, you will develop an understanding of wiring diagrams and the symbols used in them. (See Figure 1.10.)

Basic Electrical Skills An automotive electrician must develop certain skills that are basic to all electrical work. Those skills include splicing and

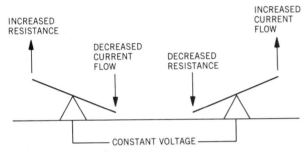

Figure 1.5 The relationships of current flow, electrical pressure, and resistance are formulated in Ohm's law. The basic concept of Ohm's law is illustrated by the resistance/current "see-saw." This text will help you to gain a working knowledge of those relationships.

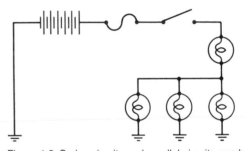

Figure 1.8 Series circuits and parallel circuits can be combined to form series-parallel circuits. In this text you will learn how various circuits are selected and wired for the jobs that they must perform.

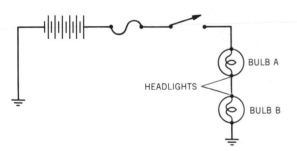

Figure 1.6 Here, a pair of headlights is connected in a series circuit. Current must flow through both bulbs before it can return to the battery. If either bulb A or bulb B burns out, the circuit will be broken. The remaining bulb then cannot operate.

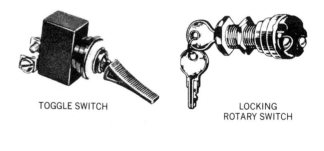

Figure 1.9 The switches shown here are just a few of the many different control devices in electrical circuits.

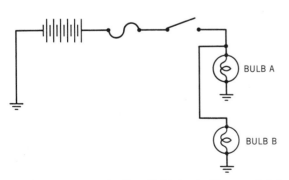

Figure 1.7 Here, a pair of headlights is wired in a parallel circuit. If either bulb A or bulb B burns out, the remaining bulb will not be effected.

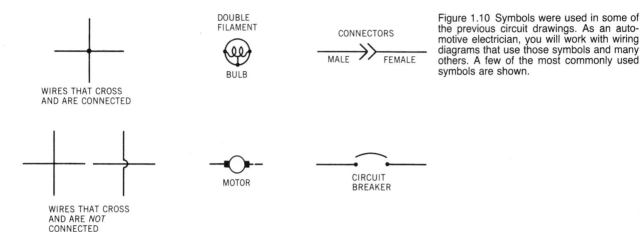

WIRES THAT CROSS
AND ARE CONNECTED

DOUBLE
FILAMENT

BULB

CONNECTORS

MALE   FEMALE

Figure 1.10 Symbols were used in some of the previous circuit drawings. As an automotive electrician, you will work with wiring diagrams that use those symbols and many others. A few of the most commonly used symbols are shown.

WIRES THAT CROSS
AND ARE *NOT*
CONNECTED

MOTOR

CIRCUIT
BREAKER

Figure 1.11 An automotive electrician often uses an ohmmeter to measure the resistance of various circuit components (courtesy of Snap-on Tools Corporation).

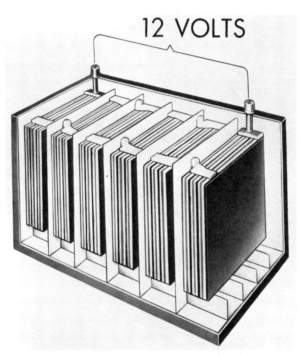

12 VOLTS

Figure 1.12 A 12-V battery is composed of six cells in a partitioned case (courtesy Chevrolet Motor Division).

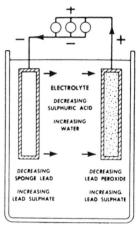

ELECTROLYTE

DECREASING
SULPHURIC ACID

INCREASING
WATER

DECREASING
SPONGE LEAD

DECREASING
LEAD PEROXIDE

INCREASING
LEAD SULPHATE

INCREASING
LEAD SULPHATE

DISCHARGING

Figure 1.13 A battery provides electrical energy through electrochemical action. This figure depicts the electrochemical action of a battery that is delivering energy (courtesy of Exide Corporation).

soldering wires and installing terminals and connectors. In this chapter, you will start to develop some of those skills by assembling jumper wires and a test light. Your knowledge of electricity will be reinforced when you learn how the *gauge,* or size, of wire is selected for a particular circuit. And you will be introduced to the use of *voltmeters, ammeters,* and *ohmmeters* as you take measurements in various circuits. (See Figure 1.11.)

The Battery—Operation, Construction, and Basic Services
The battery, shown in Figure 1.12, can be considered the primary energy source in the automotive electrical system. The mainte-

nance of the battery, its terminals, and related mounting parts requires an understanding of how a battery functions. In this chapter, you will learn about the construction of a lead-acid battery and the *electrochemical* action that takes place within a battery. (See Figure 1.13.) You also will perform basic battery services and you will learn the correct method of using a booster battery to start an engine. (See Figure 1.14.)

The Battery — Testing, Charging, and Replacement
As the battery must be considered in the diagnosis of all electrical problems, battery testing is a basic diagnostic procedure. This chap-

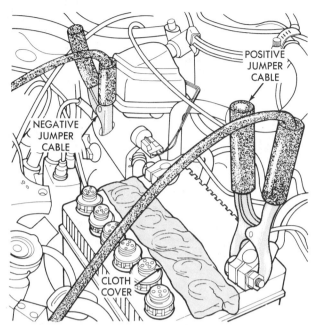

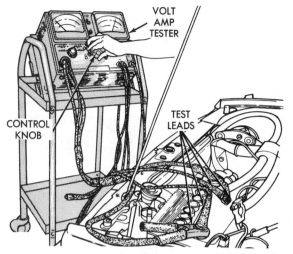

Figure 1.16 Load testing a battery (courtesy of Chrysler Corporation).

Figure 1.14 When attempting to jump start a car, the vent caps of both the discharged battery and the booster battery should be removed. The openings to the cells should be covered with cloth. The positive jumper cable should be connected to the positive terminal of the discharged battery. The negative cable should be connected to a good ground on the engine, *not* to the battery negative terminal (courtesy of Chrysler Corporation).

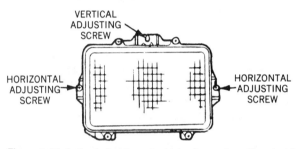

Figure 1.17 A simple but important job of an automotive electrician is the aiming of headlights. The procedures for this job are given in this text (courtesy of Chevrolet Motor Division).

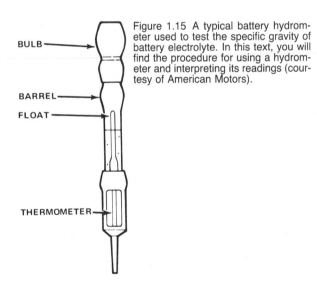

Figure 1.15 A typical battery hydrometer used to test the specific gravity of battery electrolyte. In this text, you will find the procedure for using a hydrometer and interpreting its readings (courtesy of American Motors).

ter will explain the need for different battery tests and provides procedures for testing the *specific gravity* of battery *electrolytes*. (See Figure 1.15.) As shown in Figure 1.16, you also will learn how to *load test* a battery and how to perform a *three-minute charge test.* Each test procedure is followed by an explanation of how the test results can be interpreted so that you can evaluate the condition of a battery and determine whether it should be recharged or replaced. Information on charging methods and rates is also provided.

The Lighting System — Basic Services In terms of safety, the lighting system is the most important electrical system. The lighting system consists of many different circuits, each serving a different function. The material in this chapter introduces you to some of those circuits and the components they contain, and provides procedures for performing some basic services. You will replace bulbs, lenses, flashers, and certain protection devices, and you will adjust the aim of headlights. (See Figures 1.17 and 1.18.)

The Lighting System — Advanced Services Building on the knowledge and skills you have already developed, you will now learn how to determine the causes of problems in lighting circuits. (See Figure 1.19.) In this chapter, you will further develop your diagnostic and repair skills by testing and replacing various switches, replacing *fuse links,* and working with *wiring harnesses.* (See Figures 1.20 and 1.21.)

Figure 1.18 A typical headlight aiming kit. Notice that the various adapters allow the aimers to be used on all types and sizes of headlights (courtesy of American Motors).

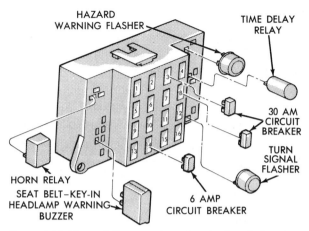

Figure 1.19 Many of the controls and protection devices in a lighting system are incorporated in a fuse block and relay module of the type shown (courtesy of Chrysler Corporation).

The Starting System — Basic Services In this chapter, you will learn how the starting system operates. You will learn about the circuits used and their various components. (See Figure 1.22.) Based on this knowlege, you will perform certain checks and diagnostic tests and replace defective parts. (See Figure 1.23.) Since the starter motor on many cars must be removed and installed from under the car, procedures for raising and supporting a car are included for your safety.

Starter Motor Construction and Overhaul The diagnosis and repair of starter motors requires a knowledge of how a motor converts electrical energy into mechanical energy. (See Figure 1.24.) This chapter provides the means by which you can gain an understanding of the principles of motor operation and of the function of starter motor components. (See Figure 1.25.) You will use this knowledge to inspect starter motor parts and to overhaul the starter motors most commonly used on domestic and imported cars. (See Figure 1.26.)

The Charging System — Basic Services The charging system provides the energy required to operate the various electrical systems while the engine is running, and maintains the state of charge of the battery. This chapter explains how the charging system operates and how each of the components in the system function. (See Figure 1.27.) Pro-

cedures are provided so that you can perform some basic diagnostic and repair services. In addition to performing basic charging system checks, you will learn how to adjust and replace drive belts and how to replace alternators and regulators. (See Figures 1.28 and 1.29.)

The Charging System — Testing and Alternator Overhaul Problems in the charging system can cause either undercharging or overcharging. In this chapter, you will learn how to perform several tests in the charging circuit that will enable you to isolate the causes of problems so that you can perform the needed adjustments or repairs. (See Figure 1.30.) As shown in Figures 1.31 and 1.32, you also will learn how to perform various inspections and tests of the internal parts of alternators to determine whether or not those parts are defective. You then will use your knowledge and skills to overhaul some of the most commonly used alternators.

Instrumentation — Warning Light and Gauge Circuit Service Warning lights and gauges are used to monitor the operation of a car. (See Figure 1.33.) This chapter will introduce you to the basic circuitry used to operate many of those lights and gauges. (See Figures 1.34 and 1.35.) You will learn about the various switches, sending units, and other controls used in those circuits. And you will learn how the most commonly used gauges function. (See Figure 1.36.) You will further develop your diagnostic and repair skills by testing warning light and gauge circuits and by replacing certain components.

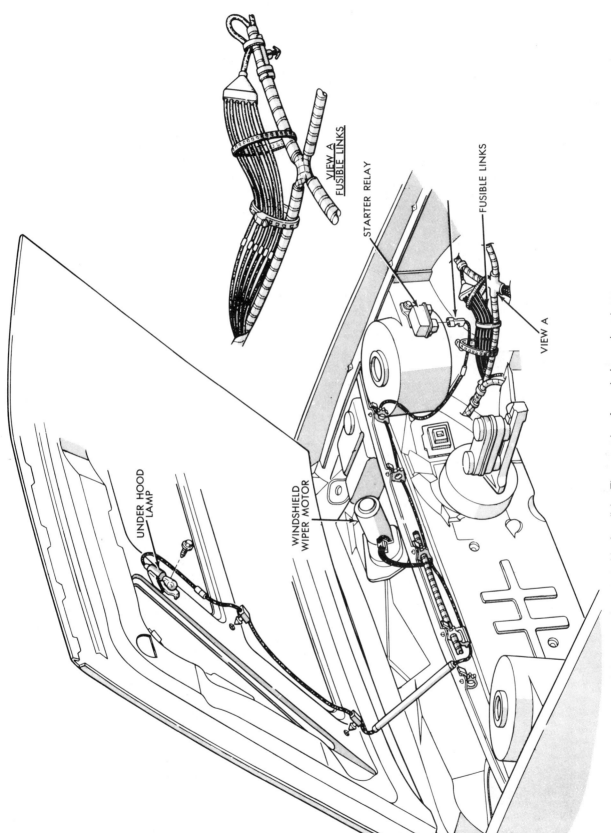

VIEW A
FUSIBLE LINKS

STARTER RELAY

FUSIBLE LINKS

VIEW A

UNDER HOOD LAMP

WINDSHIELD WIPER MOTOR

Figure 1.20 On many cars, the lighting system is protected by fuse links. The procedure for replacing various types of fuse links is included in this text (courtesy of Chrysler Corporation).

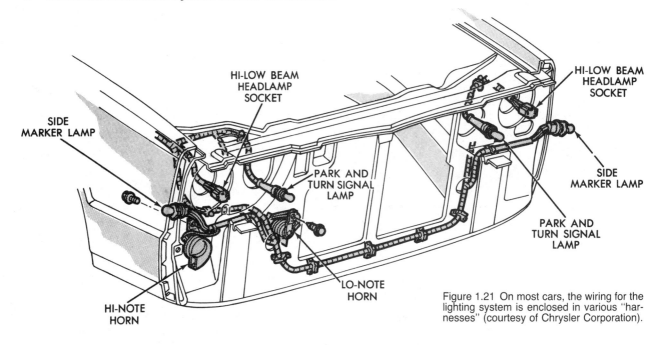

SIDE MARKER LAMP

HI-LOW BEAM HEADLAMP SOCKET

HI-LOW BEAM HEADLAMP SOCKET

PARK AND TURN SIGNAL LAMP

SIDE MARKER LAMP

PARK AND TURN SIGNAL LAMP

LO-NOTE HORN

HI-NOTE HORN

Figure 1.21 On most cars, the wiring for the lighting system is enclosed in various "harnesses" (courtesy of Chrysler Corporation).

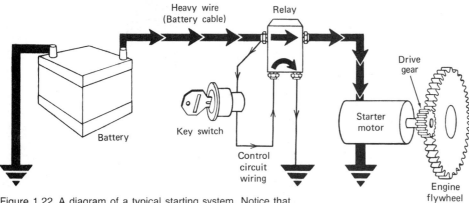

Heavy wire (Battery cable)

Relay

Drive gear

Battery

Key switch

Control circuit wiring

Starter motor

Engine flywheel

Figure 1.22 A diagram of a typical starting system. Notice that the system contains two circuits (courtesy of Chevrolet Motor Division).

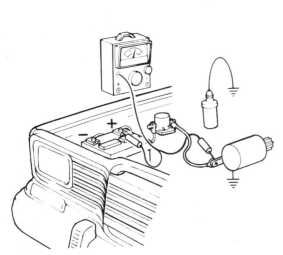

Figure 1.23 Meter connections for performing a voltage drop test of the "hot" side of the starting system motor circuit (courtesy of American Motors).

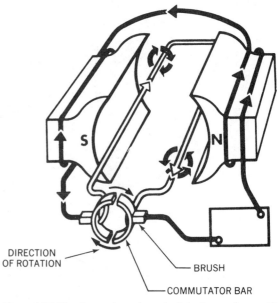

DIRECTION OF ROTATION

BRUSH

COMMUTATOR BAR

Figure 1.24 The diagnosis and repair of starting system problems requires a knowledge of motor operation. This text will explain how electrical energy is converted to mechanical energy through magnetism (courtesy of Chevrolet Motor Division).

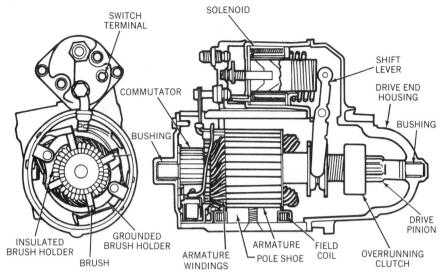

Figure 1.25 The main components of a starter motor (courtesy of Chevrolet Motor Division).

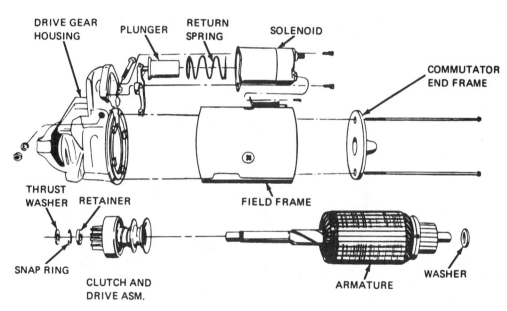

Figure 1.26 A disassembled view of a typical starter motor (courtesy of Chevrolet Motor Division).

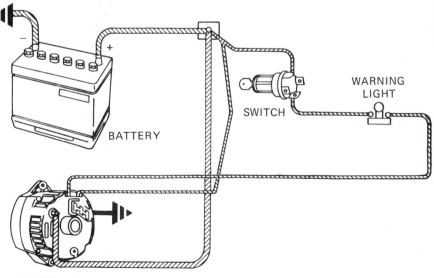

Figure 1.27 A typical charging system circuit used when the regulator is housed within the alternator (courtesy of Pontiac Motor Division, General Motors Corporation).

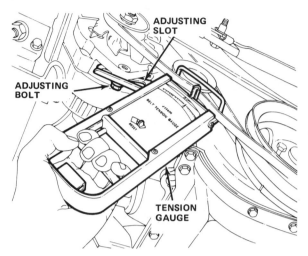

Figure 1.28 Checking alternator belt tension is an important step in the diagnosis of charging system problems (courtesy of American Motors).

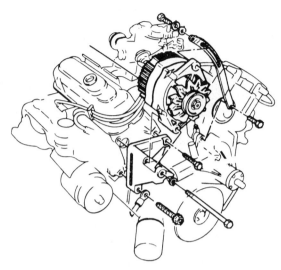

Figure 1.29 A typical alternator mounting on a V-type engine (courtesy of Chevrolet Motor Division).

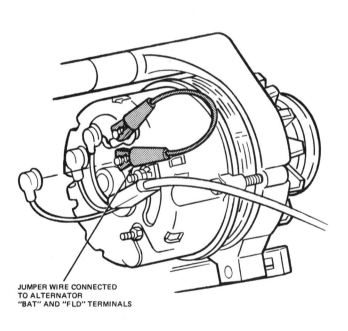

Figure 1.30 During certain alternator tests, a jumper wire is used to bypass the regulator. The procedures for those tests are provided in this book (courtesy of Ford Motor Company).

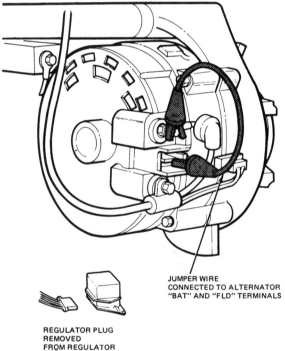

Accessories — Basic Services In recent years, car makers have made available a vast assortment of electrically operated accessories. This chapter covers the basic services required for those accessories most commonly installed as standard equipment. The circuits and the components used are described, their functions are explained, and diagnostic and repair procedures are provided. (See Figure 1.37.) You will gain additional skills by performing tests and making repairs in selected circuits. (See Figure 1.38.)

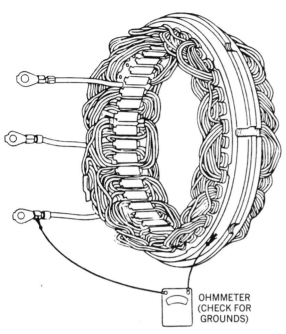

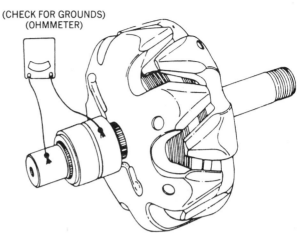

OHMMETER
(CHECK FOR
GROUNDS)

(CHECK FOR GROUNDS)
(OHMMETER)

Figure 1.31 Checking an alternator stator for grounds (courtesy of Chevrolet Motor Division).

Figure 1.32 During the overhaul of an alternator, an ohmmeter can be used to check the rotor for grounds (courtesy of Chevrolet Motor Division).

Figure 1.33 Warning lights found on a typical instrument panel cluster (courtesy of Chevrolet Motor Division).

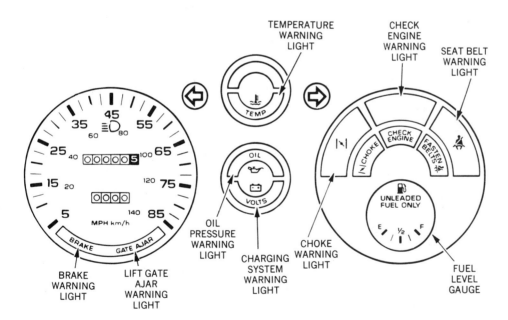

TEMPERATURE
WARNING
LIGHT

CHECK
ENGINE
WARNING
LIGHT

SEAT BELT
WARNING
LIGHT

BRAKE
WARNING
LIGHT

LIFT GATE
AJAR
WARNING
LIGHT

OIL
PRESSURE
WARNING
LIGHT

CHARGING
SYSTEM
WARNING
LIGHT

CHOKE
WARNING
LIGHT

FUEL
LEVEL
GAUGE

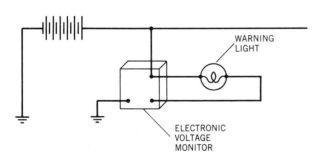

WARNING
LIGHT

ELECTRONIC
VOLTAGE
MONITOR

Figure 1.34 On some cars, a warning light is used to indicate low system voltage. An electronic unit monitors system voltage and sends current to the light when the system voltage drops below a predetermined value.

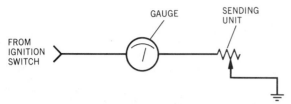

Figure 1.35 Many gauges are operated by a sending unit that contains a variable resistor. The amount of current that flows in the circuit determines the position of the pointer on the gauge face.

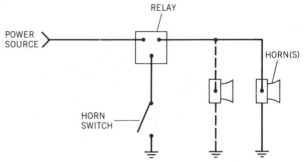

Figure 1.37 A simple horn circuit that contains a relay.

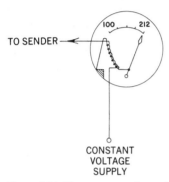

Figure 1.36 Many cars use thermal gauges. When current flows through a thermal gauge, a bimetallic strip bends, causing the pointer to move (courtesy of Chrysler Corporation).

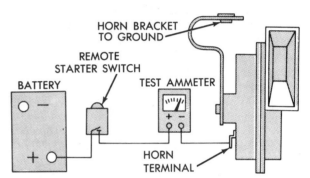

Figure 1.38 Connected as shown, an ammeter and a remote starter switch can be used to adjust a horn to the current draw specification of the manufacturer (courtesy of Chrysler Corporation).

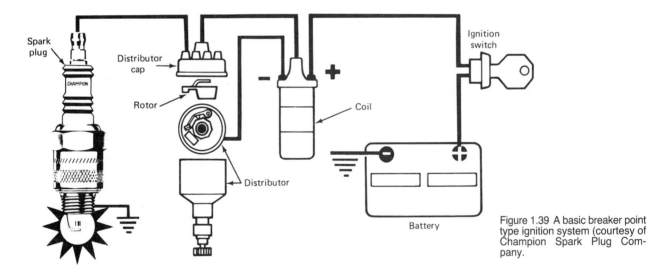

Figure 1.39 A basic breaker point type ignition system (courtesy of Champion Spark Plug Company.

The Ignition System—The Secondary Circuit   The ignition system traditionally has been considered part of the automotive electrical system. However, the use of electronic components and computerized circuitry has caused the latest ignition systems to be considered part of an "engine performance" system.

The conventional breaker-point ignition system, shown in Figure 1.39, was almost universally used in the past. In most instances, it now has been replaced by various electronic systems. A typical electronic system is shown in Figure 1.40. To provide acceptable performance from smaller engines while meeting mandated fuel economy and exhaust

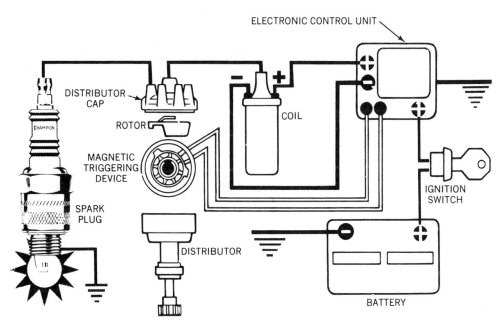

ELECTRONIC CONTROL UNIT

DISTRIBUTOR CAP

ROTOR

MAGNETIC TRIGGERING DEVICE

SPARK PLUG

COIL

DISTRIBUTOR

IGNITION SWITCH

BATTERY

Figure 1.40 A basic electronic ignition system (courtesy of Champion Spark Plug Company).

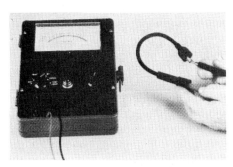

Figure 1.41 Using an ohmmeter to check the resistance of a spark plug wire (courtesy of American Motors).

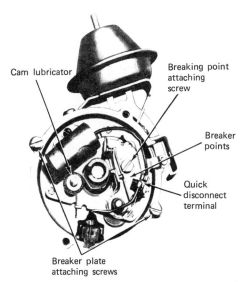

Cam lubricator

Breaking point attaching screw

Breaker points

Quick disconnect terminal

Breaker plate attaching screws

Figure 1.42 A conventional distributor that uses breaker points (courtesy of Chevrolet Motor Division).

emission standards, each car maker has developed systems that differ even among cars in their own product line.

This chapter will introduce you to basic ignition system circuitry and operation. You will learn about components that are used in most conventional and electronic systems. As shown in Figure 1.41, you will perform inspections and tests on those components used in the secondary circuit, and you will learn how to replace them.

The Ignition System—The Primary Circuit
The primary circuit is the control circuit of the ignition system. It provides current to the primary winding in the coil, it interrupts the flow of that current, and it times the occurrence of those two functions so that high voltage is induced in the secondary circuit at the exact instant that it is required.

Primary circuits using breaker points, shown in Figure 1.42, have similar circuitry and components. Common procedures are used in servicing those circuits. Primary circuits using electronic components, shown in Figure 1.43, are designed to meet the requirements of a particular application. Therefore, the components and the circuitry differ greatly, even in cars built by the same maker. When service is required in those circuits, a detailed procedure established by the manufacturer must be followed. Those procedures are not within the scope of this

ADVANCE WEIGHT (2)

MODULE

SPRING (2)

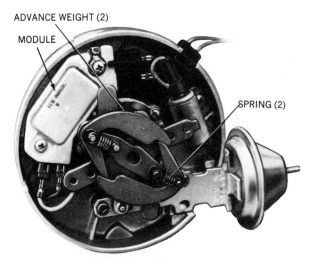

Figure 1.43 A distributor used in an electronic ignition system. The breaker points have been replaced by a module and a pickup coil assembly (courtesy of Chevrolet Motor Division).

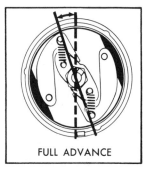

NO ADVANCE   FULL ADVANCE

Figure 1.44 Many distributors contain a centrifugal spark advance mechanism consisting of weights that can act to move the distributor cam. As engine speed increases, the weights move outward, "advancing" the position of the cam and thus causing the spark to be produced earlier (courtesy of Chevrolet Motor Division).

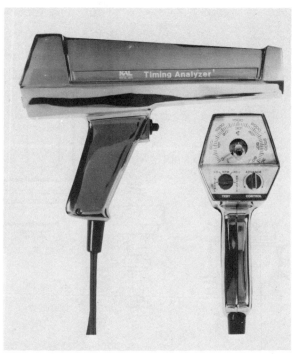

Figure 1.45 A timing light combined with a spark advance meter and a tachometer. This instrument will check initial spark timing and measure both centrifugal and vacuum spark advance (courtesy of Kal-Equip Company).

text and should be obtained from appropriate factory manuals.

This chapter covers basic primary circuit operation. You will learn how the components operate to create and collapse a magnetic field in the coil, and how the quality of that field is affected. And you will learn about spark advance and some of the various ways it is provided. (See Figures 1.44 and 1.45.) Procedures for inspecting and testing the components are given, and you will develop additional skills by repairing, replacing, and adjusting those parts.

TOOLS OF THE TRADE   The performance of automotive electrical system services requires more than knowledge, diagnostic skills, and repair skills. You also need tools. As an automotive electrician, you will require two different types of tools. The first type includes hand tools, power tools, in-

struments, and equipment. These tools enable your hands to perform diagnostic and repair procedures. The second type consists of service manuals and other reference materials. These tools tell you how those procedures should be performed and provide the specifications that must be met.

Service Manuals   Throughout this text, you will be advised to consult manuals to determine specific procedures and specifications. The various electrical systems used in all cars follow the same basic operating principles. However, those systems are not always the same. Many variations are used, even among cars built by the same manufacturer. Different systems require different diagnostic and repair procedures.

Service manuals provide drawings and photos that show where and how to perform certain procedures on the particular car on which you are working. At times, a special tool or instrument is required, and its use is shown and explained. Wiring diagrams help you to trace the various paths for current flow through a particular circuit. Sometimes a special sequence of steps for diagnostic and repair procedures is given that will save you time. Precautions are given so that you may perform the procedures without damaging parts or injuring yourself. (See Figure 1.46.)

IMPORTANT SAFETY NOTICE

Appropriate service methods and proper repair procedures are essential for the safe, reliable operation of all motor vehicles as well as the personal safety of the individual doing the work. This Shop Manual provides general directions for accomplishing service and repair work with tested, effective techniques. Following them will help assure reliability.

There are numerous variations in procedures, techniques, tools, and parts for servicing vehicles, as well as in the skill of the individual doing the work. This Manual cannot possibly anticipate all such variations and provide advice or cautions as to each. Accordingly, anyone who departs from the instructions provided in this Manual must first establish that he compromises neither his personal safety nor the vehicle integrity by his choice of methods, tools or parts.

NOTES, CAUTIONS, AND WARNINGS

As you read through the procedures, you will come across NOTES, CAUTIONS, and WARNINGS. Each one is there for a specific purpose. NOTES give you added information that will help you to complete a particular procedure. CAUTIONS are given to prevent you from making an error that could damage the vehicle. WARNINGS remind you to be especially careful in those areas where carelessness can cause personal injury. The following list contains some general WARNINGS that you should follow when you work on a vehicle.

- Always wear safety glasses for eye protection.
- Use safety stands whenever a procedure requires you to be under the vehicle.
- Be sure that the ignition switch is always in the OFF position, unless otherwise required by the procedure.
- Set the parking brake when working on the vehicle. If you have an automatic transmission, set it in PARK. If you have a manual transmission, it should be in FIRST. Place wood blocks of a 4″ x 4″ size or larger to the front and rear surfaces of the tires to provide further restraint from inadvertent vehicle movement.
- Operate the engine only in a well-ventilated area to avoid the danger of carbon monoxide.
- Keep yourself and your clothing away from moving parts, when the engine is running, especially the fan and belts.
- To prevent serious burns, avoid contact with hot metal parts such as the radiator, exhaust manifold, tail pipe, catalytic converter and muffler.
- Do not smoke while working on a vehicle.
- To avoid injury, always remove rings, watches, loose hanging jewelry, and loose clothing before beginning to work on a vehicle.
- If it is necessary to work under the hood, keep hands and other objects clear of the radiator fan blades! The electric cooling fan can start to operate any time by an increase in underhood temperature. For this reason care should be taken to ensure that the electric cooling fan motor is completely disconnected when working under the hood.

Figure 1.46 A typical safety notice as found in a service manual. A safe vehicle and a safe mechanic are of prime importance (courtesy of Ford Motor Company).

Even if you understand the circuitry in a particular system and know the correct procedure for a particular job, you still need a service manual. A service manual provides specifications. Specifications are measurements that must be taken before, during, and after assembling and connecting the components in an electrical system. Only by working to specifications can you be sure that the completed job has been performed correctly.

Two types of service manuals are in common use. One type includes the manuals published by car manufacturers. The other type includes comprehensive manuals published privately.

Manufacturers' Service Manuals Most car makers publish a complete manual or set of manuals for each model year of the cars they build. These manuals provide the best source of information for the cars they cover. The information in these manuals is arranged in groups or sections. As shown in Figure 1.47, each group or section covers a different system or area of repair. Each group or section has

1984 PONTIAC FIERO SERVICE MANUAL

This manual applies to all 1984 Pontiac Fiero models. It contains service information on all components of the car. Other information pertaining to the operation of the car is contained in the Owner's Manual which accompanies each vehicle.

All information, illustrations and specifications contained in this manual are based on the latest product information available at the time of publication approval. The right is reserved to make changes at any time without notice.

Any reference to brand names in this manual is intended merely as an example of the types of lubricants, tools, materials, etc., recommended for use in servicing 1984 Pontiac Fiero models. In all cases, an equivalent may be used.

PONTIAC MOTOR DIVISION
GENERAL MOTORS CORPORATION
PONTIAC, MICHIGAN 48053

S-8410P   © 1984 General Motors Corp.   Litho in U.S.A.

TABLE OF CONTENTS	
SECTION NAME	SECT. #
GENERAL INFORMATION A. General Information B. Maintenance and Lubrication C. Metrics	0
HEATING AND AIR CONDITIONING A. Heating and Ventilation B. Air Conditioning D2. Compressor Overhaul	1
FRAME AND BUMPERS A. Frame (Cradle) and Mounts B. Bumpers C. Body Panel Repair	2
STEERING, SUSPENSION, WHEELS AND TIRES A. Front and Rear End Alignment B2. Rack and Pinion Steering B4. Steering Wheels and Columns C. Front Suspension D. Rear Suspension E. Wheels and Tires	3
BRAKES	5
ENGINES 6. Engine Diagnosis A. 2.5 Liter L-4 B. Engine Cooling System C. Engine Fuel C13. Throttle Body Injection (Overhaul) D. Engine Electrical E2. Driveability and Emission F. Exhaust System	6
TRANSAXLE A. Automatic 125C. Automatic Transaxle Overhaul B. Manual C. Clutch D. Drive Axle	7
CHASSIS AND BODY ELECTRICAL A. Electrical Diagnosis B. Chassis Electrical C. Instrument Panel	8
ACCESSORIES	9
BODY	END OF MANUAL

Figure 1.47 The table of contents of a typical service manual published by a car manufacturer. Note that the material has been organized in sections (courtesy of Pontiac Motor Division, General Motors Corporation).

its own index to enable you to locate specific information. The pages within each section are numbered in a separate sequence.

Although all manufacturers' manuals are similar, each car maker uses different groupings and different methods of presentation. You should look through some of the manuals published by different manufacturers so that you will become familiar with these differences.

Sources of manufacturers' service manuals are listed in the Appendix.

Job 1a

LOCATE SPECIFICATIONS IN MANUFACTURERS' MANUALS

SATISFACTORY PERFORMANCE
A satisfactory performance on this job requires that you do the following:

1 Locate the specifications requested below for the two cars assigned.
2 Using appropriate manufacturers' manuals, complete the job within 30 minutes.
3 Fill in the blanks under "Information."

INFORMATION

Car #1

Vehicle identification _____

Reference used _____

Specification for	Specification	Page Number
Tail, stop, and turn signal light bulbs	_____	_____
Back-up light bulbs	_____	_____
License plate light bulb	_____	_____
Dome-light bulb	_____	_____
Hazard flasher fuse	_____	_____
Turn signal fuse	_____	_____
Heater blower motor fuse	_____	_____

Car #2

Vehicle identification _____

Reference used _____

Specification for	Specification	Page Number
Tail, stop, and turn signal light bulbs	_____	_____
Back-up light bulbs	_____	_____
License plate light bulb	_____	_____
Dome-light bulb	_____	_____
Hazard flasher fuse	_____	_____
Turn signal fuse	_____	_____
Heater blower motor fuse	_____	_____

Comprehensive Service Manuals   The most commonly used specifications and procedures for the most frequently performed service operations are compiled in these manuals. Various editions are available, covering different ranges of model years for both domestic and imported cars. Published yearly, the manuals provide the best single source of reference material you will need in your daily work.

Comprehensive service manuals usually are divided into two major sections. The first section contains procedures and specifications that apply to specific makes and models of cars. This section is indexed alphabetically by car name. The second section contains general service information and procedures.

The material in manuals from different publishers is arranged and presented differently. All the manuals are easy to use when you are familiar with their differences. You should look through the manuals you have available to become acquainted with them.

The titles and publishers of the most commonly used comprehensive service manuals are listed in the Appendix. It is advisable that you obtain manuals of this type for your reference library.

Job 1b

LOCATE SPECIFICATIONS IN A COMPREHENSIVE SERVICE MANUAL

SATISFACTORY PERFORMANCE
A satisfactory performance on this job requires that you do the following:

1 Locate the specifications requested below for the two cars assigned.
2 Using a comprehensive service manual, complete the job within 30 minutes.
3 Fill in the blanks under "Information."
INFORMATION

Car #1

Vehicle identification _____

Reference used _____

Specifications for	Specification	Page Number
Battery load test (A)	_____	_____
Spark plug gap	_____	_____
Regulated charging voltage at 80°F (27°C)	_____	_____

Car #2

Vehicle identification _____

Reference used _____

Specifications for	Specification	Page Number
Battery load test (A)	_____	_____
Spark plug gap	_____	_____
Regulated charging voltage at 80°F (27°C)	_____	_____

Manuals, Catalogs, and Other Reference Material Published by Manufacturers of Parts, Tools, Instruments, and Equipment The makers of the parts, tools, instruments, and equipment that you will use publish a wide variety of reference material. Much of that material contains information not only about the items that they produce, but about the systems and circuits in which they are used. In most instances, these materials supplement the information found in manufacturers' service manuals and in comprehensive service manuals.

Manufacturers offer material free of charge upon request. You can assemble a very useful automotive reference library by taking advantage of these offerings. A partial listing of these manufacturers is provided in the Appendix.

Hand Tools It is almost universally accepted that an automotive electrician must supply personally owned hand tools. There are many reasons for this requirement. Hand tools enable you to perform tasks that exceed your physical limitations. You cannot loosen and tighten nuts, bolts, and screws with your fingers, nor can you cut and strip wire. However, you can easily perform those operations by using wrenches, screwdrivers, cutting pliers, and wire strippers.

The need for hand tools is obvious—what may not be obvious is the relationship between the tool and the hand that holds it. This relationship is one of the main reasons why automotive electricians supply their own tools. When you use a tool repeatedly, you become accustomed to that particular tool. Each tool has its own "feel." This "feel" is developed between your hands and that particular tool. It is because of that "feel" that baseball players select particular bats and musicians select particular instruments.

As you repeatedly perform a certain operation, you become skilled in that performance. However, some of the skill you develop is in the use of the particular tool you handle. If you own the tools you use, you will develop skills more rapidly and more easily. You also will have control over one of the variables that affects the application of those skills.

There are other reasons why automotive electricians supply their own tools. A mechanic must not only complete jobs to specifications, but must complete them within a reasonable amount of time. In most shops, your wages will be based not only on the quality of the work performed, but on the amount of work completed. With personally owned tools, you save the time that would be lost in attempting to borrow tools from another mechanic or from a

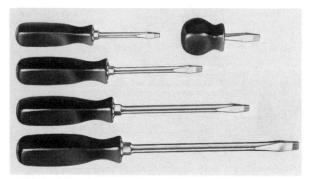

Figure 1.48 A set of screwdrivers of varying sizes in necessary for automotive electrical work (courtesy of Snap-on Tools Corporation).

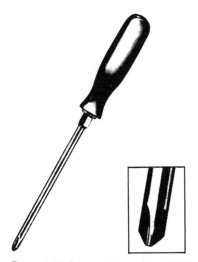

Figure 1.49 A screwdriver with a Phillips tip (courtesy of Snap-on Tools Corporation).

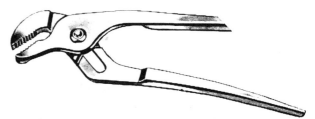

Figure 1.50 A pair of interlocking jaw pliers. Pliers of this type are also referred to as water pump pliers (courtesy of Snap-on Tools Corporation).

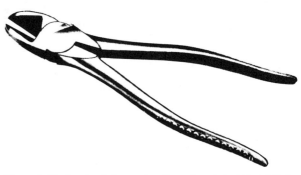

Figure 1.51 A pair of diagonal cutting pliers (courtesy of Snap-on Tools Corporation).

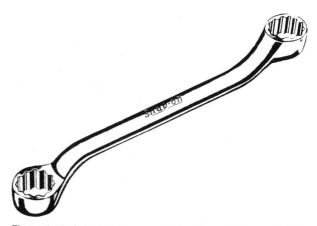

Figure 1.52 A typical box wrench (courtesy of Snap-on Tools Corporation).

toolroom. Most mechanics, having purchased their own tools, refuse to loan them to others; and most shop owners do not provide a toolroom.

Responsibility is another reason for tool ownership. When you perform a job, you must accept responsibility for it. If you are not willing to accept responsibility for supplying your own tools, you may not be ready to accept responsibility for the jobs you perform.

Pride is probably the best reason for tool ownership. You should be proud of your abilities and of the skills that you will develop. If you feel that you are a good mechanic, you will want to work with good tools. Many employers check a mechanic's toolbox before hiring the mechanic. Quality tools, kept clean and properly maintained, are considered a sign of a professional mechanic.

The Basic Tool Kit   There is no one set of tools that will meet the needs of every situation. Rather than attempt to purchase a complete set of tools, it usually is better to purchase individual tools or small sets as you need them. It is recommended that you purchase tools of high quality. Cheap tools have a short service life and often distort, slip, or break while in use. A tool that fails can damage the parts you are servicing and can cause you injury. The following tools are suggested for your consideration:

Figure 1.53 A typical open end wrench (courtesy of Snap-on Tools Corporation).

Figure 1.54 A typical combination wrench (courtesy of Snap-on Tools Corporation).

AMERICAN	METRIC	DECIMAL SIZE
⁵⁄₁₆*		0.3125
	8	0.3150
	9	0.3543
³⁄₈*		0.3750
	10*	0.3937
	11	0.4331
⁷⁄₁₆*		0.4375
	12*	0.4724
¹⁄₂*		0.5000
	13*	0.5118
	14*	0.5512
⁹⁄₁₆*		0.5625
	15	0.5906
⁵⁄₈*		0.6250
	16	0.6299
	17	0.6693
¹¹⁄₁₆		0.6875
	18	0.7087
	19*	0.7484
³⁄₄		0.7500
	20	0.7874
¹³⁄₁₆		0.8125
	21	0.8268
	22	0.8661
⁷⁄₈		0.8750
	23	0.9055
¹⁵⁄₁₆		0.9375
	24	0.9449
	25	0.9843
1		1.0000

Figure 1.55 The progression of American and metric wrench sizes. The sizes marked with an * are suggested for a basic tool kit.

Screwdrivers Screwdrivers, shown in Figure 1.48, are always needed, but two or three of the smaller sizes will be sufficient for your immediate needs. While they should have a standard tip for slotted screws, you also will need a screwdriver with a #2 Phillips tip, as shown in Figure 1.49. Additional screwdrivers can be obtained as they are needed.

Pliers There are hundreds of different types of pliers available, and eventually you will own quite a few of them. As a start, you will find that two will be used most often. A pair of *interlocking jaw pliers* that are about eight or nine inches long is especially handy. (See Figure 1.50.) Most of your cutting needs can be handled by a pair of *diagonal cutters* that are about 7 in. long. (See Figure 1.51.)

Wrenches You eventually will find that you need both *box wrenches* and *open end wrenches*. A box wrench, shown in Figure 1.52, completely encircles

a nut or the head of a bolt. A wrench of this type is not apt to slip and cause damage or injury. An open end wrench, as its name implies, has an open end, as shown in Figure 1.53. An open end wrench often can be used where a box wrench cannot be used, but an open end wrench is more liable to slip. It is suggested that you purchase *combination wrenches*. As shown in Figure 1.54, a combination wrench has a box wrench on one end and an open end wrench on the other end.

Wrenches are sized in two systems. Most older domestic cars were built with bolts and nuts that require wrenches of American standard sizes. These wrenches are most commonly found in increments of 1/16 of an inch. Most imported cars and newer domestic cars require metric-sized wrenches. These wrenches are made in increments of 1 millimeter (mm). Figure 1.55 shows the progression of wrench sizes and indicates the sizes that you will find most

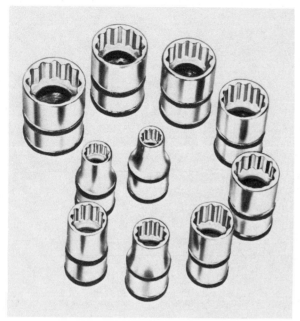

Figure 1.56 A typical set of sockets (courtesy of Snap-on Tools Corporation).

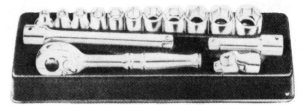

Figure 1.57 A basic 3/8-inch drive socket set complete with a ratchet, two extensions, and a universal joint (courtesy of Snap-on Tools Corporation).

Figure 1.58 A typical spark plug socket. Notice that the socket has a hex-shaped head so that it also can be turned with a box wrench or an open end wrench (courtesy of Snap-on Tools Corporation).

useful. To be able to handle the jobs in this text, you should invest in combination wrenches of the most commonly used sizes. Additional wrenches can be purchased as they become needed.

Sockets   A set of sockets, shown in Figure 1.56, combined with a ratchet handle and a few extensions, will enable you to perform many jobs that cannot be performed with combination wrenches. Although the individual tools can be purchased separately, it usually is more economical to purchase a small set similar to the one shown in Figure 1.57.

The socket set should be of 3/8-in. drive. While you eventually will need 1/4-in. drive and 1/2-in. drive sets, the purchase of those sets can be postponed.

As the socket sets usually include either American standard or metric sockets, additional sockets should be purchased so that you have the most commonly used sizes. Spark-plug sockets for 5/8-in. and 13/16-in. spark plugs should be added to the set if they are not included. A typical spark-plug socket is shown in Figure 1.58.

Additional tools should be added to your basic kit as you find that you need them. Where special tools are used in the procedures in this text, those tools are described and shown. It is suggested that you add tool catalogs to your reference library. A study of those catalogs will help you to select the tools that you need. Several manufacturers of automotive tools are listed in the Appendix.

Job 1c

IDENTIFY BASIC HAND TOOLS

SATISFACTORY PERFORMANCE
A satisfactory performance on this job requires that you do the following:

1 Identify the hand tools on page 23 by placing the number of each tool in front of the correct tool name.
2 Complete the job within 10 minutes.

PERFORMANCE SITUATION

_____ Spark plug socket   _____ Ratchet

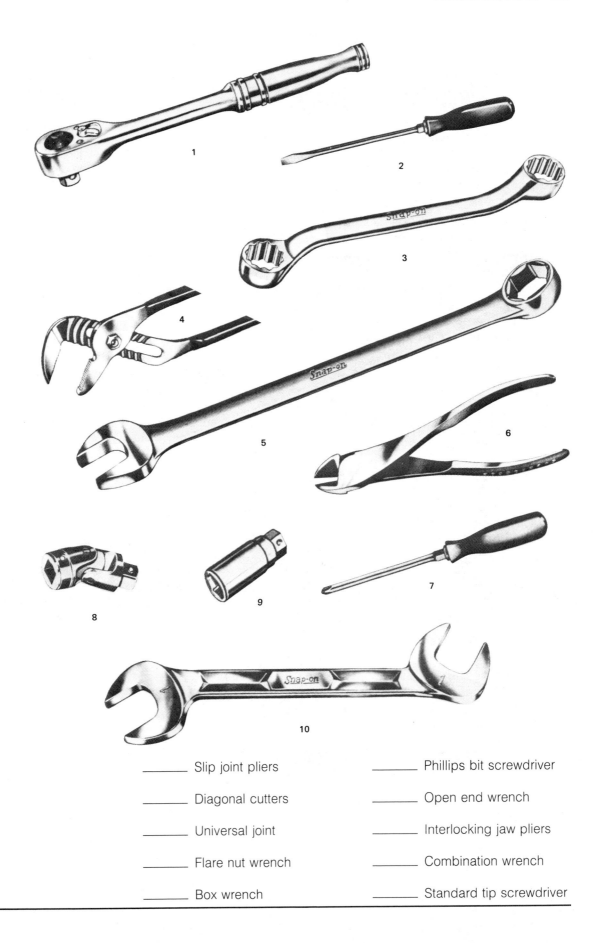

_____ Slip joint pliers

_____ Diagonal cutters

_____ Universal joint

_____ Flare nut wrench

_____ Box wrench

_____ Phillips bit screwdriver

_____ Open end wrench

_____ Interlocking jaw pliers

_____ Combination wrench

_____ Standard tip screwdriver

SUMMARY

By completing this chapter, you learned of some of the jobs that are performed by an automotive electrician. Those jobs consist of inspections, tests, adjustments, and the replacement of parts. You also learned that service manuals and other reference materials are required to provide the correct procedures for those jobs. And that even when the correct procedures are known, manuals must be used to obtain the specifications needed.

You learned that to perform those jobs correctly and within a reasonable amount of time, knowledge, diagnostic skills, and repair skills are required. That knowledge must be gained by study, and those diagnostic and repair skills must be developed through practice and performance. If you wish to become an automotive electrician, you must accept the responsibility of gaining that knowledge and developing those skills. If you accept that responsibility, you are well on your way toward your goal.

SELF-TEST

Each incomplete statement or question in this test is followed by four suggested completions or answers. In each case select the *one* that best completes the sentence or answers the question:

1 The amount of current that flows in a circuit is measured in
   A. ohms
   B. volts
   C. watts
   D. amperes

2 The electrical pressure that pushes current through a circuit is measured in
   A. ohms
   B. volts
   C. watts
   D amperes

3 The primary energy source in the automotive electrical system is the
   A. coil
   B. battery
   C. distributor
   D. starter motor

4 Two mechanics are discussing the cable connections for using a booster battery to start a car that has a discharged battery.
   Mechanic A says that the positive jumper cable should be connected to the positive battery terminal.
   Mechanic B says that the negative jumper cable should be connected to the engine away from the battery.
   Who is right?
   A. A only
   B. B only
   C. Both A and B
   D. Neither A nor B

5 A starter motor converts

   A. heat energy to electrical energy
   B. chemical energy to electrical energy
   C. mechanical energy to electrical energy
   D. electrical energy to mechanical energy

6 Two mechanics are discussing the charging system.
   Mechanic A says that the charging system provides the energy that operates the various electrical systems while the engine is running.
   Mechanic B says that the charging system maintains the state of charge of the battery.
   Who is right?
   A. A only
   B. B only
   C. Both A and B
   D. Neither A nor B

7 Two mechanics are discussing service manuals.
   Mechanic A says that manufacturers' manuals are usually published for each model year.
   Mechanic B says that comprehensive service manuals contain service procedures and specifications for many model years and for many different cars.
   Who is right?
   A. A only
   B. B only
   C. Both A and B
   D. Neither A nor B

8 This text provides a list of sources of manuals and other reference material. Where is that list located?
   A. The Appendix
   B. The Glossary
   C. The Bibliography
   D. The Introduction

9 The Phillips screwdriver most commonly used by an automotive electrician has a
   A. #1 tip
   B. #2 tip
   C. #3 tip
   D. #4 tip

10 Which of the following tools combines the functions of a box wrench and an open end wrench?
   A. A torque wrench
   B. A socket wrench
   C. An adjustable wrench
   D. A combination wrench

Chapter 2 Basic Electricity– Principles and Circuits

As an automotive electrician, you must have a basic understanding of what electricity is and how it can be used. Since it cannot be seen, electricity often is considered a mysterious form of energy. It is the purpose of this chapter to dispel some of that mystery.

Electricity follows certain natural laws. With the knowledge of some basic principles, you will understand those laws and will be able to control electricity and direct it to perform work. With that knowledge, you will form a mental picture of what cannot be seen.

This chapter will explain certain theories regarding current flow and will introduce you to voltage, amperage, and resistance. You then will work with circuits to reinforce your learning. Your specific objectives are to perform the following Jobs:

A
Identify terms relating to basic electricity
B
Identify terms relating to electrical circuitry
C
Identify circuit parts and their definitions
D
Complete wiring diagrams of basic circuits
E
Determine electrical pressure, current flow, and resistance in electrical circuits
F
Identify symbols used in automotive wiring diagrams
G
Complete automotive wiring diagrams

BASIC ELECTRICITY Your first step in becoming an automotive electrician is to gain an understanding of electricity. Once you know what electricity is and how it works, you can correctly diagnose and repair problems in electrical systems. Since you cannot see electricity, you must form a mental picture of what is happening in an electrical circuit. The clarity of that picture will depend on your knowledge of a few scientific principles.

Matter and Its Atomic Structure *Matter* often is defined as anything that has weight and occupies space. If we accept that definition, it follows that everything in the universe is made of matter. Matter can exist in three states: solid, liquid, or gas. Figure 2.1 illustrates the three states of matter.

All matter is made up of *elements*. An element can be anything that is considered "pure" in that it is not combined with anything else. Some elements

that you may be familiar with are copper, iron, hydrogen, and oxygen. While over 100 elements have been identified, most matter is made up of several elements that are combined. For example, water is made up of a combination of two elements, hydrogen and oxygen.

Elements are made up of *atoms*. An atom is the smallest part of an element that contains the identifiable properties or characteristics of that element. Although atoms are quite small—a drop of water contains billions of atoms—they are composed of still smaller particles, called *protons, electrons,* and *neutrons*. Protons have a positive (+) electrical charge and electrons have a negative (−) electrical charge. This difference in charge forms the basis of what we call electricity. Neutrons can be considered neutral because they have no electrical charge.

The structure of an atom can be likened to the structure of our solar system. Protons and neutrons form the *nucleus,* or center, of an atom in the same way that the sun forms the center of our solar system. Electrons whirl rapidly in orbits around the nucleus just as the planets rotate around the sun. (See Figures 2.2 and 2.3.)

You probably have handled magnets and noticed that opposite magnetic poles attract each other while like poles repel each other. (See Figure 2.4.) Since they have opposite charges, protons and electrons are attracted to each other in much the same way. While neutrons are electrically neutral, they form an important function in the atom. Since all protons have a positive (+) charge, protons repel each other. The presence of neutrons cancels out their repelling forces and enables a quantity of protons to

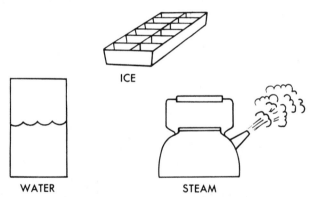

ICE

WATER   STEAM

Figure 2.1 Matter can exist in three states. As an example, water can be found as a liquid, a solid, or a gas (courtesy of General Motors Corporation).

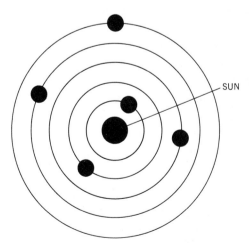

SUN

Figure 2.2 In the solar system, planets rotate around the sun in orbits.

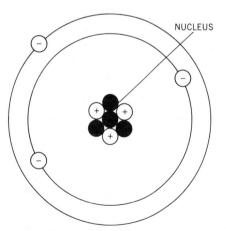

NUCLEUS

Figure 2.3 The structure of an atom. The lithium atom shown has three protons and four neutrons in the nucleus. The three electrons rotate in two orbits.

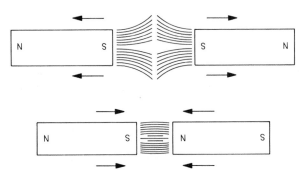

Figure 2.4 Opposite poles of magnets will attract each other while like poles will repel each other. Similar attraction and repulsion occurs with protons and electrons. Protons and electrons attract each other while protons repel protons and electrons repel electrons.

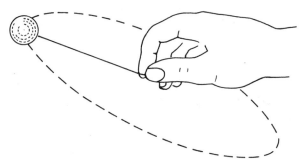

Figure 2.6 The balance of forces in an atom can be demonstrated by spinning a weight attached to a rubber band (courtesy of General Motors Corporation).

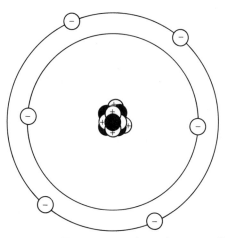

Figure 2.5 The structure of a carbon atom. The six electrons rotate on two orbit rings and are kept apart by their repulsion.

remain together in the nucleus of an atom. Electrons in an atom are not affected by neutrons, so by repelling each other they tend to position themselves alone in their orbits, as shown in Figure 2.5.

Because of the speed at which electrons are spinning around protons in the nuleus, *centrifugal force* keeps them from being pulled inward. Centrifugal force is the force that tends to pull a rotating body away from its center of rotation. It is the same force that keeps the planets in our solar system from being pulled into the sun.

The balance of forces that keeps electrons in orbit can easily be demonstrated by attaching a weight to a rubber band and spinning it, as shown in Figure 2.6. As the weight rotates around your fingers, centrifugal force pulls it outward, away from your fingers, and stretches the rubber band. The rubber band attempts to pull the weight inward, toward your fingers. If you rotate the weight at a constant speed, the two opposing forces will be balanced and the weight will remain in a constant orbit.

Increasing the speed of rotation will cause the weight to move outward until its centrifugal force is balanced by the increased force of the stretched rubber band. If that increased speed is maintained, the weight will rotate in a larger orbit. If the weight is spun so fast that the rubber band breaks, the centrifugal force will cause the weight to fly out of its orbit. In much the same manner, electrons can leave their atoms.

In most atoms, the number of protons is equal to the number of electrons. This provides an electrical balance in that the strength of the positive (+) charge of the protons is equal to the strength of the negative (−) charge of the electrons. This balance can be upset if some external force causes some of the electrons to leave their atoms. When this occurs, the atoms become unbalanced because they have more protons than electrons. They become positively (+) charged. Unbalanced atoms try to regain their balance by attracting electrons from neighboring atoms. (See Figure 2.7.) This exchange of electrons is what we call an *electric current.*

The structure of the atoms of each element is different, but not only because of the number of protons and electrons. Atoms can have from one to seven orbit rings, and the number of electrons in each ring may vary. The number of electrons in the outer orbit ring, or *valence ring,* of an atom determines the ability of an element to conduct electricity.

If the outer ring of an atom contains less than four electrons, there is room for more electrons. In addition, those electrons are held rather loosely. This allows a stray or drifting electron to join the ring and to displace another electron. Therefore, the loosely held electrons in the outer ring are called *free electrons.* Figure 2.8 illustrates the free electron of a copper atom. When the atoms of an element have free electrons, the passage of electrons from one

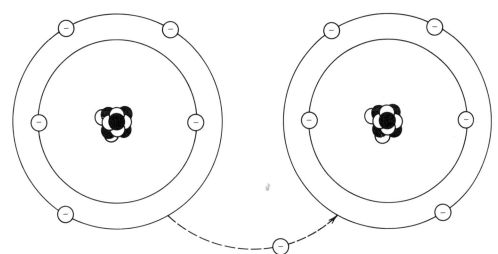

Figure 2.7 When an electron leaves an atom, the atom becomes positively charged and is thus unbalanced. An unbalanced atom will attract an electron from a neighboring atom to regain its balance.

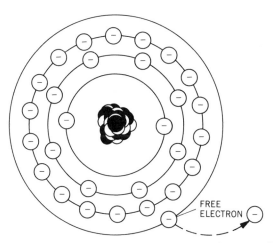

Figure 2.8 A copper atom has 29 electrons. Its outer orbit contains only one electron, which is loosely held and can easily leave its orbit ring.

atom to another is relatively easy. This ease of electron flow makes the element a good *conductor*. A conductor is a material that allows the easy flow of electrons. Copper is a good conductor, and most of the wiring in automotive electrical systems is made of that material.

When five or more electrons are present in the outer orbit ring of an atom, they are called *bound electrons*. This is because the ring is fairly full, and the electrons are held tighter. It is very difficult for a stray electron to enter the orbit ring of bound electrons. When the atoms of an element have many bound electrons, electron flow is restricted and the element is considered to be an *insulator*. An insulator is a poor conductor of electricity. Certain plastics and rubber compounds are used as insulators in automotive electrical systems.

If the outer ring of an atom contains four electrons, its element has certain properties of both conductors and insulators. Those elements are called *semiconductors*. Semiconductors are found in many automotive electronic components.

Current Flow In materials containing atoms with free electrons, random electron flow is constantly taking place. The free electrons enter and leave the outer orbit rings of the atoms as the atoms attempt to maintain an electrical balance. This is not considered a true flow of electrical current because the electron flow is neither concentrated nor directed from one point to another. Only when a concentration of electrons begins to move in one direction can that action be called a flow of current.

For electrons to flow, they must be able to move from one atom to the next. This ability is provided in a conductor. Figure 2.9 illustrates current flow in a copper wire. To cause the electrons to move from one atom to the next, a positive (+) charge must be present at one end of the conductor and a negative (−) charge at the opposite end. In Figure 2.9, the charges are supplied by a flashlight battery.

The free electron of the atom nearest the positive end of the wire is attracted to the positive charge and leaves the atom. As the atom becomes unbalanced, it attracts the free electron from its adjacent atom. This atom in turn attracts a free electron from the next atom in line, and the give-and-take action continues through the conductor. At the negative end of the wire, the negative charge repels the free electrons, and they in turn move through the wire passing from atom to atom, making up for the electrons that leave from the positive end of the wire. This electron flow can be likened to a chain passing through a pipe. As one link leaves the pipe, another

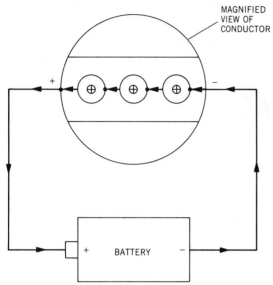

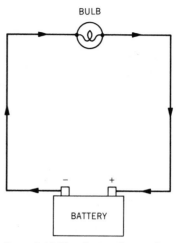

Figure 2.11 The electron theory of current flow states that current flows from negative (−) to positive (+).

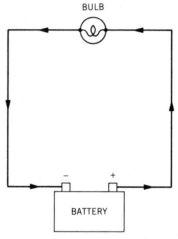

Figure 2.12 The conventional theory of current flow states that current flows from positive (+) to negative (−).

Figure 2.9 Current flow through a conductor. The positive (+) charge attracts electrons from the atoms which in turn attract electrons from adjoining atoms. The negative (−) charge repels electrons, restoring those given up by the atoms.

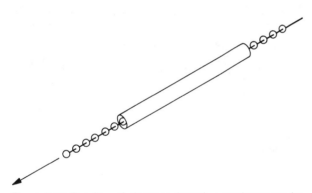

Figure 2.10 The flow of electrons through a conductor can be likened to a chain being pulled through a pipe.

link enters. (See Figure 2.10.) The flow of electrons through a conductor will continue as long as the positive (+) and negative (−) charges are present at the ends of the conductor.

The Electron Theory   The method of current flow explained in the previous paragraphs is known as the *electron theory*. This theory is relatively new, having been proposed less than 100 years ago. In essence, it states that current flows from negative (−) to positive (+), as shown in Figure 2.11.

The Conventional Theory   While the electron theory is the basis of many electronic circuits, most automotive electricians "think through" various circuits by using the *conventional theory* of current flow. The conventional theory holds that current flows

from positive (+) to negative (−), as shown in Figure 2.12.

When working on automotive electrical systems, you can use either theory as long as you use it consistently. As you will discover, the negative battery terminal is grounded on most cars. Because of this, it is usually easier to use the conventional theory in your work.

ELECTRICAL CIRCUITRY   As previously stated, electrons will not move through a conductor unless the conductor provides a complete circular path for their flow. (Refer to Figure 2.9.) That complete circular path is called a *circuit*. Since an electric current is only a flow of electrons, a complete, unbroken circuit is required if current is to flow. Figure 2.13 shows a simple circuit that contains a source

Job 2a

IDENTIFY TERMS RELATING TO BASIC ELECTRICITY

SATISFACTORY PERFORMANCE

A satisfactory performance on this job requires that you do the following:

1 Identify terms relating to basic electricity by placing the number of each term in front of the phrase that best describes it.

2 Identify all the terms correctly within twenty minutes.

PERFORMANCE SITUATION

1 Electron	9 Centrifugal force
2 Matter	10 Free electron
3 Proton	11 Semiconductor
4 Element	12 Conventional theory
5 Electron theory	13 Valence ring
6 Atom	14 Neutron
7 Nucleus	15 Conductor
8 Insulator	16 Bound electron

_____ The assumption that current flows from positive (+) to negative (−)

_____ Anything that is "pure" in that it is not combined with anything else

_____ Anything that has weight and occupies space

_____ A positively charged atomic particle

_____ The outer orbit ring of an atom

_____ An atom that has no nucleus

_____ An element whose atom contains less than four electrons in its outer orbit ring

_____ A tightly held electron in the outer orbit ring of an atom

_____ The assumption that current flows from negative (−) to positive (+)

_____ A negatively charged atomic particle

_____ A force that tends to pull a rotating body away from its center of rotation

_____ An electron that has changed polarity

_____ The smallest part of an element that contains the identifiable properties or characteristics of that element

_____ A loosely held electron in the outer orbit ring of an atom

_____ An element whose atom contains four electrons in its outer orbit ring

_____ An element whose atom contains more than four electrons in its outer orbit ring

_____ The protons and neutrons that form the center of an atom

_____ An atomic particle that has no electrical charge

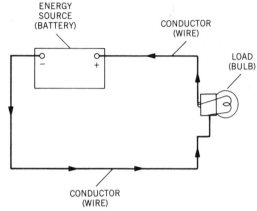

Figure 2.13 A simple circuit. Using the electron theory of current flow, current flows from the negative (−) battery terminal, through a wire to the bulb. After flowing through the filament of the bulb, the current flows to the positive (+) battery terminal.

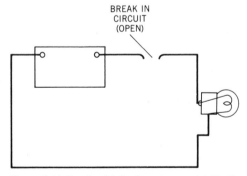

Figure 2.14 If a circuit is broken at any point, the flow of electrons and thus the flow of current stops.

of energy (battery), a load (bulb), and conductors (wires). The different charges (+ and −) at the battery cause electrons to flow through the conductors and through the load. If the circuit is broken at any point, the flow of electrons and thus the flow of current stops. The circuit can be broken by disconnecting a wire, as shown in Figure 2.14.

Electrical circuits are the means by which we direct and control current flow so that it can be used to perform work. That work may be mechanical, as in

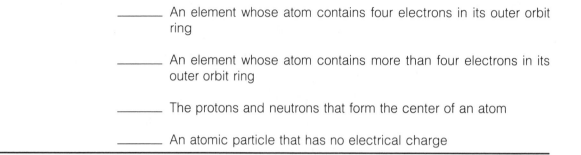

Figure 2.15 As a water meter measures water flow through a pipe, an ammeter measures current flow through a conductor. Notice that all the current flows through the ammeter as all the water flows through the water meter.

wiping a windshield or cranking an engine. It may be in the production of light for illuminating the road, or it may be for heating a window to remove frost.

Current flowing through a circuit often is compared to water flowing through a system of pipes. The amount of water flowing, the pressure in the system, and any restrictions in the various pipes are of importance to a plumber. Similar factors are of importance to an electrician. The flow of current through the circuit, the pressure applied to the circuit, and the resistance in the circuit must be understood if you are to work with electrical systems.

Current Flow—Amperes　The amount of water flowing through a pipe usually is measured in gallons per minute. The amount of current flowing through a circuit is measured in _amperes_. (See Figure 2.15.) An ampere is a measurement of the number of electrons that pass a given point in 1 second. Since electrons are very small, use of the actual count of electrons would require numbers that could be considered astronomical. One ampere is actually the flow of 6.28 billion billion electrons per second. (See Figure 2.16.)

The amount of current that flows through a circuit is dependent on two other factors: the pressure that is applied to the circuit and the resistance that is in the circuit. Those factors and their relationships will be covered later in this chapter.

Figure 2.16 The flow of electrons through a circuit is measured in amperes (courtesy of General Motors Corporation).

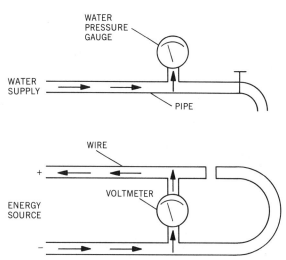

Figure 2.17 A water pressure gauge measures the difference in pressure between the water in the pipe and the atmospheric pressure outside the pipe. A voltmeter measures the difference in pressure between the positive (+) and negative (−) charges of the energy source. Notice that the water valve is closed and that the circuit is broken. Pressure exists even if nothing flows in the system or circuit.

As an example of the amount of current that flows in various automotive circuits when they are in operation, about 10 amperes flows in the windshield wiper circuit, about 12 amperes in the headlight high-beam circuit, and from 200 to 300 amperes in the starter circuit.

Electrical Pressure—Volts Water will not flow through a system of pipes unless there is a force or pressure that pushes it. Current will not flow in a circuit unless there is a force or pressure pushing it. Water pressure usually is measured in pounds per square inch. Electrical pressure, often called *electromotive force,* is measured in *volts.* (See Figure 2.17.)

Voltage actually is a measurement of a *difference in potential,* or a difference in the strength of opposite charges. You know that atoms are composed of particles with positive (+) charges and particles with negative (−) charges. And you also know that when an atom loses an electron it becomes positively charged and will pull an electron from a neighboring atom to regain its electrical balance. A battery or other source of electrical energy provides a large positive (+) charge at one terminal and a large negative (−) charge at the other terminal. Voltage is a measurement of the difference in strength between those two charges. That difference is present whether or not the battery or energy source is in a circuit.

As an example of the voltages that you will find in automotive circuits, most automobile batteries have

a difference of potential of about 12 volts, and operative charging circuits may provide a pressure of about 16 volts. Some automotive ignition systems are capable of producing electrical pressures in excess of 30,000 volts.

Resistance—Ohms Water flow through a system of pipes can be limited by a restricted pipe, an undersized pipe, or a partially closed valve. The current that flows through a circuit is also limited by certain factors that may restrict its flow. (See Figure 2.18.) Anything that restricts the flow of current in a circuit is termed *resistance.* Resistance in a device or load is designed into the device and, if within specifications, is a normal condition. Resistance in the conducting wires of a circuit should be kept to an absolute minimum if the device is to function properly.

Just as there are several causes for the restriction of water flow through a system of pipes, there are several causes of resistance in electrical circuits. As excessive resistance is a major contributor to electrical system problems, you must be aware of those causes and consider them in your diagnostic procedures.

Conductor Material As you know, copper is used for the wiring in most automotive electrical systems. Copper is considered a good conductor, but copper is not a perfect conductor. It offers a slight resistance to the flow of current. Iron also is a good

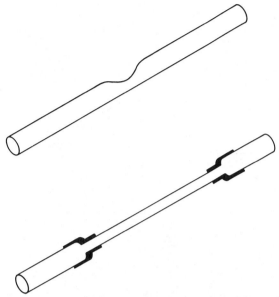

Figure 2.20 Cut or broken strands of a wire reduce its useful size and increase its resistance.

Figure 2.18 In a system of pipes, a damaged or undersized pipe can restrict the flow of water. In an electrical circuit, a damaged or undersized wire offers resistance to the flow of current.

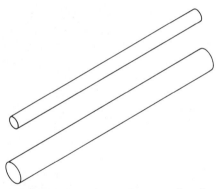

Figure 2.19 Just as more water can flow through a larger pipe, more electrons can flow through a larger wire. Wire size is selected to handle the current required by the loads in a circuit.

conductor, but it offers more than six times the resistance of copper.

Conductor Cross-sectional Area A thicker or heavier wire has more atoms per unit of length than does a wire of smaller diameter. Since more atoms provide more free electrons, wire of a larger diameter (and thus a larger cross-sectional area) provides a less restrictive path for current flow, and thus has less resistance. (See Figure 2.19.) This means that the size of the wire used in a circuit must be chosen to handle the current required by the load in the circuit.

Conductor Length The resistance of a conductor increases with the length of the conductor. There-

fore, the length of the conductor must be considered when the wire size for a circuit is chosen.

Conductor Condition If some of the strands of a wire are cut or broken, the resistance of the wire increases. Even if the damage is confined to one small area, the effect is the same as if those strands were removed from the entire length of wire. (See Figure 2.20.) Cut or broken wire strands actually reduce the useful diameter of the wire. Loose, dirty, and corroded connections also cause high resistance for the same reason.

Conductor Temperature When the temperature of a wire increases, so does its resistance. This is because the free electrons tend to move faster in random directions, and the applied pressure or voltage has less effect in moving them in a directed flow. The resistance in a heated wire produces more heat, which in turn produces more resistance, compounding the problem.

Resistance is measured in *ohms*. The Greek letter *omega* (Ω) is often used in place of the word ohms. A material that offers practically no resistance to the flow of current (there is no "perfect" conductor) is said to have no resistance, and therefore has 0-Ω resistance. A material that allows no current flow (an insulator) is said to have a resistance of *infinity* (∞). As an example of the resistances you may find in automotive electrical circuits, a battery cable should have a resistance of 0 Ω, and the winding of an alternator rotor may have a resistance of 3 Ω. Some wires used in the secondary circuit of an ignition system may have a resistance of from 5,000 to 15,000 Ω.

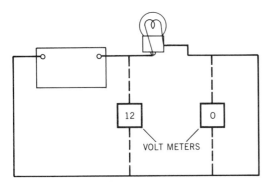

Figure 2.21 If two voltmeters were connected in a circuit as shown, they would show that the voltage in the circuit is "used up" in pushing current through the load.

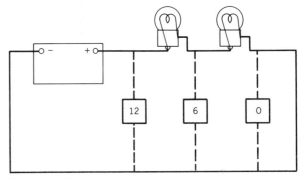

Figure 2.22 The three voltmeters show the voltage drop when a second similar load is added to the circuit in series. Each load "uses up" half the voltage, but the total voltage drop is equal to the voltage applied to the circuit.

While the resistance of a wire or an electrical device can be measured with an ohmmeter, the conductor or part must be removed from the circuit for testing. This is because an ohmmeter cannot be used in a live circuit, and may give a false reading if used in a circuit that is not energized. As you will discover later in this text, there is an easier method of locating resistance in a circuit. This procedure involves the use of a *voltmeter* to measure *voltage drop* across a wire, connection, or load.

Voltage Drop Every load in an electrical circuit converts electrical energy into another form of energy. In effect, a load "uses up" electrical energy in producing motion, light, or heat. The difference between the voltage entering a load and the voltage leaving a load is called the *voltage drop*. As shown in Figure 2.21, the voltage in a circuit with a single load is "used up" pushing current through the load. In a circuit with two or more loads connected in series, as shown in Figure 2.22, a portion of the voltage is used by each load, but the total voltage used is equal to the voltage applied to the circuit.

BASIC AUTOMOTIVE CIRCUITS On most cars, the negative (−) terminal of the battery is *grounded,* or connected to the engine block and the metal parts of the body and frame. Because of this, it is easier to visualize the flow of current from the positive (+) battery terminal to the negative (−) terminal, and the conventional theory of current flow will be used from this point on. As you learned, the conventional theory of current flow states that current flows from the positive (+) terminal to the negative (−) terminal of a power source.

Circuits A circuit is nothing more than a path for the flow of current. When current flows in a circuit,

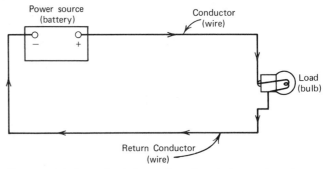

Figure 2.23 A simple circuit. Current flows from the battery through a wire to the bulb. After flowing through the filament of the bulb, the current flows back to the battery through the return wire.

that circuit is said to be *closed,* or complete. A closed circuit provides a route for current to flow from a power source to a load. And it also must provide a return route so the current can return to the power source.

A circuit must have a minimum of three components:

1 *A Power Source.* This can be the battery, or when the engine is running, the alternator.
2 *Conductors.* These can be wires and the metal parts of the car.
3 *A Load.* This can be a light, motor, or other device that converts electrical energy to another form of energy.

Figure 2.23 shows how those components can be connected to form a simple lighting circuit. In that circuit, a wire conducts current from a battery to a bulb. The current flows through the *filament* of the bulb. The filament is a thin resistance wire that gets white hot, or *incandescent,* when current flows through it. After passing through the filament, the current flows back to the battery through the return wire. The electrical energy of the battery is converted to light energy because the circuit is closed.

Job 2b

IDENTIFY TERMS RELATING TO ELECTRICAL CIRCUITRY
SATISFACTORY PERFORMANCE
A satisfactory performance on this job requires that you do the following:

1 Identify terms relating to electrical circuitry by placing the number of each term in front of the phrase that best describes it.
2 Identify all the terms correctly within 10 minutes.
PERFORMANCE SITUATION

1 Circuit	5 Ampere
2 Volt	6 Electromotive force
3 Resistance	7 Ohm
4 Voltage drop	8 Difference in potential

_____ A unit of measurement of electrical pressure

_____ The difference in strength between the positive (+) and negative (−) terminals of an energy source

_____ A unit of measurement of wire cross-section

_____ The difference between voltage entering a load and voltage leaving a load

_____ A unit of measurement of resistance

_____ A term used to describe electrical pressure

_____ A unit of measurement of current flow in a circuit

_____ A complete unbroken path for the flow of current

_____ Anything that restricts the flow of current

Circuit Control   The circuit shown in Figure 2.23 has several disadvantages. One is that it cannot be controlled. The light will continue to burn until all of the electrical energy is converted to light energy. By adding a control that will *open,* or break, the circuit, the light can be turned on and off. A control that opens and closes a circuit is called a *switch.* Figure 2.24 shows how a switch can be installed in a circuit.

Ground Return   Another disadvantage of the circuit shown in Figure 2.23 is that it requires too much wire. Figure 2.25 shows how most of the return wire can be eliminated. Since the frame and body of the car are made of metal, those parts can form the return conductor. When the metal parts of a car form the return conductor, that part of the circuit is referred to as the *ground.* Most automotive circuits use the ground as a return conductor. If an automobile had only a few circuits, several extra lengths of wire would be of no concern. However, an automobile has hundreds of separate circuits. If the ground was not used as the return conductor, the amount of wire used in a car would be doubled.

Circuit Protection   Another disadvantage of the circuit shown in Figure 2.23 is that it has no protection against overload. If the load in the circuit is increased, or if an additional load is added, the wires

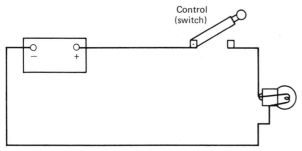

Figure 2.24 A simple circuit with a switch as a control. Opening the switch opens the circuit and stops the flow of current.

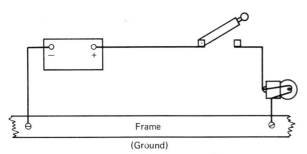

Figure 2.25 A simple circuit using the automobile frame and body as the return conductor.

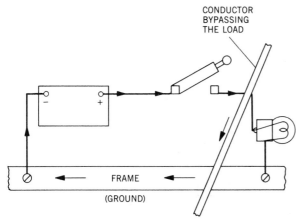

Figure 2.26 If any conductor is placed in the circuit so that current can return to the battery without passing through the load, that "shorter path" for current flow is called a short.

and the switch may be unable to carry the extra current that is required.

An extreme overload can be caused by a *short circuit.* As shown in Figure 2.26, a short circuit, or *short,* provides a path for current to return to the battery without passing through a load. As the resistance of the load is no longer in the circuit, the battery is allowed to push an excessive amount of current through the wires and parts that remain in the circuit. The overloaded wires and switch will overheat and can easily cause an electrical fire.

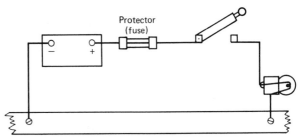

Figure 2.27 A simple circuit protected by a fuse. Notice that the fuse is placed "first" in the circuit directly after the power source.

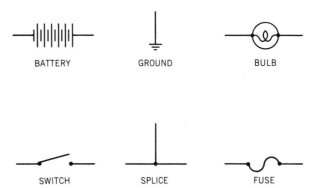

Figure 2.28 Symbols commonly used in wiring diagrams.

One method of providing circuit protection is to install a *fuse* in the circuit. A fuse is a "sacrifice" conductor that will "blow" or burn out when the current that flows through it exceeds a certain limit. Since a fuse protects only those parts of the circuit "downstream" from its location, it usually is located as close as possible to the power source. Figure 2.27 shows how a fuse is placed in a circuit.

Symbols The circuits shown in Figures 2.23 through 2.27 are in pictorial form. The parts of the circuit are drawn to resemble the actual parts. This method often is used for simple circuits with few parts. However, it has limited usage in showing automotive circuits.

Most automotive circuits are shown by *wiring diagrams.* The parts in those diagrams often are represented by *symbols.* Symbols are codes or signs that have been adopted by various auto makers to indicate certain parts and conditions. Symbols are a part of the written language of the automotive trades. Figure 2.28 shows some of the most commonly used symbols. Those symbols are used in Figure 2.29 to diagram the same circuit shown in Figure 2.27. Compare Figures 2.27 and 2.29. You will see that a diagram with symbols is not difficult to understand.

Job 2c

IDENTIFY CIRCUIT PARTS AND THEIR DEFINITIONS

SATISFACTORY PERFORMANCE
A satisfactory performance on this job requires that you do the following:

1 Identify the symbolized circuit parts in the diagram below by placing the number of each part in front of the correct part name.
2 Identify the definition of each part by placing the number of each part in front of its correct definition.
3 Identify all the parts and their definitions correctly within 10 minutes.

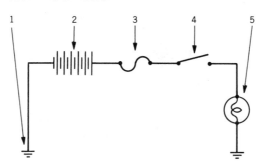

PERFORMANCE SITUATION

_____ Fuse   _____ Switch
_____ Ground   _____ Rheostat
_____ Bulb   _____ Battery

_____ A connection to the metal parts of a car

_____ A power source

_____ A device that boosts voltage in a circuit

_____ A control device that opens and closes a circuit

_____ A device that converts electrical energy to light energy

_____ A "sacrifice" part that protects a circuit

While the symbols shown in Figure 2.28 are almost universally used, not all manufacturers have agreed on a common symbolic language. For this reason, all car makers provide a legend of wiring diagram symbols in their service manuals. The Appendix contains a supplementary legend of wiring diagram symbols.

Types of Circuits   There are only two major types of circuits — *series circuits* and *parallel circuits*. Both types are used in automotive electrical systems. At

times, both are combined. A circuit of this type is called a *series-parallel* circuit.

Series Circuits   The simple circuit that was developed in Figures 2.23 through 2.29 is a series circuit. When a circuit contains only one load and one power source, it can only be a series circuit. A series circuit has one continuous path for the flow of current. Therefore, current flows through all the parts in a sequence. If any of the parts fail, or if the circuit is broken at any point, current stops flowing in the

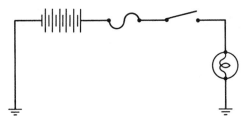

Figure 2.29 A wiring diagram for a simple circuit. Notice that the switch is shown in the "open" or OFF position.

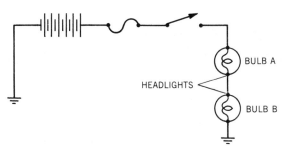

HEADLIGHTS — BULB A / BULB B

Figure 2.30 Here, a pair of headlights is connected in a series circuit. If either bulb A or bulb B burns out, the circuit will be broken. The remaining bulb then cannot operate.

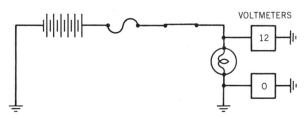

Figure 2.31 Voltage in a series circuit. Two voltmeters connected as shown will indicate that full source voltage is present before the load, and that there is no voltage after the load. Notice that the switch is shown in the "closed" or On position.

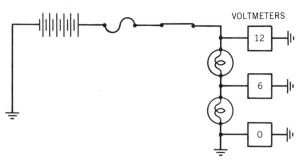

Figure 2.32 Voltage in a circuit with two loads wired in series. The voltmeters show the effects of the increased resistance. Notice that the voltage is "split" and shared by the two loads.

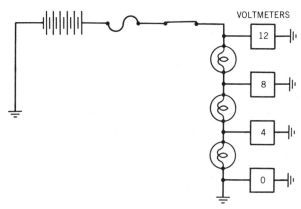

Figure 2.33 Voltage in a circuit with three loads wired in series. The voltmeters show the further effects of increased resistance.

entire circuit. Some examples of series circuits used in automobiles include the starter circuit and the primary circuit of breaker point ignition systems.

A series circuit provides only one path for the flow of current. Therefore, it is seldom used for a lighting circuit containing two or more bulbs or loads. Figure 2.30 shows a pair of headlights connected in a series circuit. As shown, the failure of one headlight would result in an open circuit. Current flow would stop and the other headlight would go out.

Series circuits have other disadvantages that relate to voltage and current flow. You know that most cars are equipped with a 12-V battery. You also know that voltage is a unit of electrical pressure. A 12-V battery provides a 12-V pressure to push current through a circuit. The load in the circuit is designed to operate at that voltage. The load is built so that its resistance allows a pressure of 12 V to push sufficient current through the load so that it functions properly. In effect, the load "uses up" the voltage in the circuit as it converts electrical energy to energy of another form. This is called voltage drop and was mentioned previously in this chapter.

Assuming no losses through bad connections, the voltage in a circuit just before the load is equal to the full voltage provided by the power source. The voltage just after the load is 0. This can be proven by using a voltmeter, as shown in Figure 2.31.

If a second similar load were placed in the circuit, the resistance in the circuit would be doubled. The

12-V pressure of the battery would be split and used up by both the loads. A voltmeter test in a circuit of this type would reveal the results shown in Figure 2.32. The split voltage would be insufficient to make both of the loads operate correctly. The reduced voltage would push only half as much current through the loads.

Adding a third similar load to the circuit, as shown in Figure 2.33, further increases the resistance. This causes a greater drop in voltage and in current flow. As you can see, series circuits containing two

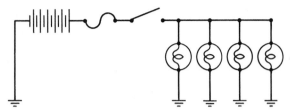

Figure 2.34 A parallel circuit. Notice that each load has a separate return wire. If one load fails, the remaining loads will still operate.

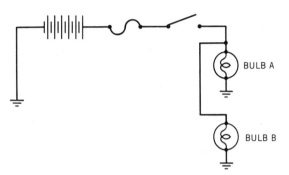

Figure 2.35 A pair of headlights wired in a parallel circuit. If either bulb A or bulb B burns out, the remaining bulb will not be affected.

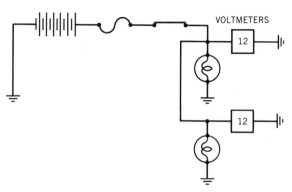

Figure 2.36 Voltage in a parallel circuit. Two voltmeters connected as shown will indicate that full source voltage is available for each load.

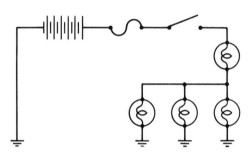

Figure 2.37 A true series-parallel circuit.

or more loads do not allow efficient use of electrical energy.

Parallel Circuits   A parallel circuit, shown in Figure 2.34, is one in which two or more loads are connected so that each is provided with its own return path to the power source. Figure 2.35 shows a pair of headlights connected in a parallel circuit. Compare that illustration with Figure 2.30. When wired in parallel, the failure of either bulb will not affect the operation of the remaining bulb. This is one reason why parallel circuits are used in automotive lighting circuits.

A parallel circuit offers another advantage: Each load receives full-system voltage, as shown in Figure 2.36. In addition, when loads are connected in parallel, the total circuit resistance drops. This is because of the additional return paths provided. Therefore, more current can flow in the circuit and the loads can operate at full efficiency. This is another reason why parallel circuits are used in automotive lighting systems.

Series-Parallel Circuits   A true series-parallel circuit is one in which there is at least one load in series with two or more loads in parallel. (See Figure 2.37.) Since full-source voltage is not available to all of the loads, such a circuit is not often used.

OHM'S LAW   The current that flows in a circuit, the electrical pressure that causes the current to flow, and the resistance in the circuit have definite relationships. Those relationships are defined by Ohm's law, which states that it requires a pressure of 1 volt to push a current of 1 ampere through a resistance of 1 ohm.

While it may not appear to be of much importance, a working knowledge of Ohm's law will be of great help to you. It will form the basis of the mental picture of current flow that you must develop. Without that mental picture, your ability to diagnose electrical problems will be severely limited.

Ohm's law is usually presented in a mathematical equation using three letters. In that equation,

$$I = \text{Current (amperes)}$$
$$E = \text{Electromotive force (volts)}$$
$$R = \text{Resistance (ohms)}$$

Ohm's law is most often written as:

$$I = \frac{E}{R} \quad \text{or} \quad \text{amperage} = \frac{\text{volts}}{\text{ohms}}$$

It also can be written in two other forms:

$$E = I \times R \quad \text{or} \quad \text{volts} = \text{amperes} \times \text{ohms}$$

Job 2d

COMPLETE WIRING DIAGRAMS OF BASIC CIRCUITS

SATISFACTORY PERFORMANCE

A satisfactory performance on this job requires that you do the following:

1 Complete the wiring diagrams in the Performance Situation by drawing in the connecting wires so that series, parallel, and series-parallel circuits are shown.

2 Using straight lines, 90° bends and splices, and ground symbols, draw in all the connecting wires so that each circuit is correctly wired.

3 Complete the job within 15 minutes.

PERFORMANCE SITUATION

Series Circuit

Parallel Circuit

Series-Parallel Circuit

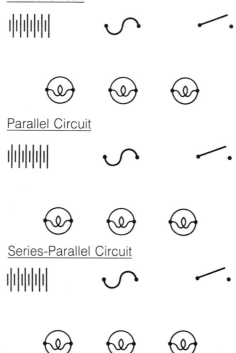

and

$$R = \frac{E}{I} \quad \text{or} \quad \text{ohms} = \frac{\text{volts}}{\text{amperes}}$$

In working with electricity, you will be concerned with voltage, amperage, and resistance. If you know the value of any two of those factors in any circuit, Ohm's law can be used to determine the third. Figure 2.38 shows the Ohm's law circle, which provides an easy way of remembering the law and how it can be used.

As an automotive electrician, you usually will not be required to perform mathematical calculations to diagnose a problem or to install a new device or circuit. However, by performing some of the calculations now, you will become aware of the relationships of voltage, amperage, and resistance. You also will better understand how, when one of those factors is changed, that change will affect one of the others.

Ohm's Law in Series Circuits In your study of series circuits, you learned that they provide only

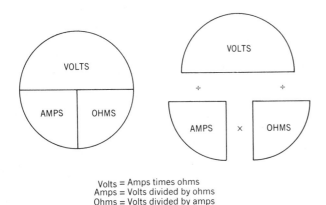

Volts = Amps times ohms
Amps = Volts divided by ohms
Ohms = Volts divided by amps

Figure 2.38 The Ohm's Law Circle. When two factors are known, the third can be found by multiplying or dividing the known factors as shown.

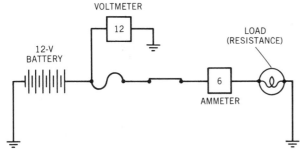

Figure 2.39 In this circuit, the ammeter indicates a current flow of 6 A. As the power source is a 12-V battery, Ohm's law $\left(R = \dfrac{E}{I}\right)$ enables you to determine that the load has a resistance of 2 Ω.

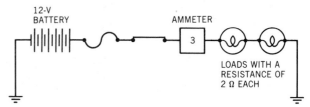

Figure 2.40 When the resistance in the circuit is doubled, the current flowing in the circuit is halved.

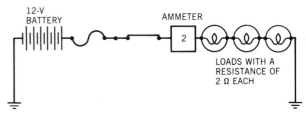

Figure 2.41 When the resistance in the circuit is tripled, the current flowing in the circuit is cut to one third.

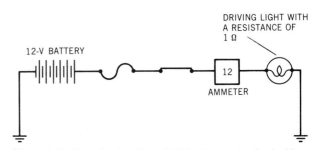

Figure 2.42 In a circuit with a 12-V battery and a load with a resistance of 1 Ω, the wires, fuse, and switch must be capable of handling at least 12 A of current.

one path for the flow of current, and that the total voltage drop in a circuit is equal to the source voltage. This information can be expanded to form four basic rules of series circuits:

1 Since there is only one path for the flow of current in a series circuit, the current flow must be the same in all parts of the circuit.
2 When two or more loads are wired in a series circuit, the voltage drop across each load will vary with the resistance of each load.
3 The total voltage drop in a series circuit is equal to the source voltage.
4 The total resistance in a series circuit is the sum of the resistance of each load in that circuit.

As an example of how Ohm's law is used in a series circuit, consult the diagram in Figure 2.39. The voltmeter indicates that the battery exerts a pressure of 12 V. The ammeter shows that 6 A is flowing in the circuit. Figure 2.38 shows that the resistance of the circuit can be found by dividing the voltage by the amperage. The number 12 divided by 6 equals 2; therefore, the circuit has a resistance of 2 Ω.

If a second load with a resistance of 2 Ω is added to the circuit, as shown in Figure 2.40, the total resistance of the circuit is doubled. Since the voltage applied to the circuit remains the same, the current flow is halved. The addition of a third similar load, shown in Figure 2.41, further increases the resistance, decreasing the current flow proportionally.

As an example of how a working knowledge of Ohm's law is necessary, consider this situation. A car owner wants a high-powered driving light installed. You must determine the size of the wire, the size of the fuse, and the rating of the switch to be used. The car has a 12-V battery, and an ohmmeter shows that the driving light has a resistance of 1 Ω. The current that will flow in that circuit can be determined by dividing the voltage by the resistance, and 12 divided by 1 equals 12. Your selection of the wire, the fuse, and the switch can be based on the knowledge that those parts must handle at least 12 A of current. (See Figure 2.42.)

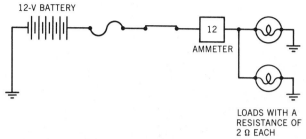

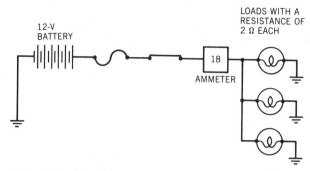

Figure 2.43 When loads are connected in parallel, the total resistance in the circuit decreases with each load added. Compare this diagram with the ones in Figures 2.38 and 2.39. Notice that twice as much current flows in the circuit when the same loads are connected in parallel.

Figure 2.44 Adding a third similar load further decreases the total resistance in the circuit. Since the 12-V pressure of the battery pushes 6 A of current through each load, the total current flow is 18 A.

Ohm's Law in Parallel Circuits When you studied parallel circuits, you learned that they provide separate paths for current flow for each of the loads in the circuit. You also learned that full-source voltage is available at each load and that the voltage drop across each load is equal to the source voltage. That information can be expanded to form four basic rules of parallel circuits:

1 Since there is more than one path for the flow of current in a parallel circuit, the current flow may vary in each path or branch.
2 The total current flow in a parallel circuit is equal to the sum of the current flow in all the branches.
3 The full source voltage is applied to each load in a parallel circuit.
4 As each load in a parallel circuit has its own path for current flow, the total resistance is less than the lowest load resistance.

To calculate the total resistance in a parallel circuit, the following formula can be used:

$$R_t = \cfrac{1}{\cfrac{1}{R_1} + \cfrac{1}{R_2} + \cfrac{1}{R_3} + \cdots}$$

In that formula,

$$R_t = \text{Total resistance}$$
$$R_1, R_2, R_3, \text{etc.} = \text{The resistance of the individual loads}$$

Figure 2.43 shows a diagram of a parallel circuit containing two loads, each having a resistance of 2 Ω. The total resistance in that circuit can be found as follows:

$$R_t = \cfrac{1}{\cfrac{1}{2} + \cfrac{1}{2}} = \frac{1}{1} = 1 \ \Omega$$

The current flow in that circuit can be found by ap-

plying Ohm's law $\left(I = \dfrac{E}{R} \right)$, which states that the source voltage is divided by the total resistance. Since 12 divided by 1 equals 12, current flow in that circuit is 12 A.

Connecting a third similar load in parallel provides a third path for current flow and further decreases the resistance in the circuit. The total resistance in that circuit, shown in Figure 2.44, can be found as follows:

$$R_t = \cfrac{1}{\cfrac{1}{2} + \cfrac{1}{2} + \cfrac{1}{2}} = \frac{1}{1.5} = 0.666$$

Applying Ohm's law $\left(I = \dfrac{E}{R} \right)$, 12 divided by 0.666 equals 18; therefore, the current flow in the circuit is 18 A.

Since most automobiles use a 12-V battery, the voltage in automotive circuits can be considered to be relatively constant. In working with Ohm's law, you may have noticed that when the voltage in a circuit remains the same, any change in the resistance causes an opposite change in the current flow. If the resistance in a circuit increases, the current flow decreases. If the resistance decreases, the current flow increases. (See Figure 2.45.)

AUTOMOTIVE CIRCUITRY The circuits you studied in the previous paragraphs are basic to all automotive circuitry. The most elaborate and complicated automotive electrical system is merely a combination of those basic circuits. And they all operate through the application of the same basic principles.

Some of the more commonly used circuit variations incorporate different types of controls, protection

Job 2e

DETERMINE ELECTRICAL PRESSURE, CURRENT FLOW, AND RESISTANCE IN ELECTRICAL CIRCUITS

SATISFACTORY PERFORMANCE

A satisfactory performance on this job requires that you do the following:

1 Using Ohm's law, determine the missing values (volts, amperes, or ohms) in the problems provided in the Performance Situation.
2 Determine correctly 8 of the 10 missing values within 20 minutes.

PERFORMANCE SITUATION

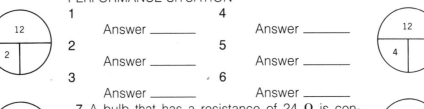

1   4
     Answer _____   Answer _____
2   5
     Answer _____   Answer _____
3   . 6
     Answer _____   Answer _____

7 A bulb that has a resistance of 24 Ω is connected to a 12-V battery. How much current will flow in that circuit?
Answer _____

8 A motor that operates an electric window has a resistance of 0.5 Ω. When installed in a circuit powered by a 12-V battery, how much current will flow in that circuit?
Answer _____

9 Two bulbs with a resistance of 2 Ω each are wired in series with a 12-V battery. How much current will flow in that circuit?
Answer _____

10 Two bulbs with a resistance of 2 Ω each are wired in a parallel circuit with a 12-V battery. How much current will flow in that circuit?
Answer _____

devices, and loads. And, as shown in Figure 2.46, different symbols are used to represent those parts. As you will service those components in various Jobs throughout this text, you must be aware of those differences.

Switches There are many types of switches and they are found in many forms. Manually operated switches, such as those shown in Figure 2.47, are described as toggle switches, push-pull switches, slide switches, rotary switches, and push-button switches. Other switches are remotely operated and may appear strange until they are handled and studied. Some switches of that type are shown in

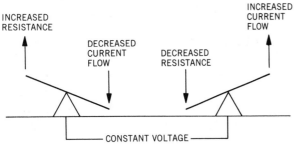

Figure 2.45 Since the voltage in most automotive circuits remains relatively constant, the resistance/current "see-saw" illustrates the basic concept of Ohm's law.

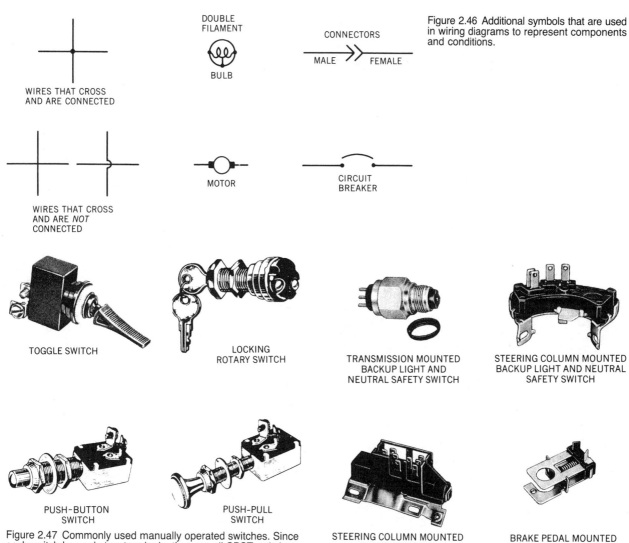

WIRES THAT CROSS
AND ARE CONNECTED

DOUBLE
FILAMENT

BULB

CONNECTORS

MALE   FEMALE

Figure 2.46 Additional symbols that are used in wiring diagrams to represent components and conditions.

WIRES THAT CROSS
AND ARE *NOT*
CONNECTED

MOTOR

CIRCUIT
BREAKER

TOGGLE SWITCH

LOCKING
ROTARY SWITCH

TRANSMISSION MOUNTED
BACKUP LIGHT AND
NEUTRAL SAFETY SWITCH

STEERING COLUMN MOUNTED
BACKUP LIGHT AND NEUTRAL
SAFETY SWITCH

PUSH-BUTTON
SWITCH

PUSH-PULL
SWITCH

Figure 2.47 Commonly used manually operated switches. Since each switch has only two terminals, they are all SPST switches.

STEERING COLUMN MOUNTED
IGNITION SWITCH

BRAKE PEDAL MOUNTED
STOP LIGHT SWITCH

Figure 2.48 Commonly used switches that are remotely operated.

Figure 2.48. Still other switches, shown in Figure 2.49, are operated by a change in temperature or a change in pressure. Regardless of their appearance or how they are actuated, all switches serve the same purpose — they open and close a circuit.

The simple switch symbolized in the diagrams in Figures 2.29 through 2.44 is called a *single-pole single-throw* (SPST) switch. That switch has only two positions, on (closed) and off (open), and can control only one circuit.

Figure 2.50 shows the symbols for a switch used to alternately control two circuits. A switch of that type is referred to as a *single-pole double-throw* (SPDT) switch. The switch used to operate the high and low beams of the headlights, usually called a dimmer switch, is of this type. Figure 2.51 shows

how a SPDT switch is wired in a headlight circuit so that current can be directed alternately to either the high-beam filament or the low-beam filament of the headlights.

At times, the function of two single-pole switches is obtained by combining multiple sets of contacts in a single switch. As symbolized in Figure 2.52, those switches are usually referred to as *double-pole single-throw* (DPST) switches and *double-pole double-throw* (DPDT) switches. Figure 2.53 diagrams a circuit where a DPST switch is used.

When one of several loads must be selected, a *rotary switch,* shown in Figure 2.54, is often used. Figure 2.55 shows how a rotary switch can be wired so that any one of four different lights can be operated.

TEMPERATURE—ACTIVATED SWITCHES

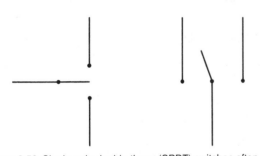

PRESSURE—ACTIVATED SWITCHES

Figure 2.49 Switches such as these are used to operate the coolant temperature and oil pressure warning lights on instrument panels. On some cars, they also are used in various engine system circuits.

Figure 2.50 Single pole double throw (SPDT) switches often are symbolized in this manner. Switches of this type allow current to be directed to either one of two circuits.

Switches in the Ground Side of a Circuit   In the circuits previously diagrammed, the switch was placed in the wire between the battery and the load. In many automotive circuits, the switch will be found between the load and the ground. As shown in Figures 2.56 and 2.57, the switches that operate the engine warning lights on the instrument panel usually are wired in that manner.

On many cars, the dome light and the courtesy lights are controlled by switches that are turned on when

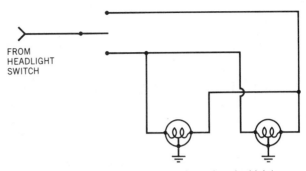

Figure 2.51 A SPDT switch is often used to select the high beam or low beam headlights. Notice that the battery or power source is not shown, but that the power source is identified as the headlight switch. This practice is followed in many wiring diagrams.

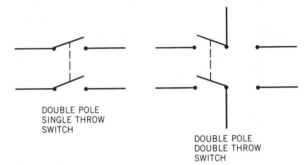

Figure 2.52 Symbols commonly used for DPST and DPDT switches. On many diagrams, the switch blades are shown connected by a dashed or dotted line to indicate that they operate in unison.

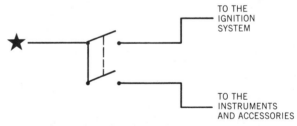

Figure 2.53 In some cars, the ignition switch is a DPST switch. This provides separate circuits for the ignition system and for the instruments and accessories. Notice that the power source is neither shown nor identified, but is indicated by a star. This practice is followed by some manufacturers in their wiring diagrams.

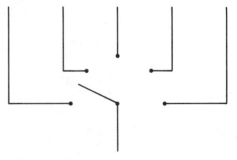

Figure 2.54 A symbol for a rotary switch. A rotary switch often is used when one of several circuits must be selected.

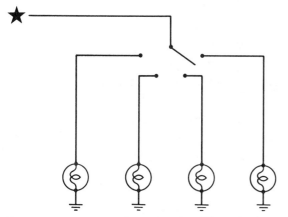

Figure 2.55 A rotary switch wired so that different loads can be selected.

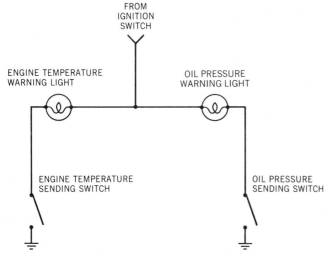

Figure 2.56 In this circuit, a temperature-actuated switch controls the engine temperature warning light, and a pressure-actuated switch controls the oil pressure warning light.

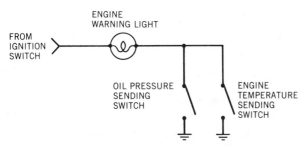

Figure 2.57 On some cars, a single warning light is used to indicate overheating and low oil pressure. When wired in this manner, either switch can complete the circuit and cause the light to go on.

the doors of the car are opened. On some of those cars, those *doorjamb switches* are wired so that they open and close the circuit between the bulbs and ground. Figure 2.58 diagrams a circuit of that type.

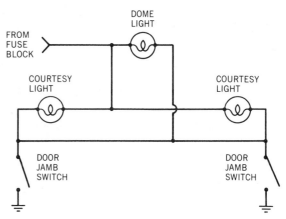

Figure 2.58 On some cars, the door jamb switches are wired so that they complete the ground circuit for the courtesy lights and dome light.

Figure 2.59 Symbols used for fixed and variable resistors.

Resistors   At times, it is necessary to add resistance to a circuit to reduce current flow intentionally. The required resistance usually is provided by a *resistor*. A resistor is considered a load as it converts electrical energy to heat and results in a voltage drop in the circuit.

Resistors are made of material that has a high natural resistance to the flow of current. While some resistors are made of carbon, most resistors used in automotive circuits are made of a metal alloy that is a poor conductor. As some resistors get very hot, they often are shielded or mounted in an enclosure.

Two types of resistors are used in automotive circuits. Some have a *fixed* or constant resistance, while others are adjustable or *variable*. The symbols for those resistors are shown in Figure 2.59.

Fixed Resistors   Fixed resistors, similar to the one shown in Figure 2.60 often are used to reduce current flow through a motor and thus reduce its speed. In some instances, resistors are wired so that they can be switched in and out of a circuit. A circuit of that type, shown in Figure 2.61, often is used to control the speed of a heater blower motor. Fixed resistors also are used in certain ignition systems to reduce current flow in the coil. On some cars, a special resistance wire is used to provide the required resistance.

Variable Resistors   On most cars, the intensity of the instrument-panel lights is adjustable. This usu-

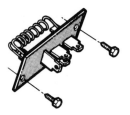

Figure 2.60 A fixed resistor. Resistors of this type are often used to control the speed of a heater blower motor (courtesy of Chrysler Corporation).

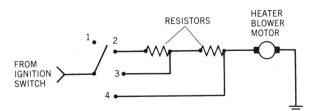

Figure 2.61 In this circuit, a rotary switch and two resistors provides three speeds for the heater blower motor. Position 1 is Off. Position 2 is Low as it places the two resistors in series with the motor. Position 3 is Medium as it places only one resistor in series with the motor. Position 4 is High as it removes both resistors from the circuit.

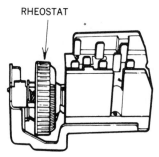

Figure 2.62 A variable resistor, or rheostat, is incorporated in many headlight switches so that the intensity of the instrument panel lights can be adjusted (courtesy of Ford Motor Company).

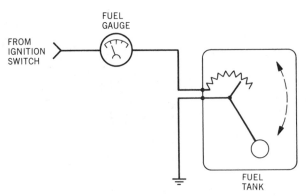

Figure 2.63 A typical fuel gauge circuit utilizing a variable resistor actuated by a float in the fuel tank.

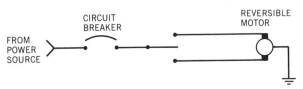

Figure 2.64 The motors used to operate power windows and seats are often subjected to overloads. Those motor circuits usually are protected by circuit breakers.

ally is accomplished by a variable resistor sometimes referred to as a *rheostat*. As shown in Figure 2.62, the variable resistor may be incorporated in the headlight switch, where it is operated by turning the knob. Variable resistors also are used to operate certain instruments and gauges. Most fuel gauges are operated by a variable resistor actuated by a float in the fuel tank, as shown in Figure 2.63.

Circuit Breakers   While a fuse usually is used to protect a circuit from overload, a fuse is not suitable for all circuits. Some circuits normally are subjected to temporary overloads. If a fuse were used in those circuits, they would melt or "blow" continually and would require frequent replacement. In the lighting system, a blown fuse could leave a driver without headlights, causing an accident.

What is required is a device that will open a circuit temporarily when current flow exceeds a certain limit and then close that circuit after a very short time. Such a device is called a *circuit breaker*. As shown in Figure 2.64, a circuit breaker is wired in a circuit in series as if it were a fuse. A circuit breaker is sensitive to heat. When excess current flows through a circuit breaker, the heat produced opens a set of contacts that breaks the circuit. When the current flow stops, the breaker cools and the contacts close again. This on—off cycle continues until the cause of the overload ceases.

Job 2f

IDENTIFY SYMBOLS USED IN AUTOMOTIVE WIRING DIAGRAMS

SATISFACTORY PERFORMANCE
A satisfactory performance on this job requires that you do the following:

1 Identify the wiring diagram symbols in the Performance Situation by placing the number of each symbol in front of the part or condition it represents.
2 Identify 13 of the 15 symbols correctly within 15 minutes.

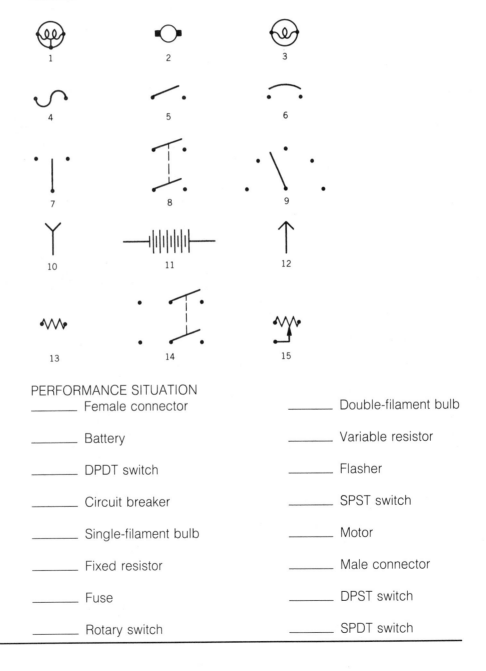

PERFORMANCE SITUATION

_____ Female connector

_____ Battery

_____ DPDT switch

_____ Circuit breaker

_____ Single-filament bulb

_____ Fixed resistor

_____ Fuse

_____ Rotary switch

_____ Double-filament bulb

_____ Variable resistor

_____ Flasher

_____ SPST switch

_____ Motor

_____ Male connector

_____ DPST switch

_____ SPDT switch

Job 2g

COMPLETE AUTOMOTIVE WIRING DIAGRAMS

SATISFACTORY PERFORMANCE
A satisfactory performance on this job requires that you do the following:

1 Complete the four wiring diagrams provided in the Performance Situation.
2 Following the instructions given, connect all the components by drawing in all of the necessary wires and ground symbols. Use straight lines and 90° bends. Splices and crossovers should meet at 90°.
3 Complete all the diagrams correctly within 30 minutes.

PERFORMANCE SITUATION
1 *Headlight Circuit*
Wire this circuit so that the dimmer switch allows the selection of high or low beams.

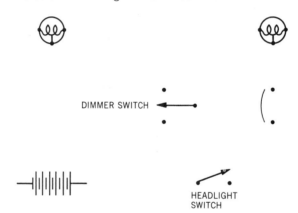

2 *Directional Signal Circuit*
Wire this circuit so that the indicator lamps light with the appropriate turn signal lamps.

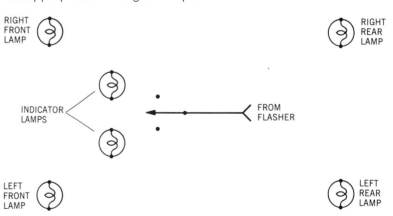

3 *Dome and Courtesy Lamp Circuit*
Wire this circuit so that the switches open and close the ground side of the circuit.

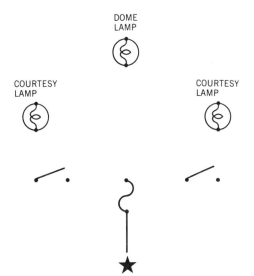

4 *Heater Blower Motor Circuit*
Wire this circuit so that the motor can be operated
at three different speeds.

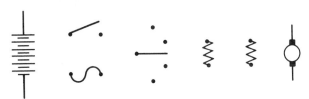

SUMMARY

In this chapter, you were introduced to some of the theories that help to explain the basic principles of electricity. You now know how the atomic structure of an element determines whether it is a conductor or an insulator, and you are aware of how current flows in a circuit. By studying the text and the illustrations, you learned about basic circuits and how they are controlled and protected.

By satisfactorily completing the Jobs provided, you have gained a working knowledge of Ohm's law and an understanding of the relationships of voltage, current, and resistance. You now can read and draw simple wiring diagrams and interpret the symbols used. You have started to form a mental picture of what happens in electrical circuitry and thus have taken an important step toward your goal.

SELF-TEST

Each incomplete statement or question in this test is followed by four suggested completions or answers. In each case select the *one* that best completes the sentence or answers the question.

1 The smallest part of an element that contains the identifiable properties of that element is called
   A. an atom
   B. a proton
   C. a neutron
   D. an electron

2 The amount of current that flows through a circuit is measured in
   A. ohms
   B. volts
   C. watts
   D. amperes

3 Electrical pressure is measured in
   A. ohms
   B. volts
   C. watts
   D. amperes

4 Resistance in an electrical circuit is measured in
   A. ohms
   B. volts
   C. watts
   D. amperes

5 The total voltage drop of a circuit is equal to the
   A. source voltage
   B. source voltage divided by the number of loads
   C. source voltage multiplied by the number of loads
   D. source voltage less the total resistance of the loads

6 Which of the following parts is NOT considered a load?
   A. A bulb
   B. A motor
   C. A switch
   D. A resistor

7 A series circuit
   A. contains only one load
   B. contains more than one load
   C. provides only one path for current flow
   D. provides more than one path for current flow

8 A parallel circuit
   A. contains only one load
   B. contains more than one load
   C. provides only one path for current flow
   D. provides more than one path for current flow

9 If 6 amperes is flowing in a circuit powered by a 12 volt battery, what is the resistance in that circuit?
   A. 1 ohm
   B. 2 ohms
   C. 3 ohms
   D. 4 ohms

10 Two mechanics are discussing electrical components.
   Mechanic A says that a fixed resistor often is called a rheostat.
   Mechanic B says that a circuit breaker is connected in series in a circuit.
   Who is right?
   A. A only
   B. B only
   C. Both A and B
   D. Neither A nor B

Chapter 3 Basic Electrical Skills

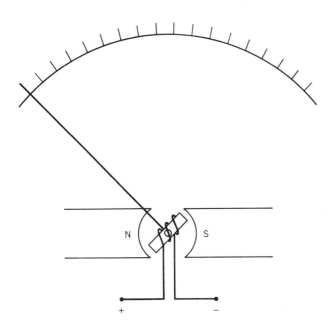

In this chapter, you will supplement your knowledge of electrical circuitry and start to develop some of the basic skills required of an automotive electrician.

In the previous chapter, you learned how the various parts of a circuit are connected by wire. You now will learn how to select the correct wire for a circuit, how to splice it, and how to install terminals and connectors.

You also learned in Chapter 2 about electrical pressure, current flow, and resistance. You now will learn how to use a voltmeter, an ammeter, and an ohmmeter to measure those factors in various automotive circuits.

Your specific objectives in this chapter are to perform the following Jobs:

A
Splice and solder wire connections
B
Install solderless terminals and connectors
C
Construct jumper wires and a test light
D
Measure open-circuit voltage and voltage under load
E
Measure current flow in electrical circuits
F
Check for continuity and measure the resistance of electrical system components.

WIRE The wire that connects the components in a circuit often is taken for granted, yet the wire is as important to the circuit as are the components. Most wire used in automotive circuits is made of copper, but occasionally you will find wire made of aluminum. As shown in Figure 3.1, most automotive wire is *stranded*, or made up of many thin strands. Solid wire often is used for internal wiring in certain components, but it rarely is used for external wiring because it is too stiff and is subject to metal fatigue and breakage.

The stranded wire used in most automotive circuits usually is insulated by a thin sheath of plastic, and often is referred to as *primary wire*. This term is used to indicate that the insulation on the wire is sufficient for the primary, or source, voltage. In most automotive applications, this voltage does not exceed 16 V, so the insulation does not have to be very thick.

Secondary wire usually is intended for use in the secondary circuit of ignition systems. The insulation on that wire may have to resist the leakage of over 30,000 V and usually is quite thick. Due to the thick insulation required, secondary wire may appear much heavier than primary wire, but the actual wire size may be much smaller (see Figure 3.2).

Wire Size As you learned in Chapter 2, the cross-sectional area, and thus the diameter, of wire must be chosen for the current it is to carry. Circuits that use large amounts of current require wire of a large diameter. Where current flow is slight, a smaller wire may be used. Wire size is measured in two systems. In most domestic cars, the wire used is measured in American Wire Gauge (AWG) sizes. The chart in Figure 3.3 lists the most commonly used AWG sizes and wire diameters. The wire used in most imported cars and in many newer domestic cars is measured in metric sizes. The conversion chart in Figure 3.4 provides the means by which

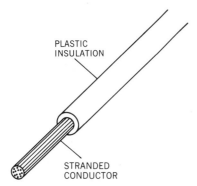

Figure 3.1 Stranded primary wire.

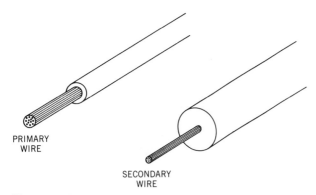

Figure 3.2 Primary wire is insulated to prevent the leakage of low voltage. Secondary wire has much heavier insulation as it may be subjected to a pressure of over 30,000 V.

GAUGE SIZE	CONDUCTOR DIAMETER (INCHES)
20	0.032
18	0.040
16	0.051
14	0.064
12	0.081
10	0.102
8	0.128
6	0.162
4	0.204
2	0.258
1	0.289
0	0.325
00	0.365
0000	0.460

Figure 3.3 Conductor diameters of the most commonly used wire gauge sizes.

AMERICAN WIRE GAUGE SIZE	METRIC SIZE (MM²)
20	0.5
18	0.8
16	1.0
14	2.0
12	3.0
10	5.0
8	8.0
6	13.0
4	19.0

Figure 3.4 This conversion chart shows the equivilant wire sizes in the AWG and metric systems.

you can select and substitute wire measured in either system.

As the resistance in a wire increases with its length, a wire capable of conducting current for a short distance may prove too small to conduct that same current for a longer distance. Figure 3.5 provides a guide for the selection of wire based on current flow and wire length.

Splices When wires must be joined, or *spliced,* it is important that both a good physical connection and a good electrical connection be made. A good physical connection is necessary to resist the strain and vibration that the splice will be subjected to in service. A good electrical connection is required to eliminate any possible resistance in the circuit.

The recommended method of splicing copper wire requires that a short length of insulation be *stripped* or removed from the wires, as shown in Figure 3.6. To avoid cutting or damaging any of the wire strands, the use of a *wire stripping tool* similar to the one shown in Figure 3.7 should be used. A razor blade can be used to slit the insulation lengthwise where necessary. The copper strands should then be

cleaned by scraping them with a knife or razor blade, using care so that none of the strands are cut. The wires then should be twisted together, as shown in Figures 3.8 and 3.9. This provides a good physical connection. A good electrical connection is obtained by soldering the spliced wires.

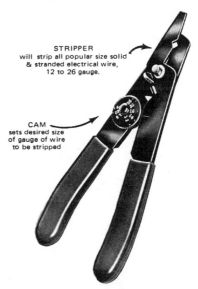

STRIPPER
will strip all popular size solid & stranded electrical wire, 12 to 26 gauge.

CAM
sets desired size of gauge of wire to be stripped

Figure 3.7 Many automotive electricians use a wire stripping tool to remove insulation without damaging the wire strands (©1985-Perfect Parts, Inc., Carlstadt, N.J.).

TOTAL CIRCUIT AMPERES @ 12 V	WIRE GAUGE REQUIRED FOR TOTAL LENGTH IN FEET					
	5'	10'	15'	20'	25'	30'
5	18	18	18	18	18	18
10	18	18	16	16	16	14
15	18	18	14	14	12	12
20	18	16	14	12	10	10
25	18	16	12	12	10	10
30	16	14	10	10	10	10
35	16	12	10	10	10	8
40	16	12	10	10	8	8
45	14	12	10	10	8	8
50	14	12	10	8	8	6

Figure 3.5 Recommended wire gauge sizes. Notice that as the current flow and/or the length of the wire increases, a heavier gauge wire is required.

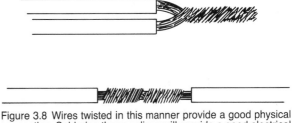

Figure 3.8 Wires twisted in this manner provide a good physical connection. Soldering these splices will provide a good electrical connection.

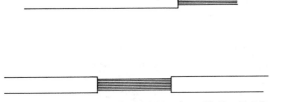

Figure 3.6 Insulation stripped from wires. Notice that the wire strands are not broken or frayed.

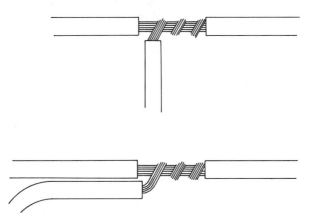

Figure 3.9 Spliced wires ready for soldering.

TIN/LEAD RATIO	MELTING TEMPERATURE
60/40	380°F (193°C)
50/50	415°F (213°C)
40/60	460°F (233°C)

Figure 3.10 Melting points of the solders most commonly used for soldering electrical wiring.

Figure 3.11 Solder in wire form is available in rolls of various types and sizes (©1985-Perfect Parts, Inc., Carlstadt, N.J.).

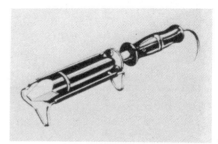

Figure 3.12 A heavy-duty soldering iron. An iron of this size is especially useful for soldering large parts where heat is dissipated rapidly (courtesy of Snap-on Tools Corporation).

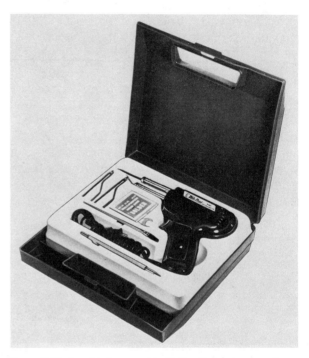

Figure 3.13 A soldering gun kit. Notice that extra gun tips, a brush, a soldering tool, and solder are included (courtesy of Snap-on Tools Corporation).

Soldering   Soldering is a method of bonding two pieces of metal together by flowing molten solder between the heated pieces and allowing the solder to cool and harden.

Solder   Solder is an alloy of tin and lead that has a relatively low melting point. While solder with a tin/lead ratio of 50/50 is most often used for electrical wiring, solders with tin/lead ratios of 60/40 and 40/60 are also used. Figure 3.10 shows the different melting points of those solders. Solder is available in many forms, but wire solder, shown in Figure 3.11, is usually used for soldering wiring when a soldering iron or soldering gun is used.

Flux   The surfaces of all metals *oxidize,* or combine with the oxygen in the air. Solder will not adhere to oxidized surfaces. Oxidation cannot always be seen, which is why wires should be scraped clean before they are spliced. When metals are heated, the rate of oxidation increases rapidly. Since the oxidation resists soldering, the surfaces of the wires and parts must be protected while they are being soldered. This protection is provided by *flux.* Flux is used to remove slight traces of oxidation and to inhibit or prevent additional oxidation.

A *rosin* or *resin* base flux is recommended for electrical work as it will not attack and corrode the parts, and it does not conduct electricity. Acid-base fluxes often are used for soldering iron and steel parts, but should never be used in electrical circuits as the acid may continue to corrode the parts and cause high resistance in the splice.

Various fluxes are available in liquid, powder, and paste form, but most automotive electricians use *rosin-core solder.* This is a hollow core wire solder that contains a rosin-base flux. When rosin-core solder is used, a separate flux usually is not required.

Soldering Irons and Soldering Guns   While a soldering iron similar to the one shown in Figure 3.12 can be used for electrical work, most mechanics prefer to use a soldering gun similar to the one shown in Figure 3.13. A soldering iron offers advantages when the parts or wires to be soldered are large and tend to dissipate heat. However, a soldering iron heats up slowly and often is too large and heavy for most jobs. A soldering gun usually will reach operating temperature in a few seconds, and its small tip is easier to work with.

Whether an iron or a gun is used, the tip must be kept clean and *tinned.* A tinned tip is one that is coated with a thin plating of solder. A tip that is not tinned will not conduct sufficient heat to the wires or parts that are to be soldered and may result in a poor connection. A dirty tip should be cleaned with a file or emery cloth and tinned by applying solder until a shiny coating is obtained.

Soldering consists of more than melting solder. The soldering iron or soldering gun should be used to heat the wires, not just the solder. To obtain a good electrical connection, the solder must flow between the wires as a liquid. This means that the wires must be hot enough to melt the solder. When using an iron or a gun, the tip should be held in contact with the wires to be soldered. The solder then should be touched to the wires and to the edge of the tip. When the wires are at the proper temperature, the solder will flow and will be drawn between the wires. A good soldered connection will look bright and shiny and have a smooth surface. A rough surface with a dull gray color indicates a *cold joint* or one in which the parts were not sufficiently heated.

Ten Steps to a Perfect Soldering Job   By following the steps listed below, after a bit of practice you will be able to obtain perfectly soldered connections.

1   Wear safety glasses while soldering. Hot solder and flux splashes and can cause serious eye injuries.
2   The wires and parts to be soldered must be clean. All traces of dirt, corrosion, oxidation, and oil or grease must be removed.
3   The wires and parts to be soldered should be held together tightly. Wherever possible, a good physical connection should be made before soldering.

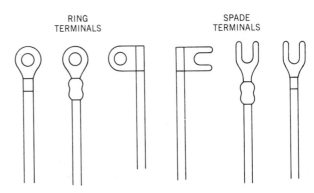

Figure 3.14 Typical ring and spade terminals.

4   The soldering iron or soldering gun should be large enough and should supply enough heat for the particular job.
5   The tip of the soldering iron or soldering gun should be clean and tinned. This improves heat transfer.
6   The tip of the soldering iron or soldering gun should be held so that there is a large area in contact with the parts to be soldered. This also improves heat transfer.
7   Use the correct solder and flux. Rosin-core solder is recommended. Acid-core solder should never be used in electrical circuits.
8   The tip of the soldering iron or soldering gun should be kept in contact with the wires or parts to be soldered until the solder flows freely. The solder should not be pasty or mushy, but should flow as a liquid.
9   Use only enough solder so that it flows between the wires or parts. Do not attempt to build up solder on the surfaces of the wires or parts.
10   After removing the soldering iron or soldering gun, do not move the wires or parts until they have cooled. Moving the wires or parts while they are still hot may break the solder bond.

TERMINALS AND CONNECTORS   The end of a wire usually is fitted with a terminal or some other type of connector. When the wire is to be connected to a stud or held by a screw, either a *ring terminal* or a *spade terminal* is used. As shown in Figure 3.14, those terminals are found in various types and sizes. At times, wires are fitted with small plugs and sockets so that they may be easily connected and disconnected. Typical connectors are shown in Figure 3.15.

Many different designs of terminals and connectors are available. Some, of the type shown in Figure 3.16, must be soldered to the wires. Others, called

Job 3a

SPLICE AND SOLDER WIRE CONNECTIONS

SATISFACTORY PERFORMANCE

A satisfactory performance on this job requires that you do the following:

1 Splice and solder four wire connections.
2 Following the steps in the Performance Outline, cut, strip, and splice 16-gauge (1.0-mm) primary wire as shown in the following illustrations.
3 Solder the splices to obtain a good electrical connection.
4 Complete the job within 30 minutes.

PERFORMANCE OUTLINE

1 Cut the wire to the dimensions shown.
2 Strip the wire to the dimensions shown.
3 Splice the wires as shown.
4 Solder the splices.

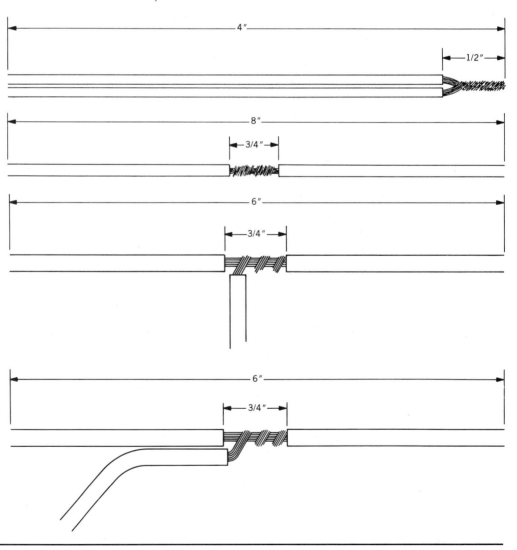

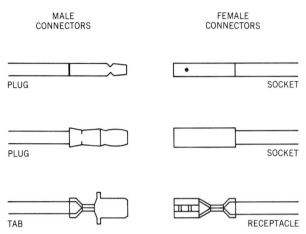

Figure 3.15 Typical wire connectors. Connectors are often used in automotive circuits so that components can easily be disconnected for testing or replacement.

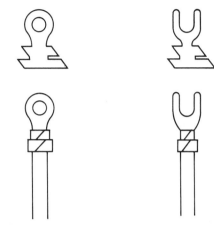

Figure 3.16 Terminals of this type must be folded around the wire and soldered.

solderless terminals and connectors, are shown in Figure 3.17. They are attached to the wires by a special crimping tool similar to the one shown in Figure 3.18.

Properly installed, solderless terminals and connectors provide both a good physical connection and a good electrical connection. For this reason, many automotive electricians avoid the use of a soldering iron or a soldering gun by using solderless terminals and connectors in most of their work.

While solderless terminals and connectors provide a satisfactory connection in most applications, they are not recommended in circuits that carry large amounts of current or where they may be subjected to road splash or moisture. When they are used under those conditions, solderless terminals and connectors should be soldered after crimping. This technique eliminates the possibility of a high-resistance connection caused by corrosion.

Installing Solderless Terminals and Connectors Solderless terminals and connectors are easily installed. As shown in Figures 3.19 and 3.20, the insulation on the wire is stripped so that the wire will enter the terminal or connector completely. A wire stripping tool (refer to Figure 3.7) should be used so that the insulation is cut square and so that none of the strands are cut. If too much insulation is removed, the wire should be trimmed. When the wire is inserted, the insulation should butt against the terminal or connector as shown. The terminal or connector should be selected to fit the wire. When the wire has a loose fit in the terminal, as shown in Figure 3.21, the wire should be doubled back to obtain a tight fit. When the wire and the terminal or

connector are properly assembled, a crimping tool (refer to Figure 3.18) is used to secure the connection.

Insulating Terminals, Connectors, and Splices Terminals, connectors, and splices should be insulated not only to avoid the possibility of a short circuit, but to protect the connection from moisture and resultant corrosion.

Insulated terminals and connectors (refer to Figure 3.17) often provide sufficient insulation. Vinyl tape, stretched and wrapped so that at least two layers cover the connection and the adjacent insulation, often is used to insulate soldered splices and uninsulated connectors. When the wires are part of a taped harness or are parallel to a taped harness, the tape can be blended into the harness so that the completed job appears to be part of the original wiring.

Many electricians use *heat-shrink tubing* to insulate connections. As shown in Figure 3.22, heat-shrink tubing is a special plastic "spaghetti" tubing that is placed over a connection. When heated with the flame from a match or cigarette lighter, the tubing shrinks to form a tight covering over the connection.

DIAGNOSTIC TOOLS Before any problem in an electrical system can be corrected, the cause of the problem must be found. At times, the cause of a problem is obvious. However, in most instances, some type of diagnosis must be performed, and that diagnosis requires the use of certain tools. All types of sophisticated diagnostic equipment are available for the automotive electrician. Yet, the cause of most problems usually can

UNINSULATED	INSULATED	TYPE	WIRE SIZE	STUD SIZE
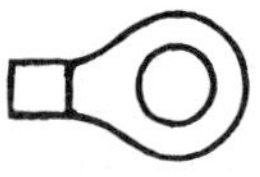	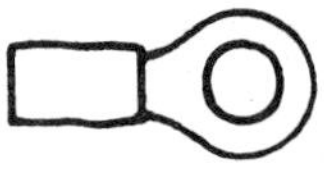	RING	16-14	10
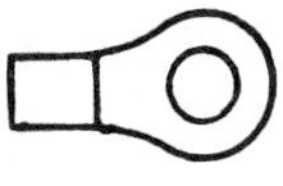	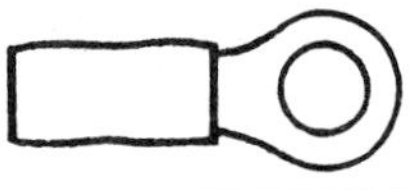	RING	12-10	10
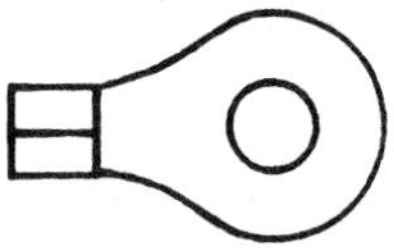	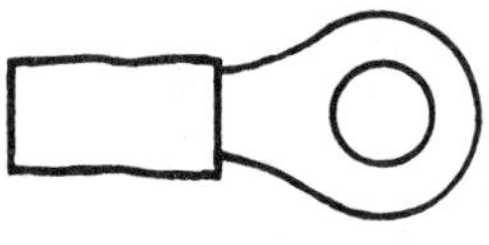	RING	12-10	1/4"
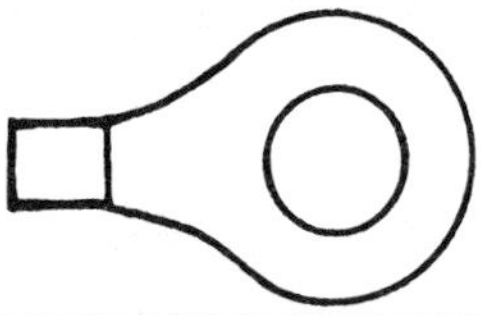	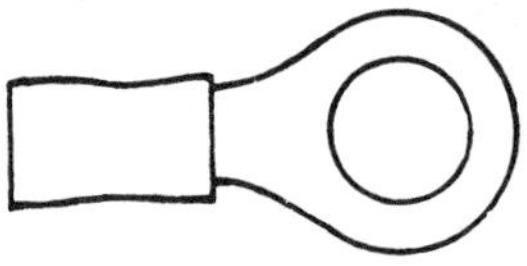	RING	12-10	3/8"
	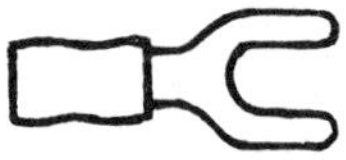	SPADE	16-14	10
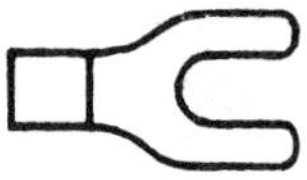	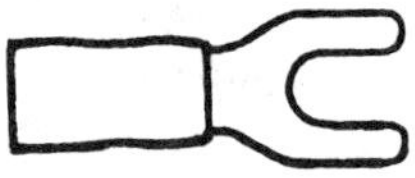	SPADE	12-10	10
		BUTT CONNECTOR	22-16	
		BUTT CONNECTOR	16-14	
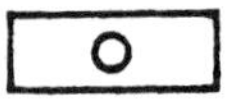		BUTT CONNECTOR	12-10	
		PLUG	16-14	
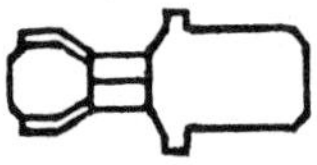		.250" TAB	18-14	
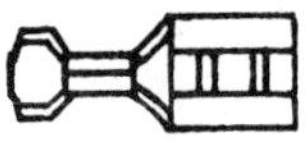		.250" RECEPTACLE	18-14	
	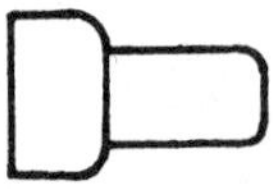	CLOSED END CONNECTOR	18-10	

Figure 3.17 Examples of some of the many types of solderless terminals and connectors used in automotive electrical service (courtesy of Easco/KD Tools, Lancaster, PA 17604).

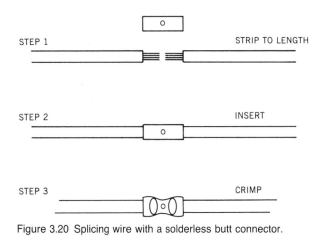

Figure 3.18 A crimping tool for installing solderless terminals. The tool shown will also cut and strip wire, and will cut and deburr small screws (©1985-Perfect Parts, Inc., Carlstadt, N.J.).

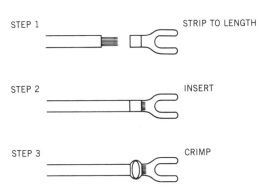

STEP 1   STRIP TO LENGTH

STEP 2   INSERT

STEP 3   CRIMP

Figure 3.19 Installing a solderless terminal.

be located with the use of simple tools and instruments.

Jumper Wires A jumper wire is probably the most commonly used diagnostic tool. As shown in Figure 3.23, a jumper wire is nothing more than a length of wire with a clip at each end. Most electricians keep several jumper wires of different lengths in their toolboxes. Used with a knowledge of basic electricity, a jumper wire can bridge or bypass switches and protection devices, complete circuits, and otherwise deliver current where it is needed. Figures 3.24 through 3.27 show some of the methods by which a jumper wire can be used for testing.

Most automotive electricians make their own jumper wires. Two lengths of 16-gauge wire, 12 in. and 24 in. long, are soldered to clips of the type shown in Figure 3.28. Jumper wires of those lengths are needed in many of the Jobs you will perform throughout this text. So that you will have them available, you will be required to make them in Job 3C.

STEP 1   STRIP TO LENGTH

STEP 2   INSERT

STEP 3   CRIMP

Figure 3.20 Splicing wire with a solderless butt connector.

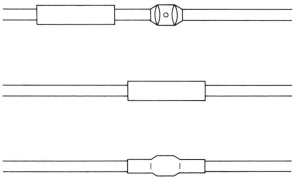

BUTT CONNECTOR TO FIT TWO WIRES

SINGLE WIRE WITH STRANDS DOUBLED BACK

Figure 3.21 A solderless butt connector can be used in this manner to splice three wires. The single wire should be doubled back so that the crimp will make a tight connection.

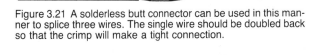

Figure 3.22 Heat shrink tubing often is used to insulate connectors. Centered over the connector, the tubing shrinks to form a tight, insulating seal when heated.

Test Lights Used almost as often as a jumper wire, a simple test light provides a quick and easy method of determining whether current is available

Job 3b

INSTALL SOLDERLESS TERMINALS AND CONNECTORS

SATISFACTORY PERFORMANCE
A satisfactory performance on this job requires that you do the following:

1 Install three solderless terminals and two solderless connectors.
2 Following the steps in the Performance Outline, cut, strip, and install the terminals and connectors on 16-gauge (1.0-mm) primary wire, as shown in the illustration below.
3 Complete the job within 15 minutes.

PERFORMANCE OUTLINE

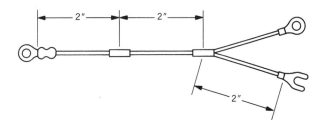

1 Cut the wire to the dimensions shown.
2 Strip the wire to fit the terminals and connectors.
3 Install the terminals and connectors.

Figure 3.23 A jumper wire is a basic diagnostic tool (courtesy of Ford Motor Company).

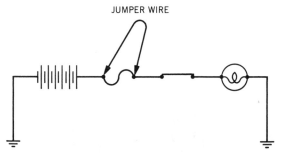

Figure 3.24 A jumper wire can be used to momentarily bridge or bypass a fuse. If the circuit operates with the jumper wire in place and does not operate when the jumper is removed, the fuse probably is defective or blown.

almost anywhere in a circuit. A typical commercially available test light, shown in Figure 3.29, consists of a bulb wired to a sharp steel probe and to a length of wire fitted with a clip. For convenience and protection, the bulb usually is shielded or enclosed in a plastic handle to which the probe is attached.

When the clip is attached to ground, the bulb will light when the probe is touched to any current-carrying conductor that is not on the ground side of

the load in a circuit. Figures 3.30 through 3.32 show how a test light can be used to check the components in a simple circuit.

Since a test light is another basic diagnostic tool that you will need, Job 3C requires that you build one. The following steps outline a procedure for making a test light:

1 Cut two 12-in. lengths of 16-gauge wire.
2 Strip ¼ in. of insulation from the ends of the wires.
3 Solder the end of one wire to the base of a #67

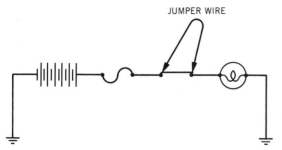

Figure 3.25 A jumper wire can be used to bridge or bypass a switch. If the circuit operates with the jumper wire in place and does not operate when the jumper switch is removed, the switch probably is defective.

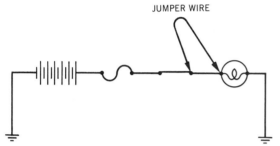

Figure 3.26 A jumper wire can be used to bridge any of the wires in a circuit. If the circuit operates with the jumper in place and does not operate with the jumper removed, the wire is broken or has a bad connection.

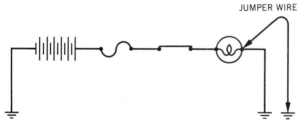

Figure 3.27 A jumper wire can be used to provide a new ground for a load. If the circuit operates with the jumper in place and does not operate when the jumper is removed, the load is improperly grounded.

bulb. (See Figure 3.33.)

Note: Other 12-V bulbs of the same base size may be substituted.

4 Bend one end of the remaining wire 90° and solder it to the contact on the base of the bulb. (Refer to Figure 3.33.)

Note: Tinning the end of the wire before attempting to solder it to the bulb will make this step easier.

5 Trim any excess solder from the bulb.
6 Slide a distributor cap nipple over the wires and push it up over the base of the bulb.
7 Solder clips to the ends of the wires.
8 Test the operation of the light by connecting it to the terminals of a battery.

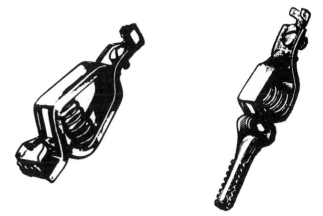

Figure 3.28 Test clips of various types are used for making jumper wires (©1985-Perfect Parts, Inc., Carlstadt, N.J.).

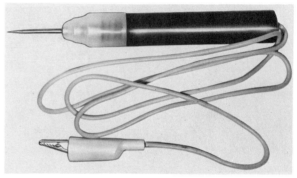

Figure 3.29 A simple test light (courtesy of American Motors).

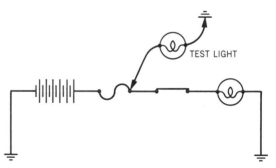

Figure 3.30 Connected as shown, the test lamp will light if the fuse is not blown.

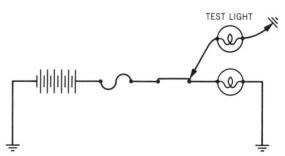

Figure 3.31 Connected as shown, the test lamp will light if the switch is in good condition. Opening and closing the circuit by turning the switch on and off should cause the test lamp to go on and off.

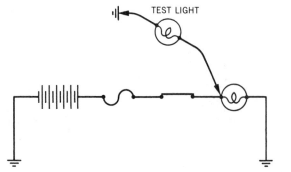

Figure 3.32 Connected as shown, a test lamp will indicate if voltage is being applied to the load.

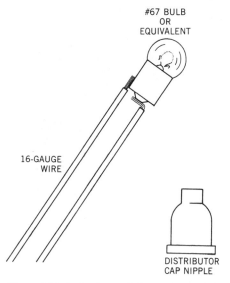

Figure 3.33 A simple test light can be made by soldering wires to the base of a bulb. The base of the bulb can then be protected and insulated with a distributor cap nipple.

Continuity Tester A continuity tester, often called a *self-powered test light,* is another simple tool that simplifies diagnosis. As shown in Figures 3.34 and 3.35, a continuity tester consists of a bulb and a battery connected so that when a complete circuit, or *continuity,* is established between the probe and the clip, the bulb will light.

Some automotive electricians do not use a continuity tester, but use a simple test light and a jumper wire connected as shown in Figure 3.36. You will use this method of testing in several Jobs throughout this text.

DIAGNOSTIC INSTRUMENTS While the simple diagnostic tools that you built will enable you to determine if voltage is available in a circuit and if current is flowing in a circuit, they cannot measure the amount of voltage and current. In many instances, it is not enough to know that voltage is

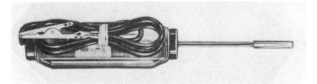

Figure 3.34 A commercial continuity tester. A small battery and bulb is enclosed in the plastic handle (©1985-Perfect Parts, Inc., Carlstadt, N.J.).

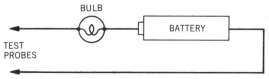

Figure 3.35 A continuity tester, or self-powered test light contains a small flashlight battery and a bulb. When continuity, or a complete circuit, is made between the test probes, the bulb will light.

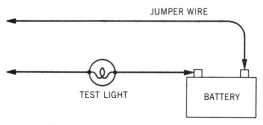

Figure 3.36 A test light and a jumper wire connected to a battery often are used to check for continuity.

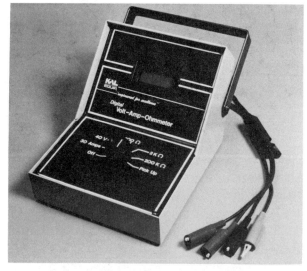

Figure 3.37 Some electronic meters do not have a typical scale. A digital readout of the value measured is provided by an LCD display (courtesy of Kal-Equip Company).

pushing current through a circuit. You also must know how much voltage is applied, how much voltage is available, and how much current is flowing. In some instances, you must know the resistance

Job 3c

CONSTRUCT JUMPER WIRES AND A TEST LIGHT

SATISFACTORY PERFORMANCE

A satisfactory performance on this job requires that you do the following:

1 Construct jumper wires and a test light.
2 Following the steps in the Performance Outline, build the tools to the specifications given within one hour.

PERFORMANCE OUTLINE

1 Cut two lengths of wire, one 12-in. long, one 24-in. long.
2 Strip ¼ in. of insulation from the ends of the wires.
3 Solder clips to the ends of the wires.
4 Cut two lengths of wire, each 12-in. long.
5 Strip ¼ in. of insulation from the ends of the wires.
6 Solder one end of each wire to a #67 bulb or a bulb with a similar base.
7 Install a distributor cap nipple on the base of the bulb.
8 Solder clips to the ends of the wires.

Figure 3.38 A typical meter that uses a pointer to indicate a reading on a numbered dial or scale (courtesy of Snap-on Tools Corporation).

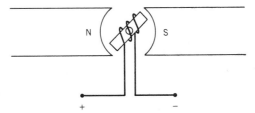

Figure 3.39 Most meters are of the d'Arsonval type. A small coil of wire is mounted on bearings between the poles of a permanent magnet. When current flows through the coil, the change in the magnetic field causes the coil to move.

of the various components in a circuit. Those measurements are made with meters.

Meters   While electronic meters that provide a digital readout as shown in Figure 3.37 are available, most automotive electricians use meters that have a pointer or needle that moves across a numbered dial or scale. Such a meter is shown in Figure 3.38.

Most meters are of the *moving coil* type. That design, usually referred to as a *D'Arsonval movement*, consists of a small coil of wire mounted on bearings between the poles of a permanent magnet. (See

Figure 3.39.) A pointer or needle is attached to the coil so that any movement of the coil is indicated on a dial or scale, as shown in Figure 3.40. A fine spiral spring holds the coil in the zero or balanced position between the magnetic poles. When current flows through the coil, a magnetic field is created. This magnetic field disturbs the balanced field between the poles of the permanent magnet and causes the coil to move, moving the pointer. (See Figure 3.41.)

The basic moving coil movement is used in all types of meters. In a voltmeter, various resistors are wired in series with the coil. In an ammeter, *shunts,* or bypasses, are wired in parallel with the coil. In an ohmmeter, the coil is wired in a circuit containing a small battery and various resistors.

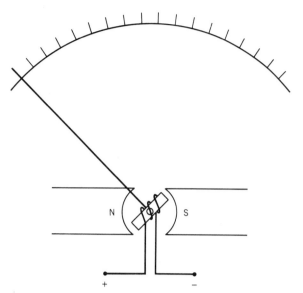

Figure 3.40 A pointer attached to the coil indicates any movement in the coil on a scale.

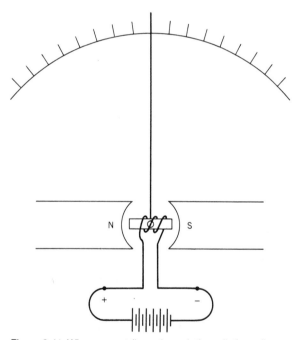

Figure 3.41 When current flows through the coil, the coil moves, causing the pointer to move on the scale.

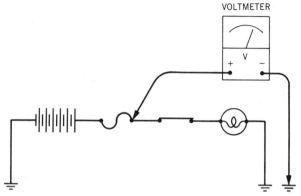

Figure 3.42 To measure the voltage applied to a circuit, a voltmeter is connected in parallel.

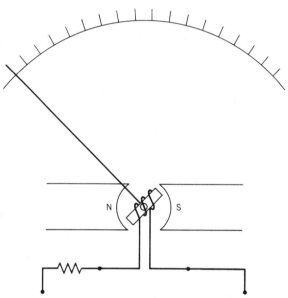

Figure 3.43 When used as a voltmeter, a moving coil movement is wired in series with a resistor. The resistor determines the range of the meter and limits the amount of current that will flow through the coil.

Voltmeters As a voltmeter measures electrical pressure, it is normally connected in parallel with the voltage source, as shown in Figure 3.42. To limit the amount of current that will flow through the coil, a resistor is wired in series with the coil. (See Figure 3.43.) The value of this resistor is chosen by the meter manufacturer so that the meter will measure voltage within a certain range. Some meters have a switch that allows the user to select different

ranges. That switch places different resistors in the meter circuit, as shown in Figure 3.44. A meter of that type will usually have a separate scale for each voltage range.

Since the combined resistance of the coil and the resistor(s) in a voltmeter is usually quite high, the instrument will allow the passage of only a very small amount of current. This allows the meter to be used in a circuit without affecting voltage drop and provides a safety feature that protects the meter from damage from excessive current.

Voltmeter Use While the high resistance of a voltmeter provides some protection, a voltmeter can be damaged by improper use. Before attempting to use a voltmeter, you must determine the correct voltage range and set the meter for that range. Many

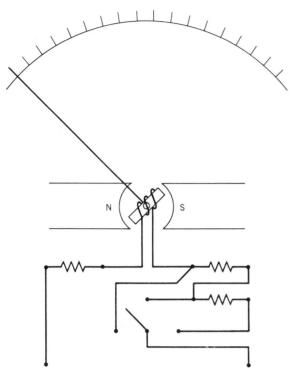

Figure 3.44 Internal circuitry of a typical voltmeter. A rotary switch places resistors in the circuit to provide different ranges.

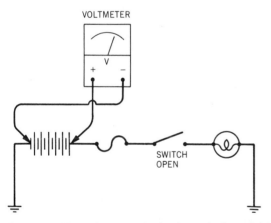

Figure 3.45 Measuring open circuit voltage. As the switch is open and there is no load on the battery, this measurement often is referred to as the "no-load" voltage. Notice the meter polarity.

automotive voltmeters have a switch that allows you to select a range appropriate to the circuit voltage. Some meters provide two ranges of 0 to 20 V and 0 to 40 V. Other meters provide a third range of from 0 to 4 or 5 V. That low voltage range is used for testing voltage drop.

When working on most automotive circuits powered by a 12-V battery, the 0- to 20-V range should be selected. When working on a 24- or 36-V system, the 0- to 40-V range should be used.

The wire leads of a voltmeter are marked as to their

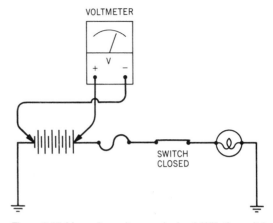

Figure 3.46 Measuring voltage under load. With the switch closed, the meter will indicate the source voltage while current is being pushed through the load.

polarity. The positive (+) lead should be connected to the positive (+) side of the circuit. The negative (−) lead should be connected to the negative (−) side of the circuit. (Refer to Figure 3.42.)

Depending on how and where a voltmeter is connected in a circuit, you can measure the source voltage, the voltage at any part in the circuit, and the voltage drop caused by any load or resistance in the circuit. In addition, a voltmeter can be used to check circuit continuity.

Measuring Source Voltage Connected across the terminals of a battery as shown in Figure 3.45, a voltmeter will indicate the source voltage of a circuit. When the circuit is *open,* or turned off, this measurement may be referred to as the *open-circuit voltage,* the *no-load voltage,* or the *base voltage.* In a 12-V system, an open-circuit voltage of less than 12 V usually indicates that the battery is discharged or defective. A measurement of open-circuit voltage is often made as the first check in the diagnosis of an electrical system problem.

As shown in Figure 3.46, source voltage also can be measured when a circuit is *closed,* or turned on. This measurement is usually referred to as the *voltage under load* and will usually be slightly lower than the open-circuit voltage. The difference between the open-circuit voltage and the voltage under load is dependent on the amount of current flowing in the circuit and the condition of the battery. For certain diagnostic procedures, that difference is often given as a specification.

Measuring Voltage in a Circuit In the previous chapter you learned that a load will function properly in a circuit only when the applied voltage is sufficient to push the required current through the

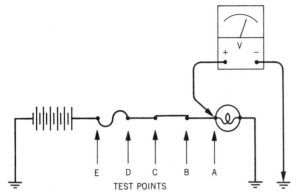

Figure 3.47 Connected in this manner, a voltmeter will measure the voltage applied to the load. If the meter indicates approximately the same as the voltage under load, there is no excessive resistance in the circuit. If the applied voltage is lower than the voltage under load, the positive (+) lead of the voltmeter should be moved to test point B.

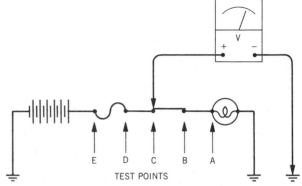

Figure 3.49 If the voltage at point C is approximately the same as the voltage under load, there is excessive resistance in the switch and its connections. If the voltage is approximately the same as the voltage indicated at point A, the excessive resistance is between point C and the battery. The test sequence should be continued.

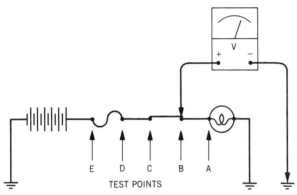

Figure 3.48 If the voltage at test point B is approximately the same as the voltage under load, there is excessive resistance in the wire and the connections between points A and B. If voltage is approximately the same as the voltage indicated at point A, move to test point C.

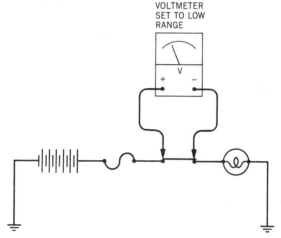

Figure 3.50 A voltmeter connected in this manner will measure the voltage drop across a component suspected of having high resistance.

load. Even though the source voltage may be sufficient, the voltage drop caused by resistance in a circuit may result in a low applied voltage at the load. Figure 3.47 illustrates how a voltmeter can be used to measure the applied voltage at the load. If the applied voltage at the load is approximately the same as the voltage under load when measured at the source, the circuit from the source to the load can be considered free of excessive resistance. If a considerable voltage drop is evident, a sequence of additional tests should be made. As shown in Figures 3.48 and 3.49, those tests consist of the measurement of voltage at various points while moving back in the circuit toward the power source. The area of high resistance will be found between the test point where the approximate voltage under load is found and the test point immediately preceding it.

Measuring Voltage Drop   When excessive resistance is discovered in a circuit, a voltmeter can be used to isolate the defective part or poor connection that is causing the problem. Connected across or in parallel with the suspected part or connection as shown in Figure 3.50, the voltmeter provides a bypass or alternate route for the flow of current around the high resistance. This current flow will cause a reading on the meter scale. In many circuits, an acceptable voltage drop will be given as a specification.

Checking Circuit Continuity   Connected in series in a circuit, as shown in Figure 3.51, a voltmeter can be used to check continuity. When the circuit is closed, a reading of the source voltage should be obtained. If no voltage is indicated, the circuit is not complete.

Job 3d

MEASURE OPEN-CIRCUIT VOLTAGE AND VOLTAGE UNDER LOAD

SATISFACTORY PERFORMANCE
A satisfactory performance on this job requires that you do the following:

1 Using a voltmeter, measure open-circuit voltage and voltage under load on the car assigned.
2 Following the steps in the Performance Outline, complete the job within 10 minutes.
3 Fill in the blanks under "Information."

PERFORMANCE OUTLINE
1 Select the appropriate range on the voltmeter.
2 Connect the voltmeter to the terminals of the battery.
3 Record the meter reading as the open circuit voltage.
4 Turn on the headlights and place the dimmer switch in the high beam position.
5 Record the meter reading as the voltage under load #1.
6 Turn on the windshield wipers.
7 Record the meter reading as the voltage under load #2.
8 Turn off the wipers and the headlights.
9 Disconnect the voltmeter.

INFORMATION
Vehicle identification _____

Voltmeter identification _____

Open-circuit voltage _____

Voltage under load #1 _____

Voltage under load #2 _____

Ammeters Since an ammeter measures the amount of current that flows in a circuit, the current must flow through the ammeter. For current to flow through an ammeter, the meter must be connected in series in the circuit, as shown in Figure 3.52.

Most of the current that flows through an ammeter passes through the shunt built into the meter. (See Figure 3.53.) The shunt is designed to pass a small percentage of the current through the coil of the meter, which indicates the total current flow on the meter scale. The shunt determines the range of the meter, and the meter scale is calibrated to be accurate with that shunt. Some ammeters contain two or more shunts that may be selected by a switch.

Meters of that type, shown in Figure 3.54, usually have several scales to provide different ranges.

As all the current that flows in the circuit must pass through the ammeter, the shunt(s) must add no resistance to the circuit. This means that an ammeter can easily be damaged if it is incorrectly connected in a circuit.

Ammeter Use Since an ammeter has very low resistance, you must be extremely careful in connecting it in a circuit. IF AN AMMETER IS CONNECTED IN PARALLEL IN A CIRCUIT, OR IF IT IS CONNECTED ACROSS THE TERMINALS OF A BATTERY, IT MAY BE DAMAGED OR DESTROYED.

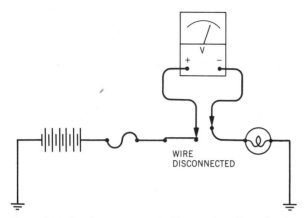

Figure 3.51 A voltmeter connected in a series with a circuit as shown will indicate source voltage if the circuit has continuity.

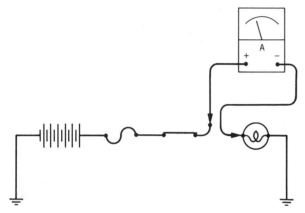

Figure 3.52 An ammeter should be connected in series in a circuit. This usually requires that a connection be disconnected as shown so that all the current that flows in the circuit flows through the ammeter. Notice the meter polarity.

Some automotive ammeters have a switch that allows you to select a range appropriate to the circuit. (See Figure 3.55.) For most automotive electrical work, a range of from 0 to 50 A is usually selected. For tests in the starting system, a range of from 0 to 500 A may be required.

The wire leads of an ammeter are marked to indicate their polarity. The positive (+) lead should be connected in the circuit so that it is closest to the positive (+) terminal of the battery. The negative (−) lead should be connected so that it is closest to the negative (−) terminal of the battery. (Refer to Figure 3.52.)

Current flow in a circuit is the same throughout the circuit; therefore, an ammeter can be connected at any point in a circuit. Since the circuit must be broken, or disconnected, to insert the ammeter in series, most mechanics find it easiest to disconnect the battery ground cable from the battery and to connect the ammeter as shown in Figure 3.56. When

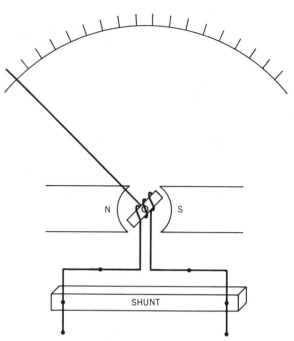

Figure 3.53 The internal circuitry of a typical ammeter. Most of the current that flows through the ammeter passes through a shunt, which allows only a small percentage of current to flow through the meter coil.

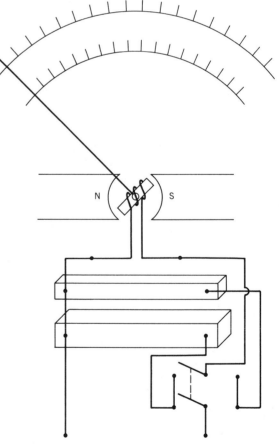

Figure 3.54 The range of an ammeter can be extended by the use of different shunts. In this example, a switch selects the shunt and scale used.

Figure 3.55 The ammeter on this tester has two ranges, −120 to +120 and −500 to +500 A. The appropriate range is selected by a switch (courtesy of Snap-on Tools Corporation).

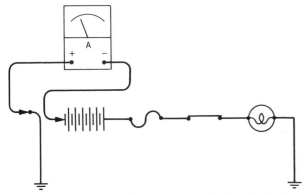

Figure 3.56 By disconnecting the ground cable from the battery terminal, an ammeter can be connected in series as shown. Connected in this manner, the current flow in almost all the electrical circuits can be measured.

connected in this manner, the current flow in almost all the circuits in the electrical system can be measured. DO NOT ATTEMPT TO START THE ENGINE WHEN AN AMMETER IS CONNECTED IN THIS MANNER. TO DO SO MAY DESTROY THE AMMETER.

Inductive Ammeters   Some shops use inductive ammeters in addition to conventional ammeters. An inductive ammeter requires no direct electrical connection in a circuit, and therefore saves the time required to disconnect and connect wires. The saving in time, however, is partially offset by a slight loss in meter accuracy, especially when small amounts of current are being measured.

An inductive ammeter is operated by the magnetic

Figure 3.57 Some ammeters measure current flow by induction. A clamp containing a sensing coil is positioned over a conductor in a circuit to measure the strength of the magnetic field created.

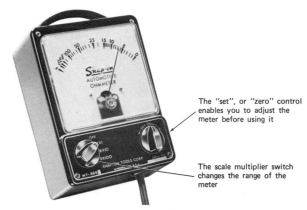

The "set", or "zero" control enables you to adjust the meter before using it

The scale multiplier switch changes the range of the meter

Figure 3.58 A typical automotive ohmmeter. Notice the controls that enable you to adjust the meter and select the resistance range (courtesy of Snap-on Tools Corporation).

field that is induced when current flows through a conductor. A special pick-up, or clamp, shown in Figure 3.57, containing a small coil is positioned over one of the wires in the circuit to be tested. The strength of the magnetic field is sensed by the coil and converted to an amperage reading on the meter scale.

As with a conventional ammeter, the polarity of an inductive ammeter must be observed when the clamp is placed over the wire. Most clamps are marked with an arrow to indicate polarity. The clamp should be placed on the wire so that the arrow points toward the positive (+) battery terminal or away from the negative (−) battery terminal.

Ohmmeters   An ohmmeter, shown in Figure 3.58, measures the resistance of any conductor. It can be used to measure the resistance of a load, a control, a safety device, and the wire that connects those parts. As shown in Figure 3.59, an ohmmeter has its own built-in power source. To provide different ranges of resistance, most ohmmeters contain resistors that can be selected by a switch. Most ohmmeters also have a small adjustable resistor that allows you to adjust, or "zero," the meter before each use. (Refer to Figure 3.58.) This adjustment provides compensation for the gradual discharge of the small battery contained in the meter.

Job 3e

MEASURE CURRENT FLOW IN ELECTRICAL CIRCUITS

SATISFACTORY PERFORMANCE
A satisfactory performance on this job requires that you do the following:

1 Using an ammeter, measure the current flow in the headlight and wiper circuits of the car assigned.
2 Following the steps in the Performance Outline, complete the job within 20 minutes.
3 Fill in the blanks under "Information."

PERFORMANCE OUTLINE
1 Select the appropriate range on the ammeter.
2 Disconnect the ground cable from the battery terminal.
3 Connect the ammeter to the battery terminal and ground.
4 Turn on the parking lights.
5 Record the meter reading.
6 Turn on the headlights and place the dimmer switch in the high beam position.
7 Record the meter reading.
8 Turn on the windshield wipers.
9 Record the meter reading.
10 Turn off the wipers and the headlights.
11 Disconnect the ammeter.
12 Connect the ground cable to the battery terminal.

INFORMATION
Vehicle identification _____

Ammeter identification _____

Current flow with parking lights on _____

Current flow with headlights on _____

Current flow with headlights and wipers on _____

An ohmmeter is calibrated so that when the test leads are held in contact with each other, the meter will indicate no resistance (0 Ω). If the test leads are separated, the meter will indicate an infinite amount of resistance (∞). This provides a range of resistance from that of a perfect conductor (0 Ω) to that of a perfect insulator (∞). The resistance of any part or wire connected between the test leads will be somewhere within that range and will be indicated on the meter scale.

An ohmmeter is a very delicate instrument in that it can safely handle only the slight current provided by its own power source. If the test leads are con-nected to any external power source, the meter may be damaged or destroyed.

Ohmmeter Use   As shown in Figure 3.60, most ohmmeters have a "backwards" scale when compared to voltmeters and ammeters. When the meter is at rest and the test leads are not in contact, the meter indicates infinity (∞). Many ohmmeters have only one set of numbers on the scale. Different ranges of resistance are provided by turning a switch that selects the ranges in multiples of 10. The switch positions are often marked X1, X10, X100, and X1000. (Refer to Figure 3.58.) Depending on the

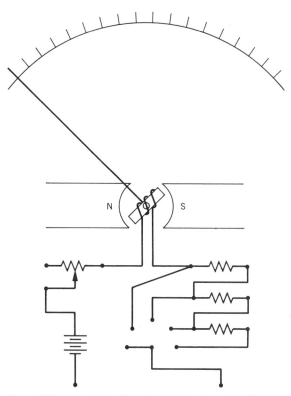

Figure 3.59 Internal circuitry of a typical ohmmeter. The instrument is powered by a small battery, and a variable resistor is used to "zero" the meter. A rotary switch is used to select different combinations of resistors to provide different ranges.

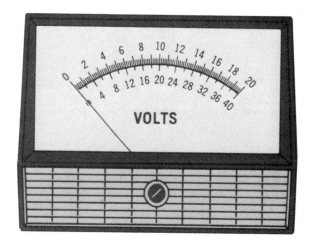

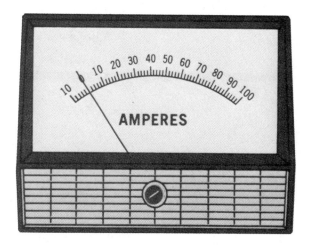

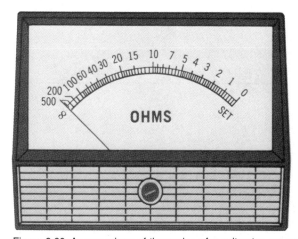

Figure 3.60 A comparison of the scales of a voltmeter, an ammeter, and an ohmmeter. Note that on the ohmmeter the scale reads from right to left (courtesy Ford Motor Company).

position of the switch, the resistance indicated on the scale is corrected by mentally adding the appropriate number of zeros.

An ohmmeter cannot be damaged if the wrong range is selected. In fact, when measuring the resistance of a part, you should start with the selector set to the lowest range. If necessary, a higher range can then be selected to obtain the most accurate meter reading. As an example, if you were measuring the resistance of the filament in a bulb, you may find that it has a resistance of 1 Ω. If the X1 range were selected, that resistance would be accurately indicated. If a higher range were selected, the resistance of 1 Ω probably could not be detected on the meter scale. In another example, a spark-plug wire with a resistance of 15,000 Ω can be most accurately checked with the meter set to the X1000 range.

Since an ohmmeter contains its own power source, the test leads usually are not marked as to their polarity. With but few exceptions, an ohmmeter will measure resistance and indicate continuity without regard to polarity. Those exceptions occur when testing diodes or components containing diodes, and will be covered in Chapter 11, where an

ohmmeter is used during the overhaul of alternators.

When using an ohmmeter, the part or wire to be

Job 3f

CHECK FOR CONTINUITY AND MEASURE THE RESISTANCE
OF ELECTRICAL SYSTEM COMPONENTS

SATISFACTORY PERFORMANCE
A satisfactory performance on this job requires that you do the following:

1 Using an ohmmeter, check the continuity and
measure the resistance of the parts supplied.
2 Following the steps in the Performance Outline,
complete the job within 15 minutes.
3 Fill in the blanks under "Information."

PERFORMANCE OUTLINE
1 "Zero" the ohmmeter.
2 Check the continuity and measure the resistance
of the parts supplied.
3 Turn off the meter.

INFORMATION
Ohmmeter identification _____

Part Tested	Continuity		Resistance
Small bulb (#)	Yes ____	No ____	_____
Sealed beam bulb (#)	Yes ____	No ____	_____
Fuse (#)	Yes ____	No ____	_____
Spark-plug wire	Yes ____	No ____	_____
Alternator rotor	Yes ____	No ____	_____
_____	Yes ____	No ____	_____
_____	Yes ____	No ____	_____

tested should be disconnected from the circuit, as shown in Figure 3.61. AN OHMMETER SHOULD NOT BE CONNECTED TO ANY PART OF A LIVE CIRCUIT, AND SHOULD NEVER BE CONNECTED ACROSS THE TERMINALS OF A BATTERY. TO DO SO MAY DAMAGE OR DESTROY THE INSTRUMENT.

The resistance of certain electrical system parts is specified in service manuals, and an ohmmeter often provides the only means to determine if those parts are serviceable. In addition to measuring the resistance of parts, most automotive electricians use an ohmmeter to check continuity and to check for unwanted resistance in controls, wires, and connections.

Multimeters Some diagnostic instruments, often called multimeters, combine two or more meters in a single housing. One such instrument, shown in Figure 3.62, contains a voltmeter and an ammeter, and incorporates a variable load called a *carbon pile*. You will use an instrument such as this when you work with batteries and starting systems in Chapters 5 and 8.

As shown in Figures 3.63 and 3.64, other instruments combine the function of several separate meters by providing different scales on the face of a single meter. You will use an instrument of this type when you work on ignition systems in Chapter 15.

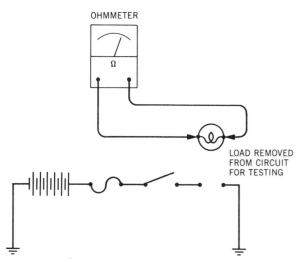

Figure 3.61 To avoid damage to an ohmmeter, the part to be tested should be removed from the circuit.

Figure 3.63 In this instrument, the functions of a voltmeter and an ammeter are combined by means of multiple scales on the same meter. The function desired is selected by a switch (courtesy of Snap-on Tools Corporation).

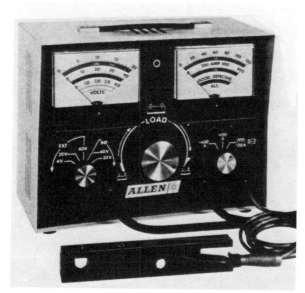

Figure 3.62 This instrument combines a voltmeter and an ammeter with a carbon pile. Instruments of this type are used to test batteries and to diagnose problems in starting systems (courtesy of Allen Group).

Although those instruments may appear different from the individual meters that you used in this chapter, they all are basically the same. They must be connected in the same manner, and the same precautions must be taken in their use.

The specific operation of the switches and controls should be obtained from the manual furnished with each instrument. Your knowledge of meter usage, combined with a few trial hookups made with the aid of the manual, will enable you to become familiar with the various meters you have available.

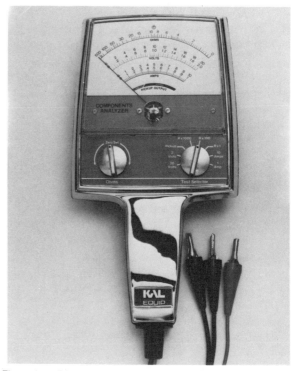

Figure 3.64 The functions of a voltmeter, ammeter, and ohmmeter are combined in this instrument. The scale desired is selected by a switch (courtesy of Kal-Equip Company).

SUMMARY

By studying the material in this chapter and completing the Jobs provided, you have gained additional knowledge and developed some of the basic skills required of an automotive electrician. You are now aware of the correct wire gauge requirements for different circuits and can select the proper wire, make splices, and install connectors and terminals. You are familiar with the use of jumper wires and test lights for quick diagnostic checks. You also have gained skills in the use of voltmeters, ammeters, and ohmmeters. That knowledge and those skills will enable you to perform many of the diagnostic tests required in the following chapters.

SELF-TEST

Each incomplete statement or question in this test is followed by four suggested completions or answers. In each case select the *one* that best completes the sentence or answers the question.

1 The resistance of a wire increases with
  A. its length
  B. its diameter
  C. its cross-sectional area
  D. the number of its strands

2 Two mechanics are discussing wire splicing.
  Mechanic A says that a good physical connection is necessary for a good splice.
  Mechanic B says that a good electrical connection is necessary for a good splice.
  Who is right?
  A. A only
  B. B only
  C. Both A and B
  D. Neither A nor B

3 Two mechanics are discussing soldering.
  Mechanic A says that flux is used to remove slight traces of oxidation and to prevent further oxidation.
  Mechanic B says that an acid-base flux should be used when soldering in electrical systems.
  Who is right?
  A. A only
  B. B only
  C. Both A and B
  D. Neither A nor B

4 Two mechanics are discussing terminals.
  Mechanic A says that a ring or spade terminal should be used when attaching a wire to a screw or stud.
  Mechanic B says that solderless terminals should be installed with special crimping pliers.
  Who is right?
  A. A only
  B. B only
  C. Both A and B
  D. Neither A nor B

5 The electrical pressure in a circuit can be measured with
  A. an ammeter
  B. an ohmmeter
  C. a voltmeter
  D. a wattmeter

6 The amount of current flowing in a circuit can be measured with
  A. an ammeter
  B. an ohmmeter
  C. a voltmeter
  D. a wattmeter

7 Two mechanics are discussing the use of an ammeter.
  Mechanic A says that an ammeter should be connected across the terminals of a battery.
  Mechanic B says that an ammeter should be connected in parallel in a circuit.
  Who is right?
  A. A only
  B. B only
  C. Both A and B
  D. Neither A nor B

8 The resistance of the components of an electrical circuit can be measured with
  A. an ammeter
  B. an ohmmeter
  C. a voltmeter
  D. a wattmeter

9 When measuring the resistance of a part, the first measurement should be made with the range selector set at
  A. X1
  B. X10
  C. X100
  D. X1000

10 Two mechanics are discussing the use of an ohmmeter.
  Mechanic A says that an ohmmeter should be connected across the terminals of a battery.
  Mechanic B says that an ohmmeter should be connected in parallel in a circuit.
  Who is right?
  A. A only
  B. B only
  C. Both A and B
  D. Neither A nor B

Chapter 4
The Battery—Operation, Construction, and Basic Services

All too often, the battery in a car is ignored until it fails to crank the engine. Annual surveys repeatedly list battery failure as one of the most common reasons for requiring road service. Failure to maintain the battery can cause more than inconvenience. Lack of proper maintenance will considerably shorten the useful life of a battery.

The most commonly performed battery services include cleaning, testing, charging, and replacement. Those services require certain knowledge, diagnostic skills, and repair skills. In this chapter, you will be given the opportunity to gain that knowledge and to develop some of those skills. Your specific objectives are to perform the following jobs:

A
Identify terms relating to battery operation and construction
B
Identify the component parts of a battery
C
Check and adjust battery electrolyte level
D
Inspect a battery, hold-down, and carrier tray
E
Clean battery terminals and cable ends
F
Remove and install a battery
G
Use a booster battery to start an engine

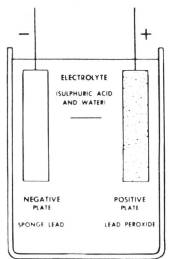

ELECTROLYTE

(SULPHURIC ACID AND WATER)

NEGATIVE PLATE

POSITIVE PLATE

SPONGE LEAD

LEAD PEROXIDE

THE BATTERY   The battery in an automobile serves four important functions:

1 It furnishes the electrical energy that cranks the engine for starting.
2 It supplies the current needed to operate the ignition system while starting.
3 It provides current to the electrical system when the engine is not running or when the demands of the system exceed the output of the charging system.
4 It acts as a stabilizer or "cushion" in the electrical system to minimize variations in system voltage.

Unless the battery performs each of these functions, the electrical system and its related systems will not operate properly.

Battery Operation   While a battery may sometimes be referred to as a "storage battery," a battery does not store electricity. A battery stores chemicals that react to produce electrical energy. A battery operates through the application of a simple scientific principle. When two different metals are immersed in an acid solution, an *electrochemical action* takes place that causes a difference in electrical pressure to exist between them.

In operation, electrons are given up by one metal and added to the electrons of the other. This transfer of electrons causes a difference of *potential*, or voltage, to exist between the two metals. The metal that has given up electrons is called the *positive plate* as it has a positive charge. The metal that has received the electrons becomes negatively charged and is referred to as the *negative plate*. Figure 4.1 shows how that difference in potential, or voltage, causes the flow of electricity.

Most automobile batteries are of the *lead-acid* type. In a lead-acid battery, the different metals are *lead peroxide* (PbO_2) and *sponge lead* (Pb). The lead peroxide forms the positive (+) plate. The sponge lead forms the negative (−) plate. The acid is *electrolyte*, a mixture of sulfuric acid (H_2SO_4) and water (H_2O). The plates and the electrolyte are contained in a *cell* or compartment. A simple lead-acid cell is shown in Figure 4.2.

A lead-acid cell can be repeatedly discharged and charged. This is possible because the electrochemical action in the cell can be reversed. If a higher external voltage is applied to the plates, the electron flow is reversed. The energy that is used to start an engine thus can be restored by the charging system when the engine is running. If such were not the case, a battery would have to be discarded after a short period of use. Throughout its life, a battery is *cycled*, or partially discharged and recharged, thousands of times. At any time in its life, a battery will be in one of the following states:

Charged   Figure 4.3 shows the chemical condition of a fully charged battery. The positive plate consists of lead peroxide, the negative plate is sponge lead, and the electrolyte is at its full strength.

Discharging   When a load is placed across the plates, the battery discharges. The difference in electrical pressure between the plates causes cur-

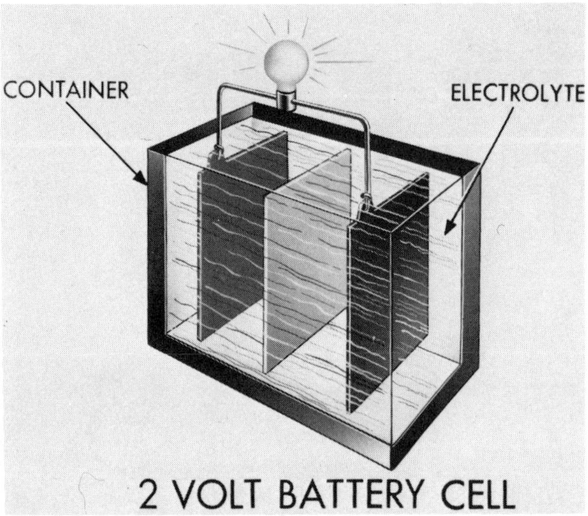

2 VOLT BATTERY CELL

Figure 4.1 The difference in electrical pressure between the plates of a simple cell causes the flow of current. In this example, the current flows through the filament of a bulb (courtesy of Chevrolet Motor Division).

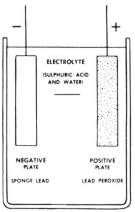

CHARGED

Figure 4.2 The composition of a simple lead-acid cell (courtesy of Exide Corporation).

rent to flow. While that current is flowing, a chemical change takes place in the plates and in the electrolyte. As shown in Figure 4.4, the electrolyte becomes diluted and the plates become *sulfated*. Actually, the electrolyte (H_2SO_4) divides into hydrogen (H_2) and a sulfate radical (SO_4). The hydrogen (H_2) combines with some of the oxygen (O) formed at the positive plate and produces additional water (H_2O). The sulfate (SO_4) combines with the lead (Pb) in both plates and forms lead sulfate ($PbSO_4$).

Discharged   If a continuous load is placed on a battery, the electrochemical action continues until the battery is discharged. As shown in Figure 4.5, the electrolyte has been diluted until it is mostly water. The plates have both become lead sulfate ($PbSO_4$), and they are no longer dissimilar.

Charging   When a battery is being charged, the electrochemical action is reversed as shown in Figure 4.6. The lead sulfate ($PbSO_4$) is broken down, restoring the plates to their original composition, lead peroxide (PbO_2) and sponge lead (Pb). The sulfate (SO_4) combines with the hydrogen (H_2) to form sulfuric acid (H_2SO_4). The sulfuric acid increases the strength of the electrolyte.

Cell Voltage   A fully charged lead-acid cell produces a maximum pressure of about 2.2 V. Internal resistance reduces that pressure slightly, and a lead-acid cell is usually referred to as a 2-V cell. Increasing the size of the plates or increasing the number of plates in a cell will not increase the voltage. In order to obtain higher voltage, cells must be connected in series, positive to negative, as shown in

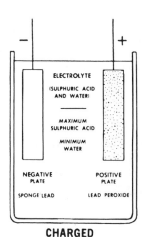

CHARGED

Figure 4.3 The chemical condition of a fully charged battery (courtesy of Exide Corporation).

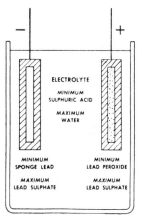

DISCHARGED

Figure 4.5 The chemical condition of a discharged battery (courtesy of Exide Corporation).

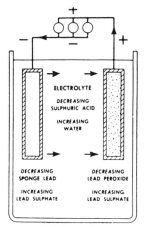

DISCHARGING

Figure 4.4 The chemical condition of a battery that is discharging (courtesy of Exide Corporation).

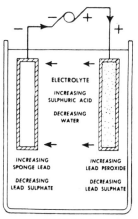

CHARGING

Figure 4.6 The chemical condition of a battery that is being charged (courtesy of Exide Corporation).

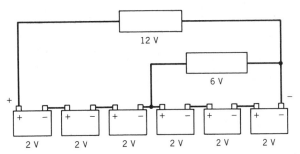

Figure 4.7 Cells are connected in series to form batteries. A 6-V battery has three cells. A 12-V battery has six cells.

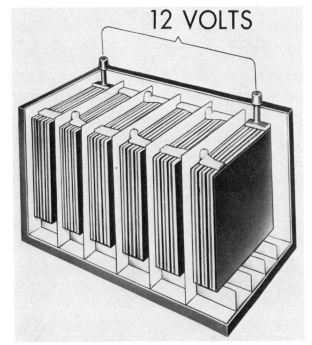

Figure 4.8 Typical arrangement of the cells in a 12-V battery (courtesy of Chevrolet Motor Division).

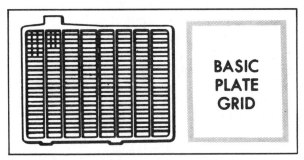

Figure 4.9 A basic plate grid is a screen or mesh made of lead alloyed with antimony (courtesy of General Motors Corporation).

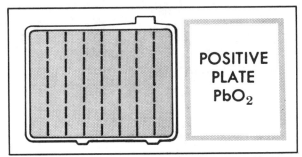

Figure 4.10 A positive plate is formed by coating a grid with lead peroxide (courtesy of General Motors Corporation).

Figure 4.11 A negative plate is formed by coating a grid with sponge lead (courtesy of General Motors Corporation).

Figure 4.7. Figure 4.8 shows how the cells are arranged in a typical 12-V battery.

Conventional Battery Construction   The internal parts of a battery are not normally visible nor can they be serviced. Yet, an understanding of those parts and how they are assembled is important if you are to interpret correctly the results of the various battery tests you will perform. There are several material and design differences between conventional batteries and those described as "maintenance-free." Those differences will be noted later in this chapter.

Grids   The plates of a lead-acid battery are built on *grids*. A grid, shown in Figure 4.9, is a screen or mesh that forms the framework of a plate. A grid usually is made of an alloy of lead and antimony as pure lead is too soft and lacks the required strength.

Positive Plates   A positive plate is built by filling all the openings in a grid with *active material*. The active material for a positive plate is a paste made of lead oxide. The lead oxide is then converted to a hard, porous, dark brown lead peroxide (PbO_2) by giving the coated plate a "forming" charge. (See Figure 4.10.)

Negative Plates   A negative plate is formed by coating a grid with a different active material. The

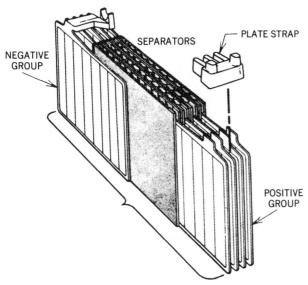

Figure 4.12 Plates are joined together by plate straps. A number of similar plates joined together is refered to as a plate group (courtesy of General Motors Corporation).

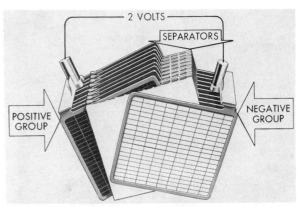

Figure 4.13 A group of positive plates and a group of negative plates are combined to form an element. Insulating separators are used to keep the plates from touching each other (courtesy of Chevrolet Motor Division).

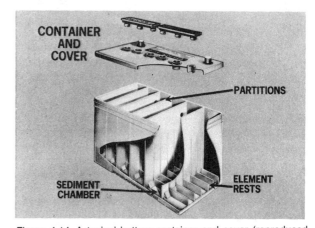

Figure 4.14 A typical battery container and cover (reproduced through the courtesy of Battery Council International).

active material used to build a negative plate consists of powdered lead (Pb) and an inert "expander" that keeps the lead porous or spongy. (See Figure 4.11.)

Separators Since the plates in a cell must not come into contact with each other, *separators* must be placed between them. In addition to being acid-resistant, separators must be porous to allow electrolyte to pass through them. Various materials are used for separators including cedar, rubber, plastic, and fiberglass.

Plate Groups A plate group is a number of similar plates (positive or negative) joined together at their tops by a *plate strap* as shown in Figure 4.12. The plate strap contains a lug or post that allows it to be connected to another plate group in the next cell. The number of plates in a group determines the *capacity* of the battery. Increasing the number of plates enables the battery to deliver more energy. When positive and negative plate groups are built for a particular battery, the negative group is built with one more plate than the positive group. By increasing the total area of the negative plates, the efficiency of the battery is increased.

Elements An *element* is comprised of a positive plate group and a negative plate group placed together in an interlocking fashion as shown in Figure 4.13. Separators are placed between each of the plates so that they do not touch. Since the negative group has one more plate than does the positive

group, the assembled element has negative plates exposed on both sides.

The Battery Case The battery case or *container* is molded of hard rubber or plastic. The case contains partitions that divide it into separate cells. The bottom of each cell has several *bridges* molded into place. Those bridges act as element rests that support the bottoms of the plates. When a battery is in service, electrochemical action and vibration combine to dislodge particles of active material from the plates. That active material eventually settles to the bottom of the battery container as sediment. Since this sediment is a conductor, it will cause a short in the cell if it contacts the bottoms of the plates. The spaces between the bridges form *sediment chambers* that collect the active material that is shed from the plates. Sufficient space is provided to collect any sediment that may accumulate during the useful life of the battery. Figure 4.14 shows the design of a typical battery container.

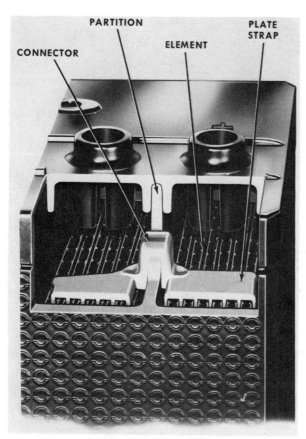

Figure 4.15 A cutaway view of a battery showing how the cell connectors pass through the partitions in the battery case (courtesy of Chevrolet Motor Division).

Figure 4.16 In some battteries, the cell connectors arch over the cell partition walls (reproduced through the courtesy of Battery Council International).

Connectors When all the elements are in place in each cell of the battery case, the plate straps of each element are joined by *connectors* so that the cells are connected in series. In some batteries, the cell connectors pass through holes in the case partitions as shown in Figure 4.15. In other designs, the connectors arch over the cell partitions as shown in Figure 4.16.

Cell Covers Most batteries have a one-piece cover that is cemented or bonded to the battery case. The cover provides a seal over each partition and contains a baffled opening for each cell as shown in

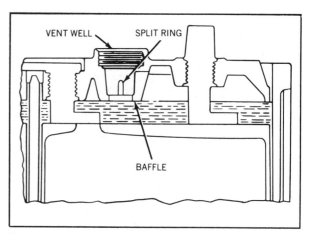

Figure 4.17 The vent holes in the cover act as baffles to deflect splashed electrolyte. The split ring provides a means of checking the electrolyte level (courtesy of General Motors Corporation).

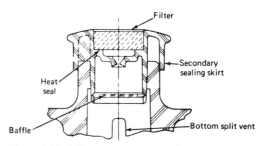

Figure 4.18 A flame arrestor vent plug. The plug shown incorporates a filter that allows the escape of gases but will not allow the entry of a hydrogen flame. A plug of this type prevents a flame or spark from entering the battery and causing an explosion (courtesy of AC-Delco, General Motors Corporation).

Figure 4.17. The opening provides access to the cell for the addition of electrolyte, for testing the electrolyte, and for adding water when it is required. The baffled opening acts to deflect any electrolyte that may be splashed against the inside of the cover and, in most instances, is formed so that the correct electrolyte level can easily be determined.

Vent Plugs The openings in the cell cover are fitted with *vent plugs*. These plugs or caps are vented to allow the escape of the gasses that are generated while the battery is working. As shown in Figure 4.18, they contain baffles to minimize the possibility of the escape of liquid electrolyte. On some batteries, the vent plugs are combined, usually in sets of three, and are friction-fitted so that they can be pulled off and pushed on. Single vent plugs may be friction-fitted or may be threaded to screw into

Figure 4.19 Single vent plugs may be threaded or friction fitted push types (courtesy of AC-Delco, General Motors Corporation).

Figure 4.20 A typical manifold or combined vent plug (courtesy of AC-Delco, General Motors Corporation).

Figure 4.21 A typical post terminal battery (courtesy of General Motors Corporation).

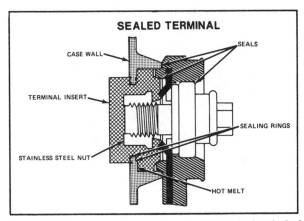

Figure 4.22 A crossection view of a sealed terminal typical of those used in side terminal batteries (courtesy of General Motors Corporation).

the cell cover. Figures 4.19 and 4.20 show some typical vent plugs.

Battery Terminals Two types of battery terminals are in common use. *Post terminals* extend through the top of the cell cover, as shown in Figure 4.21. They consist of tapered posts to which a battery cable can be clamped. The positive (+) post is $^{11}/_{16}$ in. (17.5 mm) in diameter at the top. The negative (−) post is $^5/_8$ in. (15.9 mm) in diameter at the top. The positive (+) post is larger than the negative (−) post as an aid in determining polarity. *Side terminals*, shown in Figure 4.22, are mounted on the side of the battery container and consist of a reinforced threaded hole into which a special battery cable end can be threaded.

Maintenance-Free Battery Construction
Maintenance-free batteries differ from conventional batteries in that they do not require the addition of water during their normal service life. A true maintenance-free battery is sealed and is not provided with removable vent plugs. It does, however, have a small gas vent to prevent a rise in internal pressure. Both conventional and maintenance-free batteries operate through the same electrochemical action, but they are designed and constructed in a different manner.

Most water loss in a conventional battery is through the evaporation of water vapor. The water vapor is the result of *gassing* caused by the heat generated inside a battery. The heat is produced when a bat-

tery is working, especially when it is being charged by the charging system, and actually boils the water out of the electrolyte. It has been found that the antimony used to strengthen the grids is the main cause of gassing. The antimony requires a high charging current which generates excessive heat. When the antimony in the grid alloy is replaced with calcium, less charging current is required, less heat is created, and less gassing occurs. The elimination of antimony in the alloy also increases the conductivity of the grids and thus improves the battery's capacity. Figure 4.23 shows a lead-calcium grid.

Figure 4.23 A lead-calcium grid of modern design. The grid shown is cold-worked, whereas traditional lead-antimony grids are cast (courtesy of General Motors Corporation).

Figure 4.24 Some maintenance-free batteries "bag" or encapsulate each plate in a porous plastic envelope to eliminate the need for separators (courtesy of General Motors Corporation).

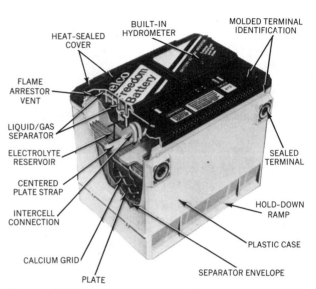

Figure 4.25 The component parts and features of a true maintenance-free battery (courtesy of General Motors Corporation).

The shedding of active material from the plates of a conventional battery requires that sediment chambers be provided at the bottom of each cell. This causes the elements to be supported high in the case and close to the top of the electrolyte. By replacing the traditional separators with microporous plastic envelopes, each plate can be separately "bagged" or encapsulated, as shown in Figure 4.24. The plastic envelopes also provide better cushioning for the plates to minimize the effects of vibration. The active material that is dislodged from the plates is trapped in the envelopes, held in contact with the plates, and is not allowed to fall to the bottom of the cell. Since no sediment is formed, no sediment chambers are required, and the elements can be positioned lower in the battery case. This provides a larger volume of electrolyte above the tops of the plates.

Figure 4.25 shows a complete battery assembly. Study this illustration to become familiar with the component parts and their location.

BATTERY SERVICE   While some batteries do not require the addition of water, all batteries require other routine maintenance services. Those services include visual inspection, cleaning, removal and installation, and cable replacement.

Safety Precautions   Battery service exposes you to certain hazards. As the electrolyte in batteries contains sulfuric acid, you should be extremely careful when working on or near a battery. Electrolyte can damage painted finishes. It can corrode metal. It can eat holes in your clothing. It can cause painful burns if spilled on your skin, and can cause blindness if splashed in your eyes.

HYDROGEN AND OXYGEN ARE PRODUCED BY A BATTERY DURING ITS NORMAL OPERATION. THOSE GASES ARE VENTED FROM THE BATTERY AND MAY EXPLODE IF IGNITED BY ANY OPEN FLAME, SPARK, OR LIGHTED CIGARETTE. THE EXPLOSION WILL USUALLY BURST THE BATTERY AND SPRAY ELECTROLYTE OVER A LARGE AREA. (SEE FIGURE 4.26.)

For your personal protection, you should observe the following safety precautions:

Job 4a

IDENTIFY TERMS RELATING TO BATTERY OPERATION AND CONSTRUCTION

SATISFACTORY PERFORMANCE

A satisfactory performance on this job requires that you do the following:

1 Identify terms relating to battery operation and construction by placing the number of each term in front of the phrase that best describes it.

2 Indentify all the terms correctly within 15 minutes.

PERFORMANCE SITUATION

1 Electrolyte	6 Lead peroxide
2 Grid	7 Sponge lead
3 Separator	8 Lead sulfate
4 Calcium	9 Plate group
5 Element	10 Antimony

_____ The active material of a positive plate in a fully charged lead-acid cell.

_____ The active material of a negative plate in a fully charged lead-acid cell.

_____ The material used in the construction of separators.

_____ A mixture of sulfuric acid and water.

_____ An insulator used between plates.

_____ The composition of the plates in a discharged cell.

_____ The metal added to the lead used to make grids for a conventional battery.

_____ The metal added to the lead used to make grids for a maintenance-free battery.

_____ An assembly of a positive plate group and a negative plate group.

_____ The meshlike framework of a plate.

_____ An assembly of similar plates connected by a plate strap.

1 Wear safety glasses any time you are working on or near a battery.

2 Remove any jewelry such as rings and watches.

3 Use fender covers to protect the finish of the car.

4 When working with tools or other metallic objects on or near a battery, use care so that they do not short across the battery terminals or across the positive terminal and ground. The resultant arc may ignite the vented gases.

5 Use a nonmetallic filler or funnel when adding water to a battery.

6 Never disconnect a battery cable or a charger cable from a battery terminal if the circuit is live. The resultant arc may ignite the vented gases.

Figure 4.26 The explosive gases given off by a battery can be ignited by a spark or a flame (courtesy of Chevrolet Motor Division).

Figure 4.28 Since each cell is a separate compartment, the electrolyte level must be checked in all cells. The caps should be placed on the battery top while checking the electrolyte level (courtesy of Ford Motor Company).

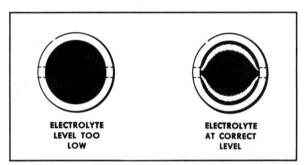

ELECTROLYTE
LEVEL TOO
LOW

ELECTROLYTE
AT CORRECT
LEVEL

Figure 4.27 Typical views of the surface of the electrolyte in a battery cell. Notice that the surface appears distorted when the electrolyte is at the correct level (courtesy of General Motors Corporation).

Checking and Adjusting Electrolyte Level

A conventional battery with removable vent plugs will normally use water at the rate of 1 to 2 oz (30 to 60 ml) per cell per month. If a car is driven extensively, especially in hot weather, the water loss may be higher. Only the water in the electrolyte is lost, not the acid. Therefore, only water should be added to adjust a low electrolyte level.

In most batteries, the correct electrolyte level is indicated by a distortion of the surface of the electrolyte. That distortion is caused by the baffled opening in the cell cover and may appear as shown in Figure 4.27. If the surface of the electrolyte is not distorted, water should be slowly added until the distortion appears. Only water that is pure, colorless, odorless, and safe for drinking should be used. Since electrolyte expands when heated, you should not overfill the cells. Overfilling the cells will cause electrolyte to leak through the vent plugs when the battery temperature rises during operation.

The following steps outline a procedure for checking and adjusting the level of electrolyte in a battery:

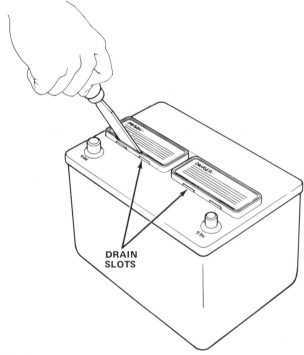

DRAIN
SLOTS

Figure 4.29 On some batteries, the cell caps must be removed by prying them up with a putty knife (courtesy of American Motors).

1 Place a fender cover on the fender near the battery.

2 Clean the battery top if it is dirty.

Note: Use wet paper towels to clean the battery top and discard the towels in a waste receptacle when finished. Use care during this step as there may be acid on the battery top.

3 Remove the vent plugs and place them carefully on the battery top as shown in Figure 4.28.

Note: Some batteries have special caps that must

Job 4b

IDENTIFY THE COMPONENT PARTS OF A BATTERY

SATISFACTORY PERFORMANCE

A satisfactory performance on this job requires that you do the following:

1 Identify the numbered parts or features on the drawing by placing the number of each part or feature in front of the correct part name or description.
2 Identify correctly 12 of the 14 parts or features within 15 minutes.

PERFORMANCE SITUATION

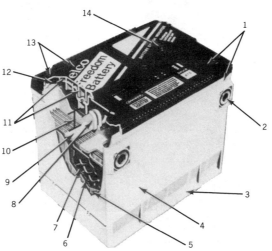

_____ Calcium grid

_____ Flame arrestor vent

_____ Terminal identification

_____ Hydrometer

_____ Intercell connector

_____ Sediment chamber

_____ Plate

_____ Hold-down ramp

_____ Plastic case

_____ Sealed terminal

_____ Liquid/gas separator

_____ Plate strap

_____ Reservoir

_____ Cover

_____ Cell partition

_____ Separator envelope

be removed with the aid of a putty knife or a similar flat-bladed tool as shown in Figure 4.29.

4 Check the electrolyte level in each cell.

Note: Use a flashlight if the surface of the electrolyte cannot be easily seen.

5 If the electrolyte level is too low in any of the cells,

use a clean syringe or a battery filler of the type shown in Figure 4.30 to add water slowly.

Note: Do not overfill the battery. Add water only until the surface of the electrolyte appears distorted.

6 Install the vent plugs.

7 If any water has spilled on the battery top, clean

Figure 4.30 A typical battery water filler. Note that the spout is fitted with a valve that helps to prevent overfilling (courtesy of Snap-on Tools Corporation).

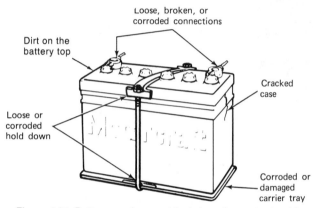

Figure 4.31 Battery service must include a thorough visual inspection of the battery, its connections, and its mounting (courtesy of Ford Motor Company).

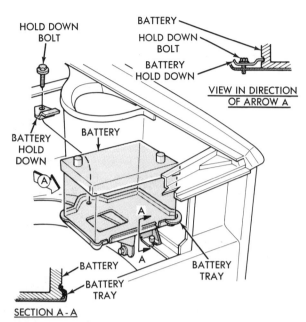

Figure 4.32 A typical battery hold-down. This type of hold-down secures the battery to the battery tray by means of grooves formed in the bottom edges of the battery case (courtesy of Chrysler Corporation).

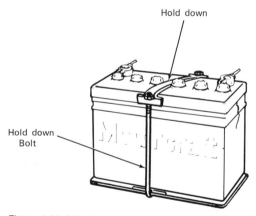

Figure 4.33 A hold-down that passes over the top of the battery (courtesy of Ford Motor Company).

and dry the top of the battery with paper towels. Discard the paper towels when finished.

8 Remove the fender cover.

Visual Inspection A visual inspection of a battery and its mounting may reveal the need for other services. Figure 4.31 illustrates some of the common faults that may be found.

A battery should be securely held in its *carrier* or *tray* by its *holddown*. Some holddowns grip the battery by its base as shown in Figure 4.32. Other holddowns grip the battery at the top, and are secured by long bolts. Such an arrangement is shown in Figure 4.33. A loose or broken holddown will allow a battery to move. If a battery is not securely mounted, movement and vibration will damage the battery. A holddown should be tightened snugly so that the battery cannot move, yet not so tight that the battery case is distorted. If a holddown is damaged so that it cannot be tightened, it should be replaced.

As shown in Figures 4.34 and 4.35, some car manu-

facturers provide heat shields to protect the battery from engine heat. Others provide air ducts that direct cool air around the battery. An inspection of the holddown should include an inspection of these devices.

A cracked case will allow electrolyte to leak from the battery. Even if that leakage is slight, the electrolyte will corrode the battery tray and its surrounding parts. A battery with a cracked case should be replaced.

Corrosion caused by spilled electrolyte should be removed and neutralized. This can be done by cleaning the parts with a solution made by dissolv-

Job 4c

CHECK AND ADJUST BATTERY ELECTROLYTE LEVEL

SATISFACTORY PERFORMANCE
A satisfactory performance on this job requires that you do the following:

1 Check the level of the electrolyte in the battery of the car assigned and adjust the level if necessary.
2 Following the steps in the "Performance Outline," complete the job within 10 minutes.
3 Fill in the blanks under "Information."

PERFORMANCE OUTLINE
1 Protect the fender of the car with a fender cover.
2 Clean the battery top.
3 Remove the vent plugs.
4 Check the electrolyte level in each cell.
5 Add water if necessary.
6 Install the vent plugs.
7 Clean the battery top.
8 Remove the fender cover.

INFORMATION
Vehicle identification _____

Was water needed? _____ Yes _____ No

Was water added? _____ Yes _____ No

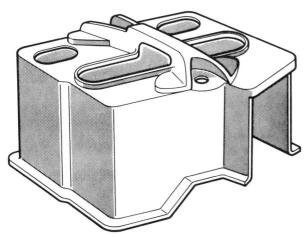

Figure 4.34 A plastic heat shield used in some vehicles to protect the battery from engine heat. Notice the openings that direct cool air around the battery (courtesy of Chrysler Corporation).

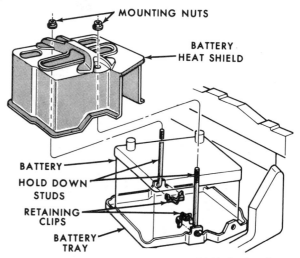

Figure 4.35 On some vehicles, the heat shield also functions as a hold-down (courtesy of Chrysler Corporation).

ing a teaspoon of baking soda in a cup of water. (A disposable coffee container is handy for this job as it can be discarded after use.) After the parts have dried, they should be protected from further corosion by a coating of oil or light grease.

Dirt on the top of a post terminal battery may hold spilled electrolyte. Since electrolyte is a conductor, that coating may cause a battery to self-discharge. The top of a battery should always be kept clean and dry.

Job 4d

INSPECT A BATTERY, HOLDDOWN, AND CARRIER TRAY

SATISFACTORY PERFORMANCE

A satisfactory performance on this job requires that you do the following:

1 Inspect the battery, holddown, and carrier tray on the car assigned.
2 Following the steps in the "Performance Outline," complete the job within 10 minutes.
3 Fill in the blanks under "Information."

PERFORMANCE OUTLINE

1 Protect the fender of the car with a fender cover.
2 Inspect the terminals, clamps, and cables for damage, looseness, and corrosion.
3 Inspect the top of the battery for the presence of dirt and electrolyte.
4 Inspect the battery case for cracks and leakage.
5 Inspect the holddown for looseness, damage, and corrosion.
6 Inspect the carrier tray for looseness, damage, and corrosion.
7 Inspect any heat shield or air duct that may be fitted for looseness, damage, and corrosion.

INFORMATION

Vehicle identification _____

Terminals, clamps, and cables:	_____ Clean	_____ Dirty
	_____ Tight	_____ Loose
Battery top:	_____ Clean	_____ Dirty
Battery case:	_____ Intact	_____ Cracked
Holddown:	_____ Clean	_____ Dirty
	_____ Tight	_____ Loose
	_____ OK	_____ Damaged
Carrier tray:	_____ Clean	_____ Dirty
Heat shield/air duct:	_____ Tight	_____ Loose
	_____ OK	_____ Damaged

Services required: _____

_____

_____

Figure 4.36 Badly corroded battery terminals and clamps often are the cause of electrical system problems (courtesy of General Motors Corporation).

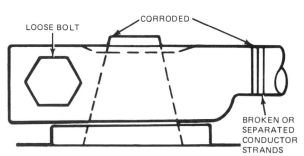

Figure 4.37 A cable clamp should be clean, tight, and exibit no broken or separated conductor strands (courtesy of Ford Motor Company).

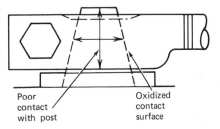

Figure 4.38 While a cable clamp may appear clean and tight, internal oxidation can create a bad connection (courtesy of Ford Motor Company).

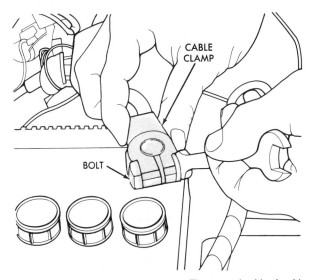

Figure 4.39 Loosening a battery clamp. The ground cable should be removed first to minimize the danger of arcing and a possible battery explosion (courtesy of Chrysler Corporation).

Figure 4.40 Typical battery pliers. This tool is desiged for the removal of cable clamp nuts that are so corroded that a wrench cannot be used (courtesy of Easco/KD Tools, Lancaster, PA 17604.)

Loose, broken, or corroded connections such as those shown in Figure 4.36 may restrict current flow. The resistance of a loose or dirty connection often causes problems that may be incorrectly blamed on the battery or on the starter motor. Broken and damaged cables should be replaced. Connections should be kept clean, tight, and in good condition.

Cleaning Battery Connections—Post Terminals A visual inspection of the connections on a battery with post terminals may reveal some of the faults shown in Figure 4.37. While those faults require correction, a more serious fault may escape detection. Oxidation and corrosion between the post and the cable clamp as shown in Figure 4.38 are a common cause of high resistance. To eliminate that resistance, the cable clamps must be removed from the posts and all parts must be cleaned thor-

oughly. The following steps outline a suggested procedure:

1 Make sure that all switches and controls are in the OFF position.

2 Loosen the clamp connecting the ground cable to the battery as shown in Figure 4.39. On most cars, the ground cable is connected to the negative ($-$) terminal.

Note: If the nut on the clamp bolt is corroded or worn so that a wrench cannot be used, a pair of battery pliers similar to those shown in Figure 4.40 will prove helpful.

Figure 4.41 A typical battery cable clamp puller (courtesy of Easco/KD Tools, Lancaster, PA 17604).

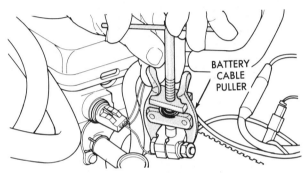

Figure 4.42 Using a puller to remove a battery cable clamp from a battery post (courtesy of Chrysler Corporation).

Figure 4.43 A typical battery post and clamp cleaning brush (courtesy of Easco/KD Tools, Lancaster, PA 17604).

3 Remove the cable clamp from the battery post.

Note: Never attempt to pry or twist a cable clamp from a battery post. To do so could damage the battery. If a clamp will not slide off easily, a cable puller similar to the one shown in Figure 4.41 should be used. Positioned over the clamp as shown in Figure 4.42, a puller will enable you to remove the clamp without damage to the battery.

4 Remove the remaining cable clamp.

5 Using a battery post and clamp cleaning tool similar to the one shown in Figure 4.43, clean the inside of the battery cable clamps. This operation is shown in Figure 4.44.

6 Clean the battery posts as shown in Figure 4.45.

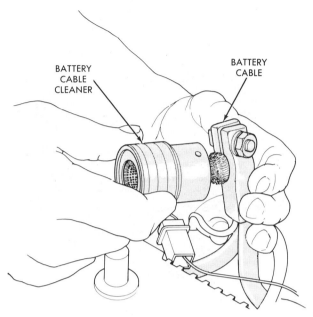

Figure 4.44 Cleaning the inside surfaces of a battery cable clamp (courtesy of Chrysler Corporation).

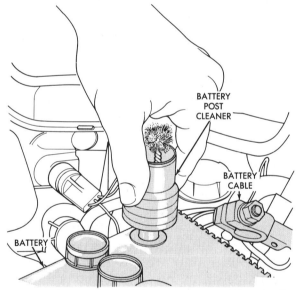

Figure 4.45 Cleaning a battery post (courtesy of Chrysler Corporation).

7 Clean the top of the battery. A solution of baking soda and water applied with a brush is ideal for cleaning as it will neutralize any spilled acid. (See Figure 4.46.)

Note: Be careful that the solution does not enter the cells. The baking soda and water solution will weaken the electrolyte.

8 Install the cable that connects the battery to the starter motor or starter solenoid. On most cars, this

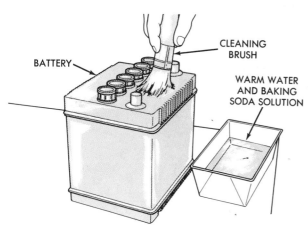

Figure 4.48 A special torque wrench for tightening the cable bolts on side terminal batteries (courtesy of Easco/KD Tools, Lancaster, PA 17604).

Figure 4.46 Cleaning the top of a battery with a solution of baking soda and water (courtesy of Chrysler Corporation).

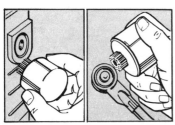

Figure 4.47 A wire brush tool for cleaning the cables and terminals of a side terminal battery (courtesy of Thexton Manufacturing Company).

cable connects to the positive (+) post of the battery.

Note: This cable should be connected first to minimize the dangers of arcing and possible battery explosion.

9 Install the ground cable. On most cars, the ground cable is connected to the negative (−) post.

Note: The ground cable should always be installed last to minimize the dangers of arcing and possible battery explosion.

10 Coat the cable clamps with a thin coating of grease.

Cleaning Battery Connections—Side Terminals

Side terminals are not as subject to corrosion as are post terminals, and the design of the

cable ends makes it difficult to check for the presense of corrosion. Because of this, their maintenance often is neglected and a high-resistance condition is not detected. The procedure for cleaning the connections of a side terminal battery differs slightly from the procedure for cleaning the connections of a post terminal battery and requires that you take certain additional precautions. A suggested procedure follows:

1 Make sure that all switches and controls are in the OFF position.
2 Using a six-point socket of the correct size, loosen the bolt holding the ground cable to the terminal. On most cars, the ground cable is connected to the negative (−) terminal.

Note: The head of the bolt is very small and is easily damaged. Do not attempt to use an open end wrench or a pair of pliers.

3 Remove the cable from the terminal.
4 Repeat steps 2 and 3 at the remaining terminal.
5 Using a cable and terminal cleaning tool of the type shown in Figure 4.47, clean the cable ends and the terminals on the battery.
6 Clean the top and side of the battery.
7 Install the cable that connects the battery to the starter motor or starter switch. On most cars, this cable connects to the positive (+) terminal, and should be installed first to minimize the dangers of arcing and possible battery explosion.

Note: Thread the cable bolt into place by hand to insure that it is properly aligned. The bolt should be tightened to the torque specification of the manufacturer (19 ft-lb—12 N•m). This may be done with a torque wrench or with a special wrench similar to the one shown in Figure 4.48. Excessive torque can damage the bolt and the terminal in the case.

8 Install the ground cable, tightening it to the torque specification as in step 7.

Removing and Installing Batteries

Battery removal may be necessary for a thorough battery inspection and cleaning, recharging, to gain access to the battery tray or other parts, or to install a replacement battery. The following steps outline

Job 4e

CLEAN BATTERY TERMINALS AND CABLE ENDS

SATISFACTORY PERFORMANCE
A satisfactory performance on this job requires that you do the following:

1 Clean the battery terminals and cable ends on the car assigned.
2 Following the steps in the "Performance Outline" and the recommendations and specifications of the manufacturer, complete the job within 30 minutes.
3 Fill in the blanks under "Information."

PERFORMANCE OUTLINE
1 Disconnect the cables from the battery.
2 Clean the terminals and the cable ends.
3 Clean the battery top.
4 Install the cables.

INFORMATION
Vehicle identification _____

Battery type: _____ Post terminal

                                                                 _____ Side terminal

Which cable was disconnected first?   _____ Positive

                                                                 _____ Negative

Which cable was connected first?   _____ Positive

                                                                 _____ Negative

If a side terminal battery is used:

What was the wrench size of the cable bolt?   _____

What was the manufacturer's torque specification?   _____

---

a procedure that will enable you to remove and install a battery safely:

Removal

1 Make sure that all switches and controls are in the OFF position.
2 Disconnect the ground cable from the battery terminal.

Note: On most cars, the ground cable is connected to the negative (−) battery terminal.

3 Disconnect the remaining cable.
4 Remove the battery holddown and any heat shield or air duct that may be present. Penetrating oil ap-

plied to the threads of the bolts will make this operation easier.

Note: On some cars, the battery holddown is located at the bottom of the battery. (Refer to Figure 4.32.) On other cars, the holddown is positioned across the top of the battery. (Refer to Figure 4.33.)

5 Lift the battery from the battery tray and remove it from the car.

Note: Take care in performing this step. Batteries are very heavy and contain acid. To avoid the possibility of dropping the battery, you should use a battery carrier. Figure 4.49 shows a battery carrier commonly used with post terminal batteries. Side

Figure 4.49 A typical battery carrier strap. This type is for use with post terminal batteries (courtesy of Easco/KD Tools, Lancaster, PA 17604).

Figure 4.50 A battery carrier strap designed for use with side terminal batteries. Knurled headed screws provide the means to attach the strap to the terminals (courtesy of Easco/KD Tools, Lancaster, PA 17604).

terminal batteries can be safely lifted with a carrier similar to the one shown in Figure 4.50. That type of carrier is attached to the battery by bolts that screw into the battery terminals. Some shops use a clamp-type carrier similar to the one shown in Figure 4.51. As shown in Figure 4.52, that type carrier can be used on both post terminal and side terminal batteries.

Some batteries have cases made of a soft, flexible plastic. Attempting to lift those batteries by hand by grasping them on their end walls may result in causing electrolyte to be forced through the vent plugs. If a battery carrier is not available, a battery should be lifted only by grasping it at diagonally opposite corners.

Installation

1 Clean the cable clamps or ends.
2 Inspect the battery tray. If any corrosion is present, clean the tray with a solution of baking soda and water. After drying the tray, it should be coated with light oil to retard future corrosion. Tighten or replace any loose or damaged bolts that secure the tray to the car. Check that no bolts or sharp edges protrude from the tray where they could contact the battery.
3 Using a battery carrier, carefully place the battery in the tray. Be sure that the terminals are in the correct position.

Note: If the original battery is to be installed, clean

Figure 4.51 A clamp-type battery carrier (courtesy of Snap-on Tools Corporation).

Place tool across middle of battery

Figure 4.52 Using a clamp-type battery carrier to lift a battery. The tool should be positioned across the center of the battery (courtesy of Buick Motor Division, General Motors Corporation).

Figure 4.53 Battery terminal spreading pliers. A tool of this type is useful for opening the clamp so that the cable can be installed on a new battery (courtesy of Easco/KD Tools, Lancaster, PA 17604).

the terminals and the battery before installing it in the tray.

4 Install the holddown assembly and any heat shield or air duct that was removed. Tighten the attaching bolt(s) to the manufacturer's torque specifications.
5 Install the battery cables, connecting the ground cable last.

Note: If the cable clamps will not fit over the terminals on a post terminal battery, they may be spread with the use of a tool similar to the one shown in Figure 4.53.

6 Tighten the clamp nuts or terminal bolts to the

Job 4f

REMOVE AND INSTALL A BATTERY

SATISFACTORY PERFORMANCE
A satisfactory performance on this job requires that you do the following:

1 Remove and install the battery in the car assigned.
2 Following the steps in the "Performance Outline" and the specifications of the car manufacturer, complete the job within 30 minutes.
3 Fill in the blanks under "Information."

PERFORMANCE OUTLINE
1 Disconnect the battery cables.
2 Remove the holddown and related parts.
3 Remove the battery.
4 Inspect and clean the cable ends and battery terminals.
5 Inspect and clean the battery tray.
6 Install the battery and secure it with the holddown.
7 Install any heat shield or air duct that was removed.
8 Connect the battery cables.

INFORMATION
Vehicle identification _____

Reference used _____ Page(s) _____

Type of battery: _____ Post terminal

                                          _____ Side terminal

Type of holddown: _____ Top mounted

                                          _____ Bottom mounted

Did the battery have a heat shield?   _____ Yes

                                          _____ No

Did the battery have an air duct?   _____ Yes

                                          _____ No

Terminal bolt (nut) torque specification _____

Holddown bolt (nut) torque specification _____

torque specifications of the manufacturer.

7 Coat the cable clamps with a thin coating of grease.

JUMP STARTING A *booster battery* is often used to "jump start" the engine of a car that has a discharged battery. The booster battery may be a separate battery brought to the disabled vehicle. It also may be the battery in another car. Jumper cables similar to those shown in Figure 4.54 are usually used to connect the booster battery to the car with the discharged battery. Properly connected, a booster battery enables you to provide an important emergency service. Improperly connected, a booster battery can cause serious damage to automotive electrical systems and may even cause a battery explosion.

The following precautions should be observed when using a booster battery:

1 Wear safety glasses.
2 Do not allow electrolyte to contact your eyes, skin, clothing, or the finish of the vehicles.
3 Do not lean over a battery when connecting or disconnecting cables.
4 Do not allow the jumper cable clamps to touch each other.
5 Keep open flame and sparks away from the batteries.
6 Do not connect a booster battery directly to a discharged battery.

Using a Booster Battery to Jump Start an Engine

The following steps outline a typical procedure for using a booster battery to jump start an engine. Some car manufacturers have established alternate procedures to protect certain electrical components used in their products. You should consult an appropriate manual for the jump starting procedures recommended by the manufacturer of the car you wish to start:

1 Check to see that all switches and other electrical controls in the car are in the OFF position.
2 If the discharged battery is a sealed, maintenance-free battery, check the charge indicator. If the charge indicator is light, as shown in Figure 4.55, *do not* attempt to jump start the engine. Replace the battery.
3 If the discharged battery has vent plugs, remove them and check the electrolyte level.

Note: During cold weather, the electrolyte in a discharged battery may freeze. If the electrolyte is not

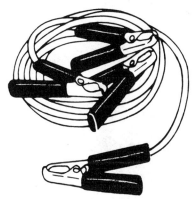
Figure 4.54 A pair of jumper cables (© 1985-Perfect Parts, Inc., Carlstadt, N.J.).

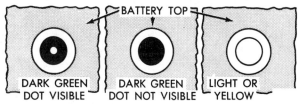
Figure 4.55 Before connecting a booster battery to a sealed, maintenance-free battery, check the built-in hydrometer. *Do not* connect battery if the indicator is light or yellow (courtesy of Chrysler Corporation).

visible, or if it appears that the electrolyte is frozen, *do not* attempt to jump start the engine. A frozen battery may rupture or explode if a booster battery is connected to it. A frozen battery should be thawed by placing it in a warm area. After the battery has been warmed, the electrolyte level should be adjusted as necessary. The battery should then be tested and charged in the normal manner.

4 If the electrolyte is not frozen, and if the level is above the tops of the plates, cover the openings with a cloth as shown in Figure 4.56.

Note: This will minimize the possibility of electrolyte spewing from the openings.

5 If the booster battery has vent plugs, remove them and cover the openings with a cloth as you did with the discharged battery.
6 Connect one of the jumper cables between the positive (+) terminal of the booster battery and the positive (+) terminal of the discharged battery as shown in Figure 4.57.

Note: Be sure that the clamps on the jumper cables are firmly connected to the battery terminals.

7 Connect one end of the remaining cable to the

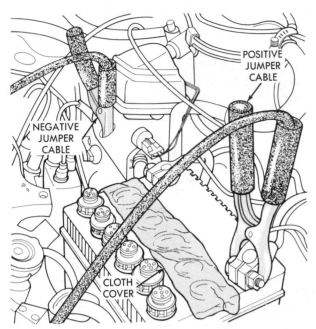

Figure 4.56 When attempting to jump start a car, the vent plugs of both the discharged battery and the booster battery should be removed. The openings in the cells should be covered with cloth (courtesy of Chrysler Corporation).

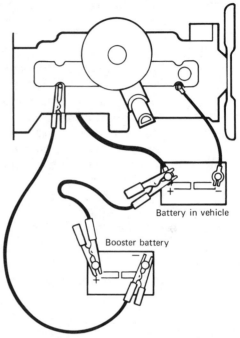

Figure 4.57 The correct placement of jumper cables when a booster battery is used to start a car. Notice that the negative (−) jumper cable is connected to the engine (courtesy of American Motors).

negative (−) terminal of the booster battery. (See Figure 4.57.)

8 Attach the remaining end of the jumper cable to a good ground on the engine, away from the battery. (See Figure 4.57.)

Note: Some manufacturers recommend that the negative (−) jumper cable be connected to the alternator bracket or to the air conditioner compressor bracket, as shown in Figure 4.58. Do not attach the cable to any part of the fuel system.

9 Making sure that the parking brake is applied and that the shift selector is in Park or Neutral, at-

tempt to start the engine in the normal manner.

10 After the engine has started (or if the engine fails to start), the jumper cables should be disconnected in the reverse order by which they were connected.

Note: The negative jumper cable should be disconnected from the ground on the engine *FIRST*.

11 Remove and discard the cloths that were used to cover the battery openings.

12 Install the vent plugs on the batteries.

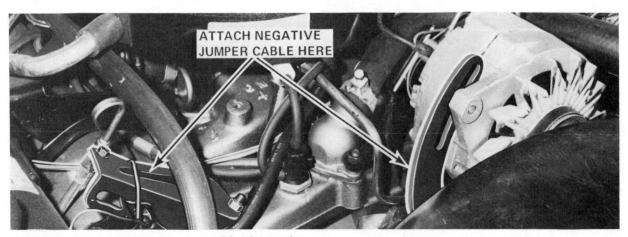

Figure 4.58 Some manufacturers recommend that the negative (−) jumper cable be connected to the alternator bracket or to the air conditioner compressor bracket as shown (courtesy of Pontiac Motor Division, General Motors Corporation).

Job 4g

USE A BOOSTER BATTERY TO JUMP START AN ENGINE

SATISFACTORY PERFORMANCE
A satisfactory performance on this job requires that you do the following:

1 Use a booster battery to start the engine of the car assigned.
2 Following the steps in the "Performance Outline" and the procedure and specifications of the car manufacturer, complete the job within 15 minutes.
3 Fill in the blanks under "Information."

PERFORMANCE OUTLINE
1 Determine that the discharged battery in the vehicle is in a condition that will allow safe starting.
2 Connect the booster battery to the car.
3 Start the engine.
4 Disconnect the booster battery.

INFORMATION
Vehicle identification: _____

Vehicle battery condition:
    Electrolyte level: _____ Above the tops of the plates

                                         _____ Below the tops of the plates

    Electrolyte temperature: _____ Above 40°F (4°C)

                                         _____ Below 40°F (4°C)

The positive (+) booster cable was connected to:

    _____ the positive (+) terminal on the car's battery

    _____ a good ground on the engine

The negative (−) booster cable was connected to:

    _____ the negative (−) terminal on the car's battery

    _____ a good ground on the engine

Which cable was connected first?   _____ Positive

                                                _____ Negative

Which cable was disconnected first?   _____ Positive

                                                _____ Negative

SUMMARY

By completing this chapter, you have gained some knowledge of the lead-acid battery. By performing the Jobs, you have developed certain skills required to service batteries. You can identify terms relating to battery operation and construction and are knowledgable of the major components of a battery. You can adjust electrolyte level and inspect a battery, its holddown, and its carrier tray. You can clean terminals to eliminate high resistance connections, and you can remove and install a battery. You also have developed skills in the correct use of a booster battery. By accomplishing your objectives, you have moved closer to your goal.

SELF-TEST

Each incomplete statement or question in this test is followed by four suggested completions or answers. In each case select the *one* that best completes the sentence or answers the question.

1 Two mechanics are discussing the construction of a lead-acid battery.
Mechanic A says that the positive plates are made of lead peroxide.
Mechanic B says that the negative plates are made of sponge lead.
Who is right?
A. A only
B. B only
C. Both A and B
D. Neither A nor B

2 Electrolyte is a mixture of water and
A. hydrogen
B. lead sulfate
C. lead peroxide
D. sulfuric acid

3 A 12-V battery consists of
A. 3 cells connected in series
B. 3 cells connected in parallel
C. 6 cells connected in series
D. 6 cells connected in parallel

4 The grids of a conventional battery are made of an alloy of lead and
A. calcium
B. antimony
C. lead sulfate
D. lead peroxide

5 The grids of a sealed, maintenance-free battery are made of an alloy of lead and
A. calcium
B. antimony
C. lead sulfate
D. lead peroxide

6 Which of the following may cause a battery to self-discharge?
A. Loose connections
B. Dirty or corroded connections
C. Frayed or broken cable strands
D. Dirt and electrolyte on the battery top

7 Two mechanics are discussing battery service.
Mechanic A says that when battery cables are removed, the ground cable should be disconnected first.
Mechanic B says that when battery cables are installed, the ground cable should be connected first.
Who is right?
A. A only
B. B only
C. Both A and B
D. Neither A nor B

8 Two mechanics are discussing battery service.
Mechanic A says that sealed batteries require no routine maintenance.
Mechanic B says that batteries with removable vent plugs require the addition of water to adjust the electrolyte level.
Who is right?
A. A only
B. B only
C. Both A and B
D. Neither A nor B

9 The level of electrolyte in the cells of a battery is considered correct when the electrolyte surface
A. appears distorted
B. is above the plates
C. is above the separators
D. is at the top of the vent plug opening

10 Two mechanics are discussing jump starting an engine by using a booster battery.

Mechanic A says that the positive (+) jumper cable should be connected first.

Mechanic B says that the negative (−) jumper cable should be connected to a good ground at the engine.

Who is right?

A. A only

B. B only

C. Both A and B

D. Neither A nor B

Chapter 5 The Battery– Testing, Charging, and Re- placement

Since the battery is a part of every automotive electrical circuit, the battery must be considered in the diagnosis of electrical problems. At times, the battery may be the cause of a problem. At other times, the problem may damage the battery or cause it to become discharged. Diagnosis often requires that the battery be tested. The results of those tests must be interpreted, and the condition of the battery evaluated before the correct repair can be made.

This chapter will make you aware of the various battery ratings commonly used. It will provide procedures by which you can perform several battery tests and interpret their results. And it will cover the procedures required to safely recharge a discharged battery and to activate a dry charged battery. The performance of the following jobs provides your specific objectives in this chapter:

A
Identify battery ratings
B
Test the specific gravity of battery electrolyte and interpret the test results
C
Load test a battery
D
Perform a three-minute charge test
E
Charge a battery
F
Activate a dry charged battery

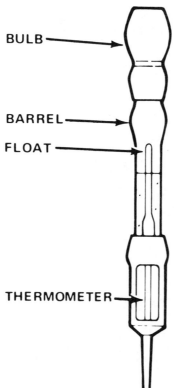

BULB

BARREL

FLOAT

THERMOMETER

BATTERY SELECTION Most cars in current production are equipped with a 12-V battery. As you learned in the previous chapter, a 12-V battery consists of 6 cells connected in series. Since a cell will produce approximately 2 V regardless of the number and size of the plates it contains, a 12-V battery can be built in just about any size and to sell for just about any price. As an automotive electrician you will examine and test many batteries and evaluate their condition. Those you find defective or worn out must be replaced. With so many different batteries available at so many different prices, how will you select the correct replacement?

When a manufacturer installs a battery in a new car, that battery is chosen to meet the requirements of that particular car. Of prime importance is the battery's ability to crank and start the engine. The current required to crank an engine can range from 150 A to over 500 A, depending on the size of the engine, the temperature, and the viscosity of the oil in the engine. Those factors are all considered in battery selection. The number and type of electrical options installed in the car are also considered. Figure 5.1 lists some typical current loads and the combined load that could be placed on a battery.

When you select a battery to install in a car, the battery you choose should have a capacity at least equal to the battery installed as original equipment. In addition, you also should consider any aftermarket electrical equipment that may have been fitted to the car, any extremes in temperature to which the car is subjected, and the type of service in which the car is used.

BATTERY RATINGS A knowledge of battery ratings will enable you to make an intelligent decision in the choice of a battery. There are several methods by which batteries are rated. They all provide a means by which you can compare the amount of electrical energy a battery can provide. These ratings are established by laboratory tests performed to the standards and specifications of the Society of Automotive Engineers (SAE) and the Battery Council International (BCI).

Ampere-Hour Rating The *ampere-hour rating* provides a measure of how much current a battery at 80°F (27°C) will deliver for a fixed period of time without the cell voltage dropping below 1.75 V (10.5 total terminal volts). Due to a specified 20 hour time period, this test is sometimes referred to as the "20 hour test." The rating number is determined by multiplying the current delivered by 20. If a battery can deliver 3 A for the 20 hour period, it receives a 60 ampere-hour rating. If a battery can deliver 5 A for the 20 hour period, it receives a rating of 100 ampere-hours.

Reserve Capacity Rating This rating was developed to replace the ampere-hour rating. The test is a measure of the amount of time a battery at 80°F (27°C) can be discharged at a rate of 25 A before the cell voltage falls below 1.75 V. Typical ratings are 70, 75, and 90 minutes. This rating was introduced to give the average car owner an easier method by which different batteries can be compared. The *reserve capacity rating* is usually explained as the amount of time the car can be driven if a charging system failure occurs. Figure 5.2 shows examples of how much time it takes for some accessories to drain a battery.

Cold Cranking Performance Rating This rating indicates the amount of current a battery is capable of delivering at 0°F (−18°C) for a period of 30 seconds while maintaining a cell voltage of at

ACCESSORY	CURRENT DRAW
STARTER	200–300 AMPS
6 WAY SEAT	20–40 AMPS
REAR WINDOW DEFOGGER	18–25 AMPS
HI BLOWER A/C	19 AMPS
HI BLOWER HEATER	16 AMPS
ELECTRIC WINDOW	12 AMPS
HEADLIGHTS HIGH BEAM	12 AMPS
LOW BEAM	9 AMPS
W/S WIPER RUNNING	10 AMPS
HORNS	12 AMPS
HAZARD FLASHER	6 AMPS

Figure 5.1 Typical loads that may be placed on a battery and which must be considered in the selection of a battery (courtesy of General Motors Corporation).

ACCESSORY	DRAIN TIME
HEADLIGHTS	1 TO 1½ HOURS
INTERIOR LIGHTS	2½ HOURS
1 SEAT BELT RETRACTOR	7 DAYS
2 SEAT BELT RETRACTORS	3½ DAYS
PARKING LIGHTS	4–6 HOURS
TRUNK LIGHT	2½ DAYS
TRUNK CLOSING MOTOR	18 HOURS

Figure 5.2 A listing of typical accessories and the amount of time in which they will discharge a battery to an unacceptable level (courtesy of General Motors Corporation).

Job 5a

IDENTIFY BATTERY RATINGS

SATISFACTORY PERFORMANCE

A satisfactory performance on this job requires that you do the following:

1 Identify battery ratings by placing the number of each rating in front of the phrase that best identifies it.

2 Identify all the ratings correctly within 10 minutes.

PERFORMANCE SITUATION

1 Ampere-hour rating
2 Reserve capacity rating
3 Cold cranking performance rating
4 Watts rating

_____ The amount of current a battery can deliver at 0°F (−18°C) for a period of 30 seconds while maintaining a cell voltage of at least 1.2 V.

_____ The amount of time a battery at 80°F (27°C) can be discharged at the rate of 200 A before the cell voltage falls below 1.2 V.

_____ The amount of current a battery at 80°F (27°C) can deliver for 20 hours before the cell voltage drops below 1.75 V.

_____ The amount of time a battery at 80°F (27°C) can be discharged at the rate of 25 A before the cell voltage falls below 1.75 V.

_____ A product of the ampere-hour capacity of a battery times its terminal voltage.

least 1.2 V (7.2 total terminal volts). Typical ratings are 315, 355, and 500. This rating was also introduced to simplify the choice of a replacement battery. It usually is held that the *cold cranking performance rating* should at least equal the cubic inch displacement of the engine it is to start. Thus, a car with an engine of 350 cubic inch displacement should have a battery with a cold cranking performance rating of at least 350.

Watts Rating   Some manufacturers provide *watts ratings* or *watt-hour ratings* for their batteries. The watts rating provides another indication of a battery's available energy or cranking power. A watt is a unit of measurement of electrical power. Watts are determined by multiplying the ampere-hour capacity of a battery by the terminal voltage. Thus, a 12-V battery with a 70 ampere-hour rating would have a watts rating of 840 watts-hours (Wh) (12 V

× 70 Ah). A 12-V battery with a 90 ampere-hour rating would have a watts rating of 1080 watts-hours (12 V × 90 Ah).

BATTERY LIFE   Properly maintained and used, the average battery will outlast the period for which it is warranted. All batteries have a limited life span, and the actual life of a battery is effected by many conditions. As an automotive electrician, you should be aware of those conditions so that you can provide appropriate service.

Electrolyte Level   Failure to maintain the correct electrolyte level will greatly shorten the life of a battery. If the electrolyte level is allowed to fall below the tops of the plates, the active material on the plates will be exposed to the air. The exposed material will dry out and harden, and will no longer react with the electrolyte. (See Figure 5.3.) Since

Figure 5.3 A plate taken from a cell that had a low electrolyte level. Notice that the active material above the water level has hardened and turned to lead sulfate. The lead sulfate is inactive and will remain so even if the electrolyte level is raised and the battery charged (reproduced through the courtesy of Battery Council International).

Figure 5.5 A buckled positive plate. Plates can be buckled by the heat created by overcharging (reproduced through the courtesy of Battery Council International).

Figure 5.6 This positive plate disintegrated when the grid, corroded by overcharging, collapsed (reproduced through the courtesy of Battery Council International).

Figure 5.4 Overcharging can cause a plate to shed its active material (reproduced through the courtesy of Battery Council International).

the remaining electrolyte has lost a large amount of water, it contains a higher percentage of acid. This acid attacks the plates and shortens their life.

Adding too much water to a battery also will effect battery life. The specific gravity of the electrolyte will be lowered, decreasing the battery's ability to deliver energy. When heated, excess electrolyte may be forced from the vent holes in the vent plugs, wetting the top of the battery and causing corrosion. Any electrolyte on the battery top may provide a conducting path between the terminals and allow the battery to self-discharge.

Overcharging   A problem in the car's charging system may result in a high charging rate. Over-

charging causes excessive heat in a battery, and results in violent gassing. This action not only causes a loss of water from the electrolyte and promotes corrosion, but dislodges particles of active material from the plates as shown in Figure 5.4. If overcharging is continuous, it can actually buckle or otherwise physically damage the grids. (See Figures 5.5 and 5.6.) Overcharging should be suspected whenever it appears that a battery is using excessive amounts of water. Damage caused by overcharging is not always the fault of the charging system. An improperly adjusted battery charger can ruin a battery as readily as it can charge it.

Undercharging   A loose drive belt or a problem in the charging system may be a cause of undercharging. Undercharging also may be caused by repeated starting of the engine without sufficient driving time to allow the charging system to restore the energy used. If a battery constantly remains in a partially discharged state, the lead sulfate on the plates will harden and resist electrochemical action. A battery that has hardened deposits of lead

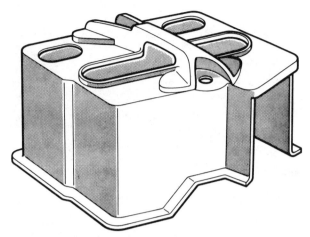

Figure 5.7 A battery heat shield. The opening in the side of the shield directs cool air around the battery (courtesy of Chrysler Corporation).

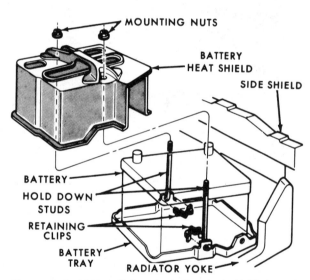

Figure 5.8 In this installation, the battery heat shield also acts as a hold-down to secure the battery to the tray (courtesy of Chrysler Corporation).

sulfate on its plates is said to be *sulfated.* (Refer to Figure 5.3.) In most instances, a sulfated battery will not accept a charge and must be replaced.

Temperature   High temperature can shorten the life of a battery. Overcharging is but one cause of high temperature and its resultant damage. Most batteries are housed in the engine compartment where they are subjected to high under-hood temperatures. In many instances, car manufacturers provide heat shields and air ducts to keep excessive heat from the battery and to direct cooler outside air around the battery. (See Figures 5.7 and 5.8.) At times, you will find these heat shields and

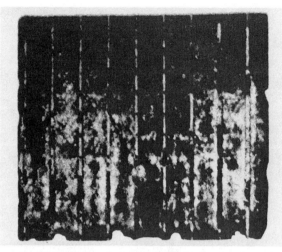

Figure 5.9 Separator damage caused by vibration (reproduced through the courtesy of Battery Council International).

ducts damaged or improperly installed. In some instances, you may find them missing. Damaged or missing heat shields and ducts should be repaired or replaced.

Movement and Vibration   A loose, corroded, or otherwise damaged carrier tray and holddown may allow the battery to move and vibrate. If the battery moves against a protruding part, a hole may be worn through the case. Vibration will shake active material from the plates and damage the separators. (See Figure 5.9.) A securely mounted battery is an important factor in battery life.

BATTERY TESTING   Routine maintenance of an electrical system requires that a battery be tested at regular intervals. In addition, the diagnosis of electrical problems, particularly in the charging and starting circuits, usually starts with a battery test. The state of charge of a battery can usually be determined by a *specific gravity test* that measures the "strength" of the electrolyte. While it indicates the percentage of charge in a battery, a specific gravity test does not provide a total picture of the battery's condition. Other tests must be made to determine the battery's ability to deliver energy.

A *high-rate discharge test* or *load test* provides a method of determining the serviceability of a battery. As its name implies, load testing consists of testing a battery while it is working under a load. A *three-minute charge test* often is made on a battery when it is suspected to be sulfated. All battery tests must be performed in a specific procedure and, in some instances, must follow certain specifications if their results are to be accurate. In addition, all

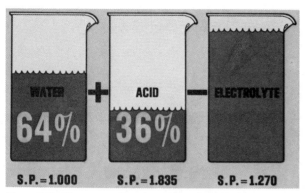

Figure 5.10 The composition, by weight, of a typical battery electrolyte (courtesy of American Motors).

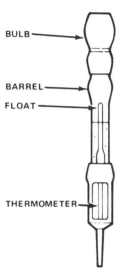

Figure 5.11 A typical battery hydrometer (courtesy of American Motors).

battery tests require the interpretation of test results.

Specific Gravity Testing *Specific gravity* may be defined as "exact weight." For use as a standard, pure water has been given the arbitrary weight or specific gravity of 1.000. The weight of all other liquids, whether heavier or lighter than water, can be determined by comparing it to the weight of water. If a liquid is lighter than water, it will have a lower specific gravity. If a liquid is heavier than water, it will have a higher specific gravity.

Sulfuric acid is much heavier than water. It has a specific gravity of 1.835. Thus, any mixture of water and sulfuric acid will have a specific gravity of more than 1.000 and less than 1.835. The specific gravity of any electrolyte is determined by the percentage of water and sulfuric acid that have been combined. One commonly used electrolyte consists of

Figure 5.12 A battery hydrometer float (courtesy of General Motors Corporation).

a mixture of 64 percent water and 36 percent sulfuric acid. As shown in Figure 5.10, that electrolyte has a specific gravity of 1.270.

New batteries usually are filled with an electrolyte that has a specific gravity of 1.265. Most manufacturers consider a battery that has been in service to be fully charged if the electrolyte in the battery has a specific gravity of from 1.250 to 1.265.

The specific gravity of the electrolyte in a battery is easily measured with a *hydrometer*. A hydrometer, shown in Figure 5.11, is a syringe-type device that enables you to withdraw a sample of electrolyte from a battery cell. As the electrolyte rises in the glass barrel of the hydrometer, it causes a calibrated float to rise. As shown in Figure 5.12, the float is marked for various specific gravities. If the electrolyte is heavy, as it is in a fully charged battery, the float will be high and will indicate a high specific gravity. If the electrolyte is light, as in a discharged battery, the float will be low and indicate a low specific gravity.

As you learned in Chapter 4, the electrolyte in a battery changes to water as the battery discharges. Thus, the electrolyte becomes lighter as the battery discharges. This change of weight, shown in Figures 5.13 through 5.16, provides a reliable indication of the battery's state of charge. The percentage of charge for various ranges of specific gravity is given in Figure 5.17.

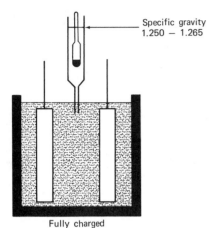

Specific gravity
1.250 – 1.265

Fully charged

Figure 5.13 The electrolyte in a fully charged battery has a high concentration of acid and thus a high specific gravity.

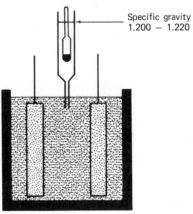

Specific gravity
1.200 – 1.220

Half discharged

Figure 5.15 A large amount of acid in the electrolyte has combined with the plates. A battery in this condition may not have sufficient energy to start an engine.

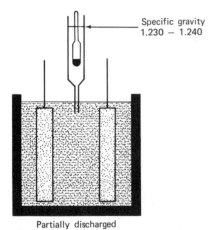

Specific gravity
1.230 – 1.240

Partially discharged

Figure 5.14 As the battery discharges, the acid combines chemically with the plates and the specific gravity drops.

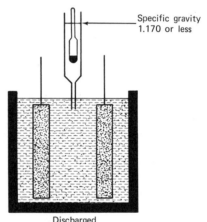

Specific gravity
1.170 or less

Discharged

Figure 5.16 Most of the acid in the electrolyte has combined with the plates. Since the electrolyte now consists mainly of water, it has a very low specific gravity.

The specific gravity of a liquid changes with its temperature. As a liquid is heated, it expands and becomes less dense. Thus, its specific gravity becomes lower. As a liquid is cooled, it contracts and becomes more dense. Thus, its specific gravity becomes higher. Because of this, a partially discharged battery may appear to be fully charged if it is tested when it is cold. For the same reason, a fully charged battery may appear to be partially discharged if it is tested when it is hot. (See Figure 5.18.)

The specific gravity readings and specifications for electrolyte are accurate only when the electrolyte temperature is 80°F (27°C). If the electrolyte is at any other temperature, the hydrometer readings must be corrected. Most hydrometers contain a thermometer that enables you to determine the temperature of the electrolyte. Electrolyte temperature must always be considered when using a hydrom-

PERCENT OF CHARGE	SPECIFIC GRAVITY RANGES
100	1.260 – 1.265
95	1.250 – 1.260
75	1.230 – 1.240
50	1.200 – 1.220
25	1.170 – 1.190

Batteries containing electrolyte with a lower specific gravity should be considered discharged.

Figure 5.17 Approximate percentages of the state of charge for various ranges of battery electrolyte specific gravity.

eter to determine the state of charge of a battery.

The temperature correction chart in Figure 5.19 provides an easy method by which you can correct specific gravity readings within a wide range of

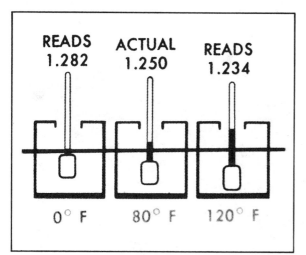

Figure 5.18 Specific gravity readings must be corrected for temperature. Conventional hydrometers are accurate only at 80°F (27°C) (courtesy of General Motors Corporation).

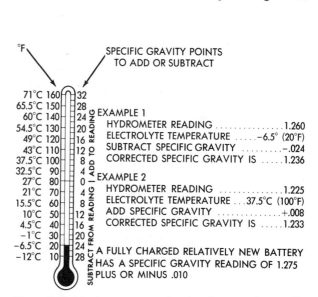

Figure 5.19 A hydrometer correction chart for correcting specific gravity readings for various electrolyte temperatures (courtesy of Chrysler Corporation).

temperatures. That chart reflects a 0.004 change in specific gravity for each 10°F (5.5°C) change in temperature. For every 10°F (5.5°C) over 80°F (27°C), 0.004 must be added to the hydrometer reading. For every 10°F (5.5°C) under 80°F (27°C), 0.004 must be subtracted from the hydrometer reading.

Measuring Specific Gravity
The following steps outline a procedure for measuring the specific gravity of the electrolyte in a battery:

1 Place a fender cover over the fender nearest the battery.

Note: Remember that electrolyte is an acid and is highly corrosive. Wear safety glasses and use care not to spill or splash any electrolyte during this procedure.

2 Remove the vent plugs or covers from the cells and place them on the battery as shown in Figure 5.20.

3 Squeeze the suction bulb of the hydrometer and insert the pick-up tube into the cell closest to the positive (+) post.

Note: Do not force the tube into the cell as you might damage the plates and separators.

4 Slowly release the bulb and draw in sufficient electrolyte to halfway fill the barrel.

Note: If there is insufficient electrolyte in the cell, an accurate test cannot be made. Water must be added to the cell and the battery must be charged so that the water will mix with the electrolyte.

5 Slowly squeeze the bulb to return the electrolyte to the cell.

Note: This step adjusts the temperature of the hydrometer to that of the electrolyte and provides for

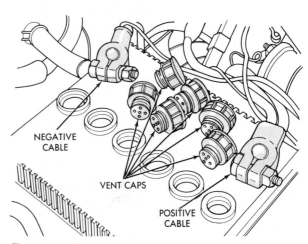

Figure 5.20 All vent plugs and covers should be removed before testing specific gravity. Placing the vent plugs on the battery minimizes the possibility of electrolyte damage to other surfaces (courtesy of Chrysler Corporation).

a more accurate first reading.

6 Slowly release the bulb again and draw in sufficient electrolyte to cause the float to rise.

Note: The hydrometer should be held in the vertical position so that the float will not drag against the inside of the barrel.

7 Read the specific gravity indicated on the float. Be sure that the float is floating free in the electrolyte and that it is not in contact with the spacer at the top of the barrel.

Note: An accurate reading requires that you bend down so that the reading is taken at eye level, as shown in Figure 5.21. Do not remove the hydrometer from the battery as that will allow electrolyte leakage.

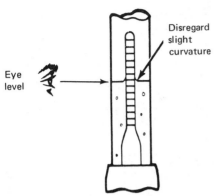

Figure 5.21 When reading a hydrometer, it should be held vertical so that the float does not touch the sides of the barrel. To obtain an accurate reading, the liquid should be at eye level (courtesy of American Motors).

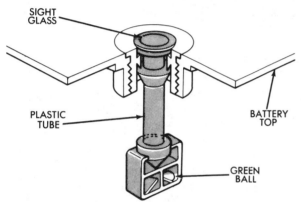

Figure 5.22 A built-in hydrometer of the type used in most sealed batteries. When the green ball floats in the electrolyte, it is visible in the sight glass. When the battery is discharged and the green ball sinks, the sight glass appears dark. When the electrolyte level drops below the hydrometer, the sight glass appears clear or light yellow (courtesy of Chrysler Corporation).

8 Note the temperature indicated by the thermometer.

9 Slowly squeeze the bulb and return all of the electrolyte to the cell.

10 Record the specific gravity and the electrolyte temperature.

11 Repeat steps 3 throgh 10 for the remaining cells.

12 Install the vent plugs or covers.

13 Clean and dry the top of the battery.

Interpreting Test Results   If the electrolyte temperature was approximately 80°F (27°C), the specific gravity readings may be considered accurate as recorded. Specific gravity readings of electrolyte at any other temperature must be corrected. Refer to Figure 5.19 to correct the specific gravity. Then refer to Figure 5.17 to determine the state of charge of the battery.

Figure 5.23 A sealed battery with a built-in temperature compensated hydrometer (courtesy of Pontiac Motor Division, General Motors Corporation).

Batteries with electrolyte whose specific gravity is less than 1.240 may be charged to increase the specific gravity. Batteries with a specific gravity reading of less than 1.200 should be charged before any further testing is attempted.

The specific gravity of all the cells in a battery should not vary more than 0.050. If the readings vary more than 0.050, the battery may be sulfated and may require replacement.

Specific Gravity Testing of Sealed Maintenance-Free Batteries   A sealed maintenance-free battery has no removable vent plugs. Therefore, a conventional hydrometer cannot be used to determine the state of charge. Most sealed batteries incorporate a built-in temperature compensated hydrometer as shown in Figure 5.22. The sight glass is located in the battery top as shown in Figure 5.23. While the built-in hydrometer does not provide an accurate measurement of the specific gravity of the electrolyte in each cell, it does provide sufficient information for diagnosis.

As shown in Figure 5.24, the most commonly used type of built-in hydrometer provides three indications:

1 *Green Dot Visible:*   Any green appearance should be considered a "green dot." This indicates that the state of charge is satisfactory and that the battery is ready for further testing.

Job 5b

TEST THE SPECIFIC GRAVITY OF BATTERY ELECTROLYTE
AND INTERPRET THE TEST RESULTS

SATISFACTORY PERFORMANCE

A satisfactory performance on this job requires that you do the following:

1 Test the specific gravity of the electrolyte in the
battery assigned.
2 Following the steps in the "Performance Outline,"
complete the job within 20 minutes.
3 Fill in the blanks under "Information."

PERFORMANCE OUTLINE

1 Remove the vent plugs.
2 Measure and record the specific gravity and the
temperature of the electrolyte in each cell.
3 Install the vent plugs.
4 Clean and dry the battery top.
5 Correct the specific gravity readings if necessary
and determine the state of charge of the battery.

INFORMATION

Vehicle identification _____

Battery identification _____

Test readings obtained:

Cell #	Specific Gravity	Temperature
1	_____	_____
2	_____	_____
3	_____	_____
4	_____	_____
5	_____	_____
6	_____	_____

Corrections and test interpretations:

Cell #	Corrected Specific Gravity	Percent of charge
1	_____	_____
2	_____	_____
3	_____	_____
4	_____	_____
5	_____	_____
6	_____	_____

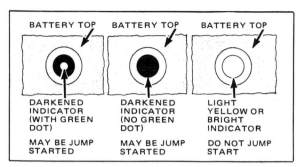

Figure 5.24 Readings provided by a built-in hydrometer (courtesy of Chevrolet Motor Division).

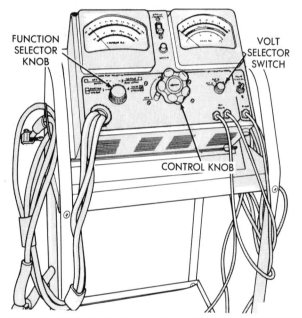

Figure 5.25 A typical tester used to load test batteries (courtesy of Chrysler Corporation).

2 *Dark—Green Dot Not Visible:* A dark appearance without a green dot indicates that the battery is partially discharged. Further testing is required to determine if the battery can be charged or if it should be replaced.

3 *Clear or Light Yellow:* A clear or light yellow appearance indicates that the electrolyte level is so low that it is below the bottom of the built-in hydrometer. This usually means that the battery is defective. If a clear or light yellow appearance is accompanied by a failure to crank the engine, the battery should be replaced. *DO NOT ATTEMPT TO CHARGE, TEST, OR JUMP START.*

Load Testing While a hydrometer can be used to determine the state of charge of a battery, it cannot measure the battery's ability to deliver energy.

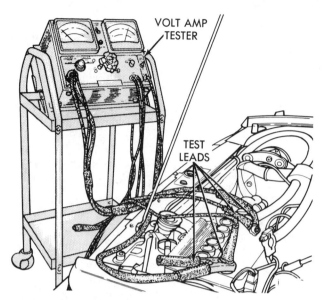

Figure 5.26 A volt-amp tester connected to perform a battery load test (courtesy of Chrysler Corporation).

A *load test* provides a method of determining the serviceability of a battery. As its name implies, load testing consists of testing a battery while it is working under a load.

Many different instruments and procedures may be used to load test a battery. All of those instruments and procedures however, require that battery voltage be measured while the battery is being discharged at a high rate. For this reason, a load test often is referred to as a *high-rate discharge test*.

A tester similar to the one shown in Figure 5.25 is most commonly used to load test a battery. In addition to various selector switches, the tester contains an adjustable resistance unit, an ammeter, and a voltmeter. The adjustable resistance unit is usually a *carbon pile*. A carbon pile is a stack of carbon discs that can be squeezed together to form a low resistance unit capable of handling large amounts of current. The ammeter measures the amount of current flowing through the carbon pile. The voltmeter measures the battery voltage.

In use, the tester is connected across the terminals of a battery as shown in Figure 5.26. The carbon pile is adjusted by means of the control knob until a specified current is shown on the ammeter. That current will vary from 150 A to over 300 A depending on the battery being tested. The voltage indicated by the voltmeter is read after the load has been maintained for 15 seconds.

The following steps outline a typical procedure for

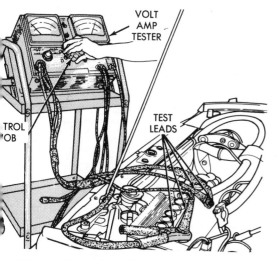

re 5.29 A load is applied to the battery by turning the control
clockwise (courtesy of Chrysler Corporation).

MINIMUM VOLTAGE UNDER SPECIFIED LOAD	BATTERY TEMPERATURE	
	°F	°C
9.6	70	21
9.5	60	16
9.4	50	10
9.3	40	4
9.1	30	−1
8.9	20	−7
8.7	10	−12
8.5	0	−18

Figure 5.30 Minimum voltages during load testing.

Maintain the desired discharge rate for 15 sec-
s and note the voltage indicated by the volt-
er.

Immediately turn the control knob to the OFF
tion.

e: It is not advisable to maintain the load for
e than 15 seconds.

Disconnect the tester from the battery.
Connect the battery cables.

e: Connect the cable that runs to the starter mo-
or starter switch first. Connect the ground cable

erpreting Test Results A battery in good
dition will maintain a voltage of at least 9.6 V
e the specified load is applied. Since the bat-
temperature will affect the indicated voltage,
chart in Figure 5.30 lists the minimum voltage
vable for temperatures lower than 70°F (21°C).

e indicated voltage during a load test drops
w those listed in Figure 5.30, the battery is de-
ve and should be replaced.

ee-Minute Charge Testing A three-minute
rge test if often performed on a battery that is
pected of being sulfated. Before this test is at-
pted, the following precautions should be ob-
ed:

his test should not be performed on a sealed
ery.

his test should not be performed on any con-
onal battery unless it has failed a load test (high-
discharge test).

3 This test should not be performed unless the battery temperature is at least 60°F (16°C).

In a three-minute charge test, a fast charger is used
to push high current through the battery for a period
of three minutes. If the sulfation is of a minor nature,
the high current flow will dislodge it from the plates.
If the sulfation is severe, it will not be removed and
the battery will usually require replacement. The following steps outline a procedure for performing a
three-minute charge test:

1 Place a fender cover on the fender near the
battery.

2 Disconnect the battery cables from the terminals.

Note: Disconnect the ground cable first. This will
minimize the possibility of arcing and a resultant battery explosion.

3 Turn the charger switch to the OFF position.

4 Connect the charger leads to the battery. The
positive (+) lead must be connected to the positive
(+) terminal. The negative (−) lead must be connected to the negative (−) terminal.

5 Turn the charger switch to the ON position. If
the charger is equipped with a timer, set the timer
for 3 minutes.

6 Adjust the charge rate control so that the battery is being charged at the highest possible rate,
but not exceeding 40 A.

7 At the end of the three-minute charge period,
connect a voltmeter across the battery terminals.

8 Turn the charger switch to the ON position and
record the voltmeter reading.

Note: The charger must be charging the battery when
the voltmeter reading is taken.

9 Turn the charger switch OFF and disconnect
the voltmeter.

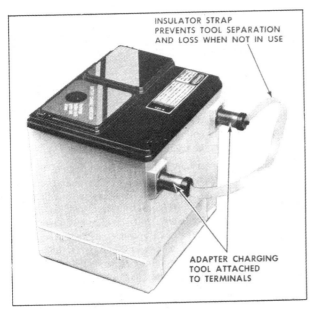

INSULATOR STRAP
PREVENTS TOOL SEPARATION
AND LOSS WHEN NOT IN USE

ADAPTER CHARGING
TOOL ATTACHED
TO TERMINALS

Figure 5.27 Adapters can be attached to side terminal batteries to facilitate testing (courtesy of Pontiac Motor Division, General Motors Corporation).

CONVENTIONAL BATTERIES	
BATTERY CAPACITY (AMPERE HOURS)	DISCHARGE (AMPERE
36	155
41	145
45	190
53	175
54	225
68	220
77	228
MAINTENANCE-FREE BATTERIES	
BATTERY CAPACITY (AMPERE HOURS)	DISCHARGE (AMPER
53	200
63	215
68	235

Figure 5.28 Discharge rates specified by on batteries of different ampere-hour ratings.

load testing a battery. You should consult an appropriate manual for the correct load to place on the battery you are testing:

1 Place a fender cover on the fender near the battery.

2 Test the specific gravity and temperature of the battery. If the specific gravity is less than 1.200, the battery should be charged before a load test is attempted. If the temperature of the battery is less than 60°F (15.5°C), allow the temperature to rise before attempting a load test.

3 Disconnect the battery cables from the terminals.

Note: Disconnect the ground cable first. This will minimize the possibility of arcing and a resultant battery explosion.

4 Turn the control knob on the battery tester to the OFF position.

5 If the tester is fitted with a function control switch, turn the switch to the Battery Test position.

6 If the tester is fitted with a volt selector switch, turn the switch so that a voltage exceeding battery voltage is selected.

7 Connect the heavy ammeter leads to the battery terminals. The positive (+) lead must be connected to the positive (+) terminal, the negative (−) lead to the negative (−) terminal.

Note: When testing side-terminal batteries, post adapters or a post adapter tool similar to the one

shown in Figure 5.27 should be use connection can be obtained.

8 Connect the voltmeter leads to minals. The positive (+) lead mus to the positive (+) terminal, the ne to the negative (−) terminal.

Note: The voltmeter leads should cc posts or adapters. If the voltmete nected to the clamps on the amme accurate voltmeter reading may be

9 Refer to an appropriate manu load to place on the battery being

Note: Some manufacturers prov charge rates for their batteries. An specific loads is shown in Figure 5 facturers recommend that the batt at a rate equal to three times the a of the battery, or at a rate equal to cranking rating.

10 Turn the control knob clock Figure 5.29 until the desired dis dicated by the ammeter. Allow the in this position for 15 seconds control to the OFF position.

Note: This initial load will rem charge" from the battery and pro test results.

11 After waiting about 15 secor control knob clockwise until the rate is indicated.

Job 5c

LOAD TEST A BATTERY

SATISFACTORY PERFORMANCE
A satisfactory performance on this job requires that you do the following:

1 Perform a load test (high-rate discharge test) on the battery assigned.
2 Following the steps in the "Performance Outline" and the specifications of the manufacturer, complete the job within 20 minutes.
3 Fill in the blanks under "Information."

PERFORMANCE OUTLINE
1 Test the specific gravity of the battery electrolyte.
2 Disconnect the battery cables.
3 Connect the tester.
4 Apply the specified load.
5 Read the voltage while the battery is under load.
6 Remove the load.
7 Disconnect the tester.
8 Connect the battery cables.

INFORMATION
Vehicle identification _____

Battery identification _____

Reference used _____ Page(s) _____

Battery ampere-hour capacity _____

Hydrometer readings: 1 _____   2 _____   3 _____

4 _____   5 _____   6 _____

Electrolyte temperature _____

Corrected hydrometer readings: 1 _____   2 _____   3 _____

4 _____   5 _____   6 _____

Specified discharge rate (test load) _____

Tester identification _____

Amount of time load was maintained _____

Voltage indicated while under load _____

Recommendations:

_____ Battery should be returned to service

_____ Battery should be charged and returned to service

_____ Battery should be replaced

Job 5d

PERFORM A THREE-MINUTE CHARGE TEST

SATISFACTORY PERFORMANCE

A satisfactory performance on this job requires that you do the following:

1 Perform a three-minute charge test on the battery assigned.
2 Following the steps in the "Performance Outline," complete the job within 10 minutes.
3 Fill in the blanks under "Information."

PERFORMANCE OUTLINE

1 Disconnect the battery cables.
2 Connect a fast charger to the battery.
3 Charge the battery at the correct rate for the proper time.
4 Read and record the battery voltage while the battery is being charged.
5 Turn the charger off.
6 Disconnect the charger.
7 Connect the battery cables.

INFORMATION

Vehicle identification _____

Battery identification _____

Charging rate _____   Charging time _____

Voltmeter reading _____

Recommendations:

The battery is: _____ badly sulfated

                       _____ not badly sulfated

The battery should be: _____ charged

                       _____ replaced

---

10 Disconnect the charger leads.
11 Connect the battery cables.

Note: Connect the ground cable last.

Interpreting Test Results   If the meter reading is less than 15.5 V, the battery is not badly sulfated and it will probably accept a charge, preferably at a low rate. If the meter reading is above 15.5 V, the battery is badly sulfated and will probably require replacement.

BATTERY CHARGING   A battery in good condition may occasionally fail. This usually is noticed by the car owner when the battery is unable to crank the engine fast enough so that it will start. The most common causes of a failure of this type are as follows:

1 The lights or other accessories were accidentally left on for an extended period of time.
2 Poor battery maintenance, including failure to maintain the electrolyte level, loose connections, dirty

connections, or an improperly secured battery.

3 Problems in the charging system, including loose drive belts, a faulty alternator, a faulty regulator, or high resistance in connections and components.

4 Trips of short duration that do not allow sufficient time for the charging system to restore the energy used for starting.

5 Defects in the electrical system such as short circuits.

6 The application of electrical loads exceeding the capacity of the alternator. This is usually caused by the use of aftermarket equipment such as radio systems, air conditioners, and special lighting systems.

7 Battery self-discharge during a long period of inactivity.

A discharged battery in good condition can be charged and returned to service. To prevent a recurrence, the causes of the discharged condition should be determined and, if possible, corrected.

Many types of battery chargers are in use, but all chargers operate on the same principle. They apply an electrical pressure that forces current through the battery to reverse the electrochemical action in the cells.

WHEN A BATTERY IS BEING CHARGED, AN EXPLOSIVE MIXTURE OF GASES INCLUDING HYDROGEN AND OXYGEN IS RELEASED FROM THE ELECTROLYTE. THOSE GASES ARE PRESENT IN THE CELLS AND IN THE AREA SURROUNDING THE BATTERY. TO AVOID THE POSSIBILITY OF IGNITING THOSE GASES YOU SHOULD OBSERVE THE FOLLOWING PRECAUTIONS:

1 Never connect or disconnect live charger leads to a battery or otherwise break a live circuit at the battery. To do so may cause arcing and possibly ignite the gases. The charger should be turned off before the leads are disconnected, connected, or otherwise disturbed.

2 Keep all open flames away from a battery.

3 Do not smoke near a battery on charge or near a battery that has recently been charged. (See Figure 5.31.)

4 If possible, charge batteries in a well ventilated area.

Charging Rates The amount of charge a battery receives is equal to the rate of charge, in amperes, multiplied by the amount of time, in hours, that the charge is applied. As an example, a battery charged at the rate of 5 A for a period of 5 hours

Figure 5.31 The explosive gases given off by a battery can be ignited by a spark or a flame (courtesy of Chevrolet Motor Division).

would receive a 25 ampere-hour charge. To bring a battery to a fully charged condition, you must restore the ampere-hours that were removed from it. Since 100 percent efficiency cannot be obtained on recharging, approximately 20 percent must be added to your ampere-hour calculations to allow for losses in the charging process.

Specific charging rates and charging times cannot be stated for the following reasons:

1 Different batteries have different electrical capacities. A fully discharged battery with a reserve capacity rating of 90 minutes will require twice as much recharging as a fully discharged battery with a reserve capacity rating of 45 minutes.

2 Electrolyte temperatures will differ. A partially discharged battery at a temperature of 80°F (27°C) will accept a higher charging rate than a battery at 20°F (−6.5°C).

3 The state of charge at the start of the charging process will vary. A battery that is completely discharged will require twice as much recharging as one that is half-charged.

4 The condition of the batteries serviced will vary. An old battery or one that has been subjected to severe service will require more recharging than a newer battery because of greater internal resistance.

Most car manufacturers specify the charging rates and charging times for their batteries, but those specifications are but guidelines for recharging relatively new batteries in good condition. Figure 5.32 shows the specifications for fully discharged batteries in the manual of one car maker.

Slow Charging If time allows, slow charging should be used to charge a discharged battery. Slow charging consists of charging a battery at a

Watt Rating	5 Amperes	10 Amperes	20 Amperes	30 Amperes	40 Amperes	50 Amperes
Below 2450	10 Hours	5 Hours	2½ Hours	2 Hours		
2450-2950	12 Hours	6 Hours	3 Hours	2 Hours	1½ Hours	
Above 2950	15 Hours	7½ Hours	3¼ Hours	2 Hours	1¾ Hours	1½ Hours

*Initial rate for constant voltage taper rate charger.
To avoid damage, charging rate must be reduced or temporarily halted if:
1. Electrolyte temperature exceeds 125° F.
2. Violent gassing or spewing of electrolyte occurs.
Battery is fully charged when over a two hour period at a low charging rate in amperes all cells are gassing freely and no change in specific gravity occurs. **For the most satisfactory charging, the lower charging rates in amperes are recommended.**
Full charge specific gravity is 1.260 - 1.280 corrected for temperature with electrolyte level at split ring.

Figure 5.32 Typical charging rates and times for fully discharged batteries. Notice that the rates and times are different for batteries with different watt ratings (courtesy of Buick Motor Division, General Motors Corporation).

Figure 5.33 A typical slow charger (courtesy of Snap-on Tools Corporation).

existing state of charge of the battery, slow charging may require from 12 to 24 hours of time. A battery that is sulfated may require even more time. During the charging period, the electrolyte temperature should not exceed 110°F (43°C). If the electrolyte temperature rises above 110°F (43°C), the charging rate should be decreased. A typical slow charger is shown in Figure 5.33.

A conventional battery with vent plugs is considered fully charged when the electrolyte is gassing freely, and when no further rise in the specific gravity is noted during three successive hydrometer readings taken at intervals of 1 hour. A sealed battery should be slow charged until the green dot appears in the built-in hydrometer. (Refer to Figure 5.24.) In some instances, a sealed battery must be tipped or slightly shaken to allow the green dot to appear.

Fast Charging If time is not available to slow charge a battery, fast charging at a high rate is permissible. While fast charging will not fully recharge a battery, it will restore the charge sufficiently to allow the battery to be used. If the charging system in the car is operating correctly, the battery will continue charging in service. Figure 5.34 shows a typical fast charger.

Fast charging consists of charging a battery at a rate of from 10 to 50 A. The exact charging rate depends on the construction of the battery, the con-

rate of about 5 A for a time sufficient to bring the specific gravity of the electrolyte to its highest reading. At no time should the charging rate exceed 1 A for each positive plate per cell. Depending on the

Figure 5.34 A typical fast charger (courtesy of Snap-on Tools Corporation).

dition of the battery, and the time available. The temperature of the electrolyte provides an indication of the correct charging rate. If the electrolyte temperature rises above 125°F (65°C), the charging rate is too high and should be reduced. Since a high charging rate and the resultant high temperature can damage a battery, a battery should be charged at the lowest possible rate.

Charging a Battery The following steps outline a procedure for charging a battery. You should consult the instructions provided with the charger you have available. The charging specifications should be obtained from an appropriate manual and used with consideration for the condition of the battery you will charge:

1 Place a fender cover over the fender nearest the battery.

2 If the battery is not sealed, check the electrolyte level in all of the cells and adjust the level if necessary.

Note: Do not attempt to charge a battery that appears to be frozen or if ice crystals are visible in the electrolyte. Allow the battery to thaw fully before charging is attempted.

3 If the battery is a sealed battery, check the built-in hydrometer. Do not attempt to charge the battery if the indicator appears clear or light yellow. (Refer to Figure 5.24.)

4 Disconnect the battery cables. This will prevent possible damage to electrical components in the car during charging.

Note: Disconnect the ground cable first to minimize the possibility of arcing and a resultant battery explosion.

5 Clean the battery terminals and the battery top.

6 Consult an appropriate manual and determine the charging rate and time for the battery.

7 Turn off the charger switch.

8 Connect the charger leads to the battery. The positive (+) lead must be connected to the positive (+) terminal. The negative (−) lead to the negative (−) terminal.

Note: If you are charging a side terminal battery, install post adapters to ensure a good connection. (Refer to Figure 5.27.)

9 Turn on the charger switch.

Note: On some chargers, the timer must be set to turn on the charger.

10 Adjust the charging rate.

11 Adjust the timer.

12 Check the charging rate and the battery temperature after the battery has been charging for about 15 minutes. Adjust the charging rate if required.

13 Continue charging until the alloted time has elapsed or until the battery is fully charged.

14 Turn off the charger switch.

15 Disconnect the charger leads from the battery.

16 Connect the battery cables.

Note: Connect the ground cable last.

Dry Charged Batteries Dry charged batteries are manufactured by special procedures that result in a battery containing dry, charged, plates. They are shipped, stocked, and sold without electrolyte in the cells. Dry charged batteries offer certain advantages over "wet" batteries. Shipping costs are lower, especially where sulfuric acid is available from local sources. Their lower weight allows easier handling and storage. However, their greatest advantage is in their long shelf life.

As you know, a "wet" battery will self-discharge, even if it is not in use. A battery starts to self-discharge as soon as electrolyte is added to the cells.

Job 5e

CHARGE A BATTERY

SATISFACTORY PERFORMANCE
A satisfactory performance on this job requires that you do the following:

1 Recharge the battery assigned.
2 Following the steps in the "Performance Outline" and the charging recommendations of the manufacturer, complete the job within 30 minutes plus the charging time.
3 Fill in the blanks under "Information."

PERFORMANCE OUTLINE
1 Check the electrolyte level and adjust the level if required.
2 Disconnect the battery cables.
3 Clean the battery terminals and the battery top.
4 Determine the charging rate and time.
5 Connect the charger and adjust the charging rate and time.
6 Check the charging rate and the battery temperature.
7 Turn off the charger when completed.
8 Disconnect the charger.
9 Connect the battery cables.

INFORMATION
Vehicle identification _____

Battery identification _____

Specific gravity of electrolyte at start _____

Reference used _____ Page(s) _____

Recommended charging rate _____ amperes for _____ hours.

Battery was charged at the rate of _____ amperes.

Battery temperature during period of charge _____

Length of time battery was charged _____

Specific gravity of electrolyte at completion _____

If a wet battery is not placed in service within a short period of time, it will require recharging before it can be used. If a wet battery remains in storage too long, the plates become sulfated. Recharging may not restore the battery to its original condition and the service life of the battery will be shortened.

Dry charged batteries are not only manufactured without electrolyte, but the cells are sealed so that moisture, even moisture in the air, cannot enter the battery. Therefore, no electrochemical action takes place in the battery, and the battery cannot self-discharge and deteriorate.

Activating Dry Charged Batteries When you replace a defective battery with a dry charged battery, you must first activate the dry charged battery. The steps that follow outline a procedure for activating a dry charged battery. The specifications given for electrolyte specific gravity, battery temperature, and charging rates may differ with batteries from different manufacturers. You should read the instructions furnished with the battery you are activating and follow the specifications provided:

1 Carefully unpack the battery and place it on a level surface.

Note: Do not install the battery and attempt to activate it in the car. It is recommended that a dry charged battery be activated on the floor near a floor drain. This will enable you to flush away any electrolyte that may be spilled.

2 Check the electrolyte supplied with the battery. It should be battery grade sulfuric acid with a specific gravity of approximately 1.265.

3 Remove the vent seals.

Note: Some batteries are sealed with disposable caps. Others have a thin plastic disc under the vent plugs. Those discs must be pushed down into the cells with the special extended vent plug provided.

4 Carefully open the electrolyte container.

Note: Use extreme care during this step and the steps that follow. Remember that electrolyte is an acid and can cause you painful burns.

5 Using a plastic or glass funnel, slowly add electrolyte to the cells.

Note: Add the electrolyte slowly and stop often to check the electrolyte level. It is easy to overfill the cells. Excess electrolyte must be removed with a syringe or with a hydrometer. Spilled electrolyte must be washed away.

6 After all the cells are filled, carefully tip the battery from side to side to release any air bubbles that may be trapped in the cells.

7 Check the electrolyte level in each cell and adjust the levels as required.

8 Check the specific gravity and the temperature of the electrolyte. The specific gravity should be at least 1.250. The temperature should be at least 80°F (27°C).

Note: Dry charged batteries manufactured by a "spin process" may provide lower specific gravity readings. Check the instructions furnished with the battery.

9 If the specific gravity or the temperature does not meet the foregoing specifications, the battery should be placed on charge at a rate of approximately 30 A for about 15 minutes.

10 If the specific gravity and the temperature meet the specifications, the battery may be installed without charging.

Note: A battery that meets the foregoing specifications is not fully charged, but is charged sufficiently for use.

11 Drain and rinse the empty acid container in running water. Crush or multilate the container so that it cannot be reused and dispose of it where it is not likely to be handled.

12 Flush away any spilled electrolyte. Discard any rags or wipers that are wet with electrolyte.

Job 5f

ACTIVATE A DRY CHARGED BATTERY

SATISFACTORY PERFORMANCE

A satisfactory performance on this job requires that you do the following:

1 Activate the dry charged battery assigned.
2 Following the steps in the "Performance Outline" and the instructions and specifications furnished with the battery, complete the job within 30 minutes.
3 Fill in the blanks under "Information."

PERFORMANCE OUTLINE

1 Place the battery where spills can be easily washed away.
2 Remove the vent seals.
3 Fill the cells with electrolyte.

4 Test the battery.
5 Charge the battery if necessary.
6 Clean up any spilled electrolyte and dispose of the empty containers.

INFORMATION

Vehicle identification: _____

Battery identification: _____

Specific gravity of electrolyte used: _____

Specific gravity of electrolyte in battery: _____

Temperature of electrolyte in battery: _____

Was a boost charge necessary?   _____ Yes   _____ No

Charging rate of boost charge: _____

Charging time of boost charge: _____

SUMMARY

By studying the text and completing the Jobs in this chapter, you have gained additional knowledge of batteries and how they are rated and tested. You have developed additional diagnostic and repair skills. You now can identify battery ratings, measure the specific gravity of electrolyte, and perform a load test. You have learned to interpret the results of your tests and to recommend the required services. You also can charge a battery to specifications and you are aware of how a dry charged battery is activated. The knowledge and skills that you gained in the area of battery service will be used throughout all areas of electrical system service.

SELF-TEST

Each incomplete statement or question in this test is followed by four suggested completions or answers. In each case select the *one* that best completes the sentence or answers the question.

1 A laboratory test that measures the amount of current a battery can deliver at 0°F (−18°C) for a period of 30 seconds while maintaining a cell voltage of at least 1.2 V is used to determine the
A. watts rating
B. ampere-hour rating
C. reserve capacity rating
D. cold cranking performance rating

2 A battery rating determined by multiplying the terminal voltage by the ampere-hour rating is known as the
A. watts rating
B. voltage rating
C. reserve capacity rating
D. cold cranking performace rating

3 Which of the following may cause a battery to self-discharge?
A. Loose connections
B. Dirty connections
C. A loose hold-down
D. Dirt and electrolyte on the battery top

4 The most probable cause of excessive water loss from a battery is
A. sulfation
B. vibration
C. overcharging
D. undercharging

5 If a battery is allowed to remain discharged or partially discharged for an extended period of time, the
A. electrolyte will corrode the terminals
B. lead sulfate will harden on the plates
C. sulfuric acid will evaporate from the electrolyte
D. concentrated acid in the electrolyte will attack the plates

6 The electrolyte in a fully charged battery has a specific gravity of approximately
   A. 1.205 to 1.220
   B. 1.220 to 1.235
   C. 1.235 to 1.250
   D. 1.250 to 1.265

7 Two mechanics are discussing the effects of temperature on electrolyte.
   Mechanic A says that as electrolyte is heated, it expands and becomes less dense.
   Mechanic B says that as electrolyte is heated, its specific gravity becomes lower.
   Who is right?
   A. A only
   B. B only
   C. Both A and B
   D. Neither A nor B

8 A test of a cold battery reveals that the electrolyte has a specific gravity of 1.240 at 30°F (−1°C). The temperature corrected specific gravity is
   A. 1.205
   B. 1.220
   C. 1.235
   D. 1.250

9 If the specific discharge rate for load testing a particular battery cannot be found, the battery may be discharged at a rate equal to
   A. 3 times the ampere-hour rating of the battery
   B. 6 times the ampere-hour rating of the battery
   C. 9 times the ampere-hour rating of the battery
   D. 12 times the ampere-hour rating of the battery

10 Two mechanics are discussing a sealed battery whose built-in hydrometer appears light yellow in color.
   Mechanic A says that the battery should be charged at the rate of 30 A for 60 minutes.
   Mechanic B says that the battery should be load tested at a rate of 60 A for 15 seconds.
   Who is right?
   A. A only
   B. B only
   C. Both A and B
   D. Neither A nor B

Chapter 6 The Lighting System— Basic Services

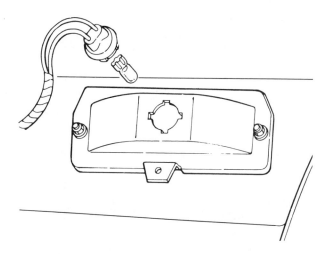

In terms of safety, the lighting system is the most important electrical system in an automobile. The lighting system uses electrical energy to illuminate the road, to signal the driver's intention to stop and turn, and to light the car so that it can be seen by other drivers. The lighting system also warns the driver of malfunctions and provides the convenience of interior illumination.

The lighting system consists of many different circuits, each serving a different function. Basic services in those circuits consist mainly of the replacement of parts such as bulbs, lenses, fuses, and protection devices. By performing the Jobs in this chapter, you will become familiar with those parts and the various service procedures required to test and replace them. Those Jobs will enable you to develop your hand skills, increase your knowledge, and provide a foundation for the development of diagnostic skills.

Your specific objectives in this chapter are to perform the following Jobs:

A
Check the operation of a lighting system
B
Replace small bulbs
C
Replace lamp lenses
D
Replace defective fuses
E
Replace circuit breakers
F
Replace flashers
G
Replace sealed beam bulbs
H
Adjust headlight aim

LIGHTING SYSTEM SERVICE The lighting system is the largest electrical system in an automobile. While it may contain miles of wire and hundreds of connectors, terminals, switches, and sockets, the lighting system is relatively simple to service, but only if you know how the system operates.

The lighting system is made up of dozens of separate circuits, each performing a specific task. When a problem appears in the lighting system, you must isolate the problem to the particular circuit involved. You must then check that circuit and its components to locate the problem and determine the appropriate repair. Only when a diagnostic procedure is followed can you be sure that the parts you replace and the repairs that you make will eliminate the problem.

Checking the Operation of a Lighting System

The first step in diagnosing a problem is to determine if a problem really exists. This can be done in a lighting system by checking each circuit individually. If a problem is found, the circuit in which it appears can then be isolated. The following steps outline a typical procedure for checking a lighting system. When a problem is found, it should be noted and the check continued so that you can be sure that all of the problems are found:

1 Turn on the headlights. Check that both headlights are operating.

2 Operate the headlight dimmer switch and check the operation of both the high and low beams. Check the operation of the high beam indicator light on the instrument panel.

3 Place the headlight switch in the parking light position. Check the operation of the parking lights, the taillights, the license plate lights, and the side marker lights.

4 Check the operation of the instrument-panel lights and other instrument lights such as the radio, transmission selector, and ash tray lights. Operate the instrument light dimmer switch to check its operation. Turn off the headlight switch.

5 Place the ignition switch in the On (Ign) position.

Note: Do not start the engine.

6 Move the transmission selector or shift lever into the Reverse position and check the operation of the back-up lights. Place the transmission selector or shift lever in Neutral or Park.

7 Check the operation of the turn signal lights and the cornering lights (if fitted). Check the operation of the turn signal indicator lights on the instrument panel. Turn off the ignition switch.

8 Turn on the hazard warning flasher switch and check the operation of the hazard warning lights. Turn off the hazard warning lights.

9 Turn on any accessory lights (driving lights, fog lights) and check their operation. Turn off the accessory lights.

10 Open the doors of the car and check the operation of any courtesy lights and dome lights.

11 Close the doors and check to see that the courtesy lights and dome lights can be operated by the switch on the instrument panel.

12 Sit in the driver's seat and turn the ignition switch to the On (Ign) position. Check the operation of the seat belt warning light.

13 With the ignition switch in the On (Ign) position, check the operation of the alternator warning light, the oil pressure warning light, the engine temperature warning light, and any other warning light fitted.

14 With the transmission selector lever in the Park or Neutral position, turn the ignition switch to the Start position and check the operation of the brake warning light. Turn off the ignition switch.

Small Bulbs A bulb consists of a *filament* enclosed in a ball-shaped glass envelope. A filament is a piece of fine wire that glows white hot when current passes through it. To prevent the filament from burning out, the air is evacuated from the bulb and it is sealed. Most automotive bulbs have a metal base that fits into a socket. That base is usually referred to as a *bayonet base* as it has *lugs,* or pins, on its side that engage with locking slots in the socket. Bulbs with threaded bases are not used in automobiles because they tend to turn loose with vibration. The construction of a typical bulb is shown in Figure 6.1

A variant bulb in common use has no metal base. Commonly called a *wedge bulb,* the glass envelope is flat on one end and the wires leading to

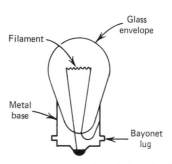

Figure 6.1 A sectioned view of a typical small bulb. The air is evacuated from the glass envelope so the filament can glow white hot without burning.

Job 6a

CHECK THE OPERATION OF A LIGHTING SYSTEM

SATISFACTORY PERFORMANCE

A satisfactory performance on this job requires that you do the following:

1 Check the operation of the lighting system on the car assigned.
2 Following the procedure in the "Performance Outline" check all the lights indicated on the Automotive Light Checklist.
3 Fill out the Automotive Light Checklist indicating any defects found and suggested repairs.

PERFORMANCE OUTLINE

1 Check the operation of all the exterior lights.
2 Check the operation of all the interior lights.

AUTOMOTIVE LIGHT CHECKLIST

Vehicle identification _____

Indicate each inoperative or defective light by placing a check in the appropriate box. Indicate the nature of the defect.

Exterior Lights	Left Front	Right Front	Left Rear	Right Rear	Defect
Headlights—high beam					
Headlights—low beam					
Parking lights					
Taillights					
License Plate lights					
Side marker lights					
Turn signal lights					
Cornering lights					
Back-up lights					
Stop lights					
Hazard warning lights					
Driving (fog) lights					

Interior Lights		Defect
Dome lights		
Courtesy lights		
Transmission selector light		
Instrument panel lights		
Radio dial lights		

Heater—A/C panel lights		
High beam indicator light		
Turn signal indicator lights		
Map light		
Ignition key light		
Cigar lighter light		
Ashtray light		
Seat belt warning light		
Brake warning light		
Alternator warning light		
Oil pressure warning light		
Engine temperature warning light		
Door ajar light		
Accessory lights		

Comments and suggested repairs: _____

_____

_____

_____

_____

Name _____ Date _____

the filament are exposed. Wedge bulbs, shown in Figure 6.2, are friction-fitted to plastic or rubber sockets and are usually found in side marker lights and other small lamps. Wedge bulbs are also used for the lights in instrument panels, where they are fitted into removable plastic sockets.

Some bulbs contain two filaments. Those bulbs are used where one lamp must serve two functions. An example of where a bulb of this type is used is in a parking light that also is used as a directional or turn signal. On many cars, the functions of the tail-lights and the stop lights are served by bulbs of this construction. A bulb with two filaments is shown in Figure 6.3.

Dozens of bulbs are used in every car, and those bulbs are usually of many different types and designs. Bulbs are identified by a universal numbering system that is used by most automobile manu-

facturers. Figure 6.4 shows some of the more commonly used bulbs and their identifying numbers.

WEDGE

Figure 6.2 While most automotive bulbs have a bayonet base, some have no base. Termed wedge bulbs, they are friction fitted to rubber or plastic sockets.

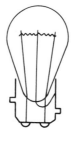

Figure 6.3 A sectioned view of a bulb with two filaments. The central wire is attached to the metal base and forms a common ground for both filaments. Notice that the lugs or pins on the base are offset, or "indexed," so that the bulb can be installed in its socket in only one direction.

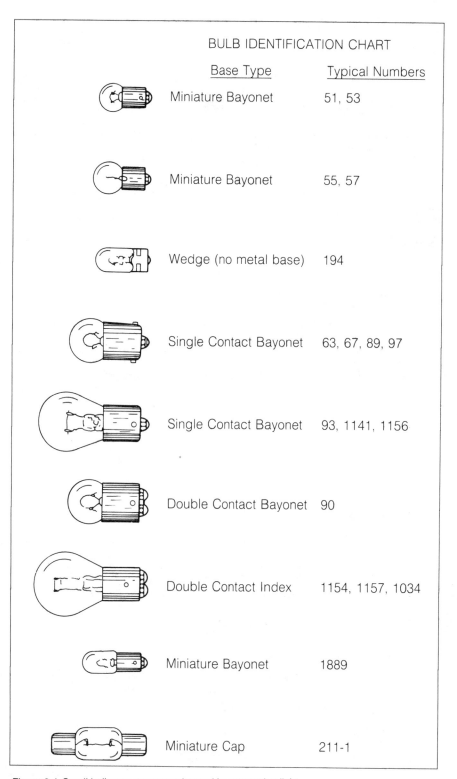

BULB IDENTIFICATION CHART

Base Type	Typical Numbers
Miniature Bayonet	51, 53
Miniature Bayonet	55, 57
Wedge (no metal base)	194
Single Contact Bayonet	63, 67, 89, 97
Single Contact Bayonet	93, 1141, 1156
Double Contact Bayonet	90
Double Contact Index	1154, 1157, 1034
Miniature Bayonet	1889
Miniature Cap	211-1

Figure 6.4 Small bulbs most commonly used in automotive lighting circuits.

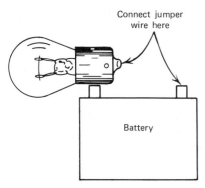

Figure 6.5 A bulb can be tested by using a jumper wire to connect the bulb across the terminals of a battery.

Figure 6.7 A pair of pliers designed for removing bulbs. The soft plastic-coated jaws grip the bulb securely and minimize breakage (courtesy of Easco/KD Tools, Lancaster PA 17604).

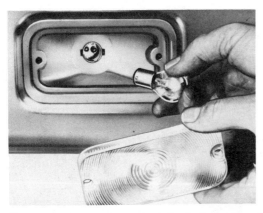

Figure 6.6 On some cars, the bulbs in the parking lights and back-up lights can be reached by removing the lens (courtesy of Ford Motor Company).

Figure 6.8 A pair of pliers designed for removing the bases of broken bulbs (courtesy of Easco/KD Tools, Lancaster, PA 17604).

Testing Bulbs In many instances, a bulb can be checked visually. If the filament is broken or missing, the bulb is defective. A bulb can be tested by using a jumper wire, as shown in Figure 6.5, to connect it across the terminals of a battery. However, since a bulb must be removed for testing, it usually is easier to test it by substitution. Simply install a replacement bulb and turn on the circuit. If the replacement bulb lights, you can assume that the old bulb was defective.

Replacing Bulbs in Exterior Lights Before you can replace a bulb, you must gain access to it. On most cars, the designs of the parking lights, taillights, and marker lights change annually with the styling of the car. Those changes often require a change in service procedure. The correct procedure for any particular car can be found in the manufacturer's service manual.

In general, you can gain access to a bulb in one of two ways. You can remove the lens, or you can

remove the socket. As shown in Figure 6.6, the lenses of some parking lights and back-up lights are secured by screws. Removing those screws will allow you to remove the lens. On some cars, the space between the bulb and the sides of the reflector is very limited and it is hard to get a grip on the bulb. In those instances, and where the bulb may be tight in its socket, you may find it helpful to use a pair of bulb gripping pliers of the type shown in Figure 6.7. Another similar pair of pliers, shown in Figure 6.8, can be used to remove the base of a broken bulb. Those tools can help you avoid cuts from broken bulbs.

On other cars, you must remove the socket from the rear of the reflector. Figure 6.9 shows two different types of removable sockets. Both types must be pushed in and turned counterclockwise for removal. Sockets of those types usually have locating keys or tabs that allow installation in only one position. When installing those sockets, you must align the tabs, push in the socket, and then turn it clockwise to lock it. Access to those sockets may be from inside the trunk or luggage compartment, behind the grill, behind the bumper, or inside the fender. Figures 6.10, 6.11, and 6.12 show some of those locations.

Replacing Bulbs in Interior Lights The bulbs used inside a car quite often outnumber those used for exterior lighting. Bulbs used to illuminate the

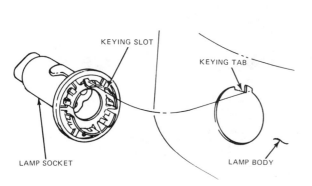

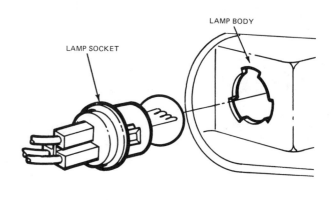

Figure 6.9 Different sockets have different types of alignment and locking keys (courtesy of Ford Motor Company)

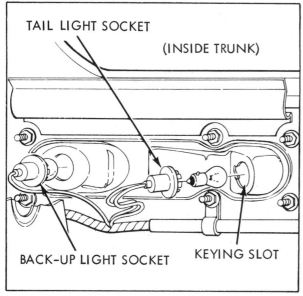

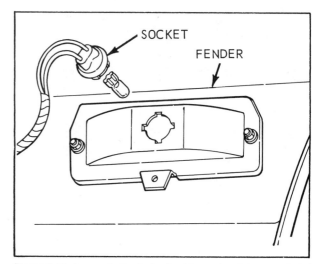

Figure 6.11 Access to the bulbs in side marker lights usually is obtained from inside the fenders (courtesy of Ford Motor Company).

Figure 6.10 On many cars, the bulbs in the rear lights are accessible from inside the trunk or luggage compartment (courtesy of Ford Motor Company).

passenger compartment will be found with all types of bases. (Refer to Figure 6.4.) In most instances, the lens of the lamp must be removed to replace the bulb. Lenses on some cars are retained by screws, as shown in Figure 6.13. On other cars, the lens snaps into place and is removed by squeezing the lens or by prying it loose, as shown in Figure 6.14. Since many different lamp designs are used, you should consult the manufacturer's manual for the correct replacement procedure.

Figures 6.15 and 6.16 show the location of the bulbs used in a typical instrument panel. Bulbs used to illuminate instruments and bulbs used in the various warning lights are usually of the wedge type. The

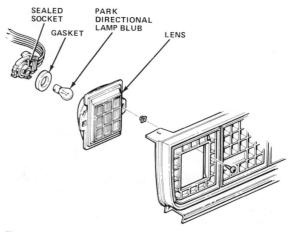

Figure 6.12 Access to the bulb in this parking lamp is from behind the grill (courtesy of American Motors).

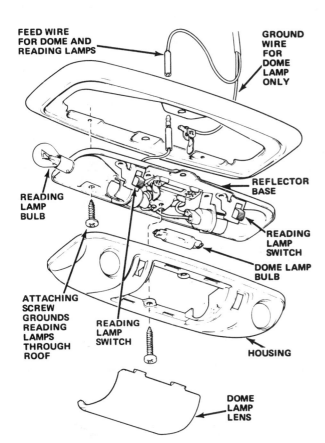

FEED WIRE FOR DOME AND READING LAMPS

GROUND WIRE FOR DOME LAMP ONLY

REFLECTOR BASE

READING LAMP BULB

READING LAMP SWITCH

DOME LAMP BULB

ATTACHING SCREW GROUNDS READING LAMPS THROUGH ROOF

READING LAMP SWITCH

HOUSING

DOME LAMP LENS

Figure 6.13 The dome lamp lens can be pried off to change the dome lamp bulb in this assembly, but the screws holding the housing must be removed to change the reading lamp bulbs (courtesy of American Motors).

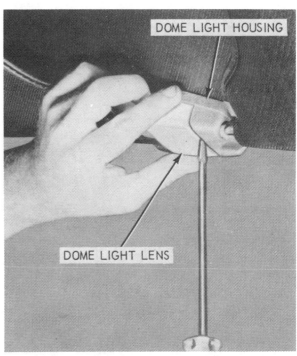

DOME LIGHT HOUSING

DOME LIGHT LENS

Figure 6.14 Some interior light lenses are removed by carefully prying the lens loose from its housing (courtesy of Ford Motor Company).

bulbs are pushed into a plastic socket that is fitted with bayonet lugs. The socket in turn is installed in the back of the instrument panel with the same push-and-twist motion used to install bayonet base bulbs. On some cars, the bulbs are accessible by reaching under the instrument panel. On most cars, however, the instrument panel must be partially or completely removed in order to change a bulb. Figure 6.17 shows the bulb locations in an instrument cluster that has been removed. The procedure for removing the instrument cluster varies greatly from car to car. A typical procedure is presented in Chapter 7, but the specific procedure for the car on which you are working should be obtained from the manufacturer's manual.

Replacing Exterior Lamp Lenses You often will be required to replace a cracked or broken lens on an exterior lamp. Where the lens is secured by screws as shown in Figure 6.18, that job is relatively simple. On some cars however, the lens is held by

an adhesive strip of butyl sealer as shown in Figure 6.19. On other cars, the lens is cemented in place or is ultrasonically welded to the reflector or to the lamp body. In some instances, the lens is not considered a separate part, and the entire lamp assembly must be replaced. (See Figure 6.20.) In other instances, the damaged lens must be removed and the replacement lens cemented in place.

Lenses Mounted in Butyl Sealer The following steps outline a typical procedure for the replacement of a lens that is mounted in butyl sealer. The manufacturer's manual should be consulted for the particular procedure that may be necessary for the car on which you are working:

1 Remove the sockets and bulbs from the lamp assembly.

2 Insert a wood dowel approximately ¾ in. (19 mm) in diameter through the socket hole(s) and carefully push the lens away from the lamp body.

Note: This method minimizes the possibility of cuts from the broken lens edges.

3 As the lens moves away from the lamp body, use a sharp knife or a single edge razor blade to cut the butyl sealer.

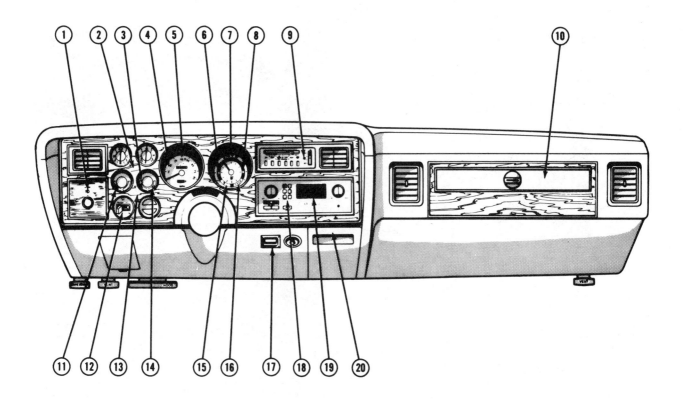

1. Headlamp Switch Illumination
2. Instrument Cluster Illumination
3. Brake System Warning Lamp Indicator
4. Instrument Cluster Illumination
5. Hi-Beam Indicator
6. Left Turn Indicator
7. Instrument Cluster Illumination
8. Right Turn Indicator
9. A/C Heater Control Illumination
10. Glove-Box Illumination
11. Washer Low Fluid Level Indicator
12. Heated Backlite Indicator
13. Cluster Illumination
14. Oil Pressure Indicator
15. Seat Belt Reminder Indicator
16. Door Ajar Warning Indicator
17. Power Amplifier Indicator
18. Radio Illumination
19. Clock Flourescent Digital Read-Out
20. Ash Tray Illumination

Figure 6.15 Bulb locations in the instrument panel of a typical automobile (courtesy of Chrysler Corporation).

4 Remove the lens.

5 Scrape and clean away all traces of the old butyl sealer from the edges of the lamp body so that the replacement lens will seat properly.

6 Clean the reflector.

7 Position the new butyl sealer or tape on the edge of the lamp body.

Note: Some manufacturers specify that a gap or gaps be left at the lower edge of the lamp body to allow for drainage of any water that may enter the lamp.

8 Install any retaining clips supplied with the replacement lens. (See Figure 6.21.)

9 Press the replacement lens in position and hold it in place for at least 10 seconds.

10 Insert the bulbs and sockets.

11 Check the operation of the lights.

Lenses Cemented or Ultrasonically Welded   The following steps outline a typical procedure for the replacement of a lens that is cemented or welded in place. You should consult the manufacturer's manual for the specific procedure that may be necessary for the car on which you are working:

1 Remove the sockets and bulbs from the lamp assembly.

2 Using a hammer, carefully break away large sections of the damaged lens.

Note: To protect yourself from flying chips, it is suggested that you cover the lens with a cloth during this step.

3 Using a hammer and a sharp chisel, carefully chip away the remaining edge of the lens from the lamp body.

Note: The edge of the lamp body must be smooth and even so that the replacement lens can be properly positioned.

4 Clean the reflector.

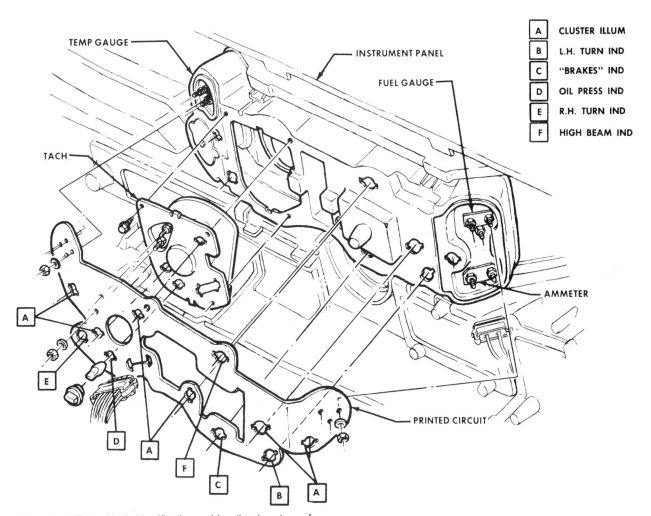

A	CLUSTER ILLUM
B	L.H. TURN IND
C	"BRAKES" IND
D	OIL PRESS IND
E	R.H. TURN IND
F	HIGH BEAM IND

Figure 6.16 Typical bulb identification and location (courtesy of Chevrolet Motor Division).

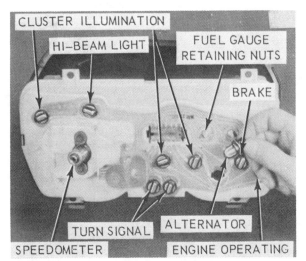

Figure 6.17 The rear of an instrument cluster that has been removed from an instrument panel. Notice the plastic caps that hold the wedge type bulbs (courtesy of Ford Motor Company).

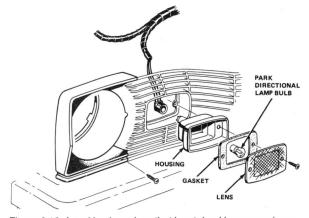

Figure 6.18 A parking lamp lens that is retained by screws (courtesy of American Motors).

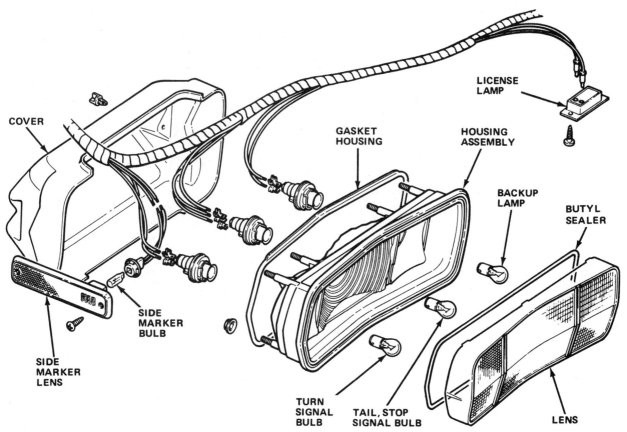

Figure 6.19 A lamp assembly with a lens that is held to the housing assembly by a strip of butyl sealer (courtesy of American Motors).

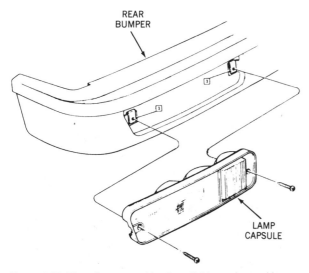

Figure 6.20 The tail, stop, and back-up lights are housed in one assembly in this type lamp. The lens and the lamp body are ultrasonically welded together to form a lamp "capsule," and if the lens is damaged, the entire assembly must be replaced (courtesy of Chevrolet Motor Division).

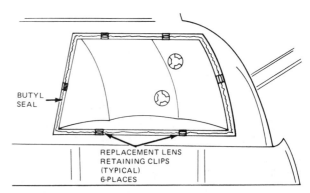

Figure 6.21 On some lamps, retaining clips are used to position the replacement lens in the butyl seal (courtesy of Ford Motor Company).

5 Position the new butyl sealer or tape on the edge of the lamp body.

Note: Some manufacturers specify that a gap or gaps be left at the lower edge of the lamp body to allow for drainage of any water that may enter the lamp.

6 Install any retaining clips supplied with the replacement lens. (Refer to Figure 6.21.)

Job 6b

REPLACE SMALL BULBS

SATISFACTORY PERFORMANCE

A satisfactory performance on this job requires that you do the following:

1 Replace the designated bulbs on the car assigned.
2 Following the steps in the "Performance Outline" and the procedures and specifications of the car manufacturer, complete the job within 200 percent of the manufacturer's suggested time.
3 Fill in the blanks under "Information."

PERFORMANCE OUTLINE

1 Remove any parts necessary to gain access to the bulbs.
2 Remove the defective bulbs.
3 Install the replacement bulbs.
4 Install any parts removed to gain access.
5 Test the operation of the bulbs.

INFORMATION

Vehicle identification _____

Reference used _____ Page(s) _____

	Bulb Location	Bulb Number
Bulb #1	_____	_____
Bulb #2	_____	_____
Bulb #3	_____	_____
Bulb #4	_____	_____

7 Press the replacement lens in position and hold it in place for at least 10 seconds.
8 Insert the bulbs and sockets.
9 Check the operation of the lamp.

Fuses A blown fuse is one of the most common causes of circuit failure. As you know, a fuse is a "sacrifice" part that blows, or burns out, when too much current flows through it. When a fuse blows, it opens the circuit and thus protects all of the other circuit parts from damage.

Glass Cartridge Fuses The most commonly used fuse consists of a strip of metal that has a low melting point. The strip is enclosed in a glass tube and is connected to a metal cap at each end. The metal strip is designed to handle a limited amount of cur-

rent. When the current flowing through the fuse exceeds that limit, the strip melts. Figure 6.22 shows a good fuse and one that has blown.

Glass cartridge fuses are available in many ratings and sizes. Figure 6.23 shows some of the more commonly used ratings and sizes. A replacement fuse should always be of the same rating and size as the one removed. If you install a fuse with a lower rating, it may not be able to handle the normal current flow in the circuit. If you install a fuse with a higher rating, it will allow too much current to flow in the circuit. The excess current could damage other components.

Most cars have a *fuse block* or *fuse panel* under the instrument panel. A fuse block provides a convenient, single location for most, if not all, of the

Job 6c

REPLACE EXTERIOR LAMP LENSES

SATISFACTORY PERFORMANCE
A satisfactory performance on this job requires that you do the following:

1 Replace the designated lens(es) on the car assigned.
2 Following the steps in the "Performance Outline" and the procedure and specifications of the car manufacturer, complete the job within 200 percent of the manufacturer's suggested time.
3 Fill in the blanks under "Information."

PERFORMANCE OUTLINE
1 Remove any parts necessary to gain access to the lens.
2 Remove the lens.
3 Install the replacement lens.
4 Install the parts removed to gain access.
5 Check the operation of the lamp.

INFORMATION
Vehicle identification _____

Location of lens(es) replaced: _____

_____

Lens(es) was secured by: _____ screws

_____ butyl sealer

_____ ultrasonic welding or cement

---

fuses in a car. Figure 6.24 shows such a fuse block. At times, a particular circuit will not have its fuse located in the fuse block. Those circuits use an *in-line fuse holder* of the type shown in Figure 6.25. In wiring diagrams, an in-line fuse holder is usually symbolized as shown in Figure 6.26. In-line fuse holders are often used in the wiring of accessories that have been added to a car. Those accessories include radios, tape decks, and special lights.

Miniaturized Fuses   Many cars use a miniaturized fuse, or *mini-fuse,* as shown in Figure 6.27. Those fuses also use a strip of metal that melts. Mini-fuses plug into special fuse blocks and cannot be interchanged with glass cartridge fuses. A fuse block for mini-fuses is shown in Figure 6.28. As with glass cartridge fuses, mini-fuses should be replaced only with fuses of the same rating. The rating of a mini-

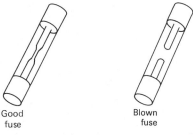

Good fuse   Blown fuse

Figure 6.22 Typical glass cartridge fuses of the type used in automotive electrical systems.

fuse is indicated by number and by color code as shown in Figure 6.29.

Testing Fuses   When it is suspected that a fuse is blown, it is easier to test the fuse in the circuit than to remove it for testing. You can make the job

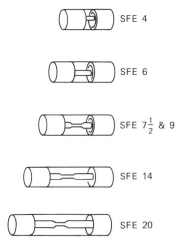

Figure 6.23 A few of the more popular fuse sizes and ratings.

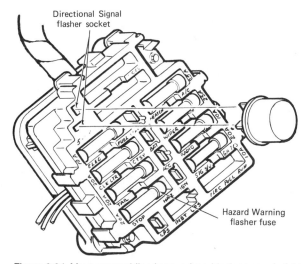

Figure 6.24 Most automobiles have a fuse block or panel of this type mounted beneath the instrument panel. Notice that the fuses, when installed, are recessed below the surface of the fuse block (courtesy of Chevrolet Motor Division).

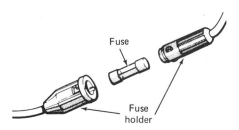

Figure 6.25 An in-line fuse holder. Fuse holders of this type often are used in accessory circuits (courtesy of Ford Motor Company).

Figure 6.26 Two symbols commonly used to represent an in-line fuse.

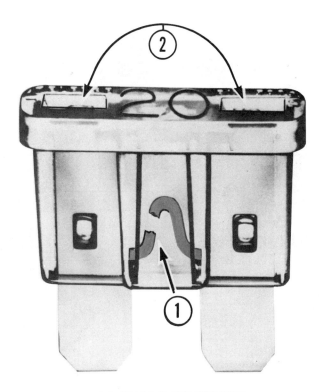

TO TEST FOR BLOWN MINI-FUSE:

1 PULL FUSE OUT AND CHECK VISUALLY.

2 WITH THE CIRCUIT ACTIVATED, USE A TEST LIGHT ACROSS THE POINTS SHOWN.

MINI-FUSE COLOR CODES

RATING	COLOR
5 AMP	TAN
10 AMP	RED
20 AMP	YELLOW
25 AMP	WHITE

Figure 6.27 An enlarged view of a mini-fuse. Notice that the fuse is blown (courtesy of Pontiac Motor Division, General Motors Corporation).

even easier by using an appropriate manual. The manual will tell you where the fuse block is located. It also will show you the location of each fuse and the correct fuse rating. Figures 6.30 and 6.31 show how that information is given.

Glass Cartridge Fuses Glass cartridge fuses can often be checked visually. With the aid of a flashlight you can often see if the fuse is blown. (Refer to Figure 6.22.) A positive test can be made by using a test lamp in the following manner:

1 Check the operation of the test lamp by connecting it across the terminals of a battery.
2 Turn on the circuit that you wish to test.

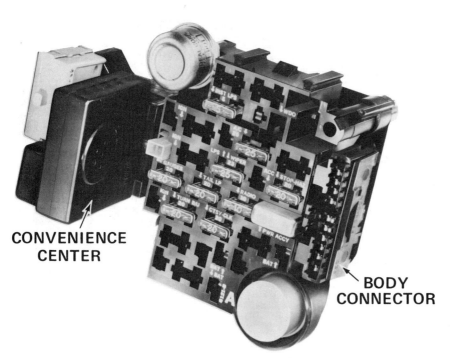

**CONVENIENCE
CENTER**

**BODY
CONNECTOR**

Figure 6.28 A fuse block for mini-fuses (courtesy of Pontiac Motor Division, General Motors Corporation).

RATING	COLOR
3 Amp	Violet
5 Amp	Tan
10 Amp	Red
20 Amp	Yellow
25 Amp	White
30 Amp	Light Green

Figure 6.29 Mini-fuses are color coded to distinguish the different ratings.

3 Connect the test lamp alligator clip to a good ground.
4 Touch the probe of the test lamp to each end of the suspected fuse. Three results are possible.
   (a) The test lamp lights when touched to either end of the fuse. That means that the fuse is good.
   (b) The test lamp lights only when touched to one end of the fuse. That means that the fuse is blown or defective.
   (c) The test lamp does not light when touched to either end of the fuse. That means that the circuit is not turned on, or that the circuit is broken between the fuse and the power source. It also can mean that you have a bad ground connection at the alligator clip.

Mini-Fuses   It is almost impossible to see if a mini-fuse is blown without removing it from the fuse block. Mini-fuses are made with access holes above their

prongs or terminals so that a test lamp can be easily used. A positive test can be made in the following manner:

1 Check the operation of the test lamp by connecting it across the terminals of a battery.
2 Turn on the circuit that you want to test.
3 Connect the test lamp alligator clip to a good ground.
4 Touch the probe of the test lamp to each of the test points exposed on the top of the fuse. (Refer to Figure 6.27.) Three results are possible.
   (a) The test lamp lights when touched to either of the test points. That means that the fuse is good.
   (b) The test lamp lights only when touched to one of the test points. That means that the fuse is blown or defective.
   (c) The test lamp does not light when touched to either of the test points. That means that the circuit is not turned on, or that the circuit is broken between the fuse and the power source. It also can mean that you have a bad ground connection at the alligator clip.

Replacing Fuses   When a glass cartridge fuse is mounted in a fuse block, it is almost impossible to remove it with your fingers. A hook made of stiff wire such as welding rod or coat hanger wire should be used. You can make that tool by following the sketch in Figure 6.32. The replacement fuse is easily pushed in place with your finger tips. (Refer to Figure 6.24.)

FUSES

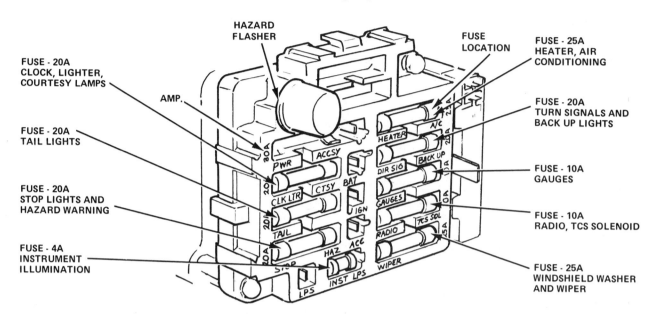

FUSE - 20A
CLOCK, LIGHTER,
COURTESY LAMPS

HAZARD
FLASHER

AMP.

FUSE
LOCATION

FUSE - 25A
HEATER, AIR
CONDITIONING

FUSE - 20A
TAIL LIGHTS

FUSE - 20A
TURN SIGNALS AND
BACK UP LIGHTS

FUSE - 20A
STOP LIGHTS AND
HAZARD WARNING

FUSE - 10A
GAUGES

FUSE - 4A
INSTRUMENT
ILLUMINATION

FUSE - 10A
RADIO, TCS SOLENOID

FUSE - 25A
WINDSHIELD WASHER
AND WIPER

DO NOT USE FUSES OF HIGHER AMPERAGE RATING THAN THOSE SPECIFIED

Figure 6.30 Fuse locations and specifications as found in a typical manufacturer's service manual (courtesy of Buick Motor Division, General Motors Corporation).

Circuit	Circuit Protection and Rating	Location
Headlamps & High Beam Indicator	22 Amp C.B.	Integral with Lighting Switch
Heated Backlite	16 GA. Fuse Link	Engine Compartment
Load Circuit	Fuse Link	In Harness
Engine Compartment Lamp	Fuse Link	In Harness
Liftgate Wiper	4.5 Amp C.B.	Instrument Panel to Right of Radio
Low Beam Headlamps①	20 Amp Fuse	In Harness②
Horn①	20 Amp Fuse	In Harness②
Tail/Park Lamps①	10 Amp Fuse	In Harness②

①For Anti-Theft System
②In Luggage Compartment

Figure 6.31 Fuses and other protection devices are not always located in the fuse block. The information shown here is typical of that found in manufacturers' service manuals (courtesy of Ford Motor Company).

Fuses contained inside in-line fuse holders are removed by grasping the ends of the fuse holder, pushing them together, and twisting the ends counterclockwise. The locking lugs will release and the ends can be pulled apart. After the replacement fuse is inserted, the ends are pushed together and twisted clockwise.

Mini-fuses are easily removed. Just grasp the top of the fuse and pull straight out. The replacement fuse is merely plugged in.

The replacement of a blown fuse may not always

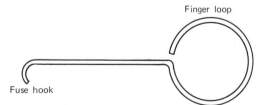

Finger loop

Fuse hook

Figure 6.32 A fuse hook. A piece of stiff wire bent to the shape shown will enable you to remove fuses from a fuse block.

solve a problem. If the replacement fuse blows or burns out, the cause of the circuit overload must be found and eliminated.

Job 6d

REPLACE DEFECTIVE FUSES

SATISFACTORY PERFORMANCE

A satisfactory performance on this job requires that you do the following:

1 Check the fuses in the vehicle assigned and replace the defective fuse.

2 Following the steps in the "Performance Outline" and the procedure and specifications of the manufacturer, complete the job within 15 minutes.

3 Fill in the blanks under "Information."

PERFORMANCE OUTLINE

1 Locate the fuse block or fuse holder.

2 Locate the defective fuse.

3 Install the replacement fuse.

4 Test the fused circuit.

INFORMATION

Vehicle identification _____

Reference used _____ Page(s) _____

Circuit in which defective fuse was found _____

Location of fuse: _____ Fuse block   _____ In-line fuse holder

Type of fuse: _____ Glass cartridge fuse   _____ Mini-fuse

Fuse identification and rating: _____

Did the circuit operate properly after the fuse was replaced?

_____ Yes   _____ No

Circuit Breakers   While fuses provide a simple method of circuit protection, they are not always desirable. Some circuits are normally subjected to temporary overloads. As an example, the switches that control the motors for windows and seats are often inadvertently held in the On position after the window or seat has moved to the limit of its travel. When that occurs, the motor stalls and the current flow in the circuit increases. If those circuits were protected by fuses, the fuses would require frequent replacement.

Another example of where the use of a fuse is not practical is in the headlight circuit. If a headlight circuit was protected by a fuse, a short that occurred in that circuit would blow the fuse. This would cause the headlights to go out and remain out until the short was eliminated and the fuse replaced. A

headlight failure of that type could be the cause of an accident.

What is needed is a device that will open the circuit when current flow exceeds a certain limit, but will close the circuit after a short length of time. Such a device is called a *circuit breaker*. Typical circuit breakers are shown in Figures 6.33, 6.34, and 6.35. A circuit breaker is wired in a circuit in series, as if it were a fuse. As is a fuse, a circuit breaker is sensitive to heat and is rated by the amperage, or amount of current, that can flow through it. The current that passes through a circuit breaker flows through a special bimetallic conductor. That conductor is termed bimetallic because it is made of two strips of different metals bonded together. When heated, as by excessive current flow, the different metals expand at different rates, causing the conductor to bend as shown in Figure 6.36. That motion opens a set of switch contacts and breaks the circuit. Since the flow of current stops, the bimetallic conductor cools and returns to its original shape. That motion closes the switch contacts and the cycle is repeated. Operating in this manner, a circuit breaker will provide the necessary protection, yet will not require replacement each time it is subjected to an overload.

In headlight circuits, a circuit breaker usually turns the lights off and then on, alternating until the short or other cause of excessive current flow is located and eliminated.

Testing a Circuit Breaker Circuit breakers rarely fail, but when they are suspected of causing trouble, they can be checked easily. The easiest way to test a circuit breaker is by substitution. A replacement circuit breaker that is known to be good is substituted for the one that is suspected of being bad. If the problem is eliminated, the breaker was bad.

Another method requires that an ammeter be connected in the circuit in place of the circuit breaker. By this method, the actual current flowing in the circuit can be measured and compared to the rating of the circuit breaker. If the actual current flow is less than the rating of the circuit breaker and the breaker has been cycling, the breaker is defective. If the current flow exceeds the rating of the circuit breaker, the breaker is functioning properly and the cause of the excessive current flow must be found and eliminated.

Replacing Circuit Breakers Replacing a circuit breaker is a simple job. On most cars, the circuit breakers are located on the fuse block as shown in Figure 6.37. They are removed by unplugging them from their socket, and the replacement breaker is merely plugged in. At times, circuit breakers are placed in other locations. These locations will vary with different makes and models of cars, and can best be found with the aid of the manufacturer's service manual.

On many cars, stud-type circuit breakers (refer to Figure 6.35) are mounted on the firewall in the engine compartment. Some window and seat motors are protected by circuit breakers at their ground wires. Those breakers are located near the motors in the doors and under the seats. On many cars, the circuit breaker for the headlight circuit is built inside the headlight switch. The breaker is not serviced as a separate part, and the headlight switch must be replaced if the breaker fails.

Figure 6.33 A typical plug-in circuit breaker.

Figure 6.34 A fuse clip type circuit breaker. Breakers of this type often are used to replace fuses since they can be plugged into standard fuse clips.

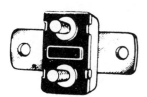

Figure 6.35 A stud type circuit breaker. Breakers of this type often are used to protect window and seat motors.

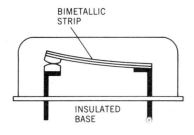

CONTACTS CLOSED

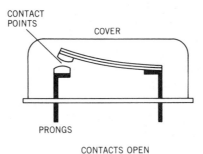

CONTACTS OPEN

Figure 6.36 The operation of a circuit breaker. When excessive current flows through the bimetallic strip, the strip is heated, causing it to bend and open the contact points. When the strip cools, the points close.

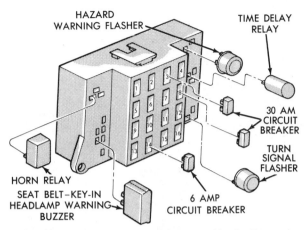

Figure 6.37 A fuse block for mini-fuses combined with a center for flashers, relays, circuit breakers, and a buzzer. An assembly of this type is called a fuse block and relay module (courtesy of Chrysler Corporation).

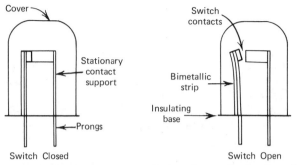

Figure 6.38 The operation of a flasher. Current flowing through the bimetallic strip causes the strip to bend, opening the contacts. When the strip cools, it returns and the contacts close.

Flashers   Most automotive lighting systems contain two flashers. The turn signal, or directional signal, circuit contains one flasher. The hazard warning circuit contains the other. Flashers are controls, or switches, that automatically open and close a circuit. Since they are switches, flashers are placed in series in a circuit. Since they are connected in series, any failure of a flasher will affect its entire circuit.

As shown in Figure 6.38, a flasher is similar to a circuit breaker in that it is operated by heat. The current in the circuit flows through a bimetallic conductor that bends and opens a set of switch contacts. The contacts close when the conductor cools. The bimetallic conductor used in a flasher is selected to bend when the normal current for the circuit flows through it.

Knowledge of how a flasher operates is very helpful in the diagnosis of problems in circuits containing flashers. Since a flasher is operated by heat, and the heat is produced by the current that flows in the circuit, the amount of current that flows determines the speed of flasher operation.

Most turn signal flashers are built to cycle at their normal rate when the front and rear turn signals on one side of a car and the indicator light on the instrument panel are in the circuit. Those three bulbs comprise the resistance in that circuit and thus determine the amount of current that flows through the flasher. If one of the front or rear bulbs burns out, the circuit resistance increases, the current flow is reduced, and the flasher may not function. The remaining turn signal bulb and the indicator bulb will remain on and not flash, or they will flash at a very slow rate. When the turn signal lights on one side of a car function properly and the turn signal lights on the other side do not, the replacement of the flasher usually is a waste of time.

At times, a flasher will cycle too rapidly. This indicates that excessive current is flowing in the circuit. This situation often occurs when aftermarket lights are installed or when the lights of a trailer are wired into the circuit. Flashers of different design, often referred to as "heavy duty" flashers, are available for the replacement of original equipment when different lights or additional lights are installed.

Testing a Flasher   The easiest method of testing a flasher is by substitution. A replacement flasher that is known to be good is substituted for the one that is suspected of being bad.

Replacing Flashers   Replacing a flasher is easy. In fact, at times it is more difficult to locate a flasher than it is to replace it. Therefore, the easiest method for replacing a flasher starts by using an appropriate manual to find where the flasher is located.

Flashers are usually mounted under the instrument panel, but the exact locations vary, even among cars built by the same maker. On some cars, both the turn signal flasher and the hazard warning flasher are located on the fuse block as shown in Figure 6.39 and 6.40. On other cars, only one flasher will be found on the fuse block. (See Figure 6.41.) The remaining flasher is mounted in some other location, and it may even be taped to part of the wiring harness. Figure 6.42 shows the location chosen by one manufacturer to mount both flashers in one model car.

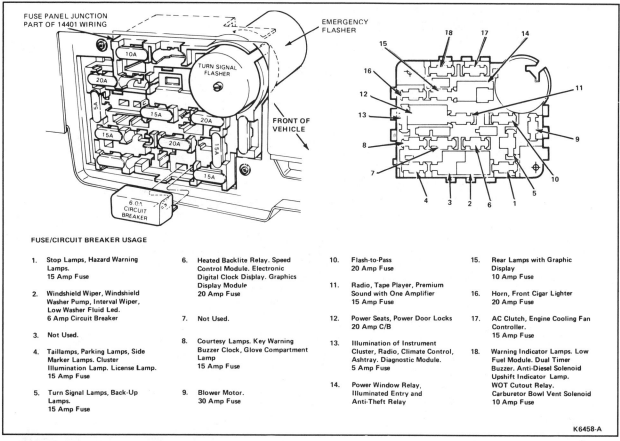

FUSE/CIRCUIT BREAKER USAGE

1. Stop Lamps, Hazard Warning Lamps. 15 Amp Fuse

2. Windshield Wiper, Windshield Washer Pump, Interval Wiper, Low Washer Fluid Led. 6 Amp Circuit Breaker

3. Not Used.

4. Taillamps, Parking Lamps, Side Marker Lamps. Cluster Illumination Lamp. License Lamp. 15 Amp Fuse

5. Turn Signal Lamps, Back-Up Lamps. 15 Amp Fuse

6. Heated Backlite Relay. Speed Control Module. Electronic Digital Clock Display. Graphics Display Module. 20 Amp Fuse

7. Not Used.

8. Courtesy Lamps. Key Warning Buzzer Clock, Glove Compartment Lamp. 15 Amp Fuse

9. Blower Motor. 30 Amp Fuse

10. Flash-to-Pass. 20 Amp Fuse

11. Radio, Tape Player, Premium Sound with One Amplifier. 15 Amp Fuse

12. Power Seats, Power Door Locks. 20 Amp C/B

13. Illumination of Instrument Cluster, Radio, Climate Control, Ashtray. Diagnostic Module. 5 Amp Fuse

14. Power Window Relay, Illuminated Entry and Anti-Theft Relay

15. Rear Lamps with Graphic Display. 10 Amp Fuse

16. Horn, Front Cigar Lighter. 20 Amp Fuse

17. AC Clutch, Engine Cooling Fan Controller. 15 Amp Fuse

18. Warning Indicator Lamps. Low Fuel Module. Dual Timer Buzzer. Anti-Diesel Solenoid. Upshift Indicator Lamp. WOT Cutout Relay. Carburetor Bowl Vent Solenoid. 10 Amp Fuse

K6458-A

Figure 6.39 Views of a typical fuse panel and the usage and locations of the fuses, circuit breakers, and flashers as provided in a manufacturer's service manual. Notice that the turn signal flasher and the hazard warning flasher are mounted back-to-back (courtesy of Ford Motor Company).

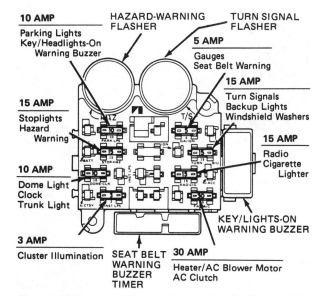

Figure 6.40 Flasher and fuse locations and applications provided by one manufacturer. Notice that the warning buzzers are included (courtesy of American Motors).

Most flashers have two prongs, as shown in Figure 6.43. Unplugging the old flasher and plugging in the replacement usually is all that is required.

Sealed Beam Bulbs A sealed beam bulb is actually an assembly of a bulb, a lens, and a reflector. In a true sealed beam bulb, the lens and the reflector are fused together into one unit. The filament is mounted in the reflector and positioned so that the light is most effectively reflected. Since the unit is sealed and all air evacuated from it, the thin glass envelope used on small bulbs is not needed. The lens has small prisms molded on its inner surface. Those prisms bend the light rays so that the desired beam shape is projected. Typical sealed beam bulbs are shown in Figure 6.44 and 6.45.

Some sealed beam bulbs have two filaments. One filament provides a *high beam* that projects a concentrated, high-intensity beam of light straight ahead.

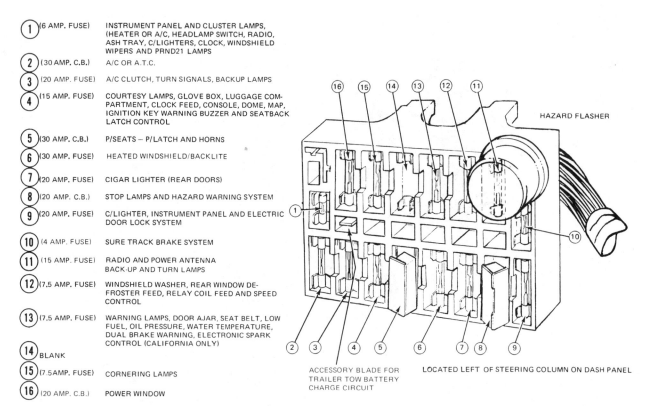

1 (6 AMP. FUSE) INSTRUMENT PANEL AND CLUSTER LAMPS, (HEATER OR A/C, HEADLAMP SWITCH, RADIO, ASH TRAY, C/LIGHTERS, CLOCK, WINDSHIELD WIPERS AND PRND21 LAMPS

2 (30 AMP. C.B.) A/C OR A.T.C.

3 (20 AMP. FUSE) A/C CLUTCH, TURN SIGNALS, BACKUP LAMPS

4 (15 AMP. FUSE) COURTESY LAMPS, GLOVE BOX, LUGGAGE COM-PARTMENT, CLOCK FEED, CONSOLE, DOME, MAP, IGNITION KEY WARNING BUZZER AND SEATBACK LATCH CONTROL

5 (30 AMP. C.B.) P/SEATS – P/LATCH AND HORNS

6 (30 AMP. FUSE) HEATED WINDSHIELD/BACKLITE

7 (20 AMP. FUSE) CIGAR LIGHTER (REAR DOORS)

8 (20 AMP. C.B.) STOP LAMPS AND HAZARD WARNING SYSTEM

9 (20 AMP. FUSE) C/LIGHTER, INSTRUMENT PANEL AND ELECTRIC DOOR LOCK SYSTEM

10 (4 AMP. FUSE) SURE TRACK BRAKE SYSTEM

11 (15 AMP. FUSE) RADIO AND POWER ANTENNA BACK-UP AND TURN LAMPS

12 (7.5 AMP. FUSE) WINDSHIELD WASHER, REAR WINDOW DE-FROSTER FEED, RELAY COIL FEED AND SPEED CONTROL

13 (7.5 AMP. FUSE) WARNING LAMPS, DOOR AJAR, SEAT BELT, LOW FUEL, OIL PRESSURE, WATER TEMPERATURE, DUAL BRAKE WARNING, ELECTRONIC SPARK CONTROL (CALIFORNIA ONLY)

14 BLANK

15 (7.5 AMP. FUSE) CORNERING LAMPS

16 (20 AMP. C.B.) POWER WINDOW

HAZARD FLASHER

ACCESSORY BLADE FOR TRAILER TOW BATTERY CHARGE CIRCUIT

LOCATED LEFT OF STEERING COLUMN ON DASH PANEL

Figure 6.41 A fuse and circuit breaker panel used on one model car. Notice that only the hazard warning flasher is mounted in this location (courtesy Ford Motor Company).

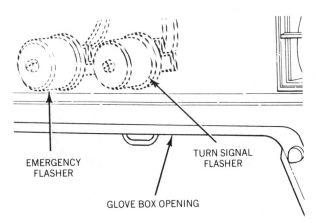

EMERGENCY FLASHER

TURN SIGNAL FLASHER

GLOVE BOX OPENING

Figure 6.42 Flasher locations vary with each model car. On this model, they are mounted above the glove box (courtesy of Ford Motor Company).

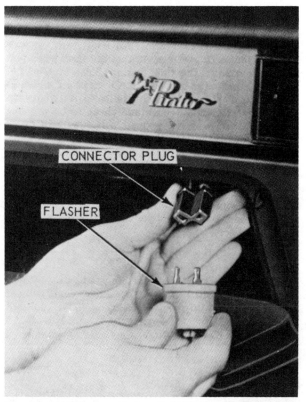

CONNECTOR PLUG

FLASHER

Figure 6.43 Replacing a flasher. Notice that the flasher has two prongs that plug into a socket or quick disconnect (courtesy of Ford Motor Company).

Job 6e

REPLACE A CIRCUIT BREAKER

SATISFACTORY PERFORMANCE
A satisfactory performance on this job requires that you do the following:

1 Replace the designated circuit breaker on the car assigned.
2 Following the steps in the "Performance Outline" and the manufacturer's procedure and specifications, complete the job within 15 minutes.
3 Fill in the blanks under "Information."

PERFORMANCE OUTLINE
1 Determine the location of the designated circuit breaker.
2 Remove the old circuit breaker.
3 Install the replacement circuit breaker.
4 Test the operation of the circuit.

INFORMATION
Vehicle identification _____

Circuit breaker replaced _____

Reference used _____ Page(s) _____

Circuit breaker location _____

Circuit breaker rating _____

Was the circuit working properly after the circuit breaker was replaced?

_____ Yes _____ No

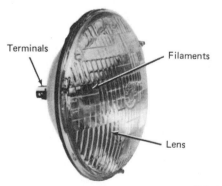

Figure 6.44 A typical sealed beam headlight bulb.

Figure 6.46 Large sealed beam bulbs with two filaments are used in systems using two headlights. The bulbs may be round (7 in.) or rectangular (5.6 × 7.9 in. − 142 × 200 mm).

Figure 6.45 A typical rectangular sealed beam bulb (courtesy of Wagner Division, Cooper Industries, Inc.).

The high beam is intended for use only when it cannot interfere with the vision of the drivers of on-coming cars. The remaining filament provides a *low beam.* The low beam provides a less concentrated distribution of light that is aimed slightly downward and toward the right. The low beam is intended for use in traffic and at other times when the high beam could disturb other drivers. Sealed beam bulbs with two filaments have three prongs, or terminals, pro-jecting from the rear of the reflector.

Some sealed beam bulbs have only one filament. That filament is used as a high beam only. Bulbs of that type are used as the inner or lower bulbs in *dual headlight systems,* systems that use four headlights. Since they have only one filament, they require only two prongs, or terminals.

Sealed beam bulbs used as automative headlights are built in two shapes, round and rectangular. Two sizes are available in each shape. The larger bulbs are for use in systems that have two headlights. The smaller bulbs are for use in dual systems. Those

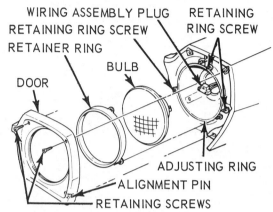

Figure 6.49 A typical headlight assembly (courtesy of Ford Motor Company).

Vertical Placement

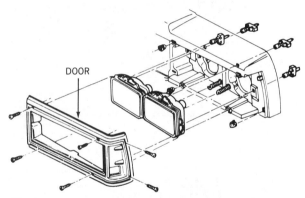

Figure 6.50 An exploded view of a typical dual headlight assembly. Notice the adjusting screws and their self-locking plastic nuts (courtesy of Chevrolet Motor Division).

Horizontal Placement

Figure 6.47 The sealed beam bulbs used in dual headlight systems are arranged so that the bulbs with two filaments are the upper bulbs or the outer bulbs. Smaller bulbs are used, and they may be round (5 in.) or rectangular (4 × 6.5 in. − 100 × 165 mm)

Figure 6.48 Identification chart for the six most commonly used automotive sealed beam headlight bulbs. The code usually is molded into the lens.

SHAPE	SIZE	NUMBER OF FILAMENTS	FUNCTION	CODE
Round	7″ (178 mm) diameter	2	High–Low	2D
Round	5¾″ (146 mm) diameter	2	High–Low	2C
Round	5¾″ (146 mm) diameter	1	High Only	1C
Rectangular	5.6″ × 7.9″ (142 × 200 mm)	2	High–Low	2B
Rectangular	4″ × 6.5″ (100 × 165 mm)	2	High–Low	2A
Rectangular	4″ × 6.5″ (100 × 165 mm)	1	High Only	1A

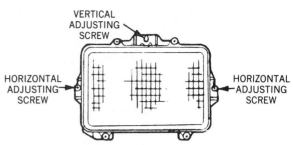

Figure 6.53 The vertical adjusting screw usually is located at the top, but the horizontal adjusting screw may be found on either side (courtesy of Chevrolet Motor Division).

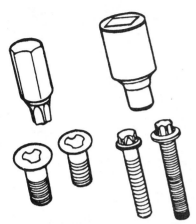

Figure 6.51 The headlight doors on some cars are secured with special screws and bolts that require special screwdriver bits and sockets (courtesy of Chevrolet Motor Division).

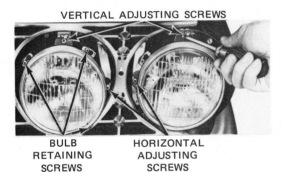

Figure 6.52 When replacing a headlight bulb, only the retaining screws should be removed. The adjusting screws should not be disturbed (courtesy of Ford Motor Company).

systems are shown in Figures 6.46 and 6.47. A number/letter code, usually molded into the lens, has been adopted to identify the most commonly used sealed beam bulbs. The number indicates the number of filaments, and the letter indicates the size and shape of the bulb. Figure 6.48 lists those bulbs and their codes.

Many cars are fitted with halogen headlight bulbs. In most instances those units are physically similar to conventional sealed beam bulbs, but they are more efficient and provide more light. While conventional and halogen sealed beam bulbs are interchangeable in most instances, they should never be mixed. All of the headlights on a car should be of the same type—all conventional or all halogen.

Replacing Sealed Beam Bulbs The following steps outline a typical procedure for replacing a sealed beam headlight bulb. It is suggested that you consult an appropriate manual for alternate procedures that may be necessary on some cars:

1 Remove the screws holding the headlight *door,* or trim. See Figures 6.49 and 6.50.

Note: While a Phillips screwdriver will remove the screws holding the headlight doors on most cars, some cars have special screws requiring tools of the type shown in Figure 6.51.

2 Remove the headlight door.

3 Remove the small screws holding the headlight retainer. Do not loosen or disturb the adjusting screws. (See Figures 6.52 and 6.53.)

Note: Some cars have a spring that holds one side of the retainer. That spring can easily be disconnected by using the tool you made to remove fuses. (Refer to Figure 6.32.)

4 Remove the retainer.

5 Carefully pull the sealed beam bulb forward and disconnect the wiring harness plug by pulling it straight off.

6 Align the prongs of the replacement bulb with the wiring harness plug and push the parts together so that the prongs are fully seated.

7 Position the bulb in the adjustment ring so that the alignment tabs on the rear edge of the bulb fit into the slots in the ring.

Note: The alignment tabs are located so that the bulb will fit properly in only one position.

8 Install the retainer.

Note: The screw holes in the edge of the retainer are usually located so that the retainer will fit properly in only one position.

9 Check the operation of the bulb.

10 Install the headlight door.

Aiming Headlights To provide effective lighting, headlights must be aimed. Proper maintenance of any vehicle should include checking the aim of the headlights. In many states, motor vehicle inspection laws require that headlight aim be

Job 6f

REPLACE FLASHERS

SATISFACTORY PERFORMANCE
A satisfactory performance on this job requires that you do the following:

1 Replace the designated flasher(s) on the car assigned.
2 Following the steps in the "Performance Outline" and the manufacturer's procedure and specifications, complete the job within 15 minutes.
3 Fill in the blanks under "Information."

PERFORMANCE OUTLINE
1 Determine the location of the designated flasher.
2 Remove the old flasher.
3 Install the replacement flasher.
4 Test the operation of the circuit.

INFORMATION
Vehicle identification _____

Flasher replaced: _____ Turn signal   _____ Hazard warning

Reference used _____ Page(s) _____

Flasher location _____

Was the circuit working properly after the flasher was replaced?

_____Yes   _____No

---

checked and adjusted periodically. The aim of the headlights can be changed by road shock, minor accidents, and changes in the suspension height. Most car manufacturers suggest that headlight aim be checked at least once each year and after the replacement of a sealed beam bulb.

Most shops use headlight aiming devices to obtain the proper adjustment. Lacking those devices, headlights can be aimed by observing the spots of light projected on a wall chart or on a marked wall.

Aiming Headlights by Using Mechanical Aimers The following steps outline a procedure for aiming headlights using mechanical aimers. The aiming device used, shown in Figure 6.54, is the type recommended by most automobile manufacturers. If the aiming device that you have available is not of that type, you should consult the manual furnished with the aimers that you have available:

Adjusting the Aimers The aimers are furnished with calibration fixtures that enable you to adjust the

Figure 6.54 A typical mechanical headlight aiming kit. Notice the various adapters furnished so that the aimers can be used on headlights of all types and sizes (courtesy of American Motors).

Job 6g

REPLACE SEALED BEAM BULBS

SATISFACTORY PERFORMANCE

A satisfactory performance on this job requires that you do the following:

1 Replace the designated sealed beam bulb(s) on the car assigned.
2 Following the steps in the "Performance Outline" and the procedure and specifications of the manufacturer, complete the job within 200 percent of the manufacturer's suggested time.
3 Fill in the blanks under "Information."

PERFORMANCE OUTLINE

1 Remove the headlight door.
2 Remove the bulb retainer.
3 Replace the bulb.
4 Install the retainer.
5 Check the bulb operation.
6 Install the headlight door.

INFORMATION

Vehicle identification _____

Location of bulb(s) replaced: _____ Right side   _____ Left side

_____ Upper   _____ Upper

_____ Lower   _____ Lower

_____ Inner   _____ Inner

_____ Outer   _____ Outer

Number marked on bulb(s) replaced: _____

Part number of replacement bulb(s): _____

Was the light working correctly after the bulb(s) was replaced?

_____ Yes   _____ No

---

aimers to compensate for any slight slope in the floor.

1 Park the vehicle in an area that appears to be level.
2 Check to see that the tires are properly inflated.
3 Attach the calibration fixtures to the aimers as shown in Figure 6.55.

Note: The calibration fixtures snap in place when properly aligned.

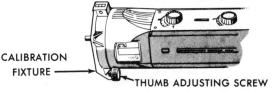

Figure 6.55 A calibration fixture attached to a headlight aimer (courtesy of Chrysler Corporation).

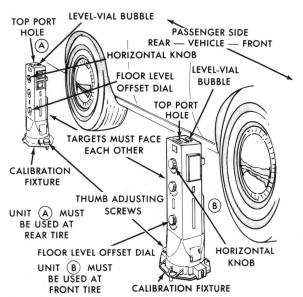

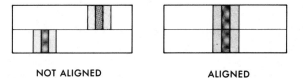

NOT ALIGNED   ALIGNED

Figure 6.58 The "split image" is aligned by turning the horizontal knob (courtesy of Chrysler Corporation).

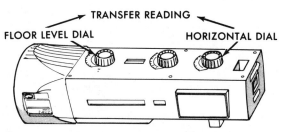

Figure 6.59 The reading obtained at the horizontal dial must be transferred to the floor level dial. this adjusts the aimer to compensate for the slope of the floor (courtesy of Chrysler Corporation).

Figure 6.56 The slope of the floor can be compensated for by measurements taken with the aimers positioned at the front and rear wheels of the car (courtesy of Chrysler Corporation).

Figure 6.57 Bubble levels are used to adjust the level of the aimers (courtesy of Chrysler Corporation).

   USE FOR FIVE INCH CIRCULAR HEADLAMP

   USE FOR SEVEN INCH CIRCULAR HEADLAMP

   USE FOR 4 x 6.5 INCH (100 x 165 MM) RECTANGULAR HEADLAMP

   USE FOR 142 x 200 MM (5.6 x 7.9 INCH) RECTANGULAR HEADLAMP

Figure 6.60 Headlight aimer adapters are provided for headlights of different shapes and sizes (courtesy of Chrysler Corporation).

4 Stand the aimers on the floor next to the right side of the car as shown in Figure 6.56. Aimer *A* must be aligned with the center of the rear wheel. Aimer *B* must be aligned with the center of the front wheel. The adjusting knobs must face outward and the targets must face each other.

5 Adjust the level of each aimer by turning the thumb adjusting screw on each calibration fixture. (Refer to Figure 6.55.) The adjustment is correct when the bubble in the top level vial is centered as shown in Figure 6.57.

6 Look down into the top port hole in aimer *A*. (Refer to Figure 6.56.) You should see an image similar to those shown in Figure 6.58.

Note: Moving your head from side to side will enable you to locate the image. At times, you will have to rotate one or both of the aimers slightly to correct their alignment.

7 While watching the image, turn the horizontal knob back and forth slowly until the split image is aligned. (Refer to Figure 6.58.)

8 Lift aimer *A* and observe the plus (+) or minus (−) reading indicated on the horizontal dial.

9 Adjust the floor level dials on both aimers to the reading found on the horizontal dial. (See Figure 6.59.)

Note: The floor level dials are self-locking. They must be pushed in slightly before they can be turned.

10 Remove the calibration fixtures from the aimers.

Mounting the Aimers   The aimers must be fitted with adapters before they can be mounted on the headlights.

1 Clean the headlight lenses with a wet paper towel.

Note: The suction cups in the aimers will not hold on dirty lenses.

2 Select the adapters required to fit the sealed beam bulbs used and install them on the aimers. Figure

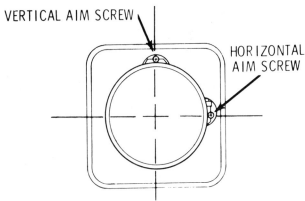

Figure 6.61 The headlight doors on some cars are notched to allow access to the adjusting screws (courtesy of Chevrolet Motor Division).

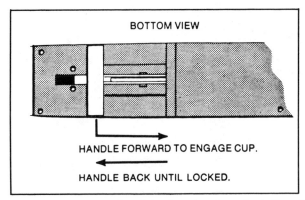

Figure 6.63 The operation of the piston handle on the bottom of the headlight aimer (courtesy of Pontiac Motor Division, General Motors Corporation).

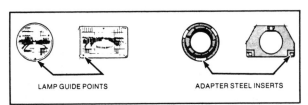

Figure 6.62 All sealed beam headlight bulbs have guide points molded into the edge of the lens. The steel inserts in the adapters must contact those points when the aimers are installed (courtesy of Pontiac Motor Division, General Motors Corporation).

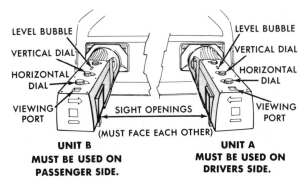

Figure 6.64 The location of the parts of the aimers when they are mounted on headlights (courtesy of Chrysler Corporation).

6.60 shows the various adapters and their applications.

3 Check the accessibility of the adjusting screws. Remove the headlight doors if necessary.

Note: On some cars, the headlight doors are notched as shown in Figure 6.61. Those notches enable you to reach the adjusting screws without removing the doors.

4 Position aimer *A* on the left (driver's side) headlight. Check to see that the steel inserts in the adapter are in contact with the guide points that project from the edge of the lens. (See Figure 6.62.)

5 Holding the aimer in position, push the piston handle on the bottom of the aimer forward to force the suction cup against the lens. Pull the handle back immediately until it locks in place. (See Figure 6.63.) The aimer should now be mounted on the lens.

6 Repeat steps 4 and 5 to mount aimer *B* on the right side headlight.

Horizontal Adjustment

1 Adjust the horizontal dial on aimer *A* to zero (0). (See Figure 6.64.)

2 Look down into the viewing port on aimer A. You should see the split image lines. (Refer to Figure 6.58.)

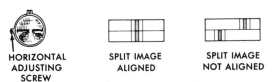

Figure 6.65 The horizontal adjusting screw is turned to align the split image (courtesy of Chrysler Corporation).

Note: If the lines are not visible, you may have to rotate one or both of the aimers slightly to shift the position of the sight openings.

3 Using a screwdriver, slowly turn the horizontal adjusting screw located at the side of the headlight. (Refer to Figures 6.52 and 6.53.) Turn the screw in or out as required until the split image is aligned. (See Figure 6.65.)

4 Repeat steps 1 through 3 at the opposite headlight.

Vertical Adjustment

1 Set the vertical dial knob on aimer *A* to 0. (Refer to Figure 6.64.)

Figure 6.66 The vertical adjusting screw is turned to center the bubble (courtesy of Chrysler Corporation).

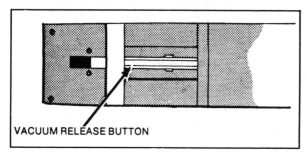

VACUUM RELEASE BUTTON

Figure 6.67 The vacuum release button on the bottom of the aimer is used to release the suction cup from the headlight lens (courtesy of Pontiac Motor Division, General Motors Corporation).

Note: The motor vehicle laws of your state may require a different setting.

2 Using a screwdriver, slowly turn the vertical adjusting screw located at the top (or bottom) of the headlight. (Refer to Figures 6.52 and 6.53.) Turn the screw in or out as required until the bubble in the level is centered. (See Figure 6.66.)

3 Repeat steps 1 and 2 at the opposite headlight.

4 Check the alignment of the split images at both aimers.

Note: A slight readjustment at one or both lights may be necessary.

5 Remove the aimers by holding them securely and pressing the vacuum release button on the piston handle. (See Figure 6.67.)

If the car is equipped with dual headlights, the steps listed under "Mounting the Aimers," "Horizontal Adjustment," and "Vertical Adjustment" must be performed on the second pair of headlights.

Aiming Headlights by Observing the Projected Light   Headlights can be aimed by observing the location of the high intensity areas of projected light. This method often is used when aiming devices are not available. A special chart can be used, or lines can be taped on a light-colored wall. A typical procedure follows:

Preparation   Since the horizontal and vertical centerlines of the headlights on various cars differ, certain measurements are used to locate lines on a chart or on a wall, as shown in Figure 6.68.

1 Position the car on a level floor so that it squarely faces a vertical wall or chart.

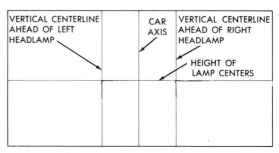

Figure 6.68 The lines that should be marked on a chart or wall to aim headlights. The locations of the lines must be determined by measurements made on the particular car (courtesy of Chrysler Corporation).

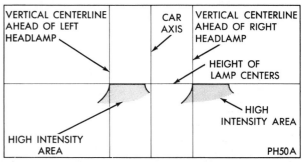

Figure 6.69 The correct low beam pattern for sealed beam bulbs with two filaments (courtesy of Chrysler Corporation).

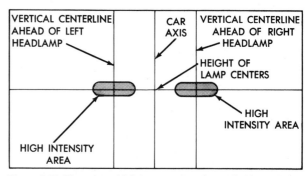

Figure 6.70 The correct high beam pattern for sealed beam bulbs with one filament (courtesy of Chrysler Corporation).

2 Adjust the position of the car so that the headlights are 25 feet (750 cm) from the chart or wall.

3 Tape a vertical line in alignment with the centerline of the car. (Refer to Figure 6.68.)

Note: The centerline of the car can be determined by standing behind the car and carefully sighting through the rear window at the chart or wall. The window trim moulding, the rear view mirror, and the hood ornament can be used as sighting aids.

4 Measure the distance between the centers of the headlights.

5 Transfer that measurement to the chart or wall and tape vertical lines representing the vertical

Job 6h

ADJUST HEADLIGHT AIM

SATISFACTORY PERFORMANCE
A satisfactory performance on this job requires that you do the following:

1 Adjust the aim of the headlights of the car assigned.
2 Following the steps in the "Performance Outline" and the procedure and specifications of the manufacturer, complete the job within 45 minutes.
3 Fill in the blanks under "Information."

PERFORMANCE OUTLINE
1 Prepare for adjustment.
2 Adjust the horizontal and vertical aim.

INFORMATION
Vehicle identification _____

Headlight system: _____ Single   _____ Dual

Reference used _____ Page(s) _____

Method used to adjust aim: _____ Mechanical aimers

_____ Projected light

Identification of instrument used _____

Was car positioned on a level floor? _____ Yes   _____ No

Was instrument adjusted for floor slope? _____ Yes   _____ No

Bulbs with two filaments were adjusted on: _____ High beam

_____ Low beam

Bulbs with one filament were adjusted on   _____ High beam

_____ Low beam

---

centerlines of the headlights. (Refer to Figure 6.68.)

Note: If the car is equipped with dual headlights, two sets of vertical centerlines must be marked.

6 Measure the distance from the floor to the center of the headlights.
7 Transfer that measurement to the chart or wall and tape a horizontal line to represent the horizontal centerline of the headlights. (Refer to Figure 6.68.)

Adjustment of Bulbs with Two Filaments   Bulbs with two filaments should be adjusted with the low beams on.

1 Turn the vertical adjusting screws (refer to Figures 6.52 and 6.53) so that the high intensity areas are located just below the horizontal centerline, as shown in Figure 6.69.
2 Turn the horizontal adjusting screws (refer to Figures 6.52 and 6.53) so that the high intensity areas are located just to the right of the vertical centerlines. (Refer to Figure 6.69.)

Adjustment of Bulbs with One Filament   Bulbs with one filament should be adjusted with the high beams on. Bulbs with two filaments used in the system should be covered while adjusting the single filament bulbs.

1 Turn the vertical adjusting screws so that the high intensity areas are centered on the horizontal centerline, as shown in Figure 6.70.

2 Turn the horizontal adjusting screws so that the high intensity areas are centered on the vertical centerlines. (Refer to Figure 6.70.)

SUMMARY

By completing the Jobs in this chapter, you gained knowledge and developed diagnostic and repair skills in the lighting system. You learned about various types of fuses and circuit breakers and can test and replace them. You learned about the different types of small bulbs, their construction, and their replacement procedures. You learned about sealed beam bulbs and can identify the various types, shapes, and sizes. You can replace sealed beam bulbs and can adjust them so that they are properly aimed. You now can perform many of the basic services in lighting systems.

SELF-TEST

Each incomplete statement or question in this test is followed by four suggested completions or answers. In each case select the *one* that best completes the sentence or answers the question.

1 A small bulb with a single contact bayonet base usually contains
A. one filament
B. two filaments
C. three filaments
D. four filaments

2 A small bulb without a metal base is usually called
A. a wedge base bulb
B. an index base bulb
C. a bayonet base bulb
D. a filament base bulb

3 Two mechanics are discussing fuses.
Mechanic A says that glass cartridge fuses are often used in fuse blocks.
Mechanic B says that glass cartridge fuses are often used in in-line fuse holders.
Who is right?
A. A only
B. B only
C. Both A and B
D. Neither A nor B

4 Two mechanics are discussing mini-fuses.
Mechanic A says that the rating of a mini-fuse is marked on the fuse.
Mechanic B says that the rating of a mini-fuse is indicated by a color code.
Who is right?
A. A only
B. B only
C. Both A and B
D. Neither A nor B

5 Two mechanics are discussing circuit breakers.
Mechanic A says that a circuit breaker is a temperature operated switch.
Mechanic B says that a circuit breaker is connected in series in a circuit.
Who is right?
A. A only
B. B only
C. Both A and B
D. Neither A nor B

6 Two mechanics are discussing flashers.
Mechanic A says that a flasher is a temperature operated switch.
Mechanic B says that a flasher is connected in series in a circuit.
Who is right?
A. A only
B. B only
C. Both A and B
D. Neither A nor B

7 Two mechanics are discussing sealed beam headlights.
Mechanic A says that a sealed beam bulb with two filaments has three terminals.
Mechanic B says that a sealed beam bulb with one filament is used as a low beam bulb.
Who is right?
A. A only
B. B only
C. Both A and B
D. Neither A nor B

8 A large (7 inch) round, sealed beam headlight has a code designation of
A. 2A
B. 2B
C. 2C
D. 2D

9 The guide points moulded into the edge of the lens of a sealed beam headlight bulb are provided for
   A. proper alignment of the headlight door
   B. accurate mounting of mechanical aimers
   C. proper positioning of the bulb in its housing
   D. accurate placement of the bulb retaining ring

10 Two mechanics are discussing aiming headlights by the light projected on a chart or wall.

Mechanic A says that bulbs with one filament should be aimed with the high beam lights on. Mechanic B says that bulbs with two filaments should be aimed with the low beam lights on. Who is right?
   A. A only
   B. B only
   C. Both A and B
   D. Neither A nor B

Chapter 7 The Lighting System— Advanced Services

In the previous chapter, you learned about some of the various circuits in the lighting system. You developed skills in replacing bulbs, fuses, and other parts that require frequent replacement. In this chapter, you will learn of some of the other components used in those circuits. You also will further develop your skills by testing and replacing those components and by making repairs in lighting system wiring.

In this chapter, your specific objectives are to perform the following Jobs:

A
Replace a dimmer switch
B
Replace a stoplight switch
C
Replace a back-up light switch
D
Replace a headlight switch
E
Replace a turn signal switch
F
Replace a fuse link
G
Repair a wiring harness
H
Replace a wiring harness

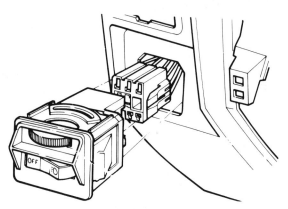

SWITCHES At times, your diagnosis of a lighting circuit problem will indicate that a switch is defective. Since the replacement of a switch can be a time-consuming operation, you should verify your diagnosis before replacing the suspected switch.

Switch Testing Most switches can be tested by (1) bypassing or "bridging" the switch with a jumper wire, (2) checking the operation of the switch with an ohmmeter or a continuity tester, and (3) substituting a replacement switch.

Bypassing a Switch Most switches are connected in a circuit by means of a plug or connector, as shown in Figures 7.1 and 7.2. Disconnecting the plug from the switch will enable you to use a jumper wire, as shown in Figure 7.3, to bridge the appropriate terminals. If the circuit operates with the jumper wire in place, the switch can be assumed to be defective.

Testing a Switch with an Ohmmeter or a Continuity Tester If access to the switch terminals can be obtained, an ohmmeter can be used to test the operation of the switch. Connected as shown in Figure 7.4, the ohmmeter should indicate 0 Ω when the switch is placed in the On position. When the switch is placed in the Off position, the ohmmeter should indicate infinity (∞). A continuity tester can be used in the same manner. The switch should turn the test lamp on and off when cycled through the On and Off positions.

Testing by Substitution If the plug or connector can be removed from a switch without removing the switch, a replacement switch often can be plugged in and operated to check the circuit. If the circuit functions properly with the replacement switch, the original switch probably is defective and can be removed. The replacement switch then can be installed.

DIMMER SWITCHES A dimmer switch is a single-pole, double-throw (SPDT) switch that directs current to either the high or low beam filaments of the headlight bulbs. On some cars, the dimmer switch is located on the floor so that it can be reached by the driver's left foot. (See Figure 7.5.) On other cars, the switch is located on the steering column so that it can be operated by

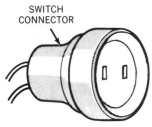

Figure 7.1 A two-terminal connector disconnected from a switch (courtesy of Chrysler Corporation).

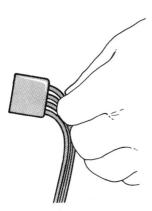

Figure 7.2 A multiterminal connector disconnected from a switch (courtesy of Chrysler Corporation).

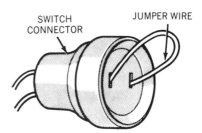

Figure 7.3 A jumper wire can be used across the terminals in a connector to act as a switch in the On position (courtesy of Chrysler Corporation).

Figure 7.4 An ohmmeter or a continuity tester can be used across the terminals of a switch to check its operation (courtesy of Chrysler Corporation).

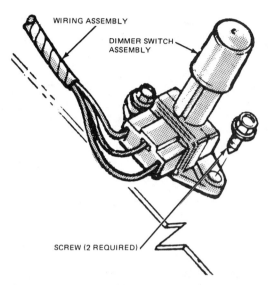

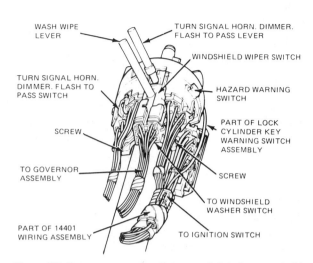

Figure 7.7 On some cars, the dimmer switch is incorporated in a switch assembly that includes the switches for the horn, turn signals, wipers, washers, hazard warning, and speed control circuits (courtesy of Ford Motor Company).

Figure 7.5 A typical floor mounted dimmer switch installation. The switch plugs into a socket on the wiring harness and is secured to the floor by screws (courtesy of Ford Motor Company).

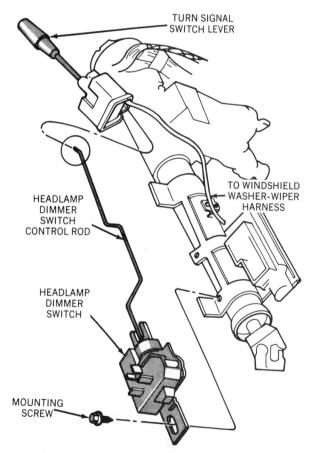

Figure 7.6 On some cars, the dimmer switch is mounted on the steering column and operated remotely by a rod connected to the turn signal switch lever (courtesy of Chrysler Corporation).

movement of the turn signal switch lever. Some steering column-mounted switches are positioned low on the column and are actuated by a rod from the turn signal lever, as shown in Figure 7.6. Other column mounted switches are combined with other switches that are positioned at the top of the steering column just under the steering wheel, as shown in Figure 7.7.

Testing Dimmer Switches Before any attempt is made to replace a dimmer switch that is suspected of being defective, the switch should be tested. The steps that follow outline procedures that can be used:

When the switch is located so that it is readily accessible (refer to Figures 7.5 and 7.6), an ohmmeter or a continuity tester can be used to test the switch.

1 Disconnect the wiring harness plug or connector from the switch.

Note: Floor-mounted switches usually require that the floor mat or carpet be lifted. Steering column-mounted switches may require that column covers be removed. (See Figure 7.8.)

2 Using an ohmmeter or a continuity tester, test for switch continuity between the battery (*B*) terminal and the high beam (*H*) terminal, and between the battery (*B*) terminal and the low beam (*L*) terminal while operating the switch. See Figure 7.9 for typical test points.

Note: Continuity (0 Ω or a lighted test lamp) should be obtained and should alternate between the high

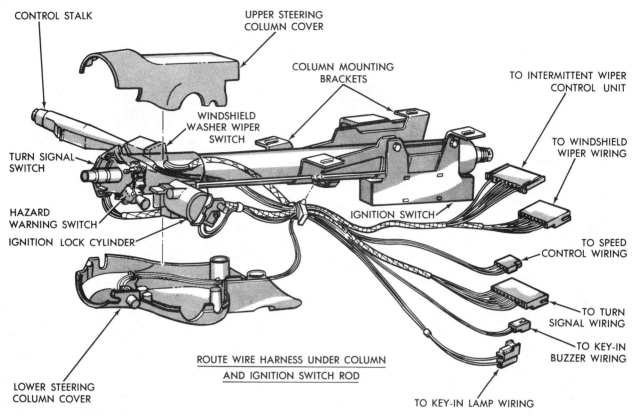

CONTROL STALK

UPPER STEERING
COLUMN COVER

COLUMN MOUNTING
BRACKETS

TO INTERMITTENT WIPER
CONTROL UNIT

WINDSHIELD
WASHER WIPER
SWITCH

TO WINDSHIELD
WIPER WIRING

TURN SIGNAL
SWITCH

HAZARD
WARNING SWITCH

IGNITION LOCK CYLINDER

IGNITION SWITCH

TO SPEED
CONTROL WIRING

TO TURN
SIGNAL WIRING

TO KEY-IN
BUZZER WIRING

ROUTE WIRE HARNESS UNDER COLUMN
AND IGNITION SWITCH ROD

LOWER STEERING
COLUMN COVER

TO KEY-IN LAMP WIRING

Figure 7.8 Typical steering column wiring. On many cars, access to the wiring and controls requires that you remove the steering column covers (courtesy of Chrysler Corporation).

beam (H) terminal and the low beam (L) terminal with each cycle of the switch. If continuity is not indicated, or if it does not alternate between the high and low beam terminals, the switch is defective.

When access to the switch is restricted, or when the dimmer switch is combined with other switches (refer to Figure 7.7), a jumper wire and a pair of test probes can be used at the plug or connector of the wiring harness to simulate the operation of the switch.

1 Disconnect the wiring harness plug from the switch or from the switch wiring.

2 Place the headlight switch in the On position.

3 Using a jumper wire and a pair of test probes, alternately bridge between the battery (B) terminal and the high beam (H) and the low beam (L) terminals in the wiring harness plug or connector. (Refer to Figure 7.9.)

Note: If the headlights function properly on high and low beams, the dimmer switch probably is defective. If the lights do not function properly, there is a problem in the wiring or in the bulb. The dimmer switch

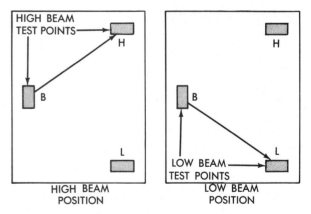

HIGH BEAM
TEST POINTS

H

B

L

HIGH BEAM
POSITION

H

B

L

LOW BEAM
TEST POINTS

LOW BEAM
POSITION

Figure 7.9 Test points for testing a dimmer switch with an ohmmeter or a continuity tester (courtesy of Chrysler Corporation).

may be in good condition. Further tests in the lighting circuit are necessary.

Dimmer Switch Replacement The following steps outline typical procedures for the replacement of dimmer switches. Since many different types of column-mounted switches are in use, you should

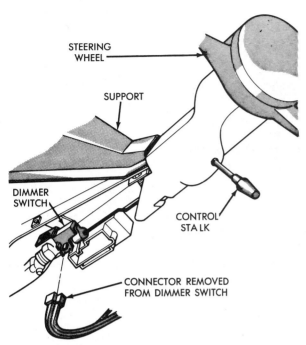

Figure 7.10 The terminals or prongs on the dimmer switch plug into a connector on the wiring harness (courtesy of Chrysler Corporation).

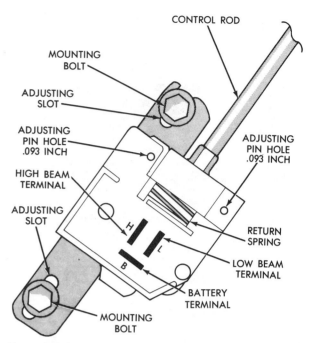

Figure 7.11 Most remotely operated dimmer switches are provided with a means for adjustment. Some switches can be locked with one pin. Others should be locked with two pins, as shown (courtesy of Chrysler Corporation).

consult an appropriate manual for the procedure and specifications required for the car on which you are working.

Floor-Mounted Dimmer Switches

1 Lift the floor mat or carpet to gain access to the switch.

2 Remove the screws holding the switch to the floor. (Refer to Figure 7.5.)

3 Unplug the switch from the harness connector.

4 Plug the replacement switch into the wiring harness connector.

5 Test the operation of the switch.

6 Install the screws holding the switch to the floor.

7 Lower the mat or carpet.

Remotely Controlled Steering Column-Mounted Dimmer Switches

1 Remove any covers or trim to gain access to the switch. (Refer to Figure 7.8.)

2 Disconnect the wiring harness connector from the switch, as shown in Figure 7.10

3 Remove the screws holding the switch to the column.

4 Remove the switch and disengage the control rod.

5 Engage the control rod on the replacement switch and install the screws securing the switch to the column. Do not tighten the screws.

Note: On most switches of this type, the mounting holes are elongated to allow for switch adjustment. A pinhole(s) is provided so that the switch can be locked in one position during the adjustment. (See Figure 7.11.)

6 Insert the pin(s) in the adjusting pinhole(s) in the switch.

7 Slide the switch upward toward the steering wheel until all play or movement is removed from between the switch and the control rod.

8 Hold the switch in this position and tighten the mounting screws.

9 Remove the pin(s) from the switch.

10 Connect the wiring harness connector to the switch.

11 Check the operation of the switch.

12 Install any covers or trim removed to gain access to the switch.

STOPLIGHT SWITCHES   On most cars, the stoplight switch is located under the instrument panel and above the brake pedal. Some switches are similar to the ones shown in Figures 7.12 and 7.13. The switch is positioned so that a plunger in the switch is in contact with the brake pedal lever. When there is no pressure on the brake pedal, its return spring holds the pedal lever up and in contact with

Job 7a

REPLACE A DIMMER SWITCH
SATISFACTORY PERFORMANCE
A satisfactory performance on this job requires that you do the following:

1 Replace the dimmer switch on the assigned car.
2 Following the steps in the "Performance Outline" and the procedure and specifications of the manufacturer, complete the job within 200 percent of the manufacturer's suggested time.
3 Fill in the blanks under "Information."

PERFORMANCE OUTLINE
1 Remove the switch.
2 Install the replacement switch.
3 Adjust the switch (if necessary).
4 Check the operation of the switch.

INFORMATION
Vehicle identification _____

Reference used _____ Page(s) _____

Type of switch replaced: _____ Floor-mounted

_____ Steering column-mounted, remote controlled

_____ Steering column-mounted, combined switch unit

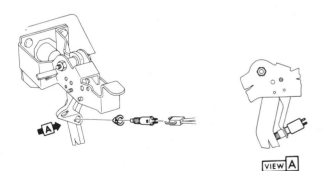

Figure 7.12 A typical plunger-type stoplight switch installation. When the brake pedal is depressed, the spring-loaded plunger moves outward, closing the switch contacts (courtesy of Chevrolet Motor Division)

the plunger on the switch, holding the switch contacts open. When the brake pedal is depressed, the plunger is allowed to move out of the switch, closing the switch contacts. Terminals or a plug on the back of the switch plug into a socket or connector on the wires of the harness.

Figure 7.14 shows another type of stoplight switch in common use. Such a switch fits on the pin that connects the brake pedal lever to the master cylinder pushrod. The switch closes the stoplight circuit when the pedal is moved the small distance allowed by the oversize hole in the eye of the pushrod. The operation of the switch is shown in Figure 7.15.

Testing Stoplight Switches A stoplight switch is easily tested by unplugging it from the wire harness and bypassing or bridging the terminals in the connector on the wire harness. (Refer to Figure 7.3.) If the stoplight circuit operates properly, the switch probably is defective. If the stoplights do not operate, there is a problem elsewhere in the circuit. The stoplight switch may be in good condition. Further tests in the stoplight circuit are required.

Stoplight Switch Replacement The following steps outline typical procedures for the replace-

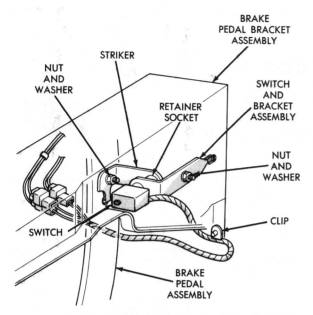

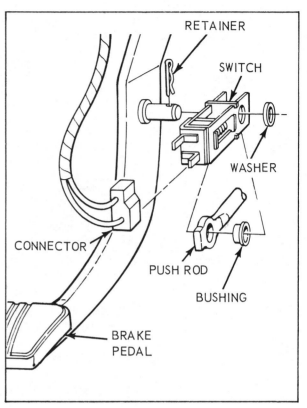

Figure 7.13 A self-adjusting plunger-type stoplight switch. This switch is held open by a striker attached to the brake pedal assembly. Pulling the brake pedal back after installing the switch causes the switch to slide back on a ratcheting mount, which provides the adjustment (courtesy of Chrysler Corporation).

Figure 7.14 A stoplight switch that is mounted on the brake pedal linkage (courtesy of Ford Motor Company).

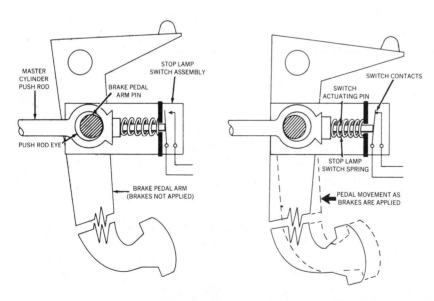

Figure 7.15 The operation of a linkage mounted stoplight switch. The switch is operated by the movement allowed by the oversize hole in the push rod eye (courtesy of Ford Motor Company).

ment of stoplight switches. As there are several variations of stoplight switches, it is suggested that you consult an appropriate manual for the procedure and specifications that may be necessary for the car on which you are working.

Plunger-Type Switches

1 Disconnect the wire connector from the switch.
2 Remove the lock nut from the switch threads (if

fitted), or remove the nut from the bolt holding the switch bracket. (Refer to Figures 7.12 and 7.13.)
3 Remove the switch.

Note: Some switches thread into the mounting bracket. This allows the switch to be threaded in or out for adjustment. Other switches do not thread into place but are adjusted by a movable bracket. Still other switches are self-adjusting.

Job 7b

REPLACE A STOPLIGHT SWITCH

SATISFACTORY PERFORMANCE
A satisfactory performance on this job requires that you do the following:

1 Replace the stoplight switch on the assigned car.
2 Following the steps in the "Performance Outline" and the procedure and specifications of the manufacturer, complete the job within 200 percent of the manufacturer's suggested time.
3 Fill in the blanks under "Information."

PERFORMANCE OUTLINE
1 Remove the switch.
2 Install the replacement switch.
3 Adjust the switch if necessary.
4 Check the operation of the switch.

INFORMATION
Vehicle identification _____

Reference used _____ Page(s) _____

Type of switch replaced: _____ Plunger-type

_____ Linkage-type

Switch closes when pedal is depressed about _____ inches (mm).

---

4 Install the replacement switch.
5 Connect the wires.
6 Adjust the switch (if required).

Note: Most manufacturers specify that the switch should be adjusted so that it closes the stoplight circuit when the brake pedal is depressed from ¼ inch (6 mm) to ⅝ inch (15 mm) from its fully released position.

Linkage-Mounted Switches

1 Disconnect the wire connector from the switch.
2 Remove the hairpin-shaped retainer from the brake pedal pin. (Refer to Figure 7.14.)
3 Remove the washer from the brake pedal pin.
4 Slide the switch and the pushrod off the brake pedal pin.

Note: Observe the position of the bushing. (Refer to Figure 7.14.)

5 Align the replacement switch with the pushrod and the bushing, and slip the assembly over the brake pedal pin.
6 Install the washer.
7 Install the hairpin-shaped retainer.
8 Connect the wires.

9 Test the operation of the switch.

Note: Switches of this type do not require adjustment.

BACK-UP LIGHT SWITCHES A back-up light switch closes the circuit to the back-up lights when the transmission selector lever is placed in the Reverse position. On some cars, the back-up light switch is combined with the neutral-safety switch in one assembly. The location of the back-up light switch varies with different cars. On some, the back-up light switch is threaded into the transmission case, as shown in Figures 7.16 and 7.17, and is operated by an internal transmission part. On other cars, the switch is mounted on the steering column beneath the instrument panel. (See Figures 7.18 and 7.19.) Back-up light switches also will be found mounted on the base of the steering column or on the transmission so as to be operated by the shift linkage.

Testing Back-up Light Switches A back-up light switch can be tested by unplugging it from the wire harness and bridging the terminals in the connector on the wire harness. (Refer to Figure 7.3.) If

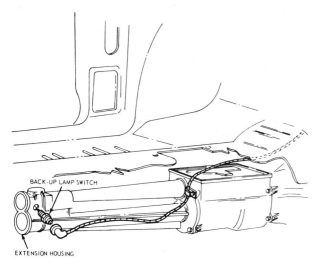

Figure 7.16 A back-up light switch mounted on the extension housing of a transmission (courtesy of Ford Motor Company).

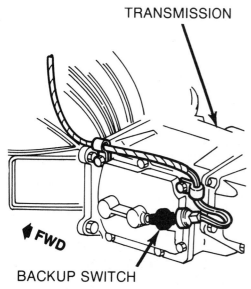

Figure 7.17 A back-up light switch mounted on the side cover of a manual transmission (courtesy of Chevrolet Motor Division).

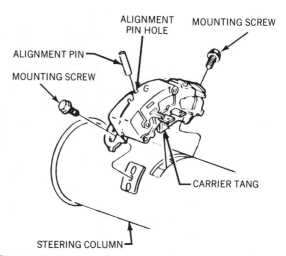

Figure 7.18 A typical column mounted back-up light switch. Notice that an alignment pin is used to position the switch during installation (courtesy of Chevrolet Motor Division).

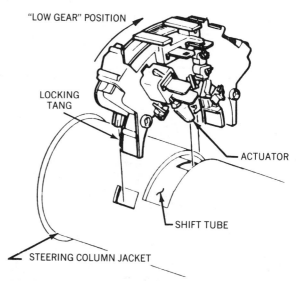

Figure 7.19 A self-adjusting back-up light switch. A ratcheting mechanism adjusts the switch when the shift lever is pulled into the Low position. Notice that no mounting screws are used as the switch is locked to the steering column jacket by tangs that snap into holes in the jacket (courtesy of Chevrolet Motor Division).

the back-up lights operate, the switch probably is defective. (Most back-up light circuits are wired so that they will operate only when the ignition switch is placed in the On position. Be sure that the ignition switch is turned on during this test.) If the back-up lights do not operate, there is a problem elsewhere in the circuit and the back-up light switch may not be defective. Further tests in the circuit are required. An ohmmeter or a continuity tester may be used across the switch terminals to verify your diagnosis. (Refer to Figure 7.4.)

Back-up Light Switch Replacement The following steps outline typical procedures for the

replacement of back-up light switches. Since many types of switches are in use, you should consult an appropriate manual for the location of the switch and any special procedure and specifications that may be required for the car on which you are working.

Column-Mounted Back-up Light Switches
1 Unplug the wiring harness connector from the switch.
2 Remove the screws holding the switch to the steering column. (Refer to Figure 7.18.)

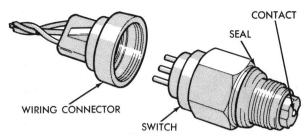

Figure 7.20 A back-up light switch combined with a neutral safety switch. A switch of this type is threaded into the side of the transmission case and actuated by a part within the transmission (courtesy of Chrysler Corporation).

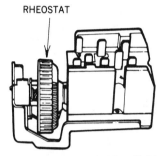

Figure 7.21 A typical headlight switch. A switch of this type incorporates a variable resistance unit to dim the instrument panel lights and a rotary switch to operate the interior lights (courtesy of Ford Motor Company).

Note: Some switches are held by barbed locking tangs that snap into holes in the column. (Refer to Figure 7.19.) Such switches must be pried loose with a screwdriver.

3 Remove the switch.

4 Place the shift selector lever in the Neutral position.

5 Position the replacement switch on the steering column so that the carrier tang or actuator is aligned with the hole in the shift tube. (Refer to Figures 7.18 and 7.19.)

Note: Some replacement switches incorporate a plastic alignment pin that holds the switch in the Neutral position for correct alignment. The pin is sheared off after installation by moving the shift lever out of the Neutral position. If the plastic pin is missing or has been sheared off during a previous installation, the switch can be held in the Neutral position by a pin fashioned from a piece of welding rod. Some switches are self-adjusting and require no alignment pin. (Refer to Figure 7.19.)

3 Install the mounting screws (if fitted).

4 Install the wiring harness connector.

5 Move the shift lever from the Neutral position to shear the pin.

Note: If a temporary pin has been installed, remove

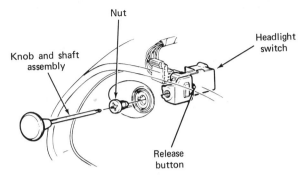

Figure 7.22 A typical headlight switch installation. Notice the release button location (courtesy of Chevrolet Motor Division).

the pin before attempting to move the shift lever. Failure to do so may damage the switch.

6 Check the operation of the switch.

Transmission Mounted Back-up Light Switches

1 Unplug the wiring harness connector from the switch. (See Figure 7.20.)

2 Using the correct size wrench, loosen and unscrew the switch from the transmission.

Note: Most switches are installed with a seal or gasket. (Refer to Figure 7.20.) Save the seal for installation on the replacement switch, as a new seal may not be supplied.

3 Install the seal or gasket on the replacement switch and thread it into the transmission case.

4 Tighten the switch with a wrench.

5 Install the wiring harness connector.

6 Check the operation of the switch.

HEADLIGHT SWITCHES A headlight switch, shown in Figure 7.21, is a multiple-pole switch that controls the headlights, parking lights, taillights, marker lights, and the instrument panel lights. In some instances, the headlight switch controls the intensity of the instrument panel lights by means of a *rheostat,* or variable resistor, and actuates the interior lights.

The headlight switch usually is mounted behind the instrument panel and is held by a nut, as shown in Figure 7.22. The knob and shaft assembly pass through the nut into the switch. In most headlight switches, the shaft is held in the switch by a spring-loaded lock that can be released by depressing a release button while the switch is in the On position. The location of the release button is shown in Figure 7.23. In some switches, the shaft is a part of the switch, and the knob must be removed as shown in Figure 7.24.

On some cars, the nut is concealed behind a *bezel,* or trim plate, that must be removed to gain access

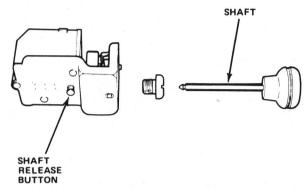

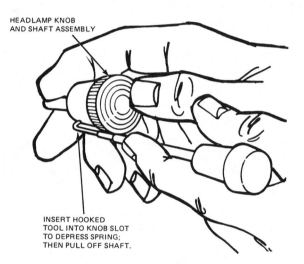

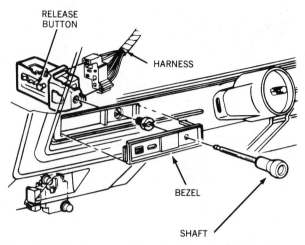

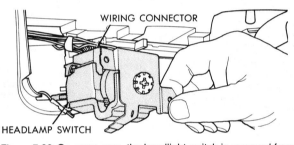

Figure 7.23 Before the nut can be removed from the switch, the shaft must be released by depressing the shaft release button (courtesy of American Motors).

Figure 7.24 On some cars, the shaft cannot be removed from the switch. The knob must be released by means of a wire hook (courtesy of Ford Motor Company).

Figure 7.25 On some cars, a bezel or trim plate must be removed to gain access to the headlight switch nut (courtesy of Chevrolet Motor Division).

Figure 7.26 On some cars, the headlight switch is removed from the front of the instrument panel after removing a bezel or trim plate (courtesy of Chrysler Corporation).

to the nut. An installation of that type is shown in Figure 7.25. While most headlight switches are removed from behind the instrument panel, they are at times removed from the front. (See Figure 7.26.)

Testing Headlight Switches   As with other switches, headlight switches can be tested with an ohmmeter or a continuity tester, and the various circuits may be checked by bypassing the switch with a jumper wire. On most cars, however, you must remove the switch to perform the tests and, unless you are familiar with the circuitry, you must have a headlight switch wiring diagram and be able to identify the switch terminals. (Figures 7.27 and 7.28 show how that information is supplied by one car maker.) Therefore, it usually is easier to test a headlight switch by substitution.

Headlight Switch Replacement   The following steps outline procedures for replacing head-

light switches. Because of the many variations of switches and switch mountings used, it is suggested that you refer to an appropriate manual for any specific procedures that may be necessary for the car on which you are working.

Headlight Switches Removed from Behind the Instrument Panel   Before attempting to remove a switch of this type, you should check to see that the switch is accessible. On many cars, certain other controls, air conditioning ducts, and brackets may have to be removed for switch access.

1 Disconnect the battery ground cable.
2 Remove the headlight switch shaft or knob. (Refer to Figures 7.23 and 7.24.)
3 Remove the shaft nut.

Note: On most cars, the nut may be removed with a screwdriver. On some cars, a special socket is required.

4 Reaching behind the instrument panel, lower the switch and disconnect the wire connector.
5 Remove the switch.

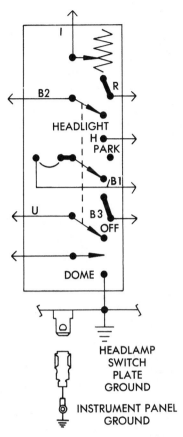

Figure 7.27 A headlight switch diagram showing the internal wiring of the switch (courtesy of Chrysler Corporation).

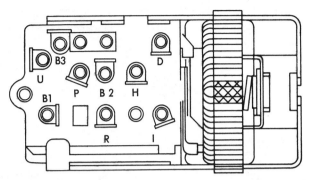

Figure 7.28 Headlight switch terminal identification. When used with a headlight switch wiring diagram, this drawing enables you to test a headlight switch with an ohmmeter or a continuity tester (courtesy of Chrysler Corporation).

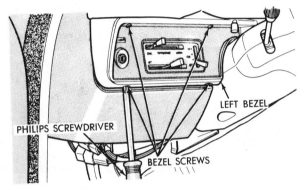

Figure 7.29 Removing the bezel screws (courtesy of Chrysler Corporation).

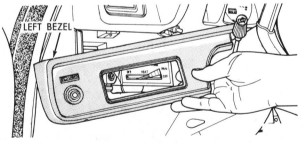

Figure 7.30 Removing the bezel (courtesy of Chrysler Corporation).

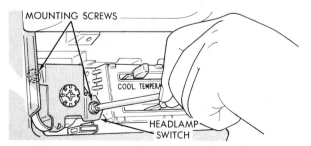

Figure 7.31 Removing the headlight switch mounting screws (courtesy of Chrysler Corporation).

6 Plug the replacement switch into the wire connector.

7 Position the switch behind the instrument panel and install the shaft nut.

Note: Most switches have a tab or protrusion that fits into a slot or notch on the instrument panel to provide the correct switch alignment.

8 Tighten the shaft nut.

9 Install the shaft or knob.

10 Connect the battery ground cable.

11 Check the operation of the switch.

Headlight Switches Removed from the Front of the Instrument Panel

1 Disconnect the battery ground cable

2 Remove the headlight switch shaft or knob. (Refer to Figures 7.23 and 7.24.)

3 Remove the screws or fasteners securing the bezel or trim plate covering the switch. (See Figure 7.29.)

4 Remove the bezel or trim plate. (See Figure 7.30.)

5 Remove the shaft nut or mounting screws that hold the switch. (See Figure 7.31.)

6 Remove the switch and disconnect the wire connector. (Refer to Figure 7.26.)

Job 7c

REPLACE A BACK-UP LIGHT SWITCH

SATISFACTORY PERFORMANCE
A satisfactory performance on this job requires that you do the following:

1 Replace the back-up light switch on the assigned car.
2 Following the steps in the "Performance Outline" and the procedure and specifications of the manufacturer, complete the job within 200 percent of the manufacturer's suggested time.
3 Fill in the blanks under "Information."

PERFORMANCE OUTLINE
1 Remove the switch.
2 Install the replacement switch.
3 Adjust the switch (if necessary).
4 Check the operation of the switch.

INFORMATION
Vehicle identification _____

Reference used _____ Page(s) _____

Type of switch replaced: _____ Column-mounted

_____ Transmission-mounted

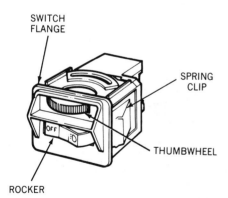

Figure 7.32 A rocker-type headlight switch. The rocker controls the headlight and parking light circuits, and the thumbwheel controls the intensity of the instrument panel lights and operates the interior lights (courtesy of Ford Motor Company).

7 Plug the replacement switch into the wire connector.
8 Position the switch in the instrument panel and install the shaft nut or mounting screws. (Refer to Figure 7.31.)
9 Install the bezel or trim plate. (Refer to Figures 7.29 and 7.30.)
10 Install the shaft or knob.

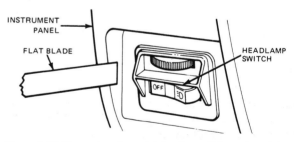

Figure 7.33 The spring clips retaining the switch can be released with a flat bladed tool (courtesy of Ford Motor Company).

11 Connect the battery ground cable.
12 Check the operation of the switch.

Rocker-type Headlight Switches Removed from the Front of the Instrument Panel (See Figure 7.32.)

1 Disconnect the battery ground cable.
2 Release the spring clips holding the switch by sliding a stiff feeler gauge or other flat bladed tool behind the flanges on the sides of the switch. (See Figure 7.33.)
3 Pull the switch out of the instrument panel. (See Figure 7.34.)

Job 7d

REPLACE A HEADLIGHT SWITCH

SATISFACTORY PERFORMANCE

A satisfactory performance on this job requires that you do the following:

1 Replace the headlight switch on the assigned car.
2 Following the steps in the "Performance Outline" and the procedure and specifications of the car manufacturer, complete the job within 200 percent of the manufacturer's suggested time.
3 Fill in the blanks under "Information."

PERFORMANCE OUTLINE

1 Disconnect the battery ground cable.
2 Remove any trim or panels blocking access to the switch.
3 Remove the switch.
4 Install the replacement switch.
5 Install any components removed to gain access.
6 Connect the battery ground cable.
7 Check the operation of the switch.

INFORMATION

Vehicle identification _____

Reference used _____ Page(s) _____

Switch was removed from: _____ Behind panel

_____ In front of panel

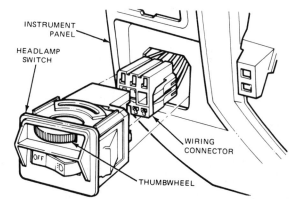

Figure 7.34 Disconnecting a rocker-type switch from the wire connector (courtesy of Ford Motor Company).

4 Disconnect the wire connector.
5 Plug the replacement switch into the wire connector.
6 Push the switch into position so that the spring clips lock it in place.

7 Connect the battery ground cable.
8 Check the operation of the switch.

TURN SIGNAL SWITCHES Most turn signal switches are mounted on the top of the steering column under the steering wheel. As shown in Figure 7.35, turn signal switches usually are combined with the hazard warning flasher switch and other switches. On most cars, the steering wheel and other related parts must be removed before the turn signal switch can be removed. Because of the involved replacement procedure, you should be sure of your diagnosis before you replace a switch thought to be defective.

Testing Turn Signal Switches If you have a wiring diagram or a continuity test diagram for the switch you wish to test, an ohmmeter or a continuity tester can be used. (See Figures 7.36 and 7.37.) A wiring diagram also will allow you to use jumper wires to bypass the switch and test the various cir-

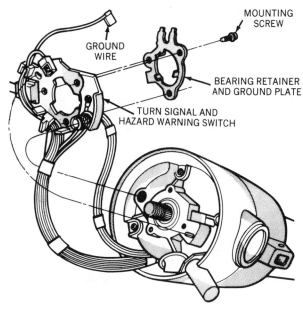

Figure 7.35 On most cars, the turn signal and hazard warning switches are combined and mounted at the top of the steering column beneath the steering wheel (courtesy of Chrysler Corporation).

cuits. Since those tests require that you gain access to and disconnect the wire connector, testing by substitution is the easiest method. Simply plug the replacement switch assembly into the connector and operate the switch by hand. If the problem in the circuit is eliminated, the old switch is defective.

Turn Signal Switch Replacement The following steps outline general procedures for the replacement of a turn signal switch. An appropriate manual should be consulted for the specific procedure and specifications required for the car on which you are working.

Removing and Installing a Steering Wheel Most cars are fitted with an energy-absorbing steering column. In the event of a frontal collision, those columns are designed to collapse or telescope to minimize driver injury. Most energy-absorbing steering columns utilize plastic retainers in their assembly. Upon impact, those retainers shear, allowing the

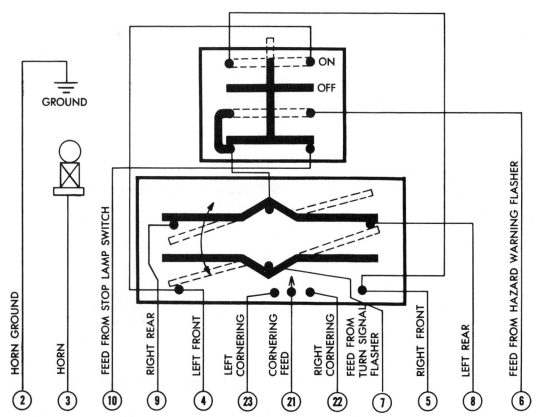

Figure 7.36 A typical turn signal switch wiring diagram. Notice that the horn switch and the hazard warning switch are incorporated in the assembly (courtesy of Chrysler Corporation).

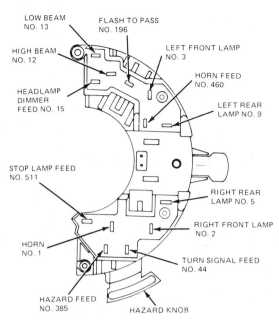

LOW BEAM
NO. 13

FLASH TO PASS
NO. 196

HIGH BEAM
NO. 12

LEFT FRONT LAMP
NO. 3

HEADLAMP
DIMMER
FEED NO. 15

HORN FEED
NO. 460

LEFT REAR
LAMP NO. 9

STOP LAMP FEED
NO. 511

RIGHT REAR
LAMP NO. 5

HORN
NO. 1

RIGHT FRONT LAMP
NO. 2

TURN SIGNAL FEED
NO. 44

HAZARD FEED
NO. 385

HAZARD KNOB

Figure 7.37 Some car manufacturers provide a chart of switch continuity test points, as shown (courtesy of Ford Motor Company).

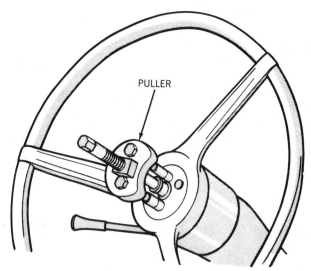

PULLER

Figure 7.38 A puller of this type will remove a steering wheel without damage to the column and shaft (courtesy of Chrysler Corporation).

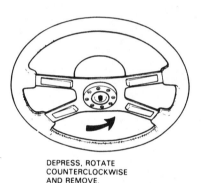

DEPRESS, ROTATE
COUNTERCLOCKWISE
AND REMOVE.

Figure 7.39 On some cars, the horn button is removed by depressing it and turning it counterclockwise (courtesy of Ford Motor Company).

column to collapse. Care must be taken while working on an energy-absorbing steering column. A sharp blow on the steering shaft or on the shift lever could shear or loosen the retainers, causing the column to lose rigidity.

The steering wheel is mounted on the steering shaft by means of a tapered spline. Because of the tight fit between the wheel and the shaft, a puller usually is needed to remove the wheel after the retaining nut has been removed. When removing a steering wheel, always use a puller of the type recommended by the car maker. Most manufacturers recommend the use of a puller similar to the one shown in Figure 7.38. Never use an impact puller or a "knock-off" type of puller. Never attempt to remove a steering wheel by hammering on the end of the steering shaft. Failure to heed these precautions may result in your shearing certain plastic retainers or otherwise damaging the column and its mounting.

The steering wheel is held on the steering shaft by a nut that is covered by the horn button or the steering wheel spoke trim. Therefore, the horn button or spoke trim must be removed to gain access to the nut. On some cars, the horn button is removed by depressing the button assembly and rotating it counterclockwise, as shown in Figure 7.39. Some horn buttons have ears or tabs that snap into notches or grooves in the steering wheel. As shown in Fig-

ure 7.40, horn buttons of that type should be carefully pried loose. When the horn button is built into the steering wheel spoke trim, the assembly usually is secured to the steering wheel by screws that are accessible from behind the wheel, as shown in Figure 7.41.

Removal

1 Disconnect the ground cable from the battery.

2 Remove the horn button or steering wheel spoke trim. (Refer to Figures 7.39, 7.40, and 7.41.)

Note: On many cars, the horn button is connected to the steering wheel by a short length of wire. Disconnect the wire carefully to avoid breaking it. (See Figure 7.42.)

3 Using a socket wrench, remove the steering wheel nut.

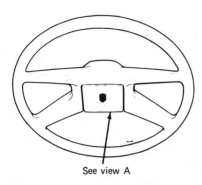

See view A

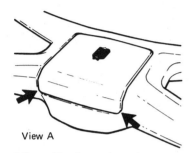

View A

Lift outside edges and remove.
Do not pry with sharp instrument.

Figure 7.40 Some horn buttons snap into place and must be carefully pried loose for removal (courtesy of Ford Motor Company).

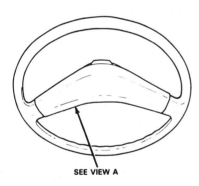

SEE VIEW A

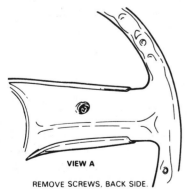

VIEW A

REMOVE SCREWS, BACK SIDE.

Figure 7.41 Some horn buttons are attached to the steering wheel by screws inserted through the rear of the spoke (courtesy of Ford Motor Company).

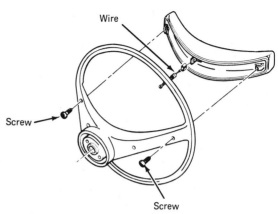

Wire

Screw

Screw

Figure 7.42 The horn button is often connected to the wheel by a short wire that must be disconnected (courtesy of Buick Motor Division, General Motors Corporation).

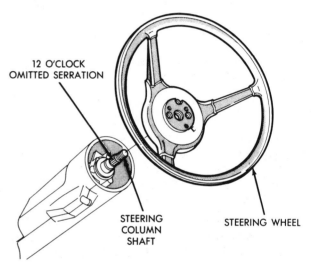

12 O'CLOCK OMITTED SERRATION

STEERING COLUMN SHAFT

STEERING WHEEL

Figure 7.43 On some cars, a tooth is omitted from the serrations in the steering wheel hub and on the steering shaft. These "master splines" ensure that the wheel is properly positioned on the shaft (courtesy of Chrysler Corporation).

4 Check the steering shaft and the steering wheel hub for reference marks that will enable you to reinstall the wheel in the same position. If no marks are present, use a sharp punch or scriber to mark both the shaft and the hub.

Note: On some cars, the steering shaft has a missing serration. That space is aligned with a "double-toothed" serration in the steering wheel hub to position the wheel. (See Figure 7.43.)

5 Install a steering wheel puller. (Refer to Figure 7.38.) Thread the attaching bolts evenly into the threaded holes in the hub approximately 0.5 in. (12.7 mm).

6 Center the puller lead screw on the steering shaft, and turn it down with a wrench until the wheel breaks loose from the shaft.

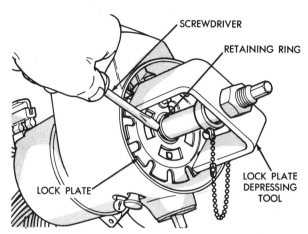

Figure 7.44 The lock plate beneath the steering wheel on many cars must be pushed down with a special tool to release its retaining ring (courtesy of Chrysler Corporation).

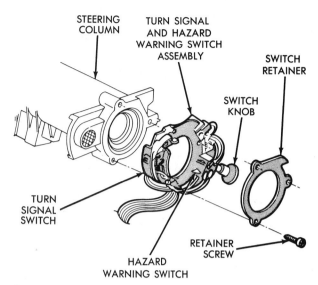

Figure 7.45 A turn signal switch removed from the steering column. Note that in this installation, a switch retainer is used between the screws and the switch (courtesy of Chrysler Corporation).

7 Remove the steering wheel from the shaft.
8 Remove the puller from the steering wheel.

Installation
1 Check to see that the turn signal switch lever is in the center or Off position.
2 Place the steering wheel over the steering shaft, guiding it over the splines or serrations so that the reference marks or "master serrations" are aligned.
3 Install the steering wheel nut.
4 Tighten the nut to the torque specification.
5 Install the horn button or steering wheel spoke trim. (Refer to Figures 7.39, 7.40, and 7.41.)
6 Connect the battery ground cable.

Removing and Installing a Turn Signal Switch Most cars are equipped with a steering wheel lock that engages when the ignition switch is turned to the Lock position. On some cars, the lock consists of a steel pin that engages a notch in a *lock plate* beneath the steering wheel. When the steering wheel is removed, the plate can be seen. To gain access to the turn signal switch, the lock plate must be removed with the aid of a special tool, as shown in Figure 7.44.

Removal
1 Remove the steering wheel. (Refer to the previous procedure.)
2 Disconnect the turn signal switch wiring at the connector.

Note: On many cars, the steering column covers must be removed. (Refer to Figure 7.8.)

3 Install a lock plate depressing tool on the steer-ing shaft so that the legs of the tool contact the lock plate. (Refer to Figure 7.44.)
4 Tighten the steering shaft nut so that the tool pushes the lock plate down against spring pressure until the retaining ring is just released.

Note: Do not overtighten the tool as the plate may be damaged.

5 Using a screwdriver, pry the retaining ring from its groove in the steering shaft. (Refer to Figure 7.44.)
6 Remove the tool, the retaining ring, and the lock plate.
7 Remove the turn signal handle or stalk assembly.

Note: On some cars, the knob on the hazard warning switch must be removed.

8 Remove the screws that hold the turn signal switch to the steering column. (See Figure 7.45.)

Note: On some cars, the screws are concealed by the switch actuating arm. In some instances, the arm can be moved to gain access to the screws. In others, the arm must be removed.

9 Carefully pull the switch up from the column while aligning and guiding the wires through their channel or opening.

Installation
1 Straighten and align the wires on the replacement switch and feed them through the channel or opening in the steering column.
2 Position the switch and install the retaining screws. (Refer to Figure 7.45.)

Job 7e

REPLACE A TURN SIGNAL SWITCH

SATISFACTORY PERFORMANCE
A satisfactory performance on this job requires that you do the following:

1 Replace the turn signal switch on the assigned car.
2 Following the steps in the "Performance Outline" and the procedure and specifications of the car manufacturer, complete the job within 200 percent of the manufacturer's suggested time.
3 Fill in the blanks under "Information."

PERFORMANCE OUTLINE
1 Remove the steering wheel.
2 Remove the turn signal switch.
3 Install the replacement switch.
4 Install the steering wheel.
5 Check the operation of the switch.

INFORMATION
Vehicle identification _____

Reference used _____ Page(s) _____

---

3 Install the handle or stalk assembly and any other knobs that may have been removed.

4 Place the lock plate and the retaining ring over the steering shaft.

Note: **The retaining ring must be a tight fit on the shaft. If the ring has been distorted, it should be squeezed to its original shape or replaced.**

5 Install the lock plate depressing tool on the steering shaft.

6 Install and tighten the steering shaft nut so that the tool pushes the lock plate down far enough so that the retaining ring groove is exposed.

7 Push the retaining ring down on the shaft so that it snaps into the groove. (Refer to Figure 7.44.)

8 Release and remove the lock plate depressing tool.

9 Connect the switch wiring.

10 Install any covers or trim that were removed.

11 Install the steering wheel. (Refer to the previous procedure.)

12 Check the operation of the switch.

FUSE LINKS Occasionally, one or more lighting circuits will be inoperative although the fuses and circuit breakers in those circuits are in good condition. The problem may be caused by a burned

fuse link. On many cars, fuse links, sometimes called *fusible links,* are built into the wiring harnesses, as shown in Figure 7.46. Those fuse links protect certain circuits from possible damage.

A fuse link is a short length of wire several gauges smaller than the wire used in the circuit that it protects. Thus, when excessive current flows in the circuit, the fuse link acts as a fuse and burns out. Figure 7.47 shows a burned-out fuse link. To prevent the possibility of a fire, fuse links are covered with a special insulation that does not burn. The insulation on some fuse links "bakes" to an ash, while others have insulation that blisters and discolors. The locations of fuse links and the circuits that they protect vary with different cars and are best found by reference to an appropriate manual.

A fuse link burns out when too much current flows in a circuit. The excessive current flow can be caused by a short in the wiring or in a circuit component, or by the incorrect use of a booster battery or battery charger. Before a fuse link is replaced, the cause of the problem must be determined and repaired.

When replacing a fuse link, you must use the correct gauge wire, and it should be cut to the correct length. You should not attempt to fabricate fuse links from ordinary wire. The insulation may not be flame-

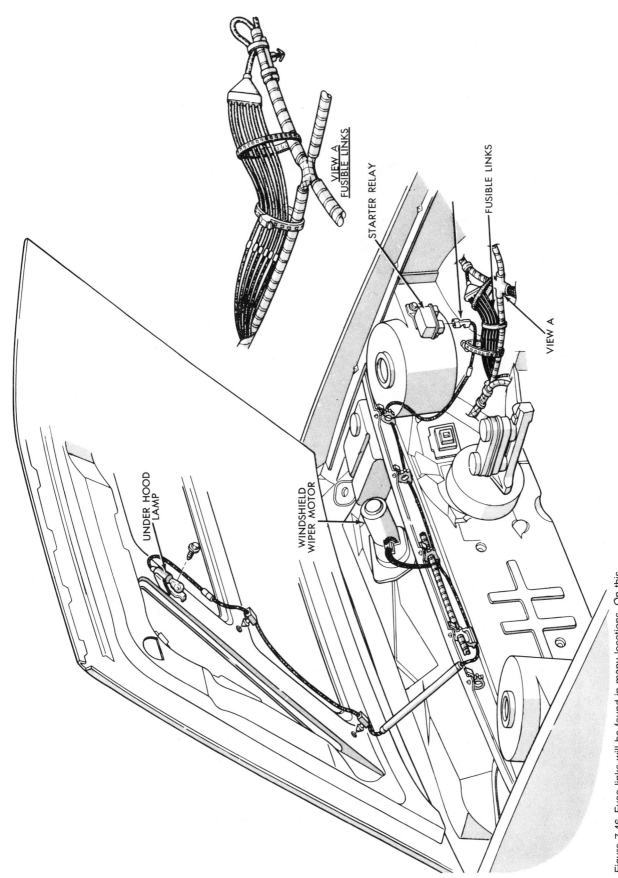

VIEW A
FUSIBLE LINKS

STARTER RELAY

FUSIBLE LINKS

VIEW A

UNDER HOOD
LAMP

WINDSHIELD
WIPER MOTOR

Figure 7.46 Fuse links will be found in many locations. On this car, multiple fuse links are located at the wiring harness over the left front wheel housing (courtesy of Chrysler Corporation).

Figure 7.47 A terminal end fuse link shown before and after a short circuit. Notice that the insulation has been burned off (courtesy of Oldsmobile Division).

proof and may be the cause of a fire. Some car makers furnish rolls of special fuse link wire in various gauges. Others provide precut lengths as shown in Figure 7.48. You usually will have to consult the car manufacturer's manual to determine the correct gauge and length.

Fuse Link Replacement The following steps outline typical procedures for the replacement of fuse links. While all connections should be soldered with rosin core solder, it is recommended that butt connectors be used where fuse links can be attached to the ends of wires. The butt connectors and the wire ends should then be soldered and insulated with vinyl tape. You should consult an appropriate manual for the correct gauge fuse link wire, its length, and any special procedure that may be necessary for the car on which you are working.

Terminal End Fuse Links

1 Disconnect the battery ground cable.

2 Remove the nut or bolt that secures the terminal end of the burned fuse link.

3 Cut off the feed wire as close as possible to the old fuse link connection.

4 Remove sufficient insulation from the end of the feed wire so that it can be inserted into a butt connector. (See Figure 7.49.)

Note: Be sure to use a butt connector sized to fit the feed wire.

5 Remove sufficient insulation from the new terminal end fuse link wire so that it can be inserted into the butt connector.

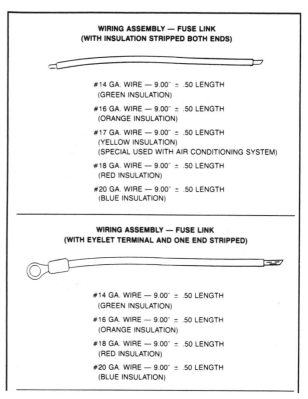

Figure 7.48 Some car manufacturers provide replacement fuse links color coded and cut to length (courtesy of Ford Motor Company).

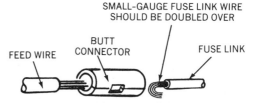

Figure 7.49 A fuse link should be attached to the feed wire with a butt connector. If the fuse link wire is a loose fit in the connector, it should be doubled back as shown so that it will make a good connection (courtesy of Ford Motor Company).

Note: If the fuse link wire fits loosely in the butt connector, it should be doubled back to obtain a tight fit. (Refer to Figure 7.49.)

6 Using a wire crimping tool, crimp the butt connector on the feed wire and on the fuse link wire to obtain a tight mechanical connection.

7 Using rosin core solder, solder the wires to the butt connector.

8 Insulate the connection with several wraps of vinyl tape, as shown in Figures 7.50 and 7.51.

9 Install the terminal end of the fuse link and tighten the nut or bolt that secures it.

10 Install the battery ground cable.

In-Line Fuse Links

1 Disconnect the battery ground cable.

2 Cut the feed wire as close as possible to the old fuse link connections.

3 Remove sufficient insulation from the ends of the feed wire so that the ends can be inserted into butt connectors.

Note: Be sure that you use butt connectors of the correct size for the gauge of the feed wire.

4 Remove sufficient insulation from the ends of the new fuse link so that the ends can be inserted into the butt connectors

Note: If the fuse link wire fits loosely in the butt connectors, it should be doubled back to obtain a tight fit. (Refer to Figure 7.49.)

5 Using a wire crimping tool, crimp the butt connectors on the feed wires and on the fuse link wire ends to obtain a tight mechanical connection.

6 Using rosin core solder, solder the wires to the butt connectors. (Refer to Figure 7.50.)

7 Insulate the connectors with several wraps of vinyl tape, as shown in Figure 7.52.

8 Connect the battery ground cable.

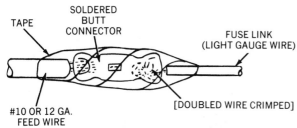

Figure 7.50 A fuse link should be installed using a crimped butt connector. The connector and the wires should then be soldered and insulated with tape (courtesy of Ford Motor Company).

Two or More Fuse Links with a Common Connector

1 Disconnect the battery ground cable.

2 Cut the feed wire as close as possible to the old fuse link connection.

3 Remove sufficient insulation from the ends of the feed wire so that it can be inserted into a butt connector.

4 Cut the fuse links as close as possible to the common connector.

5 Replace the burned-out fuse link. (Refer to the previous procedures.)

6 Remove sufficient insulation from the ends of all of the fuse link wires so that they can be inserted into a butt connector.

7 Select a butt connector that will accept all the fuse link wires.

8 Using a crimping tool, crimp the butt connector on all of the fuse link wires. (See Figure 7.53.)

9 Insert the end of the feed wire into the butt connector, and crimp the butt connector in place.

Note: If the feed wire fits loosely in the butt connector, it should be doubled back to obtain a tight fit.

10 Using rosin core solder, solder all of the wires to the butt connectors.

11 Insulate the connections with several wraps of vinyl tape. (Refer to Figure 7.53.)

12 Connect the battery ground cable.

Multiple Fuse Links

1 Disconnect the battery ground cable

2 Cut off the burned-out fuse as shown in Figure 7.54.

3 Carefully remove about 1 inch (25 mm) of the insulation from the feed wire, as shown in Figure 7.55.

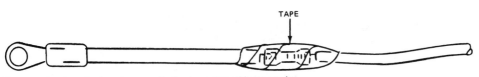

Figure 7.51 A typical replacement of a fuse link with an eyelet terminal (courtesy of Ford Motor Company).

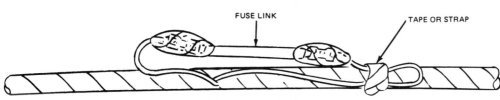

Figure 7.52 A typical repair for an in-line fuse link (courtesy of Ford Motor Company).

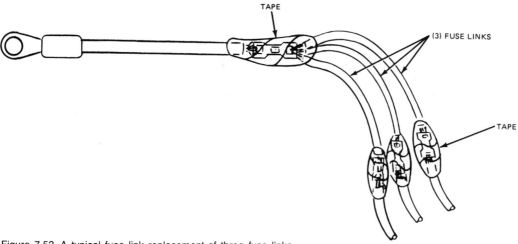

Figure 7.53 A typical fuse link replacement of three fuse links attached to a single feed wire (courtesy of Ford Motor Company).

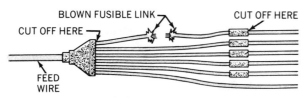

Figure 7.54 A blown or burned out fuse link in a multiple fuse link can be cut out of the circuit (courtesy of Chrysler Corporation).

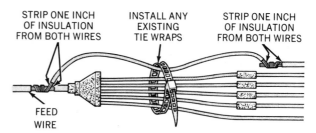

Figure 7.55 A replacement fuse link should be installed as shown (courtesy of Chrysler Corporation).

4 Remove about 1 inch (25 mm) of insulation from both ends of the replacement fuse link wire.

5 Wrap the ends of the fuse link wire around the feed wire. (Refer to Figure 7.55.)

6 Using rosin core solder, solder the connections.

7 Insulate the connections with several wraps of vinyl tape. (Refer to Figures 7.52 and 7.53.)

8 Connect the battery ground cable.

WIRING HARNESSES Your diagnosis of a lighting system problem occasionally will point to a defect in a wiring harness. Vibration may have caused a sharp edge of body sheet metal to wear through the insulation of certain wires and cause a short or a ground condition. Spilled electrolyte from an improperly maintained battery may have corroded wires and connectors. Collision damage may have cut through some wires and caused an open. Or a short or ground may have caused the insulation to burn on one or more wires.

In most instances, you can cut open the harness wrapping and repair the damaged wires by splicing in short lengths of wire of the same gauge. When the exact location of the problem cannot be found, or when access to the damaged area is restricted, you usually can bypass the defective wire or wires with new wires spliced in where the ends of the harness are accessible. When the harness is badly damaged or burned, you may have to replace the harness.

On most cars, the wiring for the front lights is contained in a harness that plugs into a *bulkhead disconnect,* sometimes called a *firewall connector,* as shown in Figures 7.56 and 7.57. On some cars, the wiring from the rear lights is also provided with a connector that allows the use of a separate harness in the trunk or behind the rear bumper, as shown in Figure 7.58. The car maker's manual will provide the identification of all of the terminals in the connectors, and the color codes of the wires in the harnesses. (See Figures 7.59 and 7.60.) Wiring location drawings similar to those shown in Figures 7.61 and 7.62 will show you how the harnesses are routed and secured.

Job 7f

REPLACE A FUSE LINK

SATISFACTORY PERFORMANCE
A satisfactory performance on this job requires that you do the following:

1 Replace the burned-out fuse link on the assigned car.
2 Following the steps in the "Performance Outline" and the procedure and specifications of the car manufacturer, replace the fuse link within 60 minutes.
3 Fill in the blanks under "Information."

PERFORMANCE OUTLINE
1 Locate the burned-out fuse link.
2 Remove the burned-out fuse link from the circuit.
3 Determine the gauge and length of the replacement fuse link.
4 Install the replacement fuse link.

INFORMATION
Vehicle identification _____

Reference used _____ Page(s) _____

Fuse link specifications: _____, Gauge _____ Length _____ Color

Checking Wiring Harnesses While some makes and models of cars may develop similar problems because of the design or routing of a particular harness, most harness problems are unique. As no procedural steps can be listed, the diagnostic procedure that you follow must be based on your understanding of electrical circuitry.

The first check of a wiring harness should be a visual inspection. Many problems blamed on a harness are caused by loose or disconnected plugs and connectors. At times, the terminals within the connectors will be found to be corroded by road splash or spilled electrolyte. Poorly performed previous repairs also are the cause of problems when connections and splices were left unsoldered or were soldered with acid core solder.

Physical damage to the harness can be suspected when the car has body damage in the harness area or when recent body repairs have been made. In many instances, a loose, cut, or frayed harness wrapping provides an indication of damaged wires. The installation of aftermarket equipment and accessories occasionally results in a harness being

pierced by a drill bit or a sheet metal screw that penetrated a panel or support under the harness.

Testing Wiring Harnesses In most instances, a test light will serve for most harness testing. If current enters a harness through a certain color-coded wire, the test light should indicate current at the other end of that wire. If a harness is unplugged from its power source, an ohmmeter or a continuity tester can be used to test the continuity of the suspected wire or wires in the harness.

Repairing Wiring Harnesses As most wiring harness problems are unique, no repair procedure can be stated. If either a short or an open is indicated in a wire in a harness and the exact location of the fault cannot be found, the wrapping on the ends of the harness can be stripped back. The ends of the defective wire then can be cut off, leaving a few inches at the connectors. Those ends now can be joined by a length of new wire. Solderless connectors can be used, but a soldered connection is preferable. The connections then can be insulated with several wraps of vinyl tape, using additional

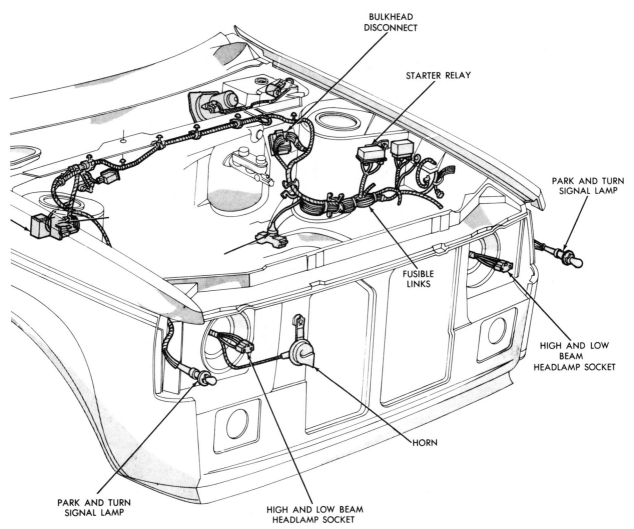

Figure 7.56 Typical underhood wiring. Notice that the harnesses plug into a bulkhead connector at the firewall (courtesy of Chrysler Corporation).

tape to replace the harness wrapping that was removed. The new wire then should be taped to the harness.

Replacing Wiring Harnesses   If a harness must be replaced, the new harness should be installed as the old harness is being removed. It usually is easier to start by unplugging the old harness from its feed connector. The new harness can then be plugged in. Before connecting any connectors, the terminals should be inspected for dirt and corrosion and cleaned if necessary. As the old harness is removed from its clamps, brackets, and straps, the new harness is installed. In this manner, the original installation and routing can be duplicated. If the old harness was damaged by sharp metal edges or by contact with exhaust system parts, the new harness should be routed so that the damage will not be repeated.

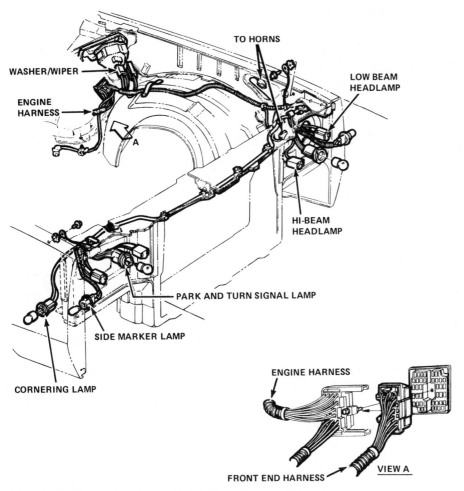

Figure 7.57 A typical wiring harness installation for the front lighting system. View A shows the connections at the firewall connector (courtesy of Pontiac Motor Division, General Motors Corporation).

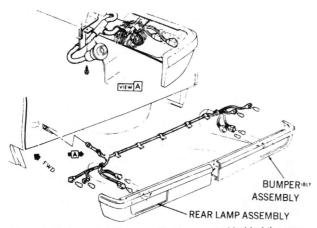

Figure 7.58 A rear light harness that is mounted behind the rear bumper. Notice the connector at view A (courtesy of Chevrolet Motor Division).

Job 7g

REPAIR A WIRING HARNESS

SATISFACTORY PERFORMANCE
A satisfactory performance on this job requires that you do the following:

1 Locate and repair a defect or defects in the assigned wiring harness.
2 Follow the steps in the "Performance Outline."
3 Fill in the blanks under "Information."

PERFORMANCE OUTLINE
1 Locate the defect(s) in the harness.
2 Repair the defect(s).
3 Insulate the connections.
4 Repair the harness wrapping.

INFORMATION
Vehicle identification _____

Harness identification _____

Reference used _____ Page(s) _____

Defect(s) found _____

_____

Repairs performed _____

_____

CAV	COLOR	DESCRIPTION
1	DB/RD*	SPEED CONTROL (OVERLAY) (PIA-LCV JET)
2	BR/RD*	SPEED CONTROL (OVERLAY) (PIA-LCV JET)
3	WT/RD*	SPEED CONTROL (OVERLAY) (PIA-LCV JET)
4	PK*	IGNITION SWITCH
5	TN*	RIGHT CORNERING LAMP (LCV JET)
6	LG	LEFT CORNERING LAMP (LCV JET)
7	GY/RD*	BRAKE WARNING LAMP (PD) (GY* LCV JET)
8	DG/RD*	HORN
9	YL	IGNITION START
10	DB	IGNITION RUN
11	VT / VT/BK*	TEMPERATURE (W/O ELECTRONIC CLUSTER) (2 WIRES-W/ELECT CLUSTER)
12	VT / GY	TEMPERATURE (W/O VOICE ALERT LCV JET) & OIL PRESSURE (2 WIRES-W/ELEC CLUSTER)
13	LG	LEFT TURN SIGNAL
14	TN	RIGHT TURN SIGNAL
15	WT/OR*	SPEED SENSOR (W/AUD MESS CTR LCV JET)
16	LB*	FUEL FLOW SENSOR (W/ELECT CLUSTER)
17	BR/YL*	LIFTGAGE (OVERLAY-PD)
18		
19	BR	WINDSHIELD WASHER
20	BK/TN*	WINDSHIELD WASHER FLUID LAMP (LCV JET)
21	WT	BACK-UP LAMP SW. TURBO BOOST LAMP (LCV JET)
22	VT/BK*	BACK-UP LAMP
23	BR*	WINDSHIELD WIPER LOW SPEED FEED
24	RD	WINDSHIELD WIPER
25	BK/YL*	PARKING LAMPS
26		
27	DG/YL*	WINDSHIELD WIPER MOTOR PARK RETURN & INTERMITTENT WIPE SIGNAL

CAV	COLOR	DESCRIPTION
28	DB	WINDSHIELD WIPER SWITCH & PARK SWITCH FEED
29	VT*	HEADLAMP LOW BEAM
30	RD	HEADLAMP HIGH BEAM (1 WIRE-W/ELECTRONIC CLUSTER)
31	PK	HAZARD FLASHER
32	DB/YL*	A/C CLUTCH
33		
34	GR/RD*	TACH W/VOICE ALERT (W/ELECTRONIC CLUSTER)
35	BK*	BATTERY FEED
36	BK/RD*	HEATED REAR WINDOW
37	RD	IGNITION SWITCH FEED
38		
39		
40	GY/YL*	OIL PRESSURE SENDING UNIT 2 WIRES (W/ELECTRONIC CLUSTER)

WIRE COLOR CODES		OR	ORANGE
BK	BLACK	PK	PINK
BR	BROWN	RD	RED
DB	DARK BLUE	TN	TAN
DG	DARK GREEN	VT	VIOLET
GY	GRAY	WT	WHITE
LB	LIGHT BLUE	YL	YELLOW
LG	LIGHT GREEN	*	WITH TRACER

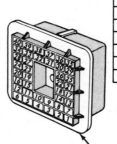

BULKHEAD DISCONNECT

Figure 7.59 A typical firewall or bulkhead disconnect. Note that the terminals are identified by cavity numbers, wire color codes, and circuits (courtesy of Chrysler Corporation).

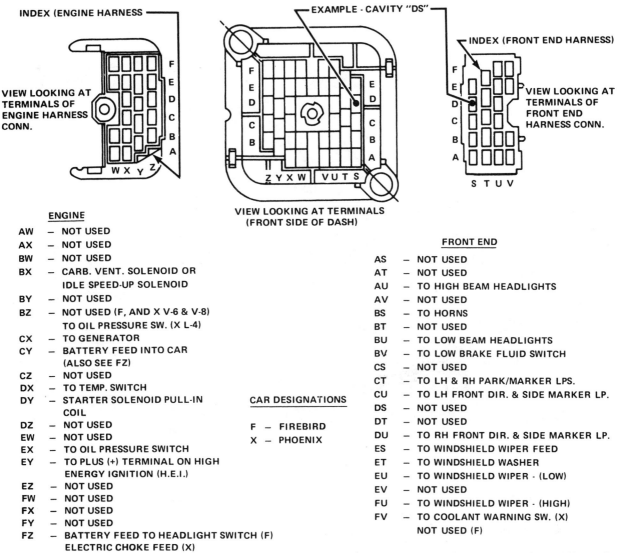

INDEX (ENGINE HARNESS

VIEW LOOKING AT TERMINALS OF ENGINE HARNESS CONN.

F E D C B A

W X Y Z

EXAMPLE - CAVITY "DS"

F E D C B

E D C B A

Z Y X W V U T S

VIEW LOOKING AT TERMINALS (FRONT SIDE OF DASH)

INDEX (FRONT END HARNESS)

VIEW LOOKING AT TERMINALS OF FRONT END HARNESS CONN.

F E D C B A

S T U V

ENGINE

AW	–	NOT USED
AX	–	NOT USED
BW	–	NOT USED
BX	–	CARB. VENT. SOLENOID OR IDLE SPEED-UP SOLENOID
BY	–	NOT USED
BZ	–	NOT USED (F, AND X V-6 & V-8) TO OIL PRESSURE SW. (X L-4)
CX	–	TO GENERATOR
CY	–	BATTERY FEED INTO CAR (ALSO SEE FZ)
CZ	–	NOT USED
DX	–	TO TEMP. SWITCH
DY	–	STARTER SOLENOID PULL-IN COIL
DZ	–	NOT USED
EW	–	NOT USED
EX	–	TO OIL PRESSURE SWITCH
EY	–	TO PLUS (+) TERMINAL ON HIGH ENERGY IGNITION (H.E.I.)
EZ	–	NOT USED
FW	–	NOT USED
FX	–	NOT USED
FY	–	NOT USED
FZ	–	BATTERY FEED TO HEADLIGHT SWITCH (F) ELECTRIC CHOKE FEED (X)

CAR DESIGNATIONS

F – FIREBIRD
X – PHOENIX

FRONT END

AS	–	NOT USED
AT	–	NOT USED
AU	–	TO HIGH BEAM HEADLIGHTS
AV	–	NOT USED
BS	–	TO HORNS
BT	–	NOT USED
BU	–	TO LOW BEAM HEADLIGHTS
BV	–	TO LOW BRAKE FLUID SWITCH
CS	–	NOT USED
CT	–	TO LH & RH PARK/MARKER LPS.
CU	–	TO LH FRONT DIR. & SIDE MARKER LP.
DS	–	NOT USED
DT	–	NOT USED
DU	–	TO RH FRONT DIR. & SIDE MARKER LP.
ES	–	TO WINDSHIELD WIPER FEED
ET	–	TO WINDSHIELD WASHER
EU	–	TO WINDSHIELD WIPER - (LOW)
EV	–	NOT USED
FU	–	TO WINDSHIELD WIPER - (HIGH)
FV	–	TO COOLANT WARNING SW. (X) NOT USED (F)

Figure 7.60 Bulkhead terminal identification as provided by one car maker (courtesy of Pontiac Motor Division, General Motors Corporation).

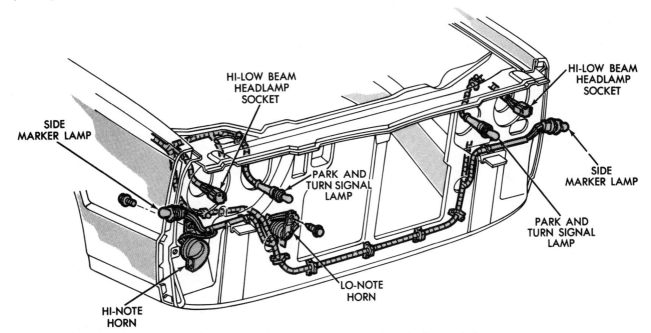

Figure 7.61 Typical harness routing for front light and horn wiring (courtesy of Chrysler Corporation).

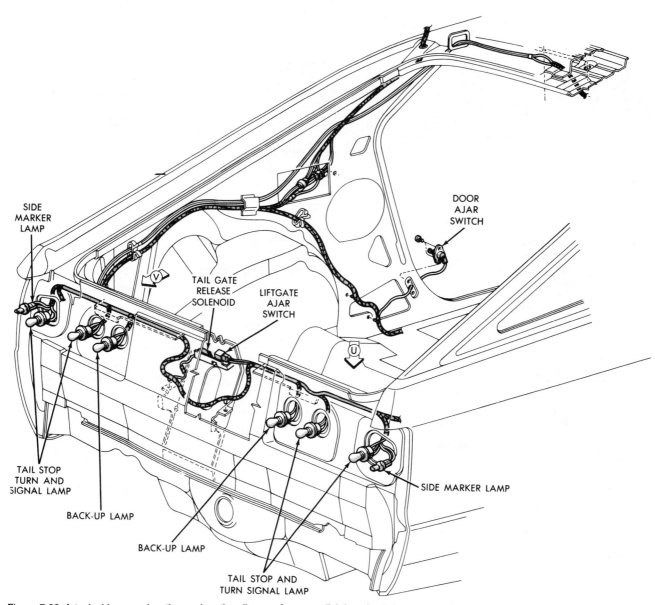

SIDE
MARKER
LAMP

DOOR
AJAR
SWITCH

TAIL GATE
RELEASE
SOLENOID

LIFTGATE
AJAR
SWITCH

TAIL STOP
TURN AND
SIGNAL LAMP

BACK-UP LAMP

BACK-UP LAMP

SIDE MARKER LAMP

TAIL STOP AND
TURN SIGNAL LAMP

Figure 7.62 A typical harness location and routing diagram for a rear lighting circuit (courtesy of Chrysler Corporation).

Job 7h

REPLACE A WIRING HARNESS

SATISFACTORY PERFORMANCE
A satisfactory performance on this job requires that you do the following:

1 Replace the assigned wiring harness.
2 Following the steps in the "Performance Outline" and the procedure of the manufacturer, complete the job within 200 percent of the manufacturer's suggested time.
3 Fill in the blanks under "Information."

PERFORMANCE OUTLINE
1 Disconnect the battery ground cable.
2 Unplug and partially remove the defective harness.
3 Connect the replacement harness and route it to follow the location of the defective harness.
4 Continue to install the replacement harness as the defective harness is removed.
5 Secure the harness.
6 Test the harness.

INFORMATION
Vehicle identification _____

Harness identification _____

Reference used _____ Page(s) _____

SUMMARY

By completing this chapter and performing the jobs presented, you have further developed your diagnostic and repair skills. You learned about various switches in the lighting system and how they can be tested and replaced. You now know about fuse links and their function. You can determine if a fuse link is burned and replace it with the correct fuse link wire. You also can inspect and test wiring harnesses, repair them, and replace them. You have advanced further toward your goal.

SELF-TEST

Each incomplete statement or question in this test is followed by four suggested completions or answers. In each case select the *one* that best completes the sentence or answers the question.

1 The operation of a switch can be tested by connecting an ohmmeter or a continuity tester
  A. across the switch terminals
  B. between the switch *B* terminal and ground
  C. between the wiring harness *B* terminal and ground
  D. across the terminals in the wiring harness plug or connector

2 A switch can be bypassed in a circuit by disconnecting the switch and using a jumper wire to bridge between the
  A. terminals on the switch
  B. switch terminals and ground
  C. terminals in the wiring plug or connector
  D. wire plug or connector terminals and ground

3 Most dimmer switches are of the
  A. single pole single throw (SPST) type
  B. single pole double throw (SPDT) type
  C. double pole single throw (DPST) type
  D. double pole double throw (DPDT) type

4 On most cars, the stoplight switch is operated by
  A. movement of the brake pedal
  B. an internal transmission part

C. master cylinder piston movement

D. hydraulic pressure in the brake system

5 A rheostat is often used to
A. dim the headlights
B. regulate switch temperature
C. adjust the speed of a flasher
D. dim the instrument panel lights

6 Two mechanics are discussing the replacement of a turn signal switch.
Mechanic A says that the steering wheel must be removed on some cars.
Mechanic B says that the lock plate must be removed on some cars.
Who is right?
A. A only
B. B only
C. Both A and B
D. Neither A nor B

7 Two mechanics are discussing fuse links.
Mechanic A says that a fuse link will burn out if excessive voltage flows in the circuit it protects.
Mechanic B says that a fuse link is a wire several gauges larger than the wire used in the circuit it protects.
Who is right?

A. A only
B. B only
C. Both A and B
D. Neither A nor B

8 Two mechanics are discussing fuse links.
Mechanic A says that fuse link wires are covered with an insulation that will not burn.
Mechanic B says that fuse links should be soldered to their feed wires.
Who is right?
A. A only
B. B only
C. Both A and B
D. Neither A nor B

9 Which of the following will NOT cause a fuse link to burn out?
A. A short circuit in the wiring.
B. An open circuit in the wiring.
C. An incorrectly connected battery charger.
D. An incorrectly connected booster battery.

10 Which of the following would NOT be used for testing a wiring harness?
A. An ohmmeter.
B. A test lamp.
C. A jumper wire.
D. A continuity tester.

Chapter 8 The Starting System— Basic Services

The starting system converts electrical energy to mechanical energy and uses that energy to crank the engine. If the starting system cannot crank the engine fast enough, or for a sufficient length of time, the engine will not start. Proper maintenance of the starting system will minimize system failure, and correct diagnostic procedures will enable you to locate the cause of any problems that may occur.

In this chapter, you will learn how the starting system operates. You will learn about the circuits used and their components. Based on this knowledge, you will perform diagnostic tests and replace defective parts. Your specific objectives are to perform the following jobs:

A
Identify the parts in a basic starting system
B
Identify the function of the parts in a basic starting system
C
Test starter current draw
D
Test the voltage drop in a starting system
E
Replace a starter relay
F
Raise a car and support it by the frame
G
Raise a car and support it by the suspension system
H
Raise and support a car with a frame contact lift
I
Raise and support a car with a suspension contact lift
J
Replace a starter motor

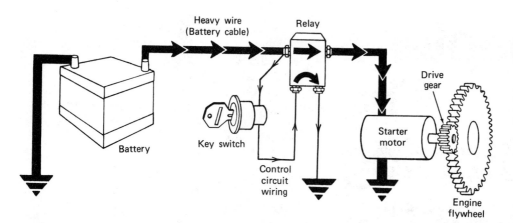

Heavy wire
(Battery cable)

Relay

Drive gear

Battery

Key switch

Control circuit wiring

Starter motor

Engine flywheel

Figure 8.1 A diagram of a typical starting system. Notice that the system contains two circuits. The motor circuit has its parts connected by heavy wires or cables. The control circuit uses small gauge wire (courtesy of Chevrolet Motor Division).

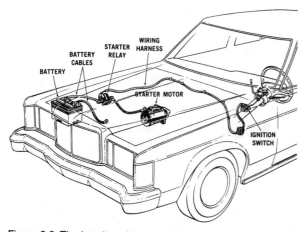

BATTERY

BATTERY CABLES

STARTER RELAY

WIRING HARNESS

STARTER MOTOR

IGNITION SWITCH

Figure 8.2 The location of the starting system components in a typical automobile. Notice that the parts in the motor circuit are close together in the engine compartment. The parts of the control circuit are widely separated and are connected by wires in a wiring harness (courtesy of Ford Motor Company).

THE STARTING SYSTEM

In most cars, the starting system consists of two circuits. The *motor circuit* provides the current that operates the starter motor. Through a gear system, the starter motor cranks the engine. The *control circuit* enables the driver to turn the starter motor on and off by means of a small switch. The switch usually is built into the ignition switch. Figure 8.1 shows the two circuits in a simple starting system.

As you learned in previous chapters, a large amount of current is needed to crank an engine. A simple switch could be used as a control in the starting system. However, the switch contacts would have to be quite large to handle the large amounts of current without being damaged. In addition, long, heavy cables would have to be used to conduct the current to and from the switch. Figure 8.2 shows the location of both circuits in a typical car and

illustrates how the use of two circuits allows the use of short battery cables.

The Motor Circuit   The motor circuit consists of the battery, the relay, and the starter motor. Those parts are connected by cables that can handle large amounts of current. (Refer to Figure 8.1.)

The Battery   The battery is the power source in the circuit. As you know from your work with batteries, a battery converts chemical energy to electrical energy. In the starting system, the starter motor converts electrical energy to mechanical energy.

While a fully charged battery can deliver large amounts of energy in the form of current, that current can be delivered for only a short amount of time. If the battery is partially discharged, it may not deliver sufficient current to crank the engine fast enough to start. Or it may not be able to deliver sufficient current for a long enough period of time. Since the battery provides the energy that cranks the engine, its state of charge is very important.

The Relay   The relay is a magnetically operated heavy duty switch. Many types of relays are used, but they all serve the same function. They allow a small amount of current in the control circuit to control a large amount of current in the motor circuit. When the driver turns the ignition switch to the Start position, a small amount of current energizes the magnet in the relay. The magnet closes a switch with large contacts, and current flows to the starter motor.

The Starter Motor   The starter motor is a heavy duty electric motor. Figure 8.3 shows one type of starter motor. The starter motor turns a small gear, called the *drive pinion,* which turns the engine fly-

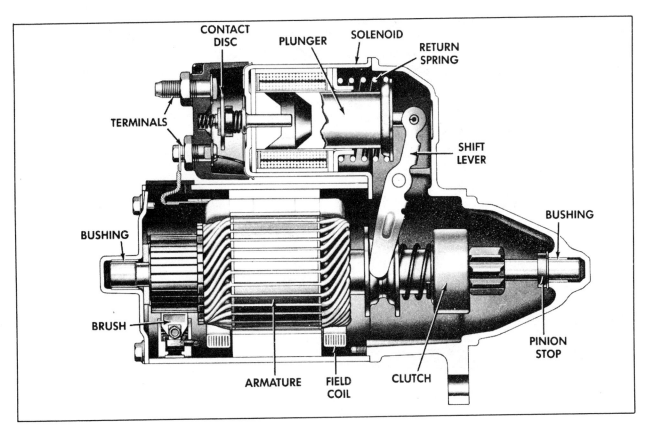

Figure 8.3 A cross section view of a typical starter motor. Notice that on this starter motor the relay is in the form of a solenoid which also acts to shift the drive pinion into mesh with the starter ring gear (courtesy of Pontiac Motor Division, General Motors Corporation).

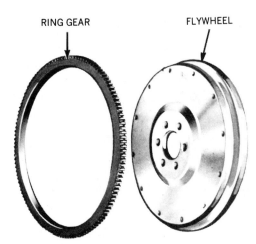

Figure 8.4 A ring gear and flywheel. The ring gear is mounted on the flywheel and is turned by the drive pinion (courtesy of Ford Motor Company).

wheel. (Refer to Figure 8.1.) The drive pinion is part of a drive system that shifts the pinion in and out of mesh with the flywheel *ring gear*. As shown in Figure 8.4, the ring gear is a toothed ring that encircles the flywheel. When the motor circuit is energized, the drive pinion is moved into mesh with the ring gear. When the starter switch is released, the drive pinion is moved back out of mesh by a spring.

The Cables   All of the parts in the motor circuit must be connected by conductors capable of handling large amounts of current. Small gauge wire offers too much resistance to the flow of current required. If small gauge wires were used, the starter motor would not be able to crank the engine, and the wires would overheat and possibly melt.

The Control Circuit   The control circuit consists of the battery, the relay, and the starter switch. On some cars, a *neutral safety switch* is used. This switch prevents the control system from working if the transmission is in gear. Since the control circuit requires very little current, the components are connected by small gauge wires. (Refer to Figure 8.1.)

The Battery   The battery is shared by both circuits. In the control circuit, it provides the energy to operate the relay.

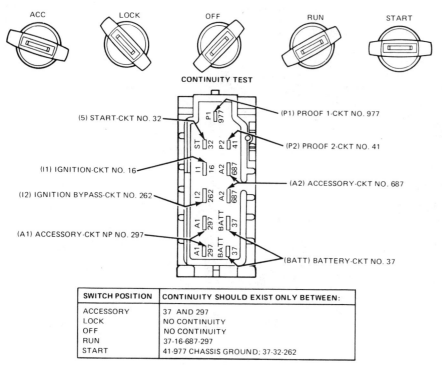

Figure 8.5 In most cars, the starter switch contacts are built into the ignition switch. Shown are typical switch positions and test points (courtesy of Ford Motor Company).

SWITCH POSITION	CONTINUITY SHOULD EXIST ONLY BETWEEN:
ACCESSORY	37 AND 297
LOCK	NO CONTINUITY
OFF	NO CONTINUITY
RUN	37-16-687-297
START	41-977 CHASSIS GROUND; 37-32-262

NOTE: CIRCUIT PAIRS 37, 687 AND 297
ARE CONNECTED TOGETHER INTERNALLY

The Relay The relay is also shared by both circuits. The control circuit delivers the current that operates the relay. Most relays used in starting systems are of the solenoid type. Some are mounted on an inner fender panel between the battery and the starter motor. (Refer to Figure 8.2.) A relay of that type acts only as a switch to close the motor circuit. Others are mounted on the starter motor. (Refer to Figure 8.3.) In addition to acting as a switch, they serve a mechanical function. The plunger of the solenoid moves a shift lever, which engages the drive pinion with the flywheel ring gear.

The Starter Switch The starter switch opens and closes the control circuit. In most cars, it is a set of contacts built into the ignition switch, as shown in Figure 8.5. When the driver turns the key to the Start position, the starter switch contacts are brought together. A spring returns the switch to the Ignition or On position and opens the starter control circuit when the key is released.

The Neutral Safety Switch To prevent the accidental starting of an engine when a car is in gear, most control circuits incorporate a neutral safety switch. The switch is placed in series with the starter switch and is in the closed position only when the transmission selector lever is in the Neutral or Park position. Some cars with manual or standard transmissions have a neutral safety switch operated by the clutch linkage. When a switch of that type is used, it is closed only when the clutch pedal is depressed.

On some cars, a neutral safety switch is placed between the starter switch and the relay, as shown in Figure 8.6. On other cars, a switch is used between the relay and ground, as shown in Figure 8.7.

STARTING SYSTEM COMPONENTS All starting systems in late-model cars operate in the same manner. However, those systems use different parts, even when they are in cars built by the same manufacturer. To service starting systems properly, you must be aware of those differences.

Relays Relays are magnetic switches. They use magnetism to close switch contacts. Two types of relays are commonly used. One type, shown in Figure 8.8, uses an electromagnet to move an armature and cause two switch contacts to come together. A relay of that type is used in the control circuit of some cars built by the Chrysler Corporation and is shown in Figure 8.9.

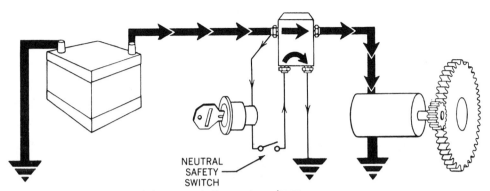

Figure 8.6 The control circuit in some starting systems incorporates a neutral safety switch between the starter switch and the relay (courtesy of Chevrolet Motor Division).

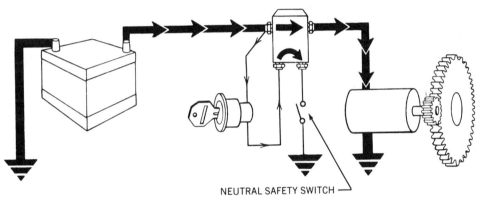

Figure 8.7 On some cars, a neutral safety switch is wired between the relay and ground (courtesy of Chevrolet Motor Division).

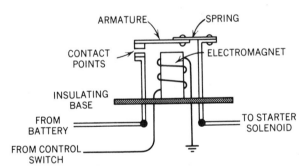

Figure 8.8 The construction of a typical relay. When current flows from the control switch, the electromagnet pulls the armature down, closing the contact points. Current then flows from the battery to the starter.

Figure 8.9 A typical starter relay of the type used on certain cars built by Chrysler Corporation.

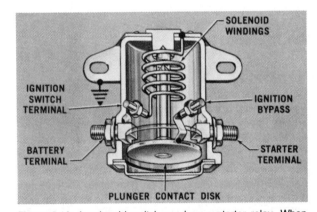

Figure 8.10 A solenoid switch used as a starter relay. When current passes through the solenoid winding, the plunger and the disc are pulled upward. The disc then contacts the two large terminals (courtesy of Ford Motor Company).

The other type of relay is commonly called a *solenoid*. A solenoid is an electromagnet with a movable core. Figure 8.10 shows a relay of that type. A large metal disc is attached to the core. When magnetism pulls the core, the disc contacts two large

Job 8a

IDENTIFY THE PARTS IN A BASIC STARTING SYSTEM

SATISFACTORY PERFORMANCE
A satisfactory performance on this job requires that you do the following:

1 Identify the numbered parts in the drawing below by placing the number of each part in front of the correct part name.
2 Complete the job by identifying all the parts correctly within 10 minutes.

PERFORMANCE SITUATION

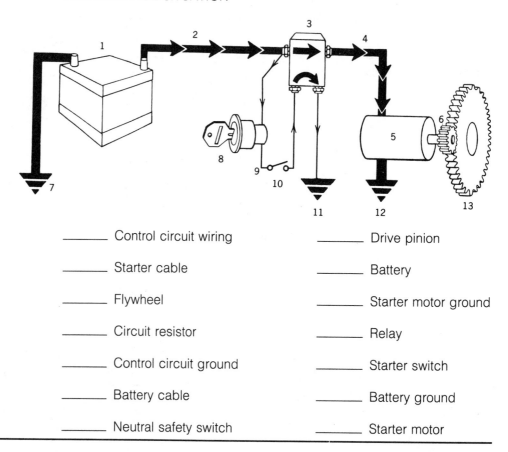

_____ Control circuit wiring

_____ Starter cable

_____ Flywheel

_____ Circuit resistor

_____ Control circuit ground

_____ Battery cable

_____ Neutral safety switch

_____ Drive pinion

_____ Battery

_____ Starter motor ground

_____ Relay

_____ Starter switch

_____ Battery ground

_____ Starter motor

terminals. A solenoid is used as a relay on most cars built by Ford Motor Company and by American Motors Corporation. It usually is mounted on the inner fender panel near the battery, as shown in Figure 8.11.

A solenoid can exert a very strong pull. On many cars, a large solenoid is mounted on the starter motor. (Refer to Figure 8.3.) The solenoid is used to shift the drive pinion into mesh with the flywheel ring gear. In some applications, the solenoid on the starter motor performs an additional task. It also

STARTER MOTOR SOLENOID

Figure 8.11 The starter motor solenoid usually is mounted on an inner fender panel or on the firewall (courtesy of American Motors).

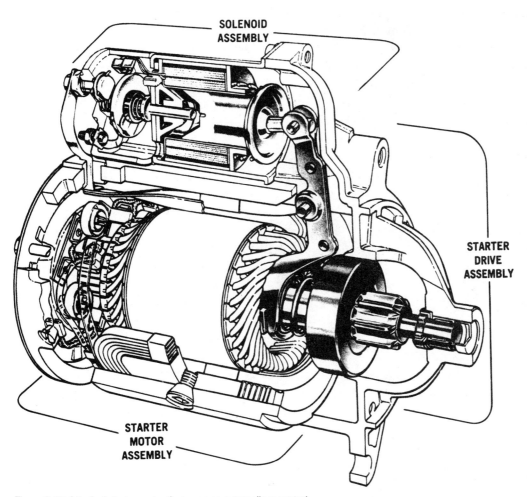

SOLENOID
ASSEMBLY

STARTER
DRIVE
ASSEMBLY

STARTER
MOTOR
ASSEMBLY

Figure 8.12 A typical starter motor that uses an externally mounted solenoid to shift the drive pinion (courtesy of Ford Motor Company).

acts as a relay, eliminating the need for a separate part. Most cars built by General Motors Corporation use the solenoid to perform both functions.

Starter Motors Many different types of starter motors are used. Some car manufacturers design and build the starter motors for their cars. Others purchase starter motors from independent makers of electrical equipment or from other car manufacturers. Because of this practice, you will find different starter motors used on apparently similar cars built by the same car maker. Regardless of the make of the starter, most starter motors now in use can be classified in three different types.

Starter Motors with the Drive Pinion Shifted by a Solenoid Figure 8.12 shows a typical starter motor of that type. An externally mounted solenoid operates a shift fork to slide out the drive pinion, as shown in Figure 8.13. Starter motors of this type are very common and will be found on various cars built by almost all manufacturers.

Gear Reduction Starter Motors with the Drive Pinion Shifted by a Solenoid As shown in Figure 8.14, starter motors of this type incorporate a gear reduction system to increase torque. Gear reduction starter motors of various designs are used on some cars built by the Chrysler Corporation and by General Motors, and on a few imported cars.

Starter Motors with the Drive Pinion Shifted by a Movable Field Pole Shoe Starter motors of this type, shown in Figure 8.15, do not use a separate solenoid to move the drive pinion. One of the field windings in the starter motor acts as a solenoid winding. A movable pole shoe takes the place of a solenoid plunger. This type of starter motor is used

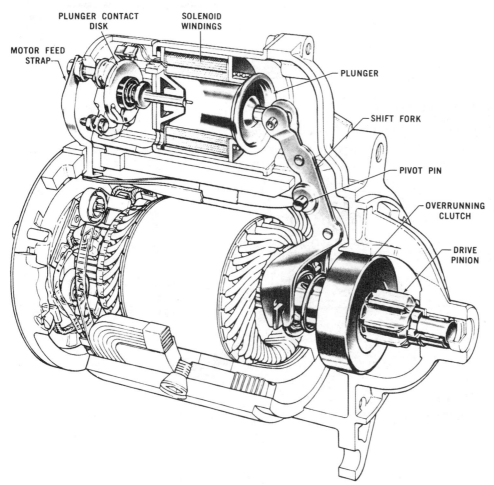

MOTOR FEED STRAP

PLUNGER CONTACT DISK

SOLENOID WINDINGS

PLUNGER

SHIFT FORK

PIVOT PIN

OVERRUNNING CLUTCH

DRIVE PINION

Figure 8.13 A starter motor that uses an externally mounted solenoid to shift the drive pinion. Notice that in this view the shift fork has moved the drive pinion outward (courtesy Ford Motor Company).

on some cars built by Ford Motor Company and on some cars built by American Motors Corporation.

STARTER DRIVE CLUTCHES If the starter motor is to crank the engine, the drive pinion must be held in mesh with the flywheel ring gear. On most cars, the solenoid serves that function as long as the ignition switch is held in the Start position. That means that when the engine starts, the flywheel turns the drive pinion until the driver releases the switch.

To enable a small starter motor to crank a large engine, the gear ratio between the starter motor and the engine must allow the starter motor to spin fast enough to develop sufficient torque. The average drive pinion has about ten teeth. Depending on the diameter of the flywheel, the number of teeth on a ring gear varies from about 150 to 200. This provides a gear ratio of from 15:1 to 20:1. When

the engine starts, the speed of the flywheel increases above its cranking speed and the flywheel now turns the drive pinion. If the engine runs at 1,000 rpm, the drive pinion is turned from 15,000 rpm to 20,000 rpm. If the starter motor armature is rotated at these speeds, centrifugal force will cause the armature windings to be torn loose, destroying the starter motor. Therefore, the armature must be instantly disconnected from the drive pinion when the engine starts. That function is performed by an *overrunning clutch.*

An overrunning clutch, shown in Figure 8.16, is a device that allows torque to be transmitted in only one direction. The clutch locks to allow the starter armature to turn the drive pinion. However, it releases when the pinion tries to turn the armature. Since it transmits torque in only one direction, an overrunning clutch is sometimes referred to as a *one-way clutch.*

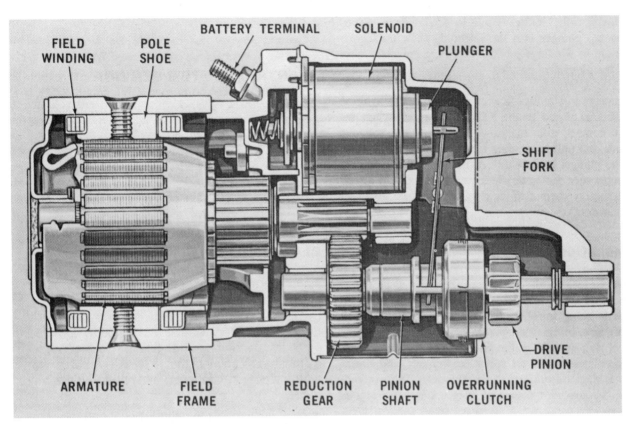

Figure 8.14 A typical gear reduction starter motor. Notice that the solenoid is enclosed (courtesy of Ford Motor Company).

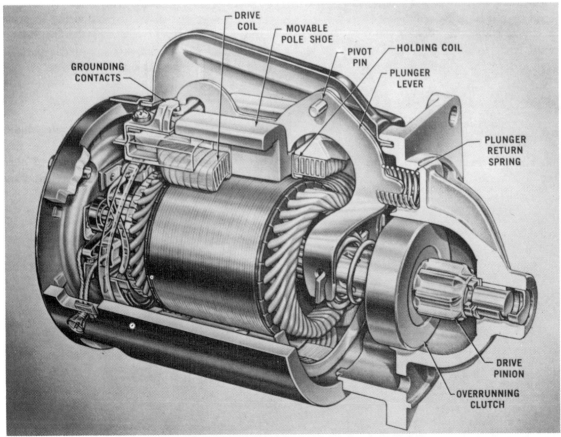

Figure 8.15 A starter motor where the drive pinion is shifted into mesh with the flywheel ring gear by means of a movable pole shoe. The field coil winding performs the function of a solenoid (courtesy of Ford Motor Company).

As shown in Figure 8.16, an overrunning clutch consists of a set of rollers housed in a ring whose inner surface is formed in the shape of ramps. Those ramps form wedge-shaped pockets for the rollers. Small springs push the rollers up the ramps and wedge them between the outer housing and an extension on the drive pinion. When the armature turns, the ramped housing turns and increases the wedging action. This locks the drive pinion to the housing and allows the armature to turn the drive pinion.

When the engine starts, the flywheel ring gear turns the drive pinion at a speed faster than the armature. This causes the extension of the drive pinion to move faster, or to *overrun,* the housing turned by the armature. The rollers are rolled down the ramps, releasing their grip on the housing and the drive pinion extension. This allows the drive pinion to rotate freely without turning the armature.

STARTING SYSTEM SERVICE

The starting system requires very little maintenance. That maintenance usually consists of (1) keeping the battery in a high state of charge and (2) keeping all of the electrical connections in the system clean and tight. When problems arise in the system, other services may be required. Those services include the replacement of parts. However, before you attempt to replace any part, you should determine if that part is at fault.

Basic Starting System Tests   The diagnosis of starting system problems is easy if you understand how the system operates and follow a logical test procedure. Many starting problems are caused by dirty or loose battery terminals or by a partially discharged battery. Cables that are badly corroded or otherwise damaged should be replaced. Remember to disconnect the negative (−) cable from the battery first before you attempt to remove any cable.

In most instances, the control circuit can easily be

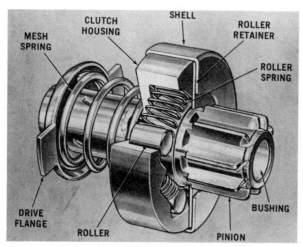

Figure 8.16 A typical overrunning clutch used in a starter drive system (courtesy of Ford Motor Company).

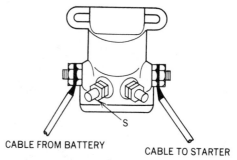

Figure 8.17 When connected between the S terminal and ground, a test lamp can be used to check the operation of the starter control circuit. If the lamp lights when the key is turned to the Start position, the starter switch and the neutral safety switch are functioning.

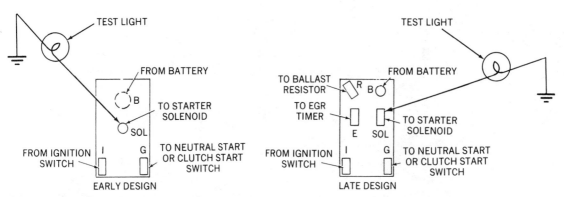

Figure 8.18 Connected as shown, a test lamp should light when the ignition switch is turned to the Start position (courtesy of Chrysler Corporation).

checked simply by attempting to start the engine. If the circuit is functioning, you usually can hear the relay or the solenoid operate. If the relay or solenoid does not appear to be functioning, a test light should be connected between the "S" terminal on the relay or solenoid and a good ground, as shown in Figures 8.17 and 8.18. The light will tell you if current is reaching the relay when the key is turned to the Start position.

If current is available and the relay or solenoid does not operate, the relay or solenoid is probably at fault. If current is not available, there is an open in the control circuit. The ignition switch or the neutral safety switch may require adjustment or replacement, or there may be a bad connection or a break in the circuit wiring. The test light may be used to check for current at the switches, or the switches may be bridged with a jumper wire for testing.

Testing Starter Current Draw A starter current draw test provides a quick check of the entire starting system. This test is performed with a battery-starter tester of the type shown in Figure 8.19. You used a tester of this type when you load tested batteries.

The following steps outline a procedure for performing a starter current draw test. You should consult the manual furnished with the tester that you have available for any special procedure that may be necessary with the instrument. The specification for starter current draw will vary with different cars and must be obtained from an appropriate manual:

1 Place a fender cover on the fender near the battery.

2 Test the specific gravity of the battery electrolyte. If the specific gravity is less than 1.200, the battery should be charged before a starter current draw test is attempted.

3 Check the battery terminals and cable clamps. Clean the terminals and clamps if necessary.

Note: **The battery connections must be clean and tight to obtain accurate test results.**

4 Turn the carbon pile control knob on the tester to the Off position.

5 If the tester is fitted with a function control switch, turn the switch to the Starter Test position.

6 If the tester is fitted with a volt selector switch, turn the switch so that a voltage exceeding battery voltage is selected.

7 Connect the heavy ammeter leads to the battery terminals. The positive (+) lead must be connected to the positive (+) terminal, the negative (−) lead to the negative (−) terminal.

8 Connect the voltmeter leads to the battery terminals. The positive (+) lead must be connected to the positive (+) terminal, the negative (−) lead to the negative (−) terminal.

Note: **The voltmeter leads should contact the battery terminals as shown in Figure 8.20. If the voltmeter leads are connected to the clamps on the ammeter leads, an inaccurate voltmeter reading may be obtained.**

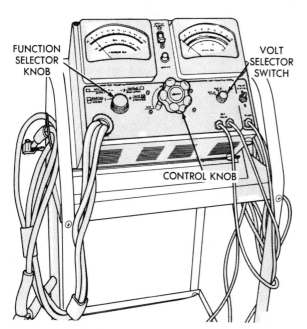

FUNCTION SELECTOR KNOB

VOLT SELECTOR SWITCH

CONTROL KNOB

Figure 8.19 A typical tester used to test batteries and starting systems (courtesy of Chrysler Corporation).

Figure 8.20 Connections for performing a starter current draw test. Notice that the voltmeter leads are not attached to the ammeter leads, but are in contact with the battery posts (courtesy of American Motors).

Job 8b

IDENTIFY THE FUNCTION OF THE PARTS IN A BASIC STARTING SYSTEM

SATISFACTORY PERFORMANCE
A satisfactory performance on this job requires that you do the following:

1 Identify the function of the listed starting system parts by placing the number of each part in front of the phrase that best describes its function.
2 Complete the job by correctly identifying all the parts within 15 minutes.

PERFORMANCE SITUATION

1 Battery
2 Control circuit ground
3 Control circuit wiring
4 Battery cable
5 Starter switch
6 Drive pinion
7 Neutral safety switch

8 Starter motor ground
9 Starter cable
10 Flywheel
11 Battery ground
12 Relay
13 Ring gear
14 Starter motor

_____ Provides a path for the return of current to the battery

_____ Provides a path for the return of control circuit current

_____ Turns the ring gear attached to the flywheel

_____ Conducts current to the relay

_____ Provides a bypass for excess starter current

_____ Conducts current to the starter motor

_____ Provides the control in the control circuit

_____ Encircles the flywheel

_____ Conducts current in the control circuit

_____ Turns the engine crankshaft

_____ Provides a path for the return of motor circuit current

_____ Provides the energy for the system

_____ Provides the control in the motor circuit

_____ Converts electrical energy to mechanical energy

_____ Opens the control circuit when the car is in gear

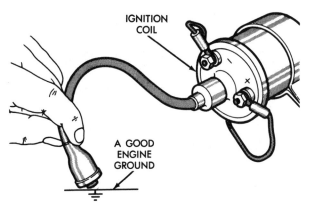

Figure 8.21 On most cars, the ignition system can be disabled by removing the coil wire from the center terminal of the distributor cap and grounding it (courtesy of Chrysler Corporation).

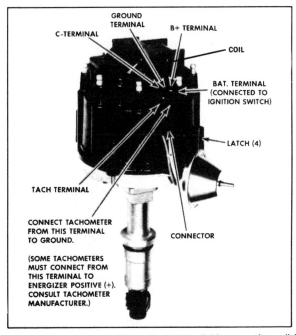

Figure 8.22 In an electronic distributor of this type, the coil is housed inside the distributor cap. Unplugging the connector will disable the system (courtesy of Pontiac Motor Division, General Motors Corporation).

Figure 8.23 A remote starter switch will enable you to crank an engine while working under the hood (courtesy of Snap-on Tools Corporation).

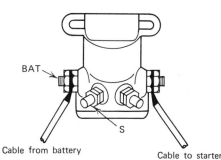

Figure 8.24 A starter solenoid of the type used by Ford Motor Company and by American Motors Corporation. Remote starter switch connections should be made at the BAT and S terminals.

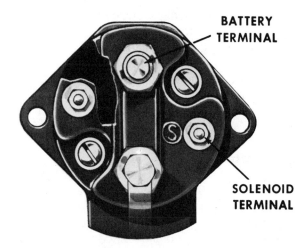

Figure 8.25 A front view of the starter solenoid used on most cars built by General Motors Corporation. When using a remote starter switch, the switch should be connected to the battery terminal and to the solenoid (S) terminal (courtesy of Chevrolet Motor Division).

9 Disable the ignition system by removing the center wire from the distributor cap and grounding it, as shown in Figure 8.21.

Note: In some ignition systems, the coil is located within the distributor cap and no center wire is used. Systems of that type may be disabled by disconnecting a connector at the distributor cap, as shown in Figure 8.22.

10 Apply the parking brake.

11 If the car has an automatic transmission, place the transmission selector lever in the Park position.

If the car has a standard or manual transmission, place the shift lever in the Neutral position.

12 Crank the engine for about 3 seconds while observing the voltmeter. Record the voltage indicated.

Note: A remote starter switch similar to the one shown in Figure 8.23 may be used. Figures 8.24, 8.25, and 8.26 show how a remote starter switch should be connected.

13 Turn the carbon pile control knob clockwise until the voltmeter indicates the voltage that you recorded in step 12. Read the current indicated by

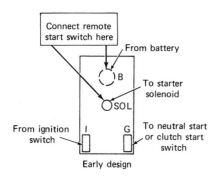

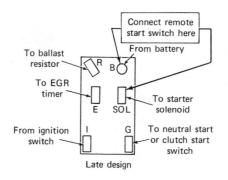

Figure 8.26 Relays of two different designs will be found on cars built by Chrysler Corporation. Remote starter switch connections should be made at the battery and solenoid terminals (courtesy of Chrysler Corporation).

CYLINDERS	AMPERES
4	120–160
6	150–180
8	160–210

Figure 8.27 Average starter current draw specifications. Current draw will vary with the temperature, oil viscosity, and condition of the engine.

the ammeter and immediately turn the control knob counterclockwise to the Off position.

14 Record the current indicated by the ammeter.

15 Compare the meter reading with the manufacturer's specification for starter current draw.

16 Disconnect and remove the tester.

17 Restore the ignition system to operating condition.

Interpretation of Test Results The specifications for starter current draw vary with different engines produced by different manufacturers. Figure 8.27 shows the specification range for cars with 12-V systems. Because of the wide range of specifications, the results of the test can be interpreted only when the specification for the particular car on which you are working is known.

Three test results are possible. The indicated starter current draw may be within specifications, above specifications, or below specifications.

Within Specifications If the ammeter reading falls within the manufacturer's specified range, the starting system is in good condition and should require no repairs.

Above Specifications If the indicated current draw exceeds the manufacturer's specifications, several problems may be indicated. In most instances, the starter motor is at fault and will require repair or replacement. High current draw also may be caused by the use of high viscosity oil, high engine tem-

perature, internal engine damage, or very cold temperatures.

Below Specifications If the starter current draw falls below the specifications of the manufacturer, it usually is an indication of high resistance in the motor circuit. All the connections, including the ground connections, should be checked, cleaned, and tightened.

Testing Voltage Drop All conductors have resistance, but their resistance is usually very low. When conductors are connected, those connections offer more resistance. If the connections are clean and tight, the additional resistance also will be slight. However, if the connections are dirty, loose, or corroded, the resistance may increase so that current flow is restricted.

The starter motor requires large amounts of current. If there is high resistance in the motor circuit, the starter motor will be unable to receive the current it requires. Any resistance in a circuit will cause a drop in voltage. Therefore, a voltage drop test can be used to detect resistance.

The steps that follow outline the procedure for a series of four voltage drop tests in a starting system. Those tests, made in the sequence given, will enable you to find the exact location of any high resistance. Since the voltmeter readings taken will be in tenths of 1 V, a voltmeter with a low scale is required for accuracy.

In performing the following tests, you will energize the starter motor to crank the engine. When working on a car with a catalytic converter, crank the engine for as short a time as possible. Extended cranking can deposit an excessive amount of fuel in the converter, causing it to overheat and pose a possible fire hazard.

Job 8c

TEST STARTER CURRENT DRAW

SATISFACTORY PERFORMANCE

A satisfactory performance on this job requires that you do the following:

1 Perform a starter current draw test on the car assigned.
2 Following the steps in the "Performance Outline" and the specifications of the manufacturer, complete the job within 15 minutes.
3 Fill in the blanks under "Information."

PERFORMANCE OUTLINE

1 Test the specific gravity of the battery electrolyte.
2 Check the battery connections and clean them, if necessary.
3 Connect the tester.
4 Disable the ignition system.
5 Read the voltage while the starter motor is cranking the engine.
6 Read the amperage while using the carbon pile to duplicate the voltage reading.
7 Compare the reading with the manufacturer's specifications.
8 Interpret the test results.

INFORMATION

Vehicle identification _____

Engine identification _____

Specification for starter current draw: _____ amperes

Starter current draw indicated by test: _____ amperes

Test interpretation and recommendations _____

_____

_____

Preparation

1 Place a fender cover on the fender near the battery.
2 Test the specific gravity of the electrolyte in the battery. If the specific gravity is less than 1.200, the battery should be charged.
3 Disable the ignition system by removing the center wire from the distributor cap and grounding it. (Refer to Figure 8.21.)

Note: In some ignition systems, the coil is located within the distributor cap and no center wire is used.

Systems of that type can be disabled by disconnecting a connector at the distributor cap. (Refer to Figure 8.22.)

4 Apply the parking brake.
5 If the car has an automatic transmission, place the transmission selector lever in the Park position. If the car has a standard or manual transmission, place the shift lever in the Neutral position.
6 Connect a remote starter switch to the relay or solenoid. (Refer to Figures 8.24, 8.25, and 8.26.)
7 Adjust a voltmeter to the low scale.

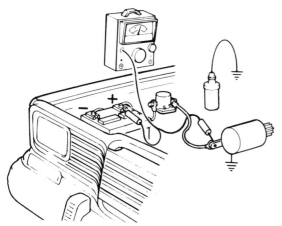

Figure 8.28 Meter connections for performing a voltage drop test of the "hot" side of the starting system motor circuit (test 1). Notice that the voltmeter leads are in contact with the battery terminal and the starter motor terminal (courtesy of American Motors).

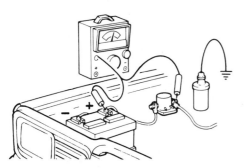

Figure 8.29 Meter connections for performing a voltage drop test of the "hot" side of the motor circuit between the battery and the starter side of the relay (test 2) (courtesy of American Motors).

Test 1   This test measures the voltage drop in the "hot" side of the motor circuit.

1 Connect the positive (+) voltmeter lead to the positive (+) battery post as shown in Figure 8.28.

Note: The voltmeter lead should contact the battery post, not the cable clamp.

2 Connect the negative (−) voltmeter lead to the terminal bolt on the starter motor. (Refer to Figure 8.28.)

Note: The voltmeter lead should contact the starter terminal bolt and not the terminal on the end of the cable.

3 Crank the engine and observe the voltmeter reading. Record the voltage indicated.
4 Disconnect the negative (−) voltmeter lead.

Test 2   This test measures the voltage drop between the battery and the starter side of the relay.

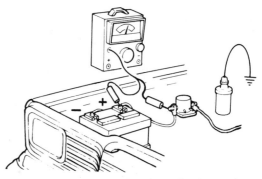

Figure 8.30 Meter connections for performing a voltage drop test of the "hot" side of the motor circuit between the battery and the battery side of the relay (test 3) (courtesy of American Motors).

1 Connect the negative (−) voltmeter lead to the terminal bolt on the starter side of the relay, as shown in Figure 8.29.
2 Crank the engine and observe the voltmeter reading. Record the voltage indicated.
3 Disconnect the negative (−) voltmeter lead.

Test 3   This test measures the voltage drop between the battery and the battery side of the relay.

1 Connect the negative (−) voltmeter lead to the terminal bolt on the battery side of the relay, as shown in Figure 8.30.
2 Crank the engine and observe the voltmeter reading. Record the voltage indicated.
3 Disconnect both voltmeter leads.

Test 4   This test measures the voltage drop in the ground side of the motor circuit.

1 Connect the positive (+) voltmeter lead to the starter motor housing as shown in Figure 8.31.
2 Connect the negative (−) voltmeter lead to the negative (−) battery post. (Refer to Figure 8.31.)

Note: The voltmeter lead must contact the battery post, not the cable clamp.

3 Crank the engine and observe the voltmeter reading. Record the voltage indicated.
4 Disconnect both voltmeter leads.
5 Restore the ignition system to operating condition.

Interpretation of Test Results

Test 1   The indicated voltage should not exceed 0.5 V.

A reading of 0.5 V or less indicates that the resistance in the "hot" side of the motor circuit is ac-

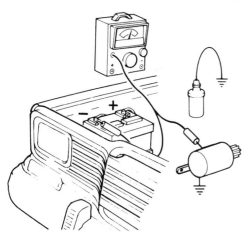

Figure 8.31 Meter connections for performing a voltage drop test of the ground side of the motor circuit (test 4) (courtesy of American Motors).

ceptable. The readings of Test 2 and Test 3 can be disregarded. Move to Test 4.

A reading of more than 0.5 V indicates excessive resistance in the "hot" side of the motor circuit. Move to Test 2.

Test 2 The indicated voltage should not exceed 0.3 V.

A reading of 0.3 V or less indicates that excessive resistance is present between the relay and the starter motor. The starter cable connections should be cleaned and tightened. The cable should be replaced if necessary.

A reading of more than 0.3 V indicates excessive resistance between the battery and the starter terminal on the relay. Move to Test 3.

Test 3 The indicated voltage should not exceed 0.2 V.

A reading of 0.2 V or less indicates that excessive resistance is present in the relay or in the relay connections. The connections should be cleaned and tightened. The relay should be replaced if necessary.

A reading of more than 0.2 V indicates excessive resistance is present between the battery and the relay. The battery cable connections should be cleaned and tightened. The battery cable should be replaced if necessary. Move to Test 4.

Test 4 The indicated voltage should not exceed 0.3 V.

A reading of 0.3 V or less indicates that the re-

sistance in the ground side of the motor circuit is acceptable.

A reading of more than 0.3 V indicates excessive resistance in the ground connections. The battery ground cable should be cleaned and tightened. The starter mounting bolts should be checked and tightened.

Verification of Repair The appropriate voltage drop test should be repeated after any repair has been made.

Replacing a Starter Relay The steps that follow outline a typical procedure for replacing starter relays and solenoids that are mounted separate from the starter motor. An appropriate manual should be consulted for any specific procedures that may be necessary for the car on which you are working:

1 Protect the fender of the vehicle with a fender cover.
2 Disconnect the ground cable from the battery terminal.
3 Disconnect the wires and cables from the relay.

Note: Arrange the wires so that they will be installed in the same position.

4 Remove the relay.
5 Install the replacement relay.
6 Install the cables and wires.
7 Connect the battery ground cable.

Replacing a Starter Motor Removing a starter motor from an engine and installing a replacement is a relatively simple job. However, when that engine is located in a crowded engine compartment, the job becomes a bit more difficult. The starter motor on some cars can be removed by working under the hood. On others, you must work under the car.

When a starter motor must be removed from under a car, the car must be raised from the floor and supported. Most of the time a car is raised by means of a floor jack similar to the one shown in Figure 8.32. Whenever you raise a car with a floor jack, you must be careful where you place the jack. An improperly placed jack may let the car fall or may damage parts under the car.

The proper placement of a jack depends on the type of frame and suspension system on the car. Since many different types of frames and suspension systems are in use, you should check the car manufacturer's service manual before you attempt to raise a car with a jack. Figures 8.33 and 8.34 are

Job 8d

TEST THE VOLTAGE DROP IN A STARTING SYSTEM

SATISFACTORY PERFORMANCE

A satisfactory performance on this job requires that you do the following:

1 Test the voltage drop in the motor circuit of the starting system of the car assigned.
2 Following the steps in the "Performance Outline" and the specifications of the manufacturer, complete the job within 20 minutes.
3 Fill in the blanks under "Information."

PERFORMANCE OUTLINE

1 Prepare the vehicle and the meter for the test.
2 Measure the voltage drop in the entire "hot" side of the motor circuit.
3 Measure the voltage drop between the battery and the starter side of the relay.
4 Measure the voltage drop between the battery and the battery side of the relay.
5 Measure the voltage drop in the ground side of the motor circuit.
6 Restore the vehicle to operating condition.
7 Compare your readings with the specifications of the manufacturer.

INFORMATION

Vehicle identification _____

Reference used _____ Page(s) _____

Meter identification _____

Meter scale used _____

Test Performed	Test Results	Specification
Voltage drop in entire "hot" side of circuit	_____ volts	_____ volts
Voltage drop between battery and starter side of relay	_____ volts	_____ volts
Voltage drop between battery and battery side of relay	_____ volts	_____ volts
Voltage drop in ground side of circuit	_____ volts	_____ volts

INTERPRETATION OF TEST RESULTS

_____ Resistance of entire motor circuit is within specifications.

Excessive resistance may be present in the

_____ Positive battery terminal and the cable clamp connection

_____ Battery cable

_____ Battery cable connection at the relay

_____ Relay

_____ Starter cable connection at the relay

_____ Starter cable

_____ Starter cable connection at the starter motor

_____ Negative battery terminal and the cable clamp connection

_____ Ground cable

_____ Ground cable connection at the engine

_____ Starter motor mounting

Job 8e

REPLACE A STARTER RELAY

SATISFACTORY PERFORMANCE

A satisfactory performance on this job requires that you do the following:

1 Replace the starter relay on the assigned car.
2 Following the steps in the "Performance Outline,"
complete the job within 20 minutes.
3 Fill in the blanks under "Information."

PERFORMANCE OUTLINE

1 Disconnect the battery ground cable.
2 Remove the relay.
3 Install the replacement relay.
4 Connect the battery ground cable.
5 Test the operation of the relay.

INFORMATION

Vehicle identification _____

Reference used _____ Page(s) _____

typical of the drawings found in service manuals. They show the areas where it is safe to place a jack. Some cars with front wheel drive require that a jack be used in special locations. Figures 8.35 and 8.36 show the locations specified by one manufacturer.

When using a jack you should always remember that a jack is made to raise cars, not to support

them. Working on or under a car that is held off the floor by a jack is very dangerous, because the car can easily slide off the jack. Anytime you raise a car, even for a few minutes, you should place _jack stands_ or _car stands_ under the car to support its weight. Typical stands are shown in Figure 8.37. The manufacturer's service manual will show you

Figure 8.32 A hydraulic jack similar to those used in most auto shops (courtesy of Hein-Werner Corporation).

Figure 8.33 A drawing of a typical frame showing where it is safe to place a jack and car stands (courtesy of Chevrolet Motor Division).

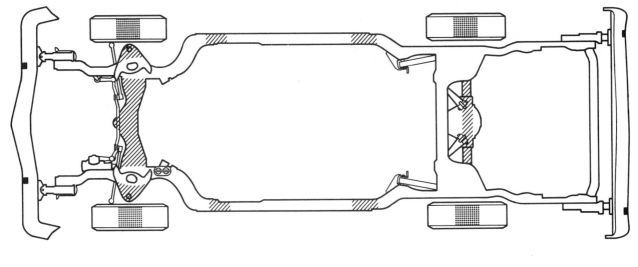

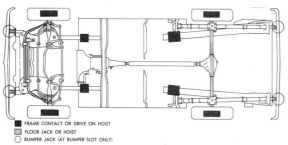

Figure 8.34 Typical lifting and support locations on a car of unit body design. Since a separate frame is not used, the jacks and car stands must be placed under reinforced areas of the floor pan (courtesy of Chrysler Corporation).

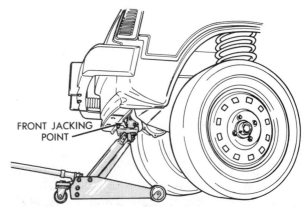

Figure 8.35 When raising the front end of some cars, the jack must be placed under a reinforced section of the body behind the front bumper (courtesy of Chrysler Corporation).

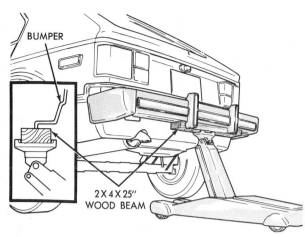

Figure 8.36 On some front wheel drive cars, the jack position for raising the rear of the car is under the rear bumper (courtesy of Chrysler Corporation).

Figure 8.37 A pair of adjustable car stands or safety stands. Stands should always be used to support a car after you raise it with a jack (courtesy of Hein-Werner Corporation).

where it is safe to place those stands. (Refer to Figures 8.33 and 8.34.)

Although there are many different areas under a car where car stands can be positioned, there are only two ways that a car can be supported. One is by the frame, the other is by the suspension systems. When a car is supported by the frame, the wheels are free to drop down to the limits of their suspension systems. On some cars, the removal of the starter motor is made easier if the car is supported by the frame. When supported in this manner, the steering linkage is held away from the starter motor, making it more accessible. When a car is supported by the suspension systems, the wheels are held up in the approximate positions they are in when the car is on the floor. The method chosen depends on the particular job to be done. The proper methods of raising and supporting a car are very important to your safety.

On many cars, the fuel tank is attached to the frame ahead of the rear bumper. In that position, it may be damaged while using a jack to raise or lower a car. When both the front and rear wheels of a car must be raised from the floor, the rear of the car should be raised and supported first. The front of the car then can be raised. When the car is lowered to the floor, the front of the car should be lowered before the rear of the car is lowered. Those procedures will minimize the possibility of fuel tank damage and the dangers of fuel leakage.

Raising a Rear-Wheel Drive Car and Supporting It by the Frame The steps that follow outline a procedure for raising a rear-wheel drive car with a jack and supporting it with jack stands

placed under the frame. You should check the manual for the car on which you are working for the correct lifting and support points.

Rear

1 Roll a jack under the rear of the car. Raise the jack slightly, and adjust its position so that it is centered under the rear axle housing. (See position C in Figure 8.38.)
2 Operate the jack to raise the car until the wheels are clear of the floor by at least 10 inches (25 cm).
3 Place jack stands under the frame side rails just ahead of the bend in the frame. (See positions D in Figure 8.38.) Raise the stands as close as possible to the frame.
4 Lower the jack so that the car is supported by the stands.
5 Remove the jack.

Front

1 Roll a jack under the front of the car. Raise the jack slightly and adjust its position so that it is centered under the front crossmember of the frame. (See position A in Figure 8.38.)
2 Operate the jack to raise the car until the wheels are clear of the floor by about 10 inches (25 cm).
3 Place jack stands under the frame side rails just behind the bend in the frame. (See positions B in Figure 8.38.) Raise the stands as close as possible to the frame.
4 Lower the jack so that the car is supported by the stands.
5 Remove the jack.

Raising a Rear-Wheel Drive Car and Supporting It by the Suspension Systems The following steps outline a typical procedure for raising a rear-wheel drive car with a jack and supporting it with jack stands placed under the suspension systems. Be sure to check the manual for the car

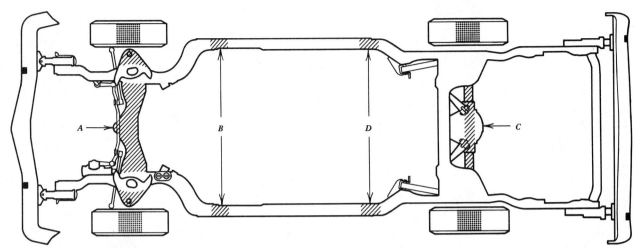

Figure 8.38 A drawing of a typical frame, showing locations for raising a car with a jack (A and C) and positions where jack stands should be placed (B and D) for supporting the car by the frame (courtesy of Chevrolet Motor Division).

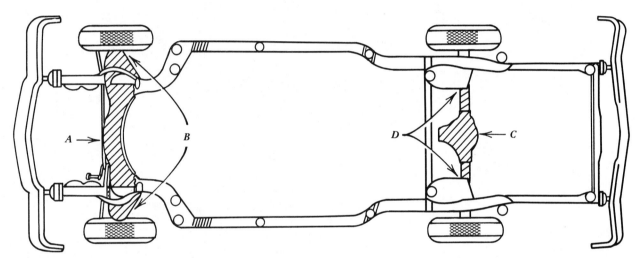

Figure 8.39 A drawing of a typical frame, showing locations for raising a car with a jack (A and C) and positions where jack stands should be placed (B and D) for supporting the car by the suspension system (courtesy of Chevrolet Motor Division).

on which you are working for the correct lift and support points.

Rear

1 Roll a jack under the rear of the car. Raise the jack slightly, and adjust its position so that it is centered under the rear axle housing. (See position C in Figure 8.39.)

2 Operate the jack to raise the car until the wheels are clear of the floor by at least 6 inches (15 cm).

3 Place jack stands under the rear axle housing as close as possible to the wheels. (See positions D in Figure 8.39.)

4 Lower the car so that the car is supported by the stands.

5 Remove the jack.

Front

1 Roll a jack under the front of the car. Raise the jack slightly and adjust its position so that it is centered under the front crossmember of the frame. (See position A on Figure 8.39.)

2 Operate the jack to raise the car until the wheels are clear of the floor by about 10 inches (25 cm).

3 Place jack stands under the lower control arms as close as possible to the wheels. (See positions B on Figure 8.39.) Raise the stands as close as possible to the lower control arms.

4 Lower the jack so that the car is supported by the stands.

5 Remove the jack.

Job 8f

RAISE A CAR AND SUPPORT IT BY THE FRAME

SATISFACTORY PERFORMANCE
A satisfactory performance on this job requires that you do the following:

1 Using a hydraulic floor jack, raise a car from the floor and support it with jack stands placed under the frame.
2 Following the steps in the "Performance Outline" and the recommendations of the car manufacturer regarding jack and stand positioning, complete the job within 15 minutes.
3 Fill in the blanks under "Information."

PERFORMANCE OUTLINE
1 Raise the rear of the car.
2 Support the rear of the car with jack stands.
3 Raise the front of the car.
4 Support the front of the car with car stands.
5 Lower the car to the floor.

INFORMATION
Vehicle identification _____

Reference used _____ Page(s) _____

Job 8g

RAISE A CAR AND SUPPORT IT BY THE SUSPENSION SYSTEM

SATISFACTORY PERFORMANCE
A satisfactory performance on this job requires that you do the following:

1 Using a hydraulic floor jack, raise a car from the floor and support it with jack stands placed under the suspension systems.
2 Following the steps in the "Performance Outline" and the recommendations of the car manufacturer regarding jack and stand positioning, complete the job within 15 minutes.
3 Fill in the blanks under "Information."

PERFORMANCE OUTLINE
1 Raise the rear of the car.
2 Support the rear of the car with stands.
3 Raise the front of the car.
4 Support the front of the car with stands.
5 Lower the car to the floor.

INFORMATION
Vehicle identification _____

Reference used _____ Page(s) _____

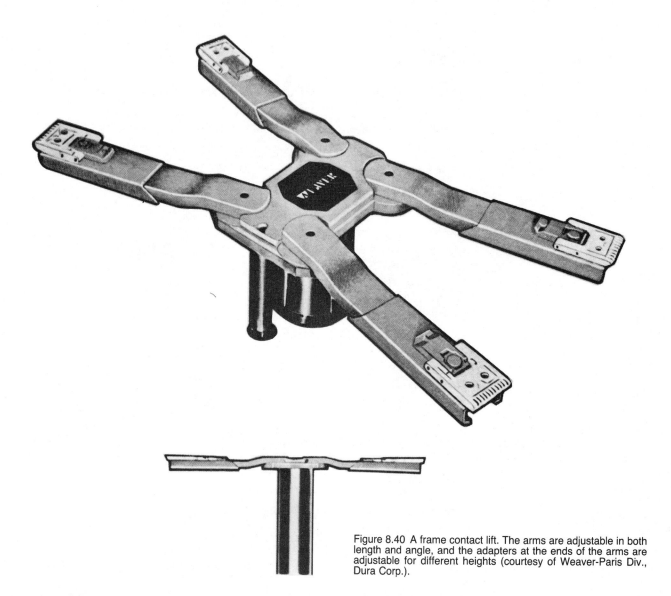

Figure 8.40 A frame contact lift. The arms are adjustable in both length and angle, and the adapters at the ends of the arms are adjustable for different heights (courtesy of Weaver-Paris Div., Dura Corp.).

Raising a Car with a Lift Many shops are equipped with hydraulic or electric lifts that raise and support cars. Most of those lifts are capable of raising a car high enough so that a mechanic can stand while working under the car. All of those lifts incorporate some type of safety system in their construction so that a car cannot drop suddenly in case of equipment failure. Some of those safety devices are automatic. Others must be set manually. Before you attempt to use a lift, you should be sure that you understand the operation of its safety device.

Although there are many different types of lifts in use, they all support a car in one of two ways. One is by the frame, the other is by the suspension system.

Since there is such a variety of lifts in use, it is beyond the scope of this book to provide specific operating instructions for all of them. Lift manufacturers provide instruction manuals for the operation of each of the designs they make. You should read the manual provided with the lift you have available. Lacking the manual, you should receive instructions from someone who is familiar with the operation of the lift you intend to use. Although the following steps do not apply to the operating of some lifts, they outline procedures that are common to most lifts.

Frame Contact Lifts A typical frame contact lift is shown in Figure 8.40. They usually are made with adjustable arms and contact pads. This allows the pads to be positioned to contact the correct lifting points on any car. As you learned when raising a car with a jack, the correct lifting points for

WAGONS

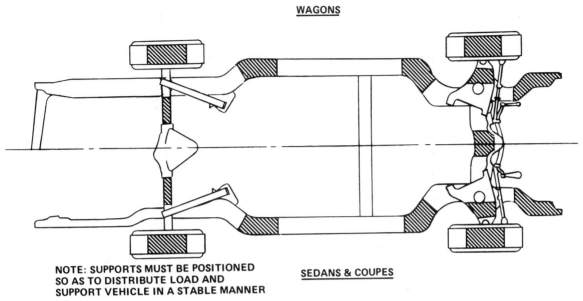

NOTE: SUPPORTS MUST BE POSITIONED
SO AS TO DISTRIBUTE LOAD AND
SUPPORT VEHICLE IN A STABLE MANNER

SEDANS & COUPES

Figure 8.41 Typical lift points specified by one manufacturer (courtesy of Pontiac Motor Division, General Motors Corporation).

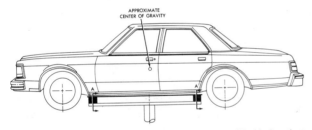

Figure 8.42 A car positioned on a frame contact lift. Notice that the approximate center of gravity is directly over the center post(s) of the lift (courtesy of Chrysler Corporation).

any car can be found in the service manual for that car. Figure 8.41 shows the lifting points specified by one manufacturer. The following steps outline a typical procedure for the operation of a frame contact lift.

Preparation
1 Check that the lift is all the way down and that the arms and pads will not come in contact with the underbody of a car driven over the lift. Lower the lift and move these parts, if necessary.
2 Slowly drive the car over the lift so that the car is centered. Position the car so that the wheels are in place in the trough or against the stops on the floor. Turn off the engine, leave the parking brake off, and place the shift selector lever in the Neutral position.
3 Check to see that the car is centered over (or between) the lift posts. Reposition the car, if necessary.

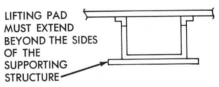

LIFTING PAD
MUST EXTEND
BEYOND THE SIDES
OF THE
SUPPORTING
STRUCTURE

Figure 8.43 A sectioned view showing how the lifting pads should be positioned under the frame or supporting structure (courtesy of Chrysler Corporation).

4 Consult an appropriate manual to determine the correct lifting points.
5 Adjust the position of the arms and pads so that they are directly under the lifting points.

Note: When positioning the pads and arms, try to keep the approximate center of gravity of the car over the post(s), as shown in Figure 8.42.

Lifting
1 Carefully operate the lift controls so that the lift rises from the floor. Stop the lift when the pads contact the car.
2 Check the position of each pad so that you are sure that they are in firm contact with the correct lifting areas.

Note: The pads should extend beyond the sides of the frame or underbody supporting structure, shown in Figure 8.43.

3 Raise the car until the tires are clear of by about 1 inch (25 mm). By pushing or ers and fenders, try to push the car

Job 8h

RAISE AND SUPPORT A CAR WITH A FRAME CONTACT LIFT

SATISFACTORY PERFORMANCE
A satisfactory performance on this job requires that you do the following:

1 Using a frame contact lift, raise and support the assigned car.
2 Following the steps in the "Performance Outline," the lift manufacturer's operating instructions, and the car maker's specifications regarding lifting points, complete the job within 15 minutes.
3 Fill in the blanks under "Information."

PERFORMANCE OUTLINE
1 Align the car on the lift.
2 Position the lift arms and pads.
3 Raise the car slightly and test the security of the points of contact.
4 Raise the car and engage the safety device.
5 Lower the car to the floor.
6 Clear the lift so that the car can be removed.

INFORMATION
Vehicle identification _____

Lift identification and type _____

Reference used _____ Page(s) _____

Note: It is better to have the car slip off the lift now, when there is little danger of damage or injury, than to have it slip off when raised several feet.

4 If the car is secure on the lift, continue to raise the car to the desired height.
5 Set and lock any manually operated safety devices that may be present.

Lowering
1 Clear the area under the car of any tools, equipment, wires, and hoses.
2 Release the safety device.
3 Carefully operate the controls to lower the car to the floor.
4 Adjust the position of the arms and pads under the car so that the car may be driven off the lift.

Suspension Contact Lifts   Figure 8.44 shows a typical suspension contact lift. The posts or heads can be adjusted so that a car is lifted by the front ower control arms and the rear axle or axle hous-

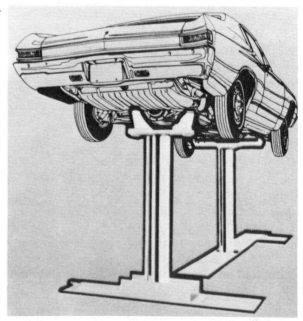

Figure 8.44 A suspension contact lift. Note that the wheels are held in the approximate position they are in when the car is on the floor (courtesy of Weaver-Paris Div., Dura Corp.).

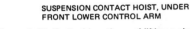

SUSPENSION CONTACT HOIST, UNDER
FRONT LOWER CONTROL ARM

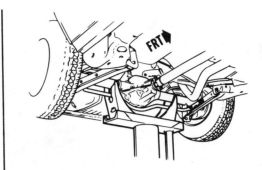

SUSPENSION CONTACT HOIST, LIFTING
ON DIFFERENTIAL HOUSING (DO NOT
LIFT ON STABILIZER BAR IF PRESENT)

Figure 8.45 Typical locations of lifting points when using a twin-post suspension contact lift (courtesy of Oldsmobile Division).

ing. Figure 8.45 shows how these lifting points are specified by one manufacturer. The following steps outline a typical procedure for the operation of a suspension contact lift:

Preparation

1 Check that the lift is all the way down and that the heads will not come into contact with the suspension system or underbody of a car driven over the lift. Lower the lift and move those parts if necessary.

2 Slowly drive the car over the lift so that the car is centered. Position the car so that the wheels are in the trough or against the stops on the floor. Turn off the engine, leave the parking brake off, and place the shift lever in the Neutral position.

3 Check to see that the car is centered over the heads. Reposition the car if necessary.

4 Consult an appropriate manual to determine the correct lifting points.

5 Adjust the position of the heads so that they are directly under the lifting points.

Lifting

1 Carefully operate the lift controls so that the lift rises from the floor. Stop the lift when the heads contact the lower control arms and the rear axle or axle housing.

2 Check the position of the heads to be sure that they are in firm contact and in the proper position. (See Figure 8.45.)

3 Raise the car until the tires are clear of the floor by about 1 inch (25 mm). By pushing on the bumpers and fenders, try to push the car off of the lift.

Note: It is better to have the car slip off of the lift

now, when there is little danger of damage or injury, than to have it slip off when raised several feet.

4 If the car is secure on the lift, continue to raise the car to the desired height.

5 Set and lock any manually operated safety devices that may be present.

Lowering

1 Clear the area under the car of any tools, equipment, wires, and hoses.

2 Release the safety device.

3 Carefully operate the controls to lower the car to the floor.

4 Adjust the position of the heads under the car so that the car may be driven off of the lift.

Starter Motor Replacement Two methods of mounting a starter motor on an engine are in common use. One uses two or three short bolts to hold the starter motor to the flywheel housing as shown in Figures 8.47 and 8.48. The other method uses two long bolts to hold the starter motor to the bottom of the flywheel housing. As shown in Figure 8.49, a reinforcement bracket often is used with that method.

On some cars, the steering linkage blocks the removal of the starter motor. Turning the wheels to the extreme left or to the extreme right may provide sufficient clearance, especially if the car is supported by the frame. In some instances, the steering linkage must be disconnected and moved.

There are many variables in engine and chassis design, even among cars built by the same manufacturer. Many of those variables require different starter motor replacement procedures. The steps

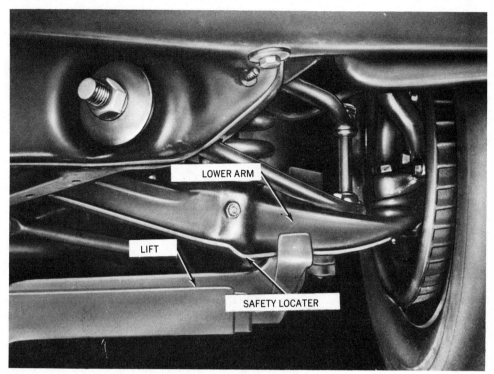

Figure 8.46 Correct positioning of a lift head at the lower control arm. As shown, some lower control arms incorporate a safety locator to guide the placement of the lift head (courtesy of Cadillac Motor Car Division).

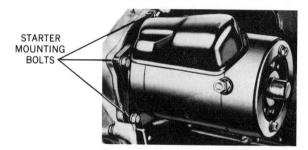

Figure 8.47 In this installation, the starter motor is held to the front of the flywheel housing by three bolts (courtesy of Ford Motor Company).

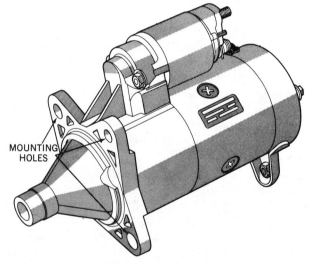

Figure 8.48 Some starter motors have a mounting flange or "ears" by which the starter is bolted to the flywheel housing (courtesy of Chrysler Corporation).

that follow outline a general procedure. You should consult an appropriate service manual for the specific procedure required for the car on which you are working.

Removal

1 Disconnect the battery ground cable.

2 If the starter motor must be removed from under the car, raise the car and support it.

Note: On some cars, supporting the car under the frame will provide more clearance at the steering linkage and allow easier starter motor removal.

3 Remove any reinforcing brackets, braces, or heat shields that may be present. (See Figure 8.50.)

4 Disconnect the wires and cables connected to the starter motor. Be sure to observe the location of each wire so that you can install it in its correct

Job 8i

RAISE AND SUPPORT A CAR WITH A SUSPENSION CONTACT LIFT

SATISFACTORY PERFORMANCE
A satisfactory performance on this job requires that you do the following:

1 Using a suspension contact lift, raise and support the car assigned.
2 Following the steps in the "Performance Outline," the lift manufacturer's operating instructions, and the car manufacturer's specifications regarding lifting points, complete the job within 15 minutes.
3 Fill in the blanks under "Information."

PERFORMANCE OUTLINE
1 Align the car on the lift and adjust the heads so that they will contact the lifting points.
2 Raise the car slightly and test the security of the points of contact.
3 Raise the car and engage the safety device.
4 Lower the car to the floor.
5 Clear the lift so that the car can be removed.

INFORMATION
Vehicle identification _____

Lift identification and type _____

Reference used _____ Page(s) _____

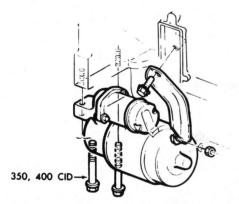

350, 400 CID →

Figure 8.49 In this installation, the starter motor is held to the bottom of the flywheel housing by two bolts. A bracket reinforces the mounting (courtesy of Chevrolet Motor Division).

location on the replacement starter motor.

Note: On some cars, it is difficult to gain access to the terminals on the starter motor or on the solenoid. On those cars, it is easier to remove the starter mounting bolts first. Then the starter motor can be positioned for better access to the terminals.

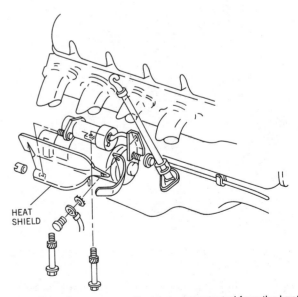

HEAT SHIELD

Figure 8.50 On some cars, the starter is protected from the heat of the exhaust manifold by a heat shield (courtesy of Chevrolet Motor Division).

Job 8j

REPLACE A STARTER MOTOR

SATISFACTORY PERFORMANCE
A satisfactory performance on this job requires that you do the following:

1 Replace the starter motor on the car assigned.
2 Following the steps in the "Performance Outline" and the procedure and specifications of the manufacturer, complete the job within 200 percent of the manufacturer's suggested time.
3 Fill in the blanks under "Information."

PERFORMANCE OUTLINE
1 Disconnect the battery ground cable.
2 Raise and support the car, if necessary.
3 Remove all of the parts necessary to gain access to the starter motor.
4 Remove the starter motor.
5 Install the replacement starter motor.
6 Install all of the parts removed to gain access.
7 Connect the battery ground cable.
8 Check the operation of the starter motor.
9 Lower the car to the floor.

INFORMATION
Vehicle identification _____

Reference used _____ Page(s) _____

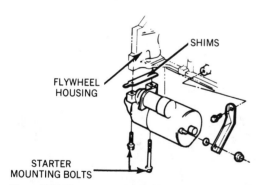

Figure 8.51 On some engines, shims are used between the starter motor and the flywheel housing to provide a means for adjusting gear mesh (courtesy of Pontiac Motor Division, General Motors Corporation).

5 Remove the bolts holding the starter to the flywheel housing.

Note: On some engines, shims are used between the starter motor and the flywheel housing, as shown in Figure 8.51. Be sure to save any shims present so that you can install them with the replacement starter motor.

6 Remove the starter motor.

Note: Starter motor removal may be blocked by other engine and chassis components. Consult an appropriate manual for the manufacturer's recommended removal procedure.

Installation
1 Clean the starter mounting surface on the flywheel housing.
2 Position the replacement starter motor against the flywheel housing and start the mounting bolts, inserting any shims that may have been removed.

Note: If you found it difficult to gain access to the terminals when you removed the original starter motor, you may find it easier to install the wires and cables before you secure the starter motor to the flywheel housing.

3 Turn all of the bolts into place gradually, holding the starter motor in position.
4 Tighten the bolts to the manfacturer's torque specification.

5 Install the wires and cables if not previously installed.

6 Install any brackets, braces, and heat shields that were removed. Tighten the attaching bolts.

7 Install any engine or chassis components that were removed to allow starter motor removal.

8 Connect the battery ground cable.

9 Check the operation of the starter motor.

10 Raise the car, remove the car stands, and lower the car to the floor.

SUMMARY

By performing the jobs in this chapter, you learned how to perform many of the services necessary to maintain starting systems. You now have knowledge of how the systems function. You are aware of the many different components used and can identify those components. You have also developed diagnostic skills. You can perform starter current draw tests and voltage drop tests. You also can interpret the results of those tests to diagnose problems in the starting system. In addition, you have gained repair skills in the replacement of defective components.

SELF-TEST

Each incomplete statement or question in this test is followed by four suggested completions or answers. In each case, select the *one* that best completes the sentence or answers the question.

1 Which of the following parts is NOT a part of the control circuit?
A. Relay
B. Battery
C. Starter motor
D. Starter switch

2 Two mechanics are discussing starting systems.
Mechanic A says that, on most cars, a relay is used to complete the motor circuit so that current can flow to the starter motor.
Mechanic B says that, on some cars, a neutral safety switch is used in the motor circuit.
Who is right?
A. A only
B. B only
C. Both A and B
D. Neither A nor B

3 On most starter motors, the drive pinion is shifted into mesh with the flywheel ring gear by the action of
A. a solenoid
B. the armature
C. a reduction gear
D. an overrunning clutch

4 When an engine starts, the drive pinion is disconnected from the starter motor armature by the action of
A. a field pole shoe
B. a torque converter
C. an overrunning clutch
D. a solenoid return spring

5 A starter current draw test should not be performed unless the specific gravity of the battery electrolyte is at least
A. 1.000
B. 1.100
C. 1.200
D. 1.300

6 When performing a starter current draw test, the voltmeter leads should be connected to the
A. relay terminals
B. battery terminals
C. ammeter lead terminals
D. starter motor terminals

7 A test of a starter motor reveals that the starter current draw is below specifications. The most probable cause of that finding is
A. low resistance in the motor circuit
B. high resistance in the motor circuit
C. low resistance in the control circuit
D. high resistance in the control circuit

8 A test of a starting system reveals that the voltage drop between the battery positive (+) post and the starter motor terminal is excessive. The most probable cause of that finding is
A. low resistance in the motor circuit
B. high resistance in the motor circuit
C. low resistance in the control circuit
D. high resistance in the control circuit

9 A test of a starting system reveals that the voltage drop between the battery negative (−) post and the starter-motor housing is 0.2 V.
Mechanic A says that the system has excessive resistance in the ground side of the motor circuit.
Mechanic B says that the system has excessive resistance in the ground side of the control circuit.

Who is right?
A. A only
B. B only
C. Both A and B
D. Neither A nor B

10 Two mechanics are discussing starter motor replacement.
Mechanic A says that, on some engines, shims are placed between the solenoid and the starter motor.
Mechanic B says that, on some engines, shims are placed between the starter motor and the flywheel housing.
Who is right?
A. A only
B. B only
C. Both A and B
D. Neither A nor B

Chapter 9 Starter Motor Construc- tion and Overhaul

As you learned in Chapter 8, a motor converts electrical energy into mechanical energy. Knowledge of how this conversion is accomplished and an understanding of the function of starter motor components will be helpful to you in overhauling starter motors.

Starter motor overhaul consists of disassembling a starter motor, cleaning, inspecting, and testing its component parts, and reassembling the starter motor with any replacement parts required. While all starter motors use the same principles of operation, there are many different designs requiring different overhaul procedures. The overhaul procedures for the most commonly used starter motors will be outlined in this chapter.

Your objectives in this chapter are to perform the following jobs:

A
Identify starter motor components
B
Inspect starter drive assemblies
C
Inspect and test starter motor armatures
D
Test starter motor field coils
E
Overhaul a Delco starter motor
F
Overhaul a Ford positive engagement starter motor
G
Overhaul a Chrysler gear reduction starter motor
H
Overhaul a Nippondenso starter motor
I
Overhaul a Bosch starter motor

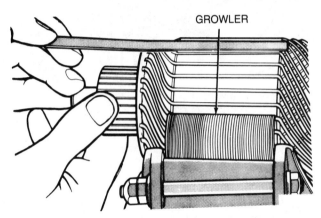

GROWLER

STARTER MOTOR OPERATION AND CONSTRUCTION

Principles of Motor Operation A motor converts electrical energy into mechanical energy through the use of magnetism. As you know, when current flows through a conductor, a magnetic field is produced around the conductor. This magnetic field is shown in Figure 9.1. Figure 9.2 illustrates another type of magnetic field, this one produced by the poles of a horseshoe magnet. Notice that the lines of force flow from the north pole to the south pole. If a straight conductor is placed between the poles of the magnet, as shown in Figure 9.3, the lines of magnetic force are strengthened or reinforced on the left side of the conductor while those on the right side of the conductor are weakened or cancelled out. The lines of magnetic force try to "straighten out" and tend to push the conductor from the strong magnetic field toward the weak magnetic field. Thus, the conductor will move from left to right, as shown in Figure 9.4.

There are two ways by which the direction of motion can be reversed. One method, shown in Figure 9.5, requires that the polarity of the magnetic poles be changed. The second method, shown in Figure 9.6, is most commonly used and requires that the current flow through the conductor be reversed. This of course results in the reversal of the magnetic field that flows around the conductor.

The "straight line" motion of the conductor can be converted to rotary motion by forming a loop in the

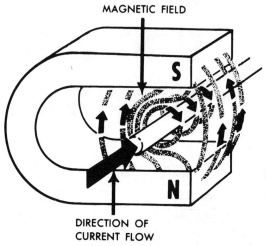

Figure 9.3 When a current-carrying conductor is placed in a magnetic field, the magnetic field on one side of the conductor will be reinforced (courtesy of Chevrolet Motor Division.)

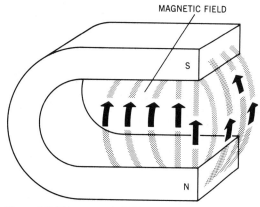

Figure 9.1 When current flows through a conductor, a magnetic field is created around the conductor.

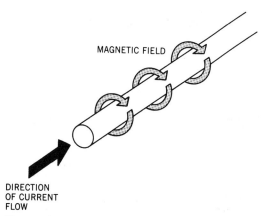

Figure 9.2 The magnetic field produced between the poles of a horseshoe magnet.

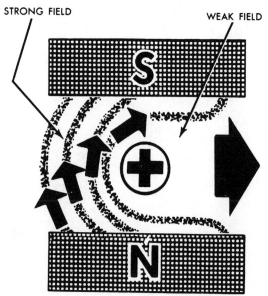

Figure 9.4 The concentration of the lines of force on one side of the conductor will push the conductor (courtesy of Chevrolet Motor Division).

conductor while providing a means of changing the direction of current flow. Figure 9.7 shows how this is done. The looped conductor is positioned between two iron poles or *pole shoes.* Those pole shoes are wound with other conductors and form the cores of electromagnets. The ends of the looped conductor are fitted with *commutator bars* which receive current through a pair of *brushes.* The brushes and the commutator bars form a rotary switch which changes the direction of current flow through the loop every half revolution.

When connected as shown in Figure 9.7, current from the battery flows through the windings on the pole shoes and creates a strong magnetic field. Through the action of the brushes and the commutator bars, the current also flows through the looped conductor before returning to the battery. The different magnetic fields around the conductor and between the poles push the looped conductor around as shown in Figure 9.8.

Starter Motor Construction   The simple motor described in the previous paragraphs will operate, but it will not develop sufficient power for any practical use. Through the application of the same principles of operation, starter motors of any de-

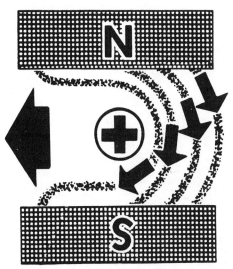

Figure 9.5 If the polarity of the magnet is reversed, the direction of motion is reversed (courtesy of Chevrolet Motor Division).

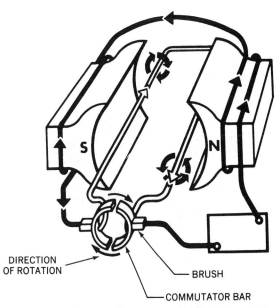

Figure 9.7 A simple motor. A current-carrying conductor is looped and positioned between magnetic poles. Brushes and commutator bars act as a rotary switch to reverse the current flow in the conductor for each 180° of rotation (courtesy of Chevrolet Motor Division).

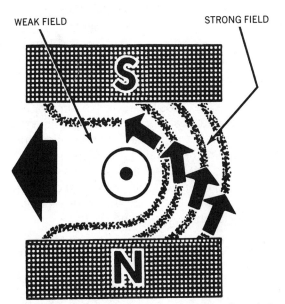

Figure 9.6 If the direction of the current flow through the conductor is reversed, the direction of motion is reversed (courtesy of Chevrolet Motor Division).

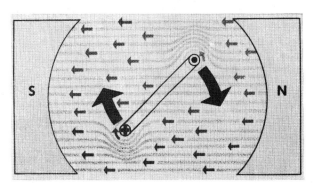

Figure 9.8 The rotational effect of the magnetic fields in a motor (courtesy of Chevrolet Motor Division).

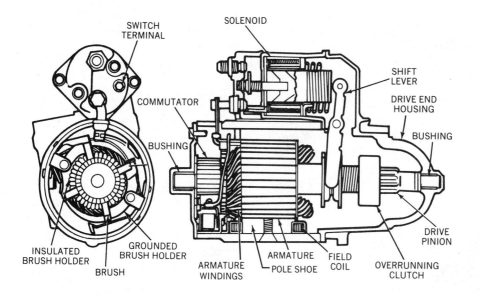

SWITCH TERMINAL
SOLENOID
COMMUTATOR
BUSHING
SHIFT LEVER
DRIVE END HOUSING
BUSHING
INSULATED BRUSH HOLDER
GROUNDED BRUSH HOLDER
BRUSH
ARMATURE WINDINGS
POLE SHOE
ARMATURE
FIELD COIL
OVERRUNNING CLUTCH
DRIVE PINION

Figure 9.9 The main components of a starter motor (courtesy of Chevrolet Motor Division).

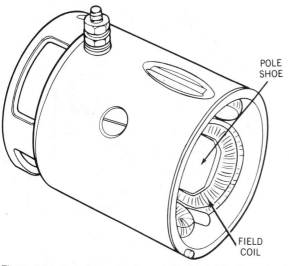

Figure 9.10 A starter motor frame with the field coils and pole shoes installed (courtesy of General Motors Corporation).

POLE SHOE
FIELD COIL

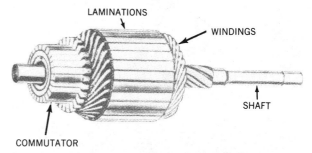

LAMINATIONS
WINDINGS
SHAFT
COMMUTATOR

Figure 9.11 A typical starter motor armature (courtesy of Chevrolet Motor Division).

sired output can be built. All starter motors operate in the same manner and, in most instances, use similar parts. The main parts of a starter motor are shown in Figure 9.9.

The Field Coils The field coils are usually made of heavy copper ribbons. They are wrapped around iron *pole pieces* or pole shoes. The pole shoes form the cores of electromagnets and concentrate the magnetic lines of force. They are fastened inside a cylindrical iron *frame* that forms the central starter housing and also acts to reinforce the magnetic field. Typical field coils are shown in Figure 9.10.

The Armature As shown in Figure 9.11, the armature contains many looped conductors. It consists of a stack of slotted iron discs called *laminations*. A steel shaft passes through the center of the laminations and protrudes from both ends so that it can be supported by bearings. The conductors, called *windings*, are positioned in the slots of the laminations. The laminations help to concentrate the magnetic lines of force that are created when current flows through the armature windings. The ends of the windings are soldered to a series of commutator bars that make up the *commutator*. Each winding is insulated both from the other windings and from the laminations and the shaft.

The Brush Holders The brush holders position the brushes and, through spring tension, hold them against the commutator. In most instances, four brushes are used. Two of the brush holders are insulated and two of the brush holders are grounded.

Job 9a

IDENTIFY STARTER MOTOR COMPONENTS
SATISFACTORY PERFORMANCE
A satisfactory performance on this job requires that you do the following:

1 Identify the numbered parts in the drawing below by placing the number of each part in front of the correct part name.
2 Complete the job by identifying all the parts correctly within 5 minutes.
PERFORMANCE SITUATION

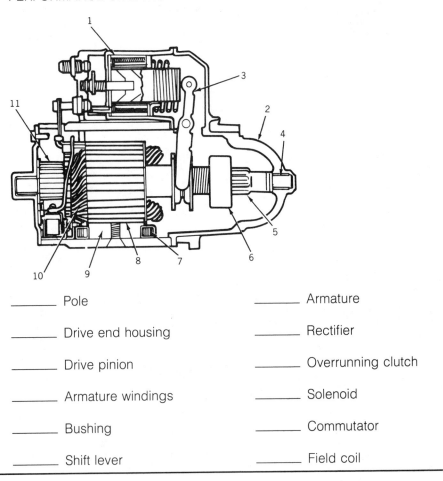

_____ Pole

_____ Drive end housing

_____ Drive pinion

_____ Armature windings

_____ Bushing

_____ Shift lever

_____ Armature

_____ Rectifier

_____ Overrunning clutch

_____ Solenoid

_____ Commutator

_____ Field coil

The brush holders may be mounted inside the frame (refer to Figure 9.9), or they may be mounted on the inside of the *end frame* or *end shield,* as shown in Figure 9.12.

The Brush End Frame   The brush end frame, also

called *commutator end frame,* provides a cover for one end of the starter frame. It contains a bushing which supports and centers the commutator end of the armature. (Refer to Figure 9.9.)

The Drive Gear Housing   The drive gear housing

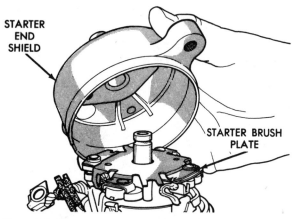

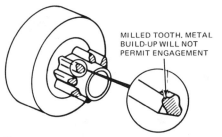

Figure 9.14 A drive pinion with chipped, worn, or milled teeth should be replaced (courtesy of Ford Motor Company).

Figure 9.12 In some starter motors, the brush holders are mounted on a brush plate inside the end frame or end shield (courtesy of Chrysler Corporation).

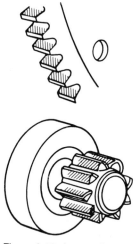

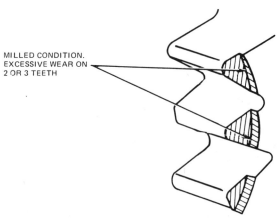

Figure 9.15 If the drive gear pinion teeth are damaged, the flywheel ring gear should be inspected for possible damage. A ring gear with broken teeth or with excessive wear on two or more teeth should be replaced (courtesy of Ford Motor Company).

Figure 9.13 A normal wear pattern on a pinion gear and a flywheel ring gear (courtesy of Ford Motor Company).

or *drive end housing* covers the other end of the starter frame. It usually houses the drive assembly and contains a bushing that centers and supports the drive end of the armature. The drive gear housing is usually formed with ears or mounting lugs by which the starter motor can be mounted on the engine.

STARTER MOTOR OVERHAUL
Common to most overhaul procedures are the steps performed in cleaning parts, checking starter drive assemblies, and in testing armatures and field coils. It is suggested that you study those steps and practice the procedures with parts known to be good and parts known to be bad so that you can familiarize yourself with the checks and tests required.

Cleaning   While the parts of a starter motor should be cleaned during an overhaul, certain parts should never be immersed in cleaning fluids. The internal parts of the drive assembly are lubricated during its initial assembly. If the drive assembly is immersed in cleaning fluids, the lubricant will be diluted and washed from those parts. The insulation on the windings of the armature, field coils, and solenoid may be damaged or softened by certain cleaning fluids. Those parts may be cleaned by the use of an air gun, a dry brush, and wipers. End frames, covers, and other small parts may be immersed in cleaning fluid without damage.

Starter Drive Inspection   The teeth on the starter drive pinion gear should exhibit a normal wear pattern as shown in Figure 9.13. If the gear "wobbles," or if the teeth are worn, chipped, or "milled," the drive assembly should be replaced. (See Figure 9.14.) When worn or milled teeth are found on the drive pinion, the flywheel ring gear

Figure 9.16 A typical armature growler. Notice the test leads that are used for checking armatures and field coils for opens and grounds (courtesy of Snap-on Tools Corporation).

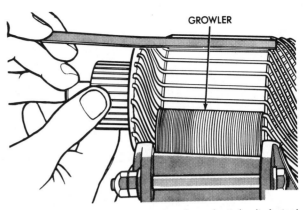

Figure 9.17 Testing a starter armature for a short circuit. A steel strip will vibrate and be attracted to the armature if a short is present (courtesy of Chrysler Corporation).

should be inspected for possible damage and the need for replacement. (See Figure 9.15.)

Check the drive assembly for slippage by holding the clutch housing while repeatedly attempting to turn the pinion gear. The gear should move freely and smoothly in one direction, but should immediately lock and refuse to turn in the opposite direction. Replace the drive assembly if the gear does not lock up or if it slips or "skips" before locking.

Armature Testing   Most shops use a *growler* for testing armatures. A growler, shown in Figure 9.16, is a device consisting of a coil wrapped around a laminated "V" block. A growler gets its name from the noise that it makes when in use. When connected to a 110-V alternating current power source, the coil creates an alternating magnetic field within the "V" block. An armature is placed in the "V" block for testing. If an armature winding is shorted, a stray magnetic field will be created above the armature. This magnetic field can be detected by holding a steel strip (a hacksaw blade works well) above the armature while it is slowly rotated. (See Figure 9.17.) If a winding is shorted, the blade will be attracted to the armature and will vibrate. If no short exists, the blade will not be attracted.

Most growlers incorporate a lamp wired in series with a pair of test probes. The lamp is used as a continuity tester to check for a grounded armature and to check for opens and grounds in field windings. On most growlers, the test lamp is wired in the 110-V circuit. Because of this, careless handling of the test probes can cause you to receive a shock. When using a growler or the test lamp, you should hold only the insulated handles of the

test probes. As an added precaution, the parts that you are testing should be placed on an insulating surface such as a wood-topped bench.

Testing an Armature for a Short

1 Place the armature in a growler.

2 Turn on the growler and hold a hacksaw blade or other thin steel strip above the armature. (Refer to Figure 9.17.)

3 Slowly rotate the armature through 360°. If the armature is shorted, the blade will be attracted to the armature and will vibrate. (A shorted armature should be replaced.)

4 Turn off the growler.

Testing an Armature for a Ground

1 Turn on the growler test lamp and touch the test probes together. The test lamp should light.

2 Touch the test probes to the commutator and to the laminations of the armature as shown in Figure 9.18. The test lamp should not light. If the test lamp lights, the armature is grounded. (A grounded armature should be replaced.)

3 Turn off the test lamp.

Alternate Test for a Grounded Armature   If the growler that you have available does not incorporate a test lamp, an armature may be tested for grounds by using a jumper wire to connect it in series with a battery and a voltmeter or test lamp as shown in Figure 9.19. If any reading is obtained on the voltmeter, or if the test lamp lights, the armature is grounded and should be replaced.

Armature Reconditioning   The armature laminations should be checked for rub marks indicat-

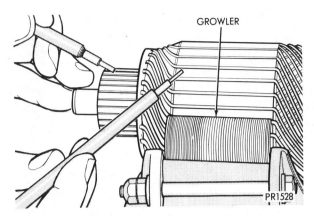

Figure 9.18 Testing a starter armature for a ground. If the armature is grounded, the test lamp will light when the probes are held as shown (courtesy of Chrysler Corporation).

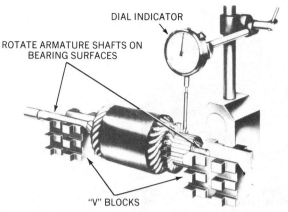

Figure 9.20 With the armature supported in "V" blocks, a dial indicator can be used to check for run-out and out-of-round. (courtesy of Ford Motor Company).

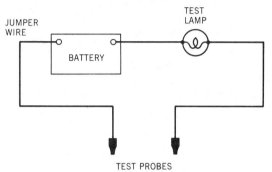

Figure 9.19 A test lamp and a jumper wire connected to a battery as shown can be used to test armatures and field coils for grounds and opens.

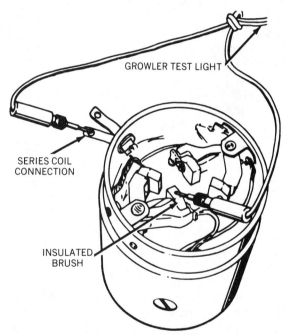

Figure 9.21 Testing a series coil for opens. The probes of the test light should be connected to the series coil connection and to each of the insulated brushes. If the lamp does not light, the coil is open (courtesy of Chevrolet Motor Division).

ing contact with the field pole shoes. If the armature has been rubbing on the pole shoes, the armature may have a bent shaft or the bushings in the starter motor may have excessive wear.

The commutator bars of an armature may be rough or burned. If the bars are not badly burned, they may be smoothed by using *commutator paper* or other abrasive paper made with a nonconductive abrasive held between your fingers while rotating the armature. A rough commutator may be turned down in a lathe so that a smooth surface is obtained. Most manufacturers do not recommend undercutting the insulation between the commutator bars.

A commutator that is out-of-round more than 0.005 inch (.127 mm) should also be turned down in a lathe. While a bent armature shaft and an out-of-round commutator may be detected in a lathe, many shops use a dial indicator and "V" blocks as shown in Figure 9.20.

Field Coil Testing   The field coils of a starter motor should be tested for opens and grounds. In most instances, the test lamp of a growler is used.

Testing Field Coils for Opens

1 Turn on the growler test lamp and touch the test probes together. The test lamp should light.
2 Hold one of the test probes to the field coil connection and the other test probe to one of the in-

Job 9b

INSPECT STARTER DRIVE ASSEMBLIES

SATISFACTORY PERFORMANCE
A satisfactory performance on this job requires that you do the following:

1 Inspect the starter drive assemblies assigned.
2 Following the steps in the "Performance Outline," complete the inspection within 15 minutes.
3 Fill in the blanks under "Information."

PERFORMANCE OUTLINE
1 Check the drive pinion for worn and damaged teeth.
2 Check the drive pinion for looseness.
3 Check the clutch for slippage or skipping.

INFORMATION
Drive identification: _____ #1   _____ #2   _____ #3

Drive #1:

_____ Teeth worn   _____ Teeth damaged   _____ Pinion loose

_____ Clutch slips or skips   _____ OK

Drive #2:

_____ Teeth worn   _____ Teeth damaged   _____ Pinion loose

_____ Clutch slips or skips   _____ OK

Drive #3:

_____ Teeth worn   _____ Teeth damaged   _____ Pinion loose

_____ Clutch slips or skips   _____ OK

sulated brushes, as shown in Figure 9.21. The test lamp should light.
3 Repeat step 2 at the remaining insulated brush. The test lamp should light.

Note: If the test lamp does not light during both of the above tests, there is an open in the field coil circuit. If the open is in the field coils, they must be replaced.

Testing Field Coils for Grounds
1 Turn on the growler test lamp and touch the test probes together. The test lamp should light.
2 Hold one of the test probes to the field coil con-nection and the other probe to one of the grounded brushes, as shown in Figure 9.22. The test lamp should not light. If the test lamp lights, the field coils are grounded. (Grounded field coils should be replaced.)

Note: On starter motors with a shunt coil, the series and the shunt coils straps must be separated as shown in Figure 9.23. The test lamp probes should then be connected to either of the insulated brushes and to one of the grounded brush holders. The test lamp should not light. If the test lamp lights, the field coils are grounded. (Grounded field coils should be replaced.)

Job 9c

INSPECT AND TEST STARTER MOTOR ARMATURES

SATISFACTORY PERFORMANCE

A satisfactory performance on this job requires that you do the following:

1 Inspect and test the starter motor armatures assigned.

2 Following the steps in the "Performance Outline," complete the tests required within 15 minutes.

3 Fill in the blanks under "Information."

PERFORMANCE OUTLINE

1 Check the armatures for shorts.

2 Check the armatures for grounds.

3 Check the armature commutator bars.

INFORMATION

Armature identification: _____ #1   _____ #2   _____ #3

Armature #1 is _____ shorted   _____ grounded   _____ OK

Armature #2 is _____ shorted   _____ grounded   _____ OK

Armature #3 is _____ shorted   _____ grounded   _____ OK

The commutator bars on

Armature #1: _____ require machining   _____ are OK

Armature #2: _____ require machining   _____ are OK

Armature #3: _____ require machining   _____ are OK

Comments _____

Alternate Field Coil Tests   If the growler you have available does not incorporate a test lamp, the field coil test procedures given above may be performed using a jumper lead and a voltmeter or a 12-V test lamp connected to a battery. (Refer to Figure 9.19.)

Overhauling a Delco Starter Motor   The following steps outline a typical procedure for the disassembly, component testing, and reassembly of a Delco starter motor. You should consult an appropriate manual for the procedure and specifications that may be required for the specific starter motor that you overhaul:

Disassembly

1 Remove the screw holding the field strap to the solenoid. (See Figure 9.24.)

2 Remove the two solenoid mounting screws.

3 Holding the solenoid in against the pressure of the return spring, rotate the solenoid about 90° to release it, and remove the solenoid and the return spring.

4 Remove the two thru-bolts (Refer to Figure 9.24.)

5 Remove the commutator end frame. (See Figure 9.25.)

Note: On starter motors used on diesel engines, the end frame insulator should also be removed. (See Figure 9.26).

6 Remove the leather washer from the end of the armature shaft.

7 Separate the field frame from the drive end housing and slide the field frame off the armature.

8 Remove the armature and drive assembly from the drive end frame.

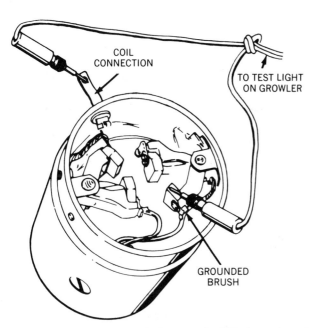

Figure 9.22 Testing field coils for grounds. With the probes of the test lamp connected as shown, the lamp should not light (courtesy of Chevrolet Motor Division).

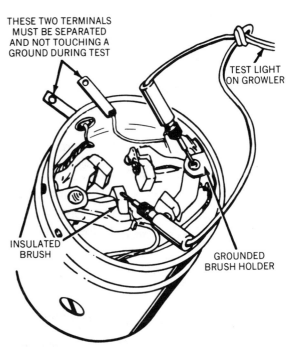

Figure 9.23 Testing a series coil for grounds. On starter motors with a shunt coil, the series and shunt coil straps must be separated. The test lamp probes should contact either of the insulated brushes and one of the grounded brush holders. The test lamp should not light (courtesy of Chevrolet Motor Division).

Note: On starter motors used on diesel engines, the shift lever pivot bolt and the two screws holding the center bearing must be removed to free the armature and the drive assembly. (See Figure 9.27.)

9 Remove the thrust washer from the end of the armature shaft and, using a socket or pipe coupling as shown in Figure 9.28, drive the retainer down off the snap ring.

10 Using a pair of lock ring pliers as shown in Figure 9.29, carefully spread the snap ring and remove it.

Note: Spread the snap ring only enough so that it can be removed. If the ring is overexpanded, it must be replaced.

11 Remove the retainer and the drive assembly from the armature shaft. (See Figure 9.30.)

Note: On starter motors used on diesel engines, the fiber washer and the center bearing should also be removed. (See Figure 9.31.)

12 Clean all of the parts.

Note: Do not immerse the armature, field coil assembly, solenoid, and drive assembly in any solvents. Those parts should be wiped clean.

13 Check the drive assembly for wear, damage, and slippage. (Refer to Figures 9.13, 9.14, and 9.15.)
14 Inspect the armature and test it for shorts and grounds. (Refer to Figures 9.17, 9.18, and 9.19.)
15 Check the fit of the armature shaft in the bushing in the drive end housing. If excessive side play or "wobble" exists, replace the bushing.

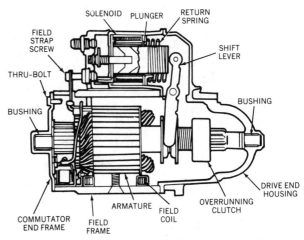

Figure 9.24 A sectioned view of a Delco starter motor (courtesy of Chevrolet Motor Division).

Note: The bushing may be driven out with a bushing driver or a tool made from the end of a defective armature shaft.

16 Check the fit of the armature shaft in the bushing in the commutator end frame. Replace the bushing if excessive side play or "wobble" exists.

Note: On most Delco starter motors, the commutator end frame bushing is not supplied as a separate part. The end frame assembly must be replaced.

Job 9d

TEST STARTER MOTOR FIELD COILS

SATISFACTORY PERFORMANCE
A satisfactory performance on this job requires that you do the following:

1 Test the starter motor field coil assemblies
assigned.
2 Following the steps in the "Performance Outline,"
complete the tests required within 15 minutes.
3 Fill in the blanks under "Information."

PERFORMANCE OUTLINE
1 Check the field coils for opens.
2 Test the field coils for grounds.

INFORMATION
Field coil identification: _____ #1   _____ #2   _____ #3

Coil assembly #1 is _____ open   _____ grounded   _____ OK

Coil assembly #2 is _____ open   _____ grounded   _____ OK

Coil assembly #3 is _____ open   _____ grounded   _____ OK

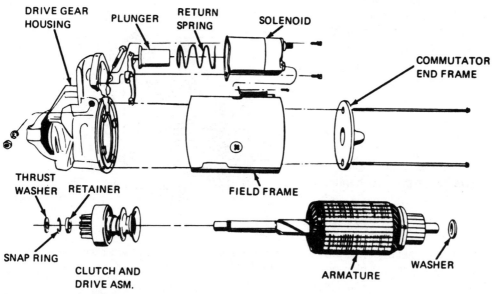

Figure 9.25 A disassembled view of a typical Delco starter motor
(courtesy of Chevrolet Motor Division).

17 Test the field coils for opens and grounds. (Refer to Figures 9.21, 9.22, and 9.23.)

Brush Replacement On most Delco starter motors, the brush holders are held by pivot pins, as shown in Figure 9.32. Pulling a pin straight out will release one grounded brush holder, one insulated brush holder, and a brush holder spring. The screws hold-ing the brushes and their leads can then be removed. It usually is easier to remove one pivot pin, replace the brushes, and reinstall the brush holders on one side at a time. This enables you to refer to the remaining side for the proper positioning of the parts and the angle of the brushes.

On the smaller 5MT starter motor shown in Figure 9.33, the brush holders slide in brush supports. Each

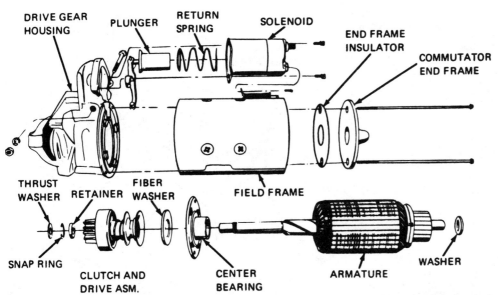

Figure 9.26 A disassembled view of a Delco starter motor of the type used on diesel engines. Note that the field frame and the armature are longer and that two screws are used to secure each pole shoe. Note also that a center bearing is used to support the armature behind the drive assembly (courtesy of Chevrolet Motor Division).

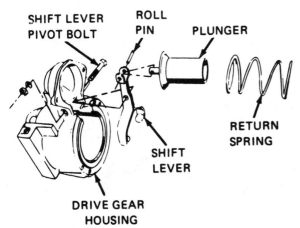

Figure 9.27 When disassembling Delco starter motors used on diesel engines, it is necessary to remove the shift lever pivot bolt before the armature and drive assembly can be removed from the drive gear housing (courtesy of Chevrolet Motor Division).

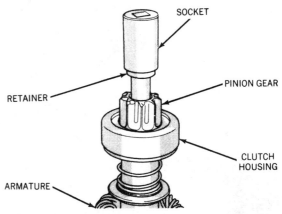

Figure 9.28 A socket or a pipe coupling can be used to drive the retainer down off the snap ring on the armature shaft (courtesy of Chrysler Corporation).

brush holder can be removed individually for brush replacement. (See Figure 9.34.)

1 Referring to Figures 9.32 and 9.34, replace one brush at a time.

Note: The face of each brush is angled for full contact with the commutator. Be sure that each brush is installed with its face angled in the proper direction. Refer to the angle of the brushes remaining in the field frame.

Solenoid Service   The solenoid is considered an assembly, and individual parts are usually not avail-

able for replacement. If the terminals and nuts appear burned, or if the cover is burned or cracked, the solenoid should be replaced. The cover may be removed for inspection, as shown in Figure 9.35. As two leads from the solenoid windings are welded to two of the terminal screws, the terminal nuts must be removed carefully so that the screws are not turned. If those terminal screws are turned, the leads usually will be broken loose and the solenoid must be discarded.

The surface of the contact disc and the contact face of the large terminal screws may be cleaned with abrasive cloth. As the battery terminal screw fits into a square hole in the cover, many mechanics

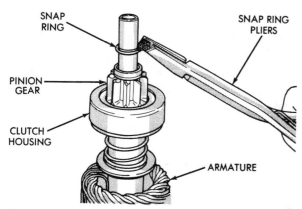

Figure 9.29 Removing the snap ring from the armature shaft (courtesy of Chrysler Corporation).

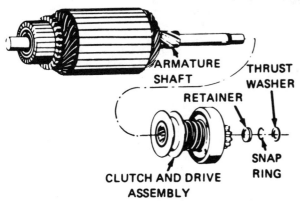

Figure 9.30 The armature, drive assembly, and related parts (courtesy of Chevrolet Motor Division).

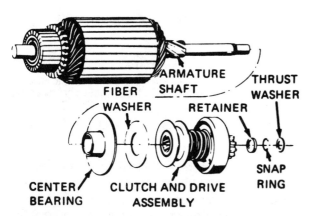

Figure 9.31 The armature, drive assembly, and related parts of a starter motor used on a diesel engine. Notice the location of the center bearing and the fiber washer (courtesy of Chevrolet Motor Division).

install the screw 180° from its original position to obtain a fresh contact surface.

The solenoid windings may be tested in the following manner:

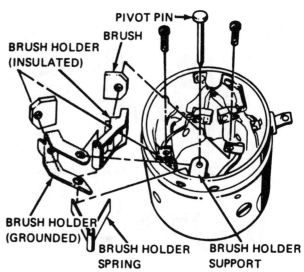

Figure 9.32 Brush holder details of a typical Delco starter motor (courtesy of Chevrolet Motor Division).

1 Connect a 12-V battery, a carbon pile, a voltmeter, an ammeter, and a heavy duty switch, as shown in Figure 9.36.

2 Check the hold-in winding by closing the switch and adjusting the carbon pile so that a voltmeter reading of 10 V is obtained. The ammeter should indicate from 14.5 to 16.5 A. Open the switch and turn the carbon pile to the Off position.

Note: Perform this test as quickly as possible. As current flows through the solenoid windings, they will overheat and their resistance will increase.

3 Check both the hold-in and pull-in windings by adding a jumper wire to the hook-up used for the previous test. As shown in Figure 9.37, the jumper wire should be connected between the motor (*M*) terminal and grounds.

4 Close the switch and adjust the carbon pile so that a voltmeter reading of 10 V is obtained. The ammeter should indicate from 41 to 47 A. Open the switch and turn the carbon pile to the Off position.

Note: Perform this test as quickly as possible. If the ammeter readings are not within the specifications given, replace the solenoid.

Assembly

1 Place a few drops of oil on the drive end of the armature shaft and install the drive assembly. (See Figure 9.38.)

Note: On starter motors used on diesel engines, install the center bearing and the fiber washer before

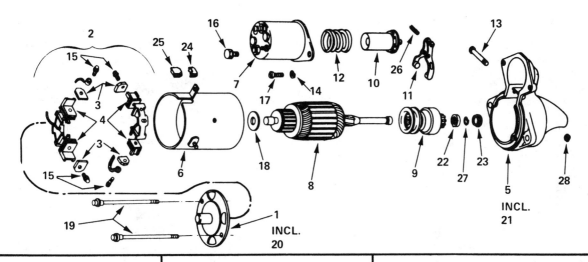

Figure 9.33 An exploded view of a Delco 5MT type starter motor. Notice that the pole shoes and field coils are not removable and that the brush holders are different (courtesy of Chevrolet Motor Division).

1. FRAME—COMMUTATOR END	10. PLUNGER	19. THRU BOLT
2. BRUSH AND HOLDER PKG.	11. SHIFT LEVER	20. BUSHING—COMMUTATOR END
3. BRUSH	12. PLUNGER RETURN SPRINGER	21. BUSHING—DRIVE END
4. BRUSH HOLDER	13. SHIFT LEVER SHAFT	22. PINION STOP COLLAR
5. HOUSING—DRIVE END	14. LOCK WASHER	23. THRUST COLLAR
6. FRAME AND FIELD ASM.	15. SCREW—BRUSH ATTACHING	24. GROMMET
7. SOLENOID SWITCH	16. SCREW—FIELD LEAD TO SWITCH	25. GROMMET
8. ARMATURE	17. SCREW—SWITCH ATTACHING	26. PLUNGER PIN
9. DRIVE ASM.	18. LEATHER WASHER—BRAKE	27. PINION STOP RETAINER RING
		28. LEVER SHAFT RETAINING RING

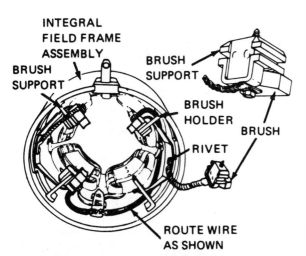

Figure 9.34 Brush holder details of the Delco type 5MT starter motor (courtesy of Chevrolet Motor Division).

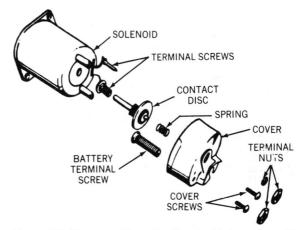

Figure 9.35 The solenoid may be disassembled as shown. Notice that two terminal screws are welded to the wires of the solenoid winding (courtesy of Chevrolet Motor Division).

installing the drive assembly. (Refer to Figure 9.31.)

2 Install the retainer on the shaft with its cupped end facing outward.

3 Install the snap ring on the armature shaft.

Note: While lock ring pliers may be used for this step (refer to Figure 9.29), an alternate method, shown in Figure 9.39, minimizes the possibility of overexpanding the snap ring.

4 Place the thrust washer on the shaft and, using pliers as shown in Figure 9.40, force the retainer up over the snap ring.

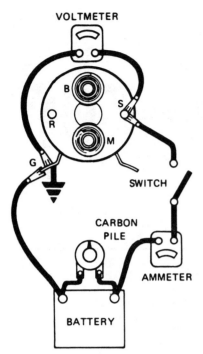

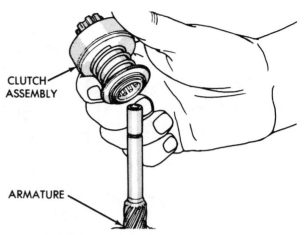

Figure 9.38 Installing the drive assembly on the armature shaft (courtesy of Chrysler Corporation).

Figure 9.36 The wire hook-up for testing the hold-in winding of a Delco starter solenoid (courtesy of Chevrolet Motor Division).

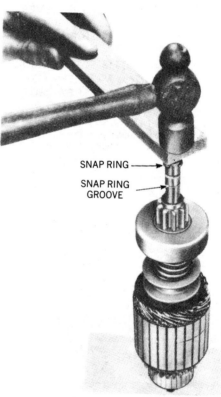

Figure 9.39 The snap ring may be installed by positioning it on the end of the shaft and starting it on by using a piece of wood and a hammer as shown. The ring can then be slid down into its groove (courtesy of Pontiac Motor Division, General Motors Corporation).

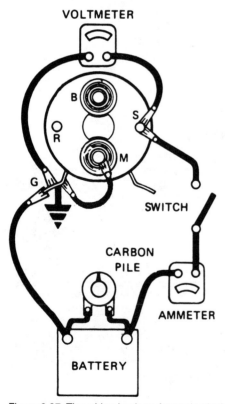

Figure 9.37 The wiring hook-up for testing both the hold-in and the pull-in windings of a Delco starter solenoid (courtesy Chevrolet Motor Division).

Note: As shown in Figure 9.41, some mechanics use a battery terminal puller to force the retainer up into place.

5 Place some lubricant in the drive end bushing and insert the armature shaft while positioning the

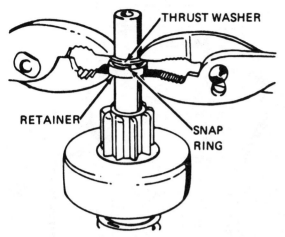

Figure 9.40 The retainer can be forced up over the snap ring by using two pliers (courtesy of Chevrolet Motor Division).

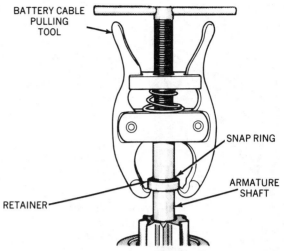

Figure 9.41 A battery terminal puller may be used to pull the retainer up over the snap ring (courtesy of Chrysler Corporation).

buttons on the shift lever in the drive assembly. (Refer to Figures 9.25 and 9.27.)

Note: On starter motors used on diesel engines, the screws holding the center bearing should now be installed.

6 Holding the brushes back against the tension of their springs, slide the field frame over the armature. Rotate the field frame so that the alignment pin on the field frame fits into the hole in the drive end housing.

7 Check the brushes for proper contact with the commutator.

Note: At times, the brushes must be shifted slightly in the brush holders so that even contact can be obtained.

8 Install the leather washer on the armature shaft.

9 Place some lubricant in the bushing in the commutator end frame bushing and slide the end frame on the shaft.

Note: On starter motors used on diesel engines, the end frame insulator should be installed before the end frame. (Refer to Figure 9.26.)

10 Align the thru-bolt holes, and install the thru-bolts.

11 Place the solenoid return spring over the plunger and install the solenoid, rotating it so that the tab on the solenoid fits into the slot at the rear of the drive end housing.

12 Install the solenoid mounting screws.

Note: Before installing the screw holding the field strap to the solenoid motor (*M*) terminal, the pinion clearance should be checked.

Checking Pinion Clearance Following an overhaul or the replacement of a solenoid, the pinion clearance should be checked. If the pinion is not pulled out far enough by the solenoid, it will not mesh properly with the gear on the flywheel. If the pinion is pulled out too far, the buttons on the yoke of the shift lever will wear rapidly. As shown in Figure 9.42, the pinion clearance is measured between the pinion and the retainer while the solenoid is energized.

1 Position a piece of cardboard or other insulating material between the field strap and the motor (*M*) terminal on the solenoid.

Note: Be sure that the field strap is well insulated. If the field strap contacts the motor (*M*) terminal, the starter motor will be energized, causing injury to your fingers during the following steps.

2 Using jumper leads, connect a 12-V battery to the solenoid switch (*S*) terminal and to the drive end frame as shown in Figure 9.43.

3 Using a jumper wire as shown in Figure 9.44, momentarily ground the motor (*M*) terminal. Remove the jumper.

Note: This will energize the solenoid and it will remain energized until the battery is disconnected.

4 With the solenoid energized, push the pinion back as far as possible and check the clearance between the pinion and the retainer. (Refer to Figure 9.42.)

Note: The clearance is nonadjustable. If the clear-

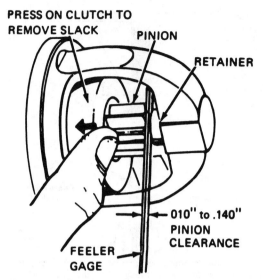

Figure 9.42 Pinion clearance should be checked with a feeler gauge while the solenoid holds the pinion in the cranking position (courtesy of Chevrolet Motor Division).

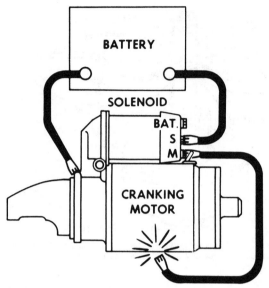

Figure 9.44 The solenoid is energized by momentarily touching a third jumper wire between the M terminal on the solenoid and the starter motor frame (courtesy of Pontiac Motor Division, General Motors Corporation).

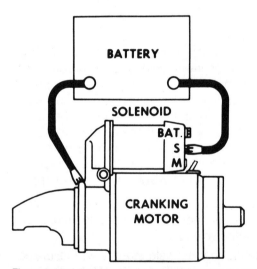

Figure 9.43 A battery should be connected to the starter motor frame and to the S terminal on the solenoid to perform the pinion clearance check (courtesy of Pontiac Motor Division, General Motors Corporation).

ance is not within the limits given, check for worn parts or improper assembly.

5 Disconnect the battery leads.

6 Install the screw holding the field strap to the solenoid motor (M) terminal.

Overhauling a Ford Positive-Engagement Starter Motor There are several variations of this starter motor in use. Figure 9.45 is representative of later designs that hold the cover in place with a screw. On earlier versions, the cover is held

by a band that covers access holes in the field frame. Figure 9.46 shows a starter motor of the earlier type.

The following steps outline a typical procedure for the disassembly, component testing, and reassembly of a Ford positive engagement starter motor. You should consult an appropriate manual for the procedure and specifications that may be required for the specific starter motor you overhaul:

Disassembly

1 Remove the cover screw and lift the cover from the top of the starter motor. (Refer to Figure 9.45.)

Note: On earlier starter motors, the band screw must be loosened and the band removed to release the cover. (Refer to Figure 9.46.)

2 On earlier designs with access holes in the field frame, use a brush hook similar to the one shown in Figure 9.47 to raise the brush springs and pull the brushes out of their holders.

3 Remove the thru-bolts and pull the drive end housing off of the front of the starter motor.

Note: This will free the shift lever return spring.

4 Push out the pivot pin holding the shift lever and remove the shift lever.

5 Remove the armature from the front of the field frame.

6 Remove the fiber washer from the armature shaft.

Job 9e

OVERHAUL A DELCO STARTER MOTOR

SATISFACTORY PERFORMANCE
A satisfactory performance on this job requires that you do the following:

1 Overhaul the starter motor assigned.
2 Following the steps in the "Performance Outline" and the procedure and specifications of the manufacturer, complete the job within 120 minutes.
3 Fill in the blanks under "Information."

PERFORMANCE OUTLINE
1 Disassemble the starter motor.
2 Inspect the parts for wear and damage.
3 Test the armature and the field coils.
4 Inspect the solenoid.
5 Assemble the starter motor, replacing the parts required.
6 Check the pinion clearance.

INFORMATION
Starter motor identification _____

Reference used _____ Page(s) _____

Armature was: _____ shorted _____ grounded _____ OK

Field coils were: _____ open _____ grounded _____ OK

Brushes were: _____ excessively worn _____ OK

Parts replaced _____

---

7 Remove the end plate from the rear of the field frame.

8 On later designs, remove the insulator. (Refer to Figure 9.45.)

9 On later designs, release the brushes from the brush holder and remove the brush holder assembly.

Note: Observe the position of the brush holder assembly in its relation to the field terminal that extends from the rear of the field frame.

10 Remove the washer and the retainer from the armature shaft.

11 Remove the stop ring from its groove in the armature shaft.

Note: A $7/16$ inch open end wrench can be used as a fork to drive the horseshoe-shaped stop ring from its groove.

12 Remove the drive assembly from the armature shaft.

13 Check the drive assembly for wear, damage, and slippage. (Refer to Figures 9.13, 9.14, and 9.15.)

14 Inspect the armature and test it for shorts and grounds. (Refer to Figures 9.17, 9.18, and 9.19.)

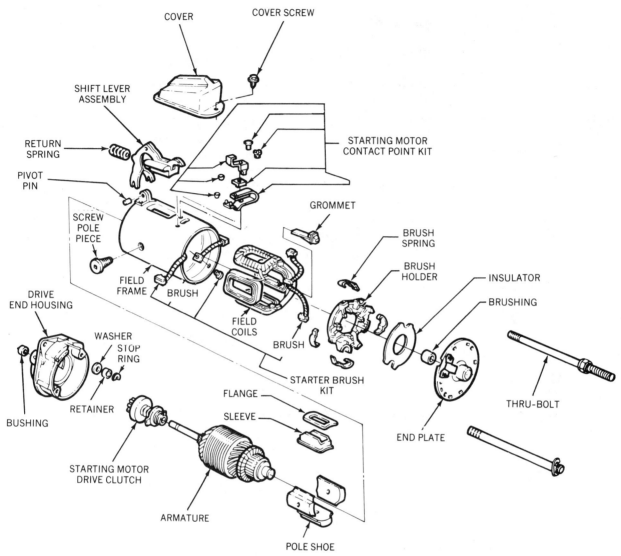

COVER

COVER SCREW

SHIFT LEVER
ASSEMBLY

RETURN
SPRING

STARTING MOTOR
CONTACT POINT KIT

PIVOT
PIN

SCREW
POLE
PIECE

GROMMET

BRUSH
SPRING

BRUSH
HOLDER

INSULATOR

BRUSHING

DRIVE
END HOUSING

FIELD
FRAME

BRUSH

FIELD
COILS

BRUSH

STARTER BRUSH
KIT

END PLATE

THRU-BOLT

WASHER
STOP
RING

FLANGE

SLEEVE

BUSHING

RETAINER

STARTING MOTOR
DRIVE CLUTCH

ARMATURE

POLE SHOE

Figure 9.45 An exploded view of a Ford positive engagement
starter motor (courtesy of Ford Motor Company).

15 Check the fit of the armature shaft in the bushing in the drive end housing. If excessive side play or "wobble" exists, replace the bushing.

Note: The bushing may be driven out with a bushing driver or a tool made from the end of a defective armature shaft.

16 Check the fit of the armature shaft in the bushing in the end frame. Replace the bushing if excessive side play or "wobble" exists.

17 Test the field coils for opens and grounds. (Refer to Figures 9.21, 9.22, and 9.23.)

Brush Replacement   In Ford positive engagement starter motors, the ground brushes are connected

to the field frame by screws. The insulated field coil brushes are welded or soldered to the field windings as shown in Figure 9.48. If the brushes are worn to less than ¼ inch (6 mm) in length, they should be replaced.

If the insulated brushes require replacement, they should be cut off as close as possible to the field windings. On those starter motors where the brush leads were originally soldered in place, they may be unsoldered. The leads of the new brushes should be soldered to the field windings using rosin core solder. As a soldering gun or small soldering iron will not provide sufficient heat, a soldering iron with a rating of at least 300 W should be used.

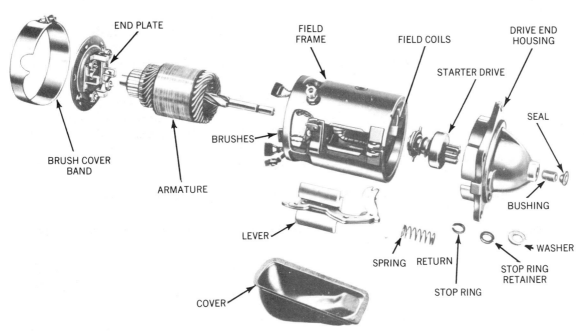

Figure 9.46 Earlier versions of the Ford positive engagement starter motor had access holes in the field frame covered by a brush cover band (courtesy of Ford Motor Company).

Figure 9.47 Fashioned as shown from a piece of welding rod or coat hanger wire, a brush hook is very useful in overhauling a starter motor.

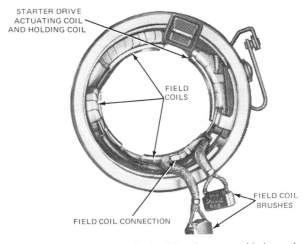

Figure 9.48 The insulated field coil brushes are welded or soldered to the field coil connections (courtesy of Ford Motor Company).

Assembly

1 Place a few drops of oil on the drive end of the armature shaft and install the drive assembly. (Refer to Figure 9.38.)

2 Install the stop ring in its groove in the shaft.

Note: The stop ring should fit tight in its groove.

3 Install the retainer so that its cupped end fits over the stop ring and holds it in place. Install the washer (if fitted). (Refer to Figure 9.45.)

4 Position the armature within the field frame.

Note: On starter motors with access holes in the field frame, each brush should be pushed out through its access hole.

5 Install the shift lever assembly so that the notches on the yoke fit over the tabs on the drive assembly.

6 Install the shift lever pivot pin.

7 Place some lubricant in the bushing in the drive end housing, and slide the housing over the armature.

8 Install the shift lever return spring in the pocket in the drive end housing, aligning it with the protrusion on the shift lever.

9 On starter motors without the access holes, install the brush holder in the field frame, aligning it with the extended field terminal.

10 On starter motors without the access holes, in-

sert the brushes in the brush holder and install the springs. Install the insulator.

11 Install the fiber washer on the armature shaft.

12 Place some lubricant in the bushing in the end plate and install the end plate.

13 Align the drive end housing, the field frame, and the end plate and install the thru-bolts.

Note: Notches in the ends of the field frame align with a pin in the drive end housing and a tab on the end plate to facilitate alignment of those parts.

14 On starter motors with access holes, use a brush hook to raise the brush springs and insert the brushes in the brush holders.

Note: Be sure that the springs rest in the notch in the center of each brush and that the leads are not pinched.

15 Install the cover over the shift lever and secure it with the screw (or band).

Overhauling a Chrysler Gear Reduction Starter Motor The following steps outline a typical procedure for the disassembly, component testing, and reassembly of a Chrysler gear reduction starter motor of the type shown in Figures 9.49 and 9.50. You should consult an appropriate manual for the procedure and specifications that may be required for the specific starter motor you overhaul.

Job 9f

OVERHAUL A FORD POSITIVE-ENGAGEMENT STARTER MOTOR

SATISFACTORY PERFORMANCE

A satisfactory performance on this job requires that you do the following:

1 Overhaul the starter motor assigned.

2 Following the steps in the "Performance Outline" and the procedure and specifications of the manufacturer, complete the job within 120 minutes.

3 Fill in the blanks under "Information."

PERFORMANCE OUTLINE

1 Disassemble the starter motor.

2 Inspect the parts for wear and damage.

3 Test the armature and the field coils.

4 Assemble the starter motor, replacing the parts required.

INFORMATION

Starter motor identification _____

Reference used _____ Page(s) _____

Armature was: _____ shorted   _____ grounded   _____ OK

Field coils were: _____ open   _____ grounded   _____ OK

Brushes were: _____ excessively worn   _____ OK

Parts replaced _____

_____

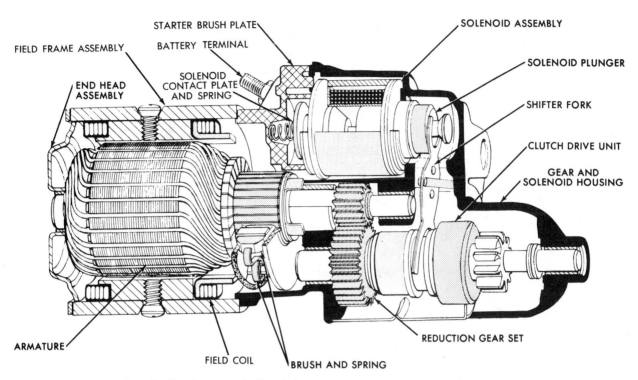

STARTER BRUSH PLATE

FIELD FRAME ASSEMBLY

BATTERY TERMINAL

SOLENOID ASSEMBLY

SOLENOID PLUNGER

END HEAD ASSEMBLY

SOLENOID CONTACT PLATE AND SPRING

SHIFTER FORK

CLUTCH DRIVE UNIT

GEAR AND SOLENOID HOUSING

ARMATURE

FIELD COIL

BRUSH AND SPRING

REDUCTION GEAR SET

Figure 9.49 A cutaway view of a Chrysler gear reduction starter motor (courtesy of Chrysler Corporation).

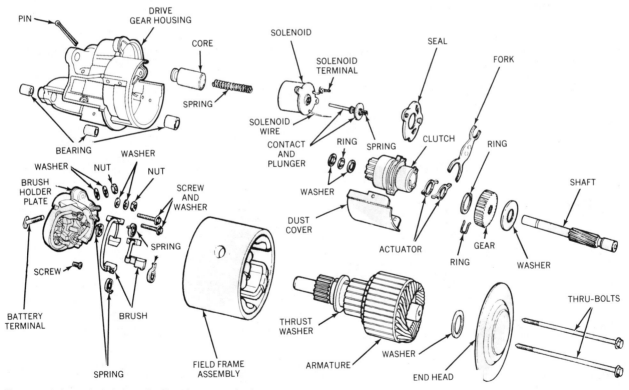

PIN

DRIVE GEAR HOUSING

CORE

SOLENOID

SOLENOID TERMINAL

SEAL

FORK

SPRING

SOLENOID WIRE

CONTACT AND PLUNGER

RING

SPRING

CLUTCH

RING

SHAFT

BEARING

WASHER

WASHER

WASHER

NUT

NUT

DUST COVER

ACTUATOR

RING

GEAR

WASHER

BRUSH HOLDER PLATE

SCREW AND WASHER

SPRING

SCREW

BATTERY TERMINAL

BRUSH

THRUST WASHER

THRU-BOLTS

FIELD FRAME ASSEMBLY

ARMATURE

WASHER

END HEAD

SPRING

Figure 9.50 An exploded view of a Chrysler gear reduction starter motor (courtesy of Chrysler Corporation).

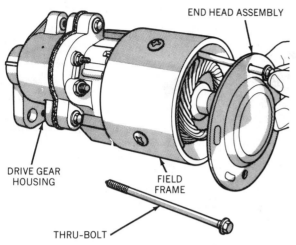

Figure 9.51 Removing the end head assembly (courtesy of Chrysler Corporation).

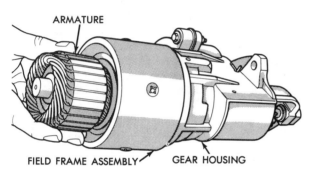

Figure 9.52 With the end head removed, the armature can be pulled from the field frame (courtesy of Chrysler Corporation).

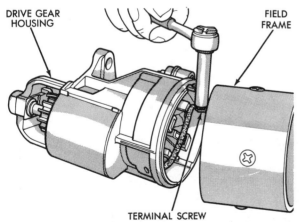

Figure 9.53 When the field frame and the drive gear housing are separated, the terminal screw can be removed (courtesy Chrysler Corporation).

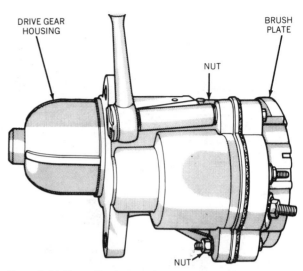

Figure 9.54 Removing the nuts from the solenoid bolts (courtesy of Chrysler Corporation).

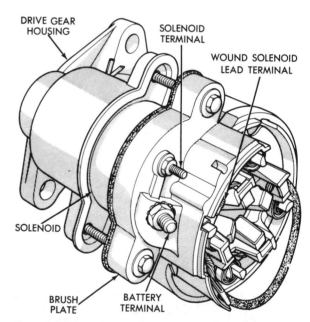

Figure 9.55 The solenoid and brush plate assembly can be separated from the drive-gear housing (courtesy of Chrysler Corporation).

Disassembly

1 Remove the two thru-bolts and the end head assembly. (See Figure 9.51.)

2 Slide the armature out of the rear of the field frame, as shown in Figure 9.52. Remove the washer from the armature shaft.

3 Separate the drive gear housing and the field frame just enough to expose the terminal screw in the field winding.

4 As shown in Figure 9.53, remove the terminal screw.

5 Remove the nuts from the bolts holding the brush plate to the drive gear housing. (See Figure 9.54.)

6 Remove the solenoid and brush plate assembly

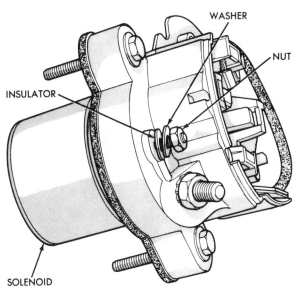

Figure 9.56 Removing the nut, washer, and insulator from the solenoid terminal (courtesy of Chrysler Corporation).

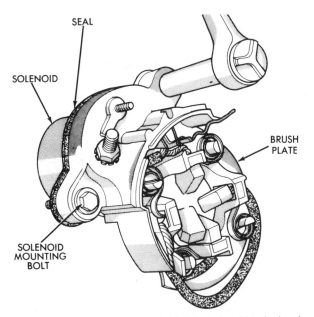

Figure 9.58 Removing the bolts holding the solenoid to the brush plate (courtesy of Chrysler Corporation).

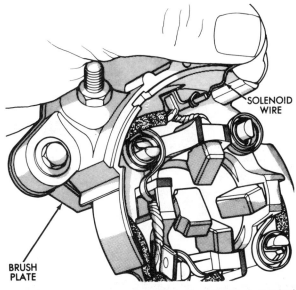

Figure 9.57 Unwinding the solenoid wire from the brush terminal (courtesy of Chrysler Corporation).

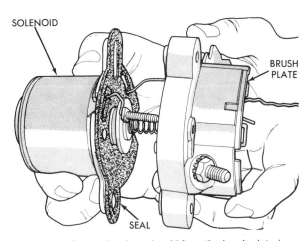

Figure 9.59 Separating the solenoid from the brush plate (courtesy of Chrysler Corporation).

from the drive gear housing, as shown in Figure 9.55.

7 As shown in Figure 9.56, remove the nut, washer, and insulator from the solenoid terminal.

8 Carefully unwind the solenoid wire from the brush terminal, as shown in Figure 9.57.

9 Remove the bolts holding the solenoid to the brush plate. (See Figure 9.58.)

10 Separate the solenoid from the brush plate as shown in Figure 9.59.

Note: Separate the parts carefully so that the seal

is not torn. Use care not to break the solenoid wire.

11 Remove the battery terminal nut and washer and remove the battery terminal from the brush plate. (See Figure 9.60.)

12 Slide the contact plunger assembly from the solenoid, as shown in Figure 9.61.

13 Remove the return spring from the solenoid core. (See Figure 9.62.)

14 Using a screwdriver as shown in Figure 9.63, remove the dust cover from the drive gear housing.

15 As shown in Figure 9.64, pry off the snap ring that positions the driven gear on the pinion shaft.

Note: As the snap ring will fly out when released,

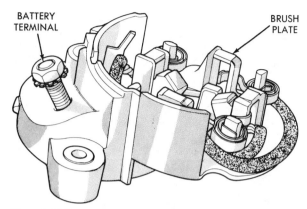

Figure 9.60 Removing the battery terminal nut and washer (courtesy of Chrysler Corporation).

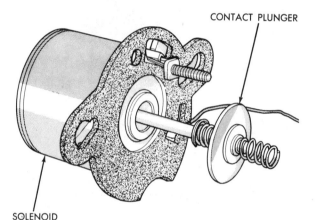

Figure 9.61 Removing the contact plunger from the solenoid (courtesy of Chrysler Corporation).

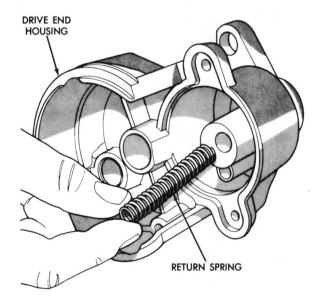

Figure 9.62 Removing the return spring from the solenoid (courtesy of Chrysler Corporation).

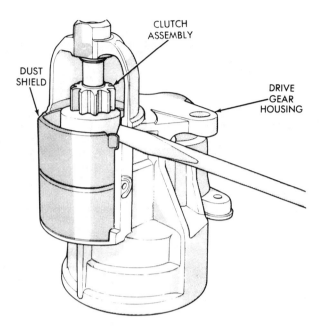

Figure 9.63 The dust cover is removed from the drive gear housing by prying it loose with a screwdriver (courtesy of Chrysler Corporation).

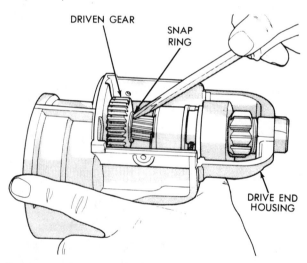

Figure 9.64 Removing the snap ring that positions the driven gear on the pinion shaft (courtesy of Chrysler Corporation).

you should hold a wiper over the assembly to catch the snap ring.

16 Using snap ring pliers as shown in Figure 9.65, remove the snap ring from its groove in the pinion shaft.

17 Using a punch or a rod as shown in Figure 9.66, push the pinion shaft toward the rear of the drive gear housing and allow the thrust washers and the lock ring to slide off the shaft.

18 Pull the pinion shaft out of the housing as shown in Figure 9.67 and remove the clutch assembly with

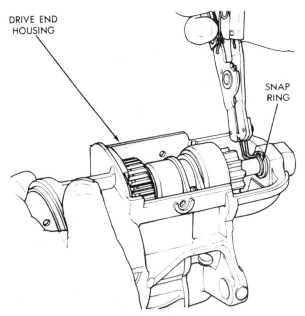

Figure 9.65 Removing the snap ring from the pinion shaft (courtesy of Chrysler Corporation).

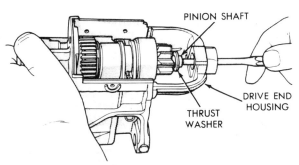

Figure 9.66 Pushing the pinion shaft through the clutch assembly (courtesy of Chrysler Corporation).

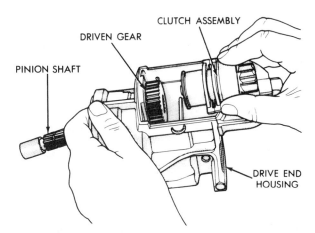

Figure 9.67 Removing the pinion shaft and clutch assembly (courtesy of Chrysler Corporation).

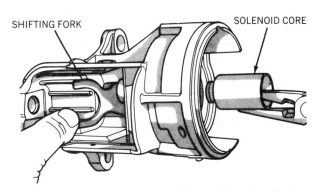

Figure 9.68 If the shifting fork is held forward, the solenoid core can be disconnected and removed (courtesy of Chrysler Corporation).

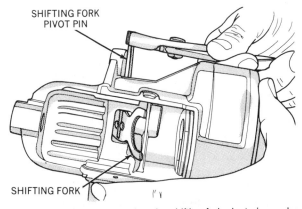

Figure 9.69 After straightening, the shifting fork pivot pin can be pulled from the housing (courtesy of Chrysler Corporation).

the two actuators. (Refer to Figure 9.50.)

19 Lift the driven gear and its washer from the housing. (Refer to Figure 9.50).

20 Push the shifting fork forward as shown in Figure 9.68, and remove the solenoid core.

21 Straighten the ends of the shifting fork pivot pin and pull it from the housing as shown in Figure 9.69.

22 Remove the shifting fork.

23 Clean all the parts.

24 Check the clutch assembly for wear, damage, and slippage. (Refer to Figures 9.13 and 9.14.)

25 Inspect the armature and test it for shorts and grounds. (Refer to Figures 9.17 and 9.18.)

26 Check the fit of the pinion shaft in the bushings in the drive gear housing. If excessive side play or "wobble" exists, replace the bushings.

27 Check the fit of the armature shaft in the bushing in the drive gear housing. If excessive side play or "wobble" exists, replace the bushing.

Note: While it is recommended that the bushings be replaced with the use of a special bushing puller

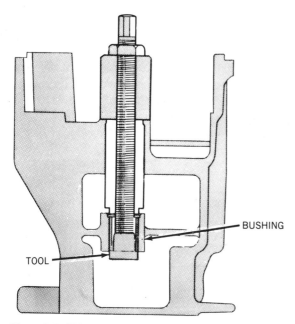

Figure 9.70 Using a special bushing puller to install the armature shaft bushing in the drive gear housing (courtesy of Chrysler Corporation).

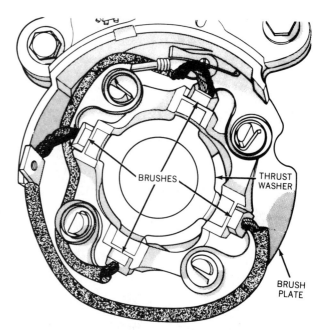

Figure 9.71 The correct position of the brushes. Note that the armature thrust washer acts as an assembly aid and holds the brushes up in the brush holders so that the armature can be installed (courtesy of Chrysler Corporation).

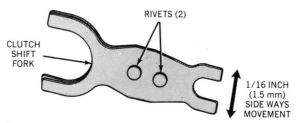

Figure 9.72 The clutch shift fork should have a slight amount of movement between the plates as shown (courtesy of Chrysler Corporation).

similar to the one shown in Figure 9.70, they may be replaced by careful use of bushing drivers.

28 Check the fit of the armature shaft in the bushing in the end head assembly. If excessive side play or "wobble" exists, replace the bushing or the end head assembly.

Note: In most instances, the bushing in the end head assembly is not supplied as a separate part. The end head assembly must be replaced.

29 Test the field coils for opens and grounds. (Refer to Figures 9.21, 9.22, and 9.23.)

30 Inspect the brushes. If they are worn to one half or less than the length of new brushes, they should be replaced.

Note: The armature thrust washer has four protrusions on its outer edge. The washer is formed in this manner so that it can be used as an assembly aid. The brushes should be placed in the brush holders as shown in Figure 9.71 with the thrust washer positioned to hold the brushes in place against the tension of the springs.

Assembly

1 Apply a couple of drops of oil between the steel plates that form the clutch shift fork. Check to see that there is about 1/16 inch (1.5 mm) side movement between the plates as shown in Figure 9.72.

2 Install the shift fork in the drive gear housing and insert the pivot pin. (Refer to Figure 9.69).

Note: The pin should straddle the shift fork between the rivets as shown in Figure 9.73.

3 Spread the ends of the pivot pin slightly so that it will be retained in the housing. Check the operation of the shift fork to see that it moves freely.

4 Hold the shift fork in the forward position and install the solenoid core so that it engages the fork. (Refer to Figure 9.68.)

5 Lubricate the pinion shaft and the bushings in the drive gear housing.

6 Slide the pinion shaft a short way into the rear of the drive gear housing and install the washer and the driven gear. (Refer to Figure 9.67.)

7 Install the clutch assembly, thrust washers, and snap ring, sliding the pinion shaft forward. (Refer to Figure 9.50 for the correct positioning of those parts.)

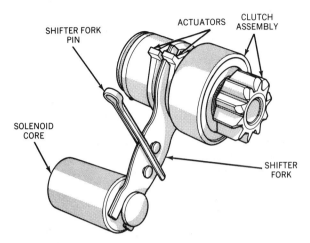

Figure 9.73 The clutch assembly, the shift fork, and the solenoid core should fit together as shown when the starter motor is assembled. Notice that the fork fits between the two actuators on the clutch assembly (courtesy of Chrysler Corporation).

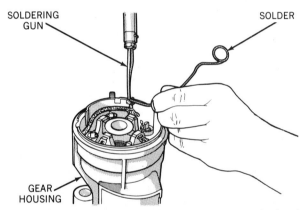

Figure 9.74 The solenoid wire must be soldered to the brush terminal during starter assembly (courtesy of Chrysler Corporation).

8 Install the pinion shaft snap ring. (Refer to Figure 9.65.)

9 Install the driven gear snap ring. (Refer to Figure 9.64.)

10 Install the dust cover. (Refer to Figure 9.63).

11 Install the return spring in the solenoid core. (Refer to Figure 9.62.)

12 Install the contact plunger assembly in the solenoid. (Refer to Figure 9.61.)

Note: If the contact disc is burned or pitted, it can be removed from the plunger and reversed.

13 Install the battery terminal in the brush plate and secure it with the nut and the washer. (Refer to Figure 9.60.)

14 Position the seal on the solenoid and assemble the solenoid to the brush plate. (Refer to Figure 9.59.)

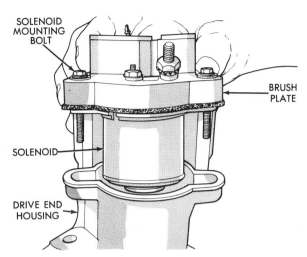

Figure 9.75 Installing the solenoid and brush plate assembly into the drive gear housing (courtesy of Chrysler Corporation).

Note: Use care so that the solenoid wires are not damaged.

15 Install the bolts holding the brush plate to the solenoid. (Refer to Figure 9.58.)

16 Install the insulator, washer, and nut on the solenoid terminal. (Refer to Figure 9.56.)

17 Carefully wrap the solenoid wire around the brush terminal. (Refer to Figure 9.57.)

18 Using a soldering gun and rosin core solder, as shown in Figure 9.74, solder the solenoid wire to the brush terminal.

19 Insert the solenoid and brush plate assembly into the drive gear housing as shown in Figure 9.75.

20 Install and tighten the nuts on the solenoid bolts. (Refer to Figure 9.54.)

21 Align the field frame with the drive gear housing assembly and install the terminal screw. (Refer to Figure 9.53.)

22 Align the field frame with the drive gear housing and slide the armature into the field frame. (Refer to Figure 9.52.) Rotate the armature slightly so that it meshes with the driven gear.

Note: As the commutator slides under the brushes, the armature thrust washer is pushed out from under the brushes, allowing them to be forced into contact with the commutator.

23 Install the washer on the armature shaft.

24 Lubricate the bushing in the end head assembly.

25 Install the end head assembly and the two thru-bolts. (Refer to Figure 9.51.)

Job 9g

OVERHAUL A CHRYSLER GEAR REDUCTION STARTER MOTOR

SATISFACTORY PERFORMANCE

A satisfactory performance on this job requires that you do the following:

1 Overhaul the starter motor assigned.
2 Following the steps in the "Performance Outline" and the procedure and specifications of the manufacturer, complete the job within 150 minutes.
3 Fill in the blanks under "Information."

PERFORMANCE OUTLINE

1 Disassemble the starter motor.
2 Inspect the parts for wear and damage.
3 Test the armature and the field coils.
4 Inspect the solenoid.
5 Assemble the starter motor, replacing the parts required.

INFORMATION

Starter motor identification _____

Reference used _____ Page(s) _____

Armature was: _____ shorted   _____ grounded   _____ OK

Field coils were: _____ open   _____ grounded   _____ OK

Brushes were: _____ excessively worn   _____ OK

Parts replaced _____

Overhauling a Nippondenso Starter Motor

The following steps outline a procedure for the disassembly, component testing, and reassembly of a typical Nippondenso starter motor. (See Figures 9.76 and 9.77.) You should consult an appropriate manual for the procedure and specifications that may be required for the specific starter motor that you overhaul.

Disassembly

1 Remove the nut holding the field coil wire to the solenoid and remove the field coil wire. (See Figure 9.78.)
2 Remove the nuts holding the solenoid to the drive end housing as shown in Figure 9.79.
3 Remove the solenoid as shown in Figure 9.80.

Note: The solenoid must be tipped to disengage the hook on the plunger from the shift lever.

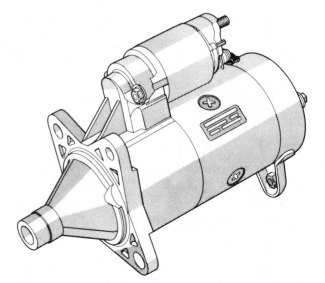

Figure 9.76 A typical Nippondenso direct drive starter motor (courtesy of Chrysler Corporation).

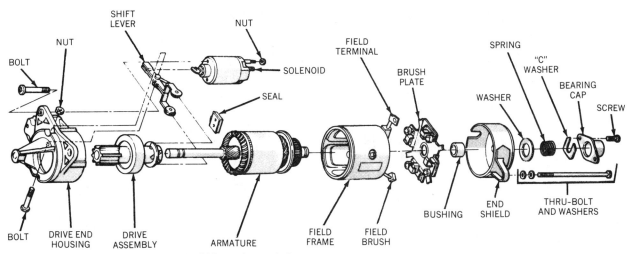

Figure 9.77 An exploded view of a typical Nippondenso starter motor (courtesy of Chrysler Corporation).

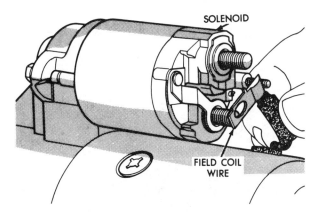

Figure 9.78 Removing the field coil wire from the solenoid terminal (courtesy of Chrysler Corporation).

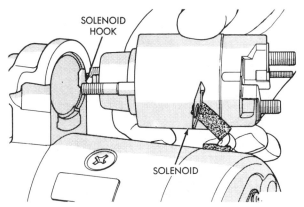

Figure 9.80 Removing the solenoid from the drive end housing (courtesy of Chrysler Corporation).

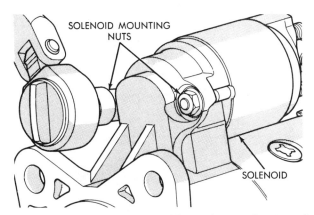

Figure 9.79 Removing the solenoid mounting nuts (courtesy of Chrysler Corporation).

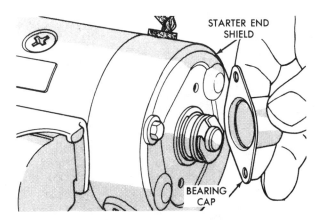

Figure 9.81 Removing the bearing cap (courtesy of Chrysler Corporation).

4 Remove the screws holding the bearing cap to the end shield and remove the bearing cap. (See Figure 9.81.)

5 Remove the "C" washer, the spring, and the flat washer from the end shield. (See Figure 9.82.)

6 Remove the two thru-bolts and remove the end shield as shown in Figure 9.83.

7 Lift the brush retaining springs and remove the

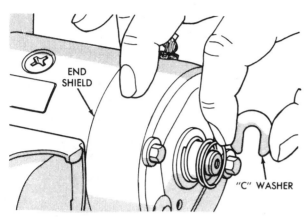

Figure 9.82 Removing the "C" washer from the rear of the armature shaft. Notice the position of the spring and the flat washer (courtesy of Chrysler Corporation).

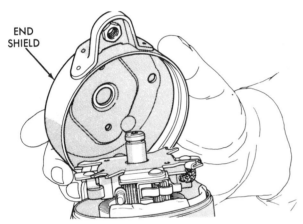

Figure 9.83 Removing the end shield (courtesy of Chrysler Corporation).

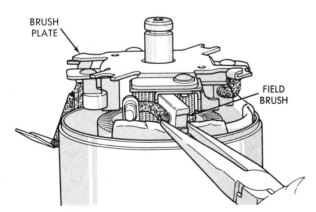

Figure 9.84 Removing the insulated field brushes from the brush plate (courtesy of Chrysler Corporation).

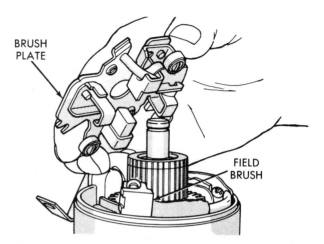

Figure 9.85 Removing the brush plate. Note that the grounded brushes remain with the brush plate (courtesy of Chrysler Corporation).

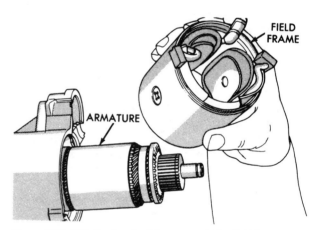

Figure 9.86 With the brush plate removed, the field frame may be removed from the armature (courtesy of Chrysler Corporation).

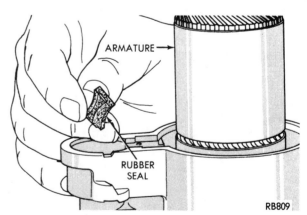

Figure 9.87 Removing the rubber seal from the drive end housing (courtesy of Chrysler Corporation).

two insulated field brushes from the brush holder. (See Figure 9.84.)

8 Remove the brush plate from the starter motor, as shown in Figure 9.85.

9 Remove the field frame. (See Figure 9.86.)

10 Lift the rubber seal from its slot in the drive end housing. (See Figure 9.87.)

11 Remove the armature and drive assembly from the drive end housing as shown in Figure 9.88.

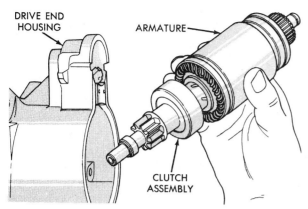

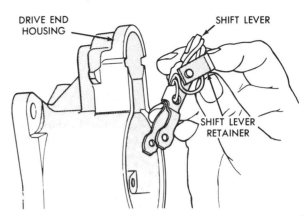

Figure 9.88 Removing the armature assembly from the drive end housing (courtesy of Chrysler Corporation).

Figure 9.90 Removing the shift lever assembly (courtesy of Chrysler Corporation).

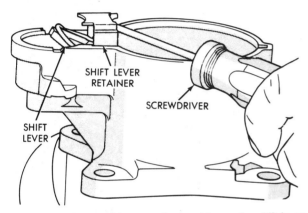

Figure 9.89 A screwdriver may be used to pry the shift lever retainer from the drive end housing (courtesy of Chrysler Corporation).

12 As shown in Figure 9.89, pry the shift lever retainer from the drive end housing.

13 Remove the shift lever assembly, as shown in Figure 9.90.

14 Using a socket or pipe coupling, drive the retainer off the snap ring on the armature shaft. (Refer to Figure 9.28.)

15 Using snap ring pliers, remove the snap ring from the armature shaft. (Refer to Figure 9.29.)

Note: Spread the snap ring only enough so that it can be removed. If the ring is overexpanded, it must be replaced.

16 Remove the retainer and the drive assembly from the armature shaft.

17 Check the drive assembly for wear, damage, and slippage. (Refer to Figures 9.13 and 9.14.)

18 Inspect the armature and test it for shorts and grounds. (Refer to Figures 9.17, 9.18, and 9.20.)

19 Check the fit of the armature shaft in the bushing in the drive end housing. If excessive side play or "wobble" exists, replace the bushing or the drive end housing.

Note: On most Nippondenso starter motors, the bushing in the drive end housing is not supplied as a separate part. The drive end housing must be replaced.

20 Check the fit of the armature shaft in the bushing in the end shield. Replace the bushing or the end shield if excessive side play or "wobble" is detected.

Note: On most Nippondenso starter motors, the end shield bushing is not supplied as a separate part. The end shield assembly must be replaced.

21 Test the field coils for opens and grounds. (Refer to Figures 9.21, 9.22, and 9.23.)

Note: On most Nippondenso starter motors, the field coils are not supplied as separate parts. The field frame assembly must be replaced if the field coils are found to be defective.

Brush Replacement The brushes should be replaced if they are worn to less than $1\frac{1}{32}$ inch (9 mm) in length. The leads of the insulated field brushes are welded to the field coils and are not supplied as separate parts. The field frame assembly must be replaced. The leads of the ground brushes are welded to the brush plate and are not supplied as separate parts. The brush plate assembly must be replaced.

Assembly

1 Place a few drops of oil on the drive end of the armature shaft and install the drive assembly. (Refer to Figure 9.38.)

2 Install the retainer on the armature shaft with its cupped end facing outward.

3 Install the snap ring on the armature shaft.

Note: While the snap ring may be installed with lock ring pliers (refer to Figure 9.29), an alternate method (refer to Figure 9.39) minimizes the possibility of overexpanding the snap ring.

4 Using a battery cable puller, pull the retainer up over the snap ring. (Refer to Figure 9.41.)

5 Install the shift lever assembly in the drive end housing. (Refer to Figure 9.90.)

6 Place some lubricant in the bushing in the drive

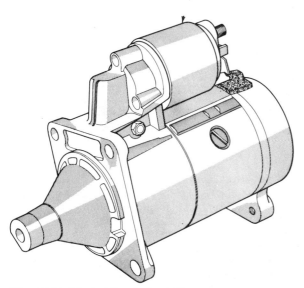

Figure 9.91 A typical Bosch direct drive starter motor (courtesy of Chrysler Corporation).

end housing and insert the armature assembly while positioning the shift lever yoke on the drive assembly. (Refer to Figure 9.88.)

7 Install the rubber seal in its slot in the drive end housing. (Refer to Figure 9.87.)

8 Position the field frame over the armature. (Refer to Figure 9.86.)

9 Install the brush plate. (Refer to Figure 9.85.)

10 Using a brush hook to hold the springs back, install the brushes in the brush plate. (Refer to Figure 9.84.)

Note: Check that the brushes make proper contact with the commutator and that they move freely.

11 Place some lubricant in the bushing in the end shield and slide the end shield on the armature shaft.

12 Install the thru-bolts.

13 Install the flat washer, spring, and "C" washer on the end of the armature shaft. (Refer to Figure 9.82.)

14 Install the bearing cap and secure it with the two screws. (Refer to Figure 9.81.)

15 Install the solenoid. (Refer to Figure 9.80.)

Note: The solenoid must be tipped and "wiggled" so that the hook on the plunger engages the shift fork.

16 Install the nuts holding the solenoid to the drive end housing. (Refer to Figure 9.79.)

17 Position the field coil wire on the solenoid terminal and install the nut. (Refer to Figure 9.78.)

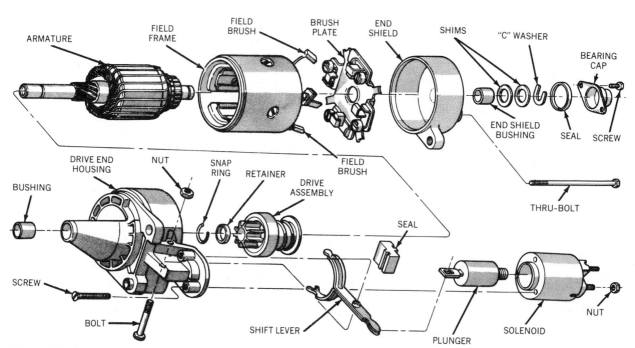

Figure 9.92 An exploded view of a typical Bosch direct drive starter motor (courtesy of Chrysler Corporation).

Job 9h

OVERHAUL A NIPPONDENSO STARTER MOTOR

SATISFACTORY PERFORMANCE

A satisfactory performance on this job requires that you do the following:

1 Overhaul the assigned starter motor.
2 Following the steps in the "Performance Outline" and the procedure and specifications of the manufacturer, complete the job within 120 minutes.
3 Fill in the blanks under "Information."

PERFORMANCE OUTLINE

1 Disassemble the starter motor.
2 Inspect the parts for wear and damage.
3 Test the armature and the field coils.
4 Assemble the starter motor, replacing the parts required.

INFORMATION

Starter motor identification _____

Reference used _____ Page(s) _____

Armature was: _____ shorted _____ grounded _____ OK

Field coils were: _____ open _____ grounded _____ OK

Brushes were: _____ excessively worn _____ OK

Parts replaced _____

Overhauling a Bosch Starter Motor The following steps outline a procedure for the disassembly, component testing, and reassembly of a typical Bosch starter motor. (See Figures 9.91 and 9.92.) You should consult an appropriate manual for the procedure and specifications that may be required for the specific starter motor that you overhaul.

Disassembly

1 Remove the nut holding the field coil wire to the solenoid and remove the field coil wire. (See Figure 9.93.)
2 Remove the screws holding the solenoid to the drive end housing, as shown in Figure 9.94.
3 Remove the solenoid, as shown in Figure 9.95.

Note: The solenoid must be tipped to disengage the plunger from the end of the shift lever.

4 Remove the screws holding the bearing cap to the end shield and remove the bearing cap. (See Figure 9.96.)

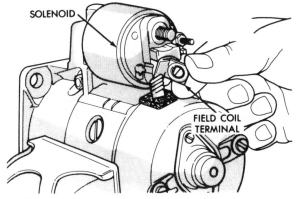

Figure 9.93 Removing the field coil lead from the solenoid (courtesy of Chrysler Corporation).

5 Remove the "C" washer and the shim washer(s) from the end shield as shown in Figure 9.97.
6 Remove the two thru-bolts and remove the end shield. (See Figure 9.98.)

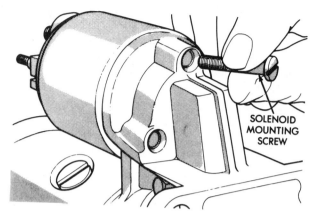

Figure 9.94 Removing the solenoid mounting screws (courtesy of Chrysler Corporation).

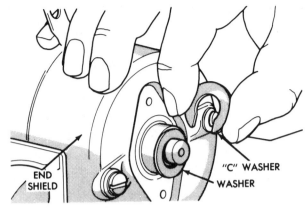

Figure 9.97 Removing the "C" washer. One or more flat shim washers may be found under the "C" washer (courtesy of Chrysler Corporation).

Figure 9.95 Removing the solenoid from the drive end housing (courtesy of Chrysler Corporation).

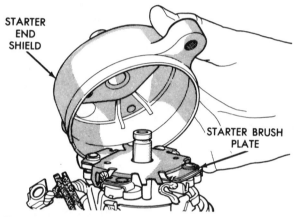

Figure 9.98 Removing the starter end shield (courtesy of Chrysler Corporation).

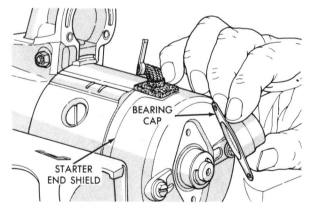

Figure 9.96 Removing the bearing cap from the end shield (courtesy of Chrysler Corporation).

7 Lift the brush retaining springs and remove the two insulated field brushes from the brush plate. (See Figure 9.99.)

8 Remove the brush plate from the starter motor as shown in Figure 9.100.

9 Remove the field frame. (See Figure 9.101.)

10 Remove the armature and drive assembly from

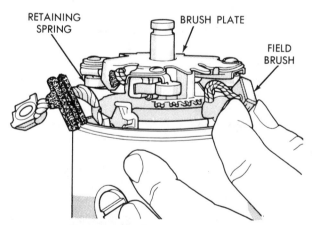

Figure 9.99 Removing the insulated field brushes from the brush plate (courtesy of Chrysler Corporation).

the drive end housing, as shown in Figure 9.102.

11 Lift the rubber seal from its slot in the drive end housing. (See Figure 9.103.)

12 Remove the nut from the shift lever pivot bolt and remove the bolt. (See Figure 9.104.)

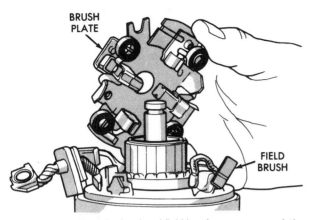

Figure 9.100 When the insulated field brushes are removed, the brush plate can be lifted from the armature (courtesy of Chrysler Corporation).

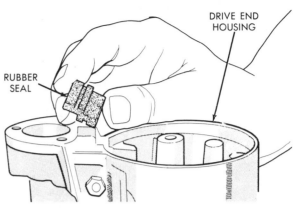

Figure 9.103 Removing the rubber seal from the drive end housing (courtesy of Chrysler Corporation).

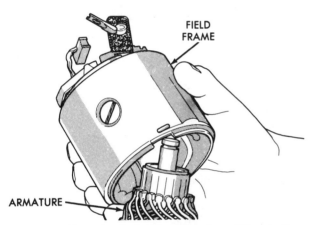

Figure 9.101 Removing the field frame (courtesy of Chrysler Corporation).

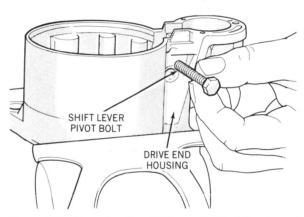

Figure 9.104 Removing the shift lever pivot bolt (courtesy of Chrysler Corporation).

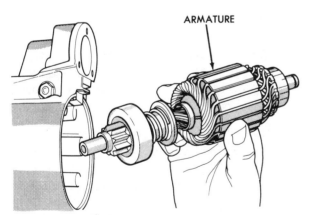

Figure 9.102 Removing the armature and drive assembly from the drive end housing (courtesy of Chrysler Corporation).

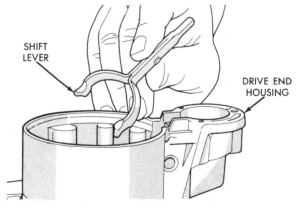

Figure 9.105 Removing the shift lever from the drive end housing (courtesy of Chrysler Corporation).

13 Remove the shift lever from the drive end housing. (See Figure 9.105.)

14 Using a socket or a pipe coupling, drive the retainer off the snap ring on the armature shaft. (Refer to Figure 9.28.)

15 Using lock ring pliers, remove the snap ring from the armature shaft. (Refer to Figure 9.29.)

Note: Spread the snap ring only enough so that it can be removed. If the ring is overexpanded, it must be replaced.

Job 9i

OVERHAUL A BOSCH STARTER MOTOR

SATISFACTORY PERFORMANCE

A satisfactory performance on this job requires that you do the following:

1 Overhaul the assigned starter motor.
2 Following the steps in the "Performance Outline" and the procedure and specifications of the manufacturer, complete the job within 120 minutes.
3 Fill in the blanks under "Information."

PERFORMANCE OUTLINE

1 Disassemble the starter motor.
2 Inspect the parts for wear and damage.
3 Test the armature and the field coils.
4 Assemble the starter motor, replacing the parts required.

INFORMATION

Starter motor identification _____

Reference used _____ Page(s) _____

Armature was: _____ shorted _____ grounded _____ OK

Field coils were: _____ open _____ grounded _____ OK

Brushes were: _____ excessively worn _____ OK

Parts replaced _____

---

16 Remove the retainer and the drive assembly from the armature shaft.

17 Check the drive assembly for wear, damage, and slippage. (Refer to Figures 9.13 and 9.14.)

18 Inspect the armature and test it for shorts and grounds. (Refer to Figures 9.17, 9.18, and 9.20.)

19 Check the fit of the armature shaft in the bushing in the drive end housing. If excessive side play or "wobble" is found, replace the bushing or the drive end housing.

Note: On most Bosch starter motors, the bushing in the drive end housing is not supplied as a separate part. The drive end housing must be replaced.

20 Check the fit of the armature shaft in the bushing in the end shield. Replace the bushing or the end shield if excessive side play or "wobble" is found.

Note: On most Bosch starter motors, the end shield bushing is not supplied as a separate part. The end shield assembly must be replaced.

21 Test the field coils for opens and grounds. (Refer to Figures 9.21, 9.22, and 9.23.)

Note: On most Bosch starter motors, the field coils are not supplied as separate parts. The field frame assembly must be replaced if the field coils are found to be defective.

Brush Replacement The brushes should be replaced if they are worn to less than $\frac{11}{32}$ inch (9 mm) in length. The leads of the insulated brushes are welded to the field coils and are not supplied as separate parts. The field frame assembly must be replaced. The leads of the ground brushes are welded to the brush plate and are not supplied as separate parts. The brush plate assembly must be replaced.

Assembly

1 Place a few drops of oil on the drive end of the

armature shaft and install the drive assembly. (Refer to Figure 9.38.)

2 Install the retainer on the shaft with its cupped end facing outward.

3 Install the snap ring on the armature shaft.

Note: While lock ring pliers may be used for this step (refer to Figure 9.29), an alternate method (refer to Figure 9.39) minimizes the possibility of overexpanding the snap ring.

4 Using a battery cable puller, pull the retainer up over the snap ring. (Refer to Figure 9.41.)

5 Position the shift lever in the drive end housing and install the pivot bolt and nut. (Refer to Figures 9.104 and 9.105.)

6 Install the rubber seal in its slot in the drive end housing. (Refer to Figure 9.103.)

7 Place some lubricant in the bushing in the drive end housing and insert the armature assembly while positioning the shift lever yoke on the drive assembly. (Refer to Figure 9.102.)

8 Position the field frame assembly over the armature. (Refer to Figure 9.101.)

9 Install the brush plate. (Refer to Figure 9.100.)

10 Using a brush hook to hold the springs back, install the brushes in the brush plate. (Refer to Figure 9.99.)

Note: Check that the brushes make proper contact with the commutator and that they move freely.

11 Place some lubricant in the bushing in the end shield and slide the end shield on the armature shaft. (Refer to Figure 9.98.)

12 Install the thru-bolts.

13 Install the flat shim washer(s) and the "C" washer on the armature shaft. (Refer to Figure 9.97.)

14 Install the bearing cap and secure it with the two screws. (Refer to Figure 9.96.)

15 Install the solenoid. (Refer to Figure 9.95.)

Note: The solenoid must be tipped and "wiggled" so that the end of the shift lever enters the slot on the solenoid plunger.

16 Install the solenoid mounting screws. (Refer to Figure 9.94.)

17 Position the field coil wire on the solenoid terminal and install the nut. (Refer to Figure 9.93.)

SUMMARY

By completing this chapter, you have gained knowledge of how starter motors operate. By performing the jobs provided, you have developed some of the skills required to overhaul starter motors. You are aware of the design and assembly differences of the most commonly used units, and can perform many of the inspections and tests required in their overhaul. By accomplishing your objectives, you have moved closer to your goal.

SELF-TEST

Each incomplete statement or question in this test is followed by four suggested completions or answers. In each case select the *one* that best completes the sentence or answers the question.

1 The windings of an armature are connected to the
   A. field coils
   B. brush holders
   C. commutator bars
   D. armature laminations

2 Most starter motors have
   A. one brush
   B. two brushes
   C. three brushes
   D. four brushes

3 A growler often is used to check
   A. armatures
   B. solenoids
   C. pole shoes
   D. end frames

4 A test lamp indicates continuity when connected between the commutator and the laminations of an armature. This indicates that the armature is
   A. OK
   B. open
   C. shorted
   D. grounded

5 A test lamp indicates continuity when connected between the field coil connection and either of the insulated brushes. This indicates that the field coils are
   A. OK
   B. open
   C. shorted
   D. grounded

6 A test lamp indicates continuity when connected between the field coil connection and the field frame. This indicates that the field coils are
   A. OK
   B. open
   C. shorted

D. grounded

7 In a typical Delco starter motor, the brush holders are mounted in the
A. solenoid
B. field frame
C. drive gear housing
D. commutator end frame

8 In a typical Ford positive engagement starter motor, the shift lever assembly is mounted in the
A. end plate
B. field frame
C. center bearing

D. drive end housing

9 In a typical Chrysler gear reduction starter motor, the solenoid is housed in the
A. end head
B. field frame
C. center bearing
D. drive gear housing

10 In a typical Nippondenso starter motor, the brushes are mounted in
A. a brush plate
B. the end shield
C. the field frame
D. the drive end housing

Chapter 10
The Charging System— Basic Services

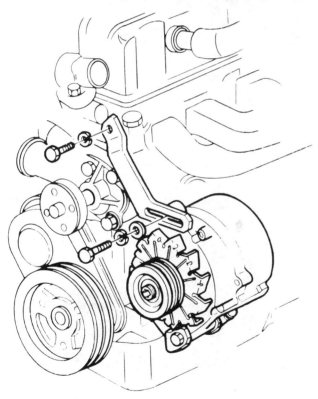

The charging system is responsible for providing the energy required to operate the various electrical systems while the engine is running and for maintaining the state of charge of the battery. If the charging system should fail, the battery can provide energy for only a limited time.

In this chapter, you will learn how charging systems operate. You will learn about the components used and how they function, and you will perform some basic diagnostic and repair services. Your specific objectives are to perform the following jobs:

A
Identify the parts in a charging system
B
Identify the function of the parts in a charging system
C
Adjust an alternator belt
D
Replace an alternator belt
E
Perform basic charging system checks
F
Replace a regulator
G
Replace an alternator

THE CHARGING SYSTEM The charging system converts mechanical energy to electrical energy. Charging system failure can be minimized by proper maintenance and, when failure does occur, can quickly be repaired after proper diagnosis. But that diagnosis must be based on a knowledge of the system, its components, and their function.

Charging System Components Most charging systems include four major components. Those components are (1) the alternator, (2) the regulator, (3) the battery, and (4) the indicating device. Figure 10.1 shows the location of those components in a typical charging system.

The Alternator An alternator, shown in Figure 10.2, is a generator that produces alternating current. For that reason, it is sometimes referred to as an *A.C. generator*. The alternator is driven by a belt from the engine crankshaft, as shown in Figure 10.3. Through magnetism, the alternator converts mechanical energy to electrical energy. Some of the electrical energy is used by the various other electrical systems in a car. Some of that energy also is converted to chemical energy and stored in the battery.

The Regulator The regulator acts as an automatic control in the charging system. Without a regulator, an alternator will operate at its highest possible out-

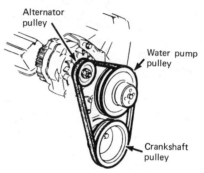

Figure 10.3 On most cars, the alternator is driven by a belt from the crankshaft (courtesy of Pontiac Motor Division, General Motors Corporation).

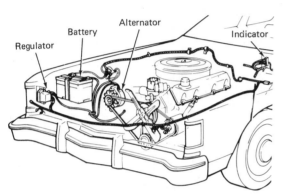

Figure 10.1 The location of the components in a typical charging system (courtesy of Ford Motor Company).

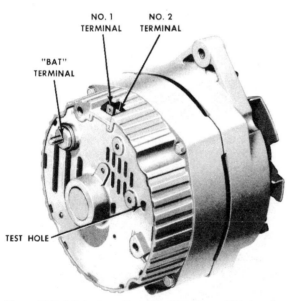

Figure 10.2 A Delcotron alternator. Alternators of this type have an integral regulator and are used on vehicles built by General Motors Corporation (courtesy of Pontiac Motor Division, General Motors Corporation).

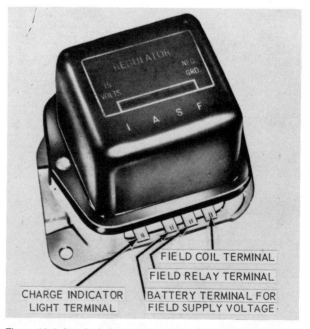

Figure 10.4 A typical alternator regulator. A regulator of this type is mounted separate from the alternator (courtesy of Ford Motor Company).

put. It will do this even when the energy it produces is not needed. Uncontrolled, alternator output voltage will exceed the limits of the other electrical systems in a car. Bulbs and other components will burn out. The battery will be damaged by overcharging. Within a very short time, the alternator will burn itself out. Figure 10.4 shows a typical regulator.

The Battery While the battery converts and stores energy through electrochemical action, it also acts as a "cushion" or "shock absorber" in the system. The battery does this by balancing out slight differences in the energy supplied by the alternator and the energy demanded by the other systems.

The Indicating Device The indicating device provides the driver with some indication of system failure. On most cars, the indicating device consists of a warning light on the instrument panel. If that light remains on while the engine is running, it indicates that the alternator is not charging the battery.

Some cars have an ammeter as an indicating device. Those meters usually are not as accurate as those you use for testing. However, they do inform the driver of the approximate amount of current that is flowing in the electrical system. Ammeters used on instrument panels show current flow in two directions. Current flowing into the battery moves the meter needle to the Charge side of the dial. When the needle moves to the Discharge side, it means that current is flowing out of the battery. Figure 10.5 shows a typical automotive ammeter.

On other cars, a voltmeter is provided as the indicating device. The voltmeter indicates system voltage, usually on a scale divided into voltage ranges. Figure 10.6 illustrates such a meter.

Charging System Circuits Many different circuits are used to connect the components in the charging system. Those circuits vary even among cars built by the same manufacturer. Figures 10.7 and 10.8 show two typical circuits. When a diagram for the charging circuit for a particular car is needed, it is best found in the car manufacturer's manual.

Figure 10.5 Some cars have an ammeter that indicates approximate charge or discharge rates (courtesy of Stewart-Warner Corporation).

Figure 10.6 Some instrument panels incorporate a voltmeter to monitor system voltage (courtesy of Stewart-Warner Corporation).

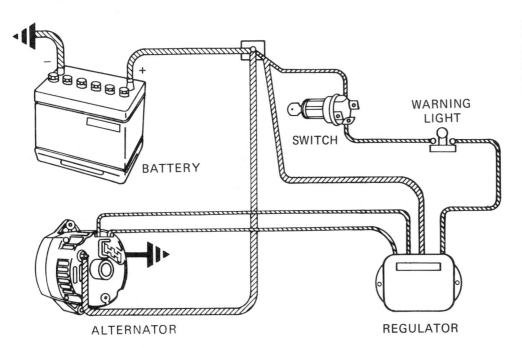
Figure 10.7 A typical charging system that uses an external regulator (courtesy of Pontiac Motor Division, General Motors Corporation).

BATTERY

SWITCH

WARNING LIGHT

ALTERNATOR

REGULATOR

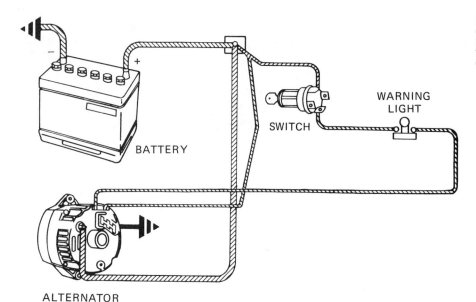

Figure 10.8 A charging system circuit used when the regulator is housed within the alternator (courtesy of Pontiac Motor Division, General Motors Corporation).

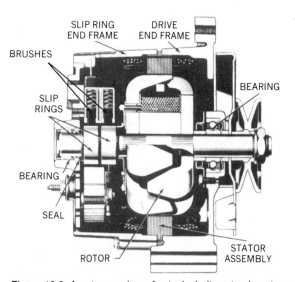

Figure 10.9 A cutaway view of a typical alternator (courtesy of Pontiac Motor Division, General Motors Corporation).

CHARGING SYSTEM COMPONENT OPERATION
Many different types of alternators and regulators are used in automotive charging sytems. Due to the different service procedures required, a knowledge of how the components of those systems operate is very important.

Alternators When a magnetic field is moved across a conductor, a current is induced in that conductor. In an alternator, an electromagnet called a *rotor* turns inside a set of stationary wire coils, or windings, called a *stator.* Those parts are shown in Figure 10.9. As the rotor turns, its magnetic lines of force move across all of the windings in the stator. This action induces current in all of the stator windings.

As the rotor turns, the poles of its magnetic field at the stator continuously alternate from north to south. Therefore, the current induced in the stator windings is *alternating current.* The polarity of alternating current continuously changes back and forth from positive (+) to negative (−). The battery and the electrical systems in a car requires *direct current.* Direct current does not change polarity.

To be of use in an automotive electrical system, alternating current must be *rectified,* or converted to direct current. A rectifier is used for this task. Figure 10.10 shows a rectifier and its location in an alternator. A rectifier consists of a group of *diodes.* A diode is an electrical check valve that conducts current in only one direction. (See Figure 10.11.) Both positive (+) and negative (−) diodes are used in pairs to "split" the alternating current. Most rectifiers contain six diodes. Half the diodes conduct positive (+) current and half the diodes conduct negative (−) current. Thus, the alternating current is converted to direct current before it leaves the alternator.

The voltage produced by an alternator depends largely on the strength of the rotor's magnetic field. That magnetic field is created by a coil of wire inside the rotor. Figure 10.12 shows that *field coil.* Battery voltage is conducted to the field coil by a pair of brushes that contact a pair of slip rings on the rotor shaft. (Refer to Figure 10.9.) By controlling the amount of current that flows through the field

Job 10a

IDENTIFY THE PARTS IN A CHARGING SYSTEM

SATISFACTORY PERFORMANCE

A satisfactory performance on this job requires that you do the following:

1 Identify the numbered parts in the drawing below by placing the number of each part in front of the correct part name.

2 Complete the job by correctly identifying all the parts within 5 minutes.

PERFORMANCE SITUATION

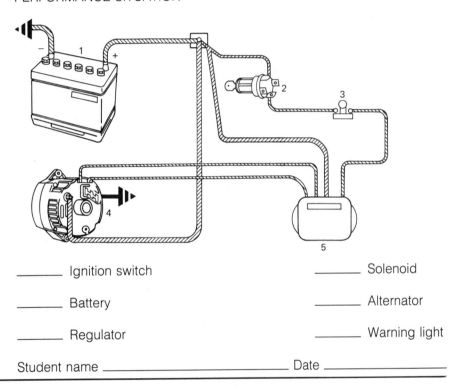

_____ Ignition switch   _____ Solenoid

_____ Battery   _____ Alternator

_____ Regulator   _____ Warning light

Student name _____ Date _____

Electro-Mechanical-Regulators   Figure 10.13 shows a typical electro-mechanical regulator. Those regulators use magnetically operated switches that function in the same manner as a relay. One switch acts as a *voltage limiter*. By directing current through various resistors, it limits the voltage that flows to the field coil winding. The remaining switch, usually called the *field relay,* opens the circuit to the field coil when the alternator is not operating. Figure 10.14 shows the internal wiring diagram of a typical electro-mechanical regulator. Since electro-mechanical regulators have moving parts that are subject to wear, and switch contacts that are subject to pitting, they usually will be found only on older model vehicles.

coil, the output voltage of the alternator is controlled. Current flow through the field coil is controlled automatically by a regulator.

Regulators   As mentioned above, regulators control alternator output by controlling the amount of current that flows in the field coil winding. However, regulators have other functions. In some systems, they turn off the field current when the engine stops running. They also are used to operate the warning light on the instrument panel.

There are many types of alternator regulators, but they can be classified in three distinct types as follows:

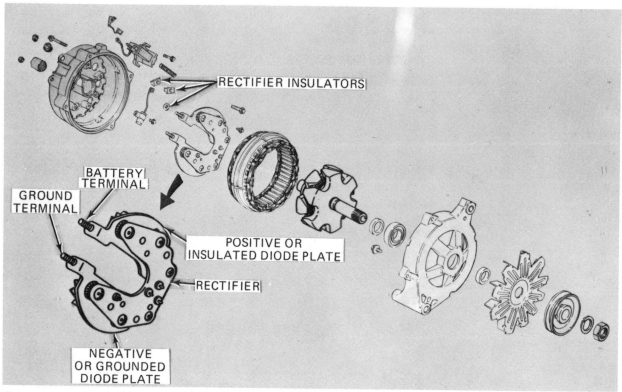

Figure 10.10 A typical rectifier assembly and its location in an alternator. (courtesy of Ford Motor Company).

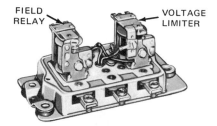

Figure 10.11 The symbol for a diode. As the symbol indicates, a diode is an electrical check valve that allows current to flow in only one direction.

Figure 10.13 A typical electro-mechanical regulator (courtesy of Ford Motor Company).

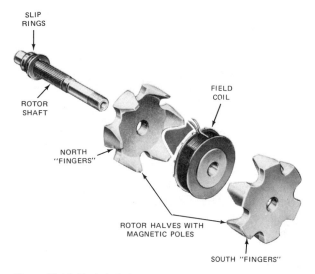

Figure 10.12 Exploded view of a rotor (courtesy of Ford Motor Company).

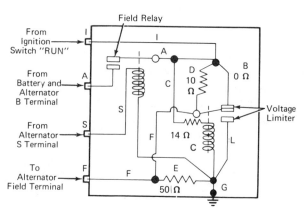

Figure 10.14 An internal wiring diagram of a typical electro-mechanical regulator (courtesy of Ford Motor Company).

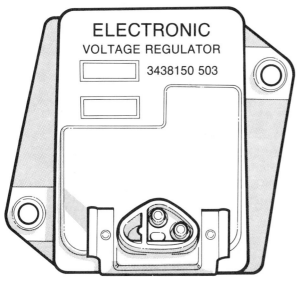

Figure 10.15 A typical transistorized regulator. The unit is sealed and has no moving parts (courtesy of Chrysler Corporation).

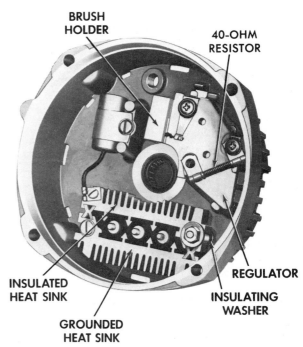

Figure 10.17 A transistorized regulator mounted inside the rear housing of an alternator (courtesy of Chevrolet Motor Division).

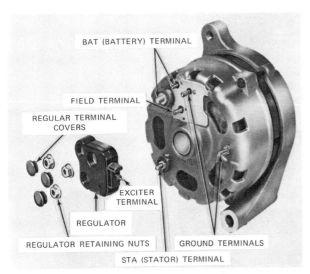

Figure 10.16 A transistorized regulator that is mounted on the rear of an alternator (courtesy of Ford Motor Company).

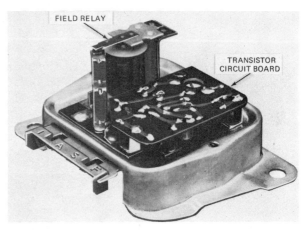

Figure 10.18 A composite regulator. A mechanical field relay is used together with a transistorized voltage limiter circuit board (courtesy of Ford Motor Company).

SWITCH	ENGINE	LAMP
Off	Stopped	Off
On	Stopped	On
On	Running	Off

Figure 10.19 The charging system warning light operates in three phases.

Transistorized Regulators   Most late-model cars have transistorized regulators. Often called electronic regulators, they have no moving parts but use transistorized circuits to replace the mechanical switches. Many types of transistorized regulators are in use. Some are mounted separate from the alternator. A regulator of that type is shown in Figure 10.15. Others are mounted externally on the rear of the alternator, as shown in Figure 10.16. Still others are mounted inside the alternator as an integral part of that assembly (see Figure 10.17).

Composite Regulators   As shown in Figure 10.18, some regulators contain both transistorized and electro-mechanical parts. Regulators of that type are mounted separate from the alternator.

Warning Lights   Most cars have an alternator warning light on the instrument panel to warn the driver of any system failure. Figure 10.19 shows the three phases of warning light operation.

Job 10b

IDENTIFY THE FUNCTION OF THE PARTS IN A CHARGING SYSTEM

SATISFACTORY PERFORMANCE

A satisfactory performance on this job requires that you do the following:

1 Identify the function of the listed charging system parts by placing the number of each part in front of the phrase that best describes its function.

2 Complete the job by correctly identifying the function of all the parts within 10 minutes.

PERFORMANCE SITUATIONS

1 Alternator
2 Stator
3 Rotor
4 Diode
5 Rectifier

6 Electro-mechanical regulator
7 Transistorized regulator
8 Battery
9 Field coil winding
10 Brushes and slip rings

_____ Conducts current to the field coil winding

_____ Uses magnetically operated switches to control voltage

_____ Converts alternating current to direct current

_____ Converts mechanical energy to electrical energy

_____ Conducts current to the stator

_____ Acts as an electrical check valve

_____ Creates a magnetic field in the rotor

_____ Uses transistorized circuits to control voltage

_____ Moves a magnetic field across the stator windings

_____ Receives current through induction

_____ Converts and stores energy

CHARGING SYSTEM SERVICES Routine maintenance of the charging system consists of

(1) keeping the alternator drive belt in adjustment, and

(2) keeping all of the electrical connections in the system clean and tight. Other services may be required when problems arise in the system. Those services include the replacement of parts. However, before you attempt to replace any part, you should determine if that part is defective.

Drive Belts The alternator and many other engine accessories are usually driven by V belts. These belts are driven by a pulley or pulleys on the engine crankshaft. There are several types of V belts in common use. (See Figure 10.20.) The type of belt is chosen by the vehicle manufacturer to meet the performance requirements of the particular application. As shown in Figure 10.21, belt arrangements are determined by the accessories in each vehicle. A failure of the belt that drives the alternator results in failure of the charging system. An inspection and adjustment of the drive belt should be considered a part of routine maintenance of the charging system and one of the first steps in problem diagnosis.

Inspecting Belts Before any attempt to adjust a belt is made, the belt should be inspected. The underside of the belt should have no cracks. If cracks are found, the belt should be replaced. The tapered sides of the belt should be smooth. A loose belt slips, and most slipping belts develop *glazing*. A belt is glazed when its sides have a shiny, glasslike surface. A glazed belt will continue to slip, even after it has been properly adjusted. For this reason, a glazed belt should be replaced. In most cases, an oil-soaked belt indicates that the seal at the end of the crankshaft is leaking. An oil-soaked belt must also be replaced, and the location of the oil leak must be found and the required repair performed.

A loose alternator belt cannot drive the alternator at the proper speed. A belt that is too tight will cause premature wear of the alternator bearings. On those cars where the alternator belt drives the water pump, the water pump bearings may also be damaged. A belt that is not cracked, glazed, or oil-soaked should be adjusted to the correct tension. Most alternator belts are adjusted by moving the alternator in or out on its mountings. (See Figure 10.22.)

Vehicle manufacturers specify the tension to which the belts on their cars should be adjusted. (See

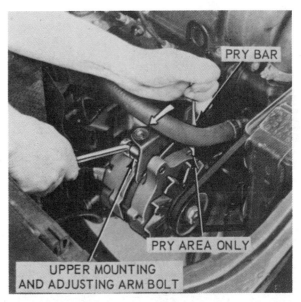

Figure 10.22 On most cars, the alternator belt tension is increased by prying the alternator outward in its mountings (courtesy of Ford Motor Company).

Figure 10.20 Commonly used drive belts (courtesy of Ford Motor Company).

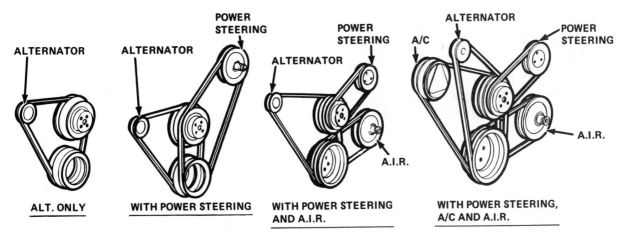

Figure 10.21 Typical belt arrangements (courtesy of Pontiac Motor Division, General Motors Corporation).

ENGINE	TENSIONING	GENERATOR	POWER STEERING	AIR CONDITIONING
L-4	NEW	650N (146 LB)	650N (146 LB)	750N (168 LB)
	USED	222-350 N (50-80 LB)	222-350 N (50-80 LB)	289-422 N (65-95 LB)

Figure 10.23 Belt tension specifications provided by one manufacturer for one particular engine. Note that the specifications for a "used" belt are approximately half those for a "new" belt. A belt that has been in service for more than 15 min can be considered "used." (courtesy of Chevrolet Motor Division).

Figure 10.24 A typical belt tension gauge (courtesy of American Motors).

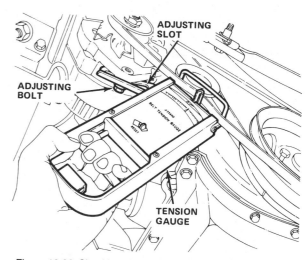

Figure 10.26 Checking alternator belt tension with a belt tension gauge. As shown, in some instances it is easier to check the belt tension from under the car (courtesy of American Motors).

Figure 10.25 Installing a belt tension gauge on an alternator belt. On this engine, the gauge is placed between the alternator and the water pump (courtesy of Chevrolet Motor Division).

Figure 10.23.) After a belt is inspected, its tension should be measured. Belt tension is easily measured by the use of a *belt tension gauge*. A typical belt tension gauge is shown in Figure 10.24. The gauge is positioned on the belt as shown in Figure 10.25. The tension indicated on the gauge should be compared to the manufacturer's specifications. If the tension is within the specified range, no adjustment is necessary. If the tension is too low or too high, an adjustment is required.

Adjusting an Alternator Belt   The following steps outline a typical procedure for adjusting an alternator belt. You should consult an appropriate manual for the particular procedure and for the tension specifications required for the car on which you are working:

1 Position a belt tension gauge on the belt as shown in Figure 10.26.
2 Loosen the adjustment nuts or bolts on the alternator slightly so that the alternator can be moved on its mountings. (See Figure 10.27.)
3 Carefully pry the alternator in the appropriate direction until the correct tension is indicated on the gauge. (Refer to Figure 10.22.)

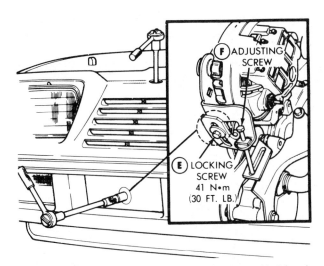

Figure 10.27 On some cars, access holes are provided for alternator belt adjustment. Note that this mounting has a separate locking screw to secure the adjustment (courtesy of Chrysler Corporation).

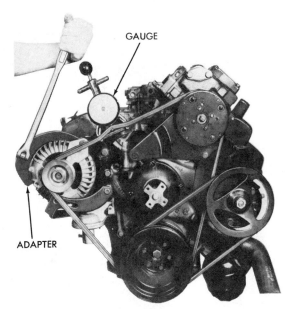

Figure 10.29 Adjusting the tension of a belt using a special tool to grasp the alternator (courtesy of Chrysler Corporation).

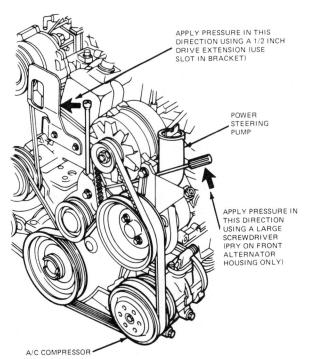

Figure 10.28 On some applications, more than one belt will require adjustment. Note that pry points for both the alternator and the idler pulley are specified (courtesy of Ford Motor Company).

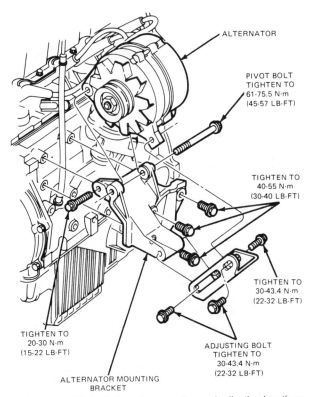

Figure 10.30 Typical alternator mounting and adjusting locations and torque specifications (courtesy of Ford Motor Company).

Note: Some manufacturers specify the pry points for particular applications and may even specify the use of a special tool. (See Figures 10.28 and 10.29.)

4 While maintaining the correct belt tension, tighten the adjustment bolts or nuts to the manufacturer's torque specifications. (See Figure 10.30.)

5 Remove the belt tension gauge.

Replacing an Alternator Belt The following steps outline a procedure for replacing an alternator belt. On many cars, other belts must be removed to gain access to the alternator belt. (Refer

Job 10c

ADJUST AN ALTERNATOR BELT

SATISFACTORY PERFORMANCE
A satisfactory performance on this job requires that you do the following:

1 Adjust the alternator belt tension on the vehicle assigned.
2 Following the steps in the "Performance Outline" and the manufacturer's procedure and specifications, complete the job within 200 percent of the manufacturer's suggested time.
3 Fill in the blanks under "Information."

PERFORMANCE OUTLINE
1 Measure the belt tension.
2 Compare the tension with the specifications.
3 Loosen the adjustment bolts or nuts.
4 Adjust the belt tension to the specifications.
5 Tighten the bolts or nuts to the torque specification.

INFORMATION
Vehicle identification _____

Reference used _____ Page(s) _____

Belt tension specification _____

Belt tension at the start of the job _____

Belt tension at the completion of the job _____

---

to Figure 10.21.) Therefore an appropriate manual should be consulted for the particular procedures and specifications that may be necessary for the car on which you are working:

1 Determine if other belts must be removed to gain access to the alternator belt.
2 Loosen the adjustment and mounting bolts or nuts necessary to relieve the tension on the belt(s).
3 Remove the belt(s).
4 Install the replacement belt(s).
5 Position a belt tension gauge on the alternator belt. (Refer to Figures 10.25 and 10.26).
6 Carefully pry the alternator in the appropriate direction until the correct tension is indicated on the gauge. (Refer to Figures 10.28 and 10.29.)
7 Tighten the adjustment and mounting bolts or nuts to the manufacturer's torque specifications while maintaining the correct belt tension. (Refer to Figure 10.30.)
8 Repeat steps 5, 6, and 7 on any remaining belts.

Preliminary Electrical Checks

Test the Battery If the battery is discharged, an accurate test of the charging system cannot be made. Check the specific gravity of the battery electrolyte and charge the battery if necessary. If the battery is defective, the alternator will be unable to keep it charged. Load test the battery and replace it if it is defective.

Check the Battery Posts and Cable Clamps Loose, dirty, and corroded connections will prevent the alternator from maintaining the state of charge. Clean and tighten the connections as necessary.

Test the Voltage Drop in the Cables and Their Connections Even though the battery cable connections may appear to be clean and tight, they may have excessive resistance. The steps that follow outline a procedure similar to the one you used when you tested for voltage drop in the starting system:

Job 10d

REPLACE AN ALTERNATOR BELT

SATISFACTORY PERFORMANCE

A satisfactory performance on this job requires that you do the following:

1 Replace the alternator belt on the vehicle assigned.

2 Following the steps in the "Performance Outline" and the manufacturer's procedure and specifications, complete the job within 200 percent of the manufacturer's suggested time.

3 Fill in the blanks under "Information."

PERFORMANCE OUTLINE

1 Loosen the necessary adjustment and mounting bolts or nuts.

2 Remove the belt(s).

3 Install the replacement belt(s).

4 Adjust the tension of the belt(s) to the specifications.

5 Tighten the adjustment and mounting bolts or nuts to the torque specifications.

INFORMATION

Vehicle identification _____

Reference used _____ Page(s) _____

Part number of replacement belt _____

Belt type: _____ conventional _____ cogged _____ multi-V

Alternator belt tension specification _____

Accessory belt tension specification(s) _____

_____

In performing the following tests, you will crank the engine with the starter motor. When working on a car with a catalytic converter, crank the engine for as short a time as possible. Extended cranking can deposit an excessive amount of fuel in the converter, causing it to overheat and pose a possible fire hazard.

1 Disable the ignition system by removing the center wire from the distributor cap and grounding it.

Note: An alternate disabling procedure may be used.

2 Apply the parking brake.

3 If the car has an automatic transmission, place the transmission selector lever in the Park position. If the car has a manual transmission, place the shift lever in the Neutral position.

4 Connect a remote starter switch to the starter relay.

5 Adjust a voltmeter to the low scale.

6 Connect the positive (+) voltmeter lead to the positive (+) battery post as shown in Figure 10.31.

Note: **The voltmeter lead must contact the battery post, not the cable clamp.**

7 Connect the negative (−) voltmeter lead to the terminal bolt on the battery side of the starter relay. (Refer to Figure 10.31.)

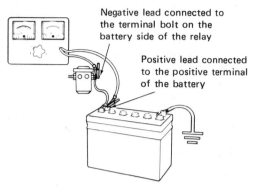

Negative lead connected to the terminal bolt on the battery side of the relay

Positive lead connected to the positive terminal of the battery

Figure 10.31 Connections for checking voltage drop in the battery cable and its connections (courtesy of Ford Motor Company).

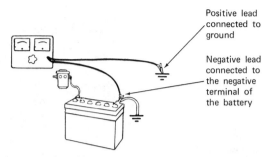

Positive lead connected to ground

Negative lead connected to the negative terminal of the battery

Figure 10.32 Connections for checking voltage drop in the battery ground cable and its connection (courtesy of Ford Motor Company).

Note: The voltmeter lead must contact the terminal bolt, not the battery cable terminal.

8 Crank the engine and observe the voltmeter reading.

Note: The voltage reading should not exceed 0.2 V. A reading of more than 0.2 V indicates excessive resistance in the cable or in its connections. Clean and tighten the connections. Replace the cable if necessary.

9 Disconnect the voltmeter.

10 Connect the positive (+) voltmeter lead to a good ground on the engine as shown in Figure 10.32.

11 Connect the negative (−) voltmeter lead to the negative (−) battery post. (Refer to Figure 10.32.)

Note: The voltmeter lead must contact the battery post, not the cable clamp.

12 Crank the engine and observe the voltmeter reading.

Note: The voltage reading should not exceed 0.2 V. A reading of more than 0.2 V indicates excessive resistance in the battery ground cable or its connections. Clean and tighten the connections. Replace the cable if necessary.

13 Remove the voltmeter.

14 Restore the ignition system to operating condition.

Check the Wiring   An open circuit or a short circuit in the wiring is a common cause of charging system problems. Check for loose connections and damaged wires. Clean and tighten loose connections, and repair or replace damaged wires.

Fuse Links   Some cars have fuse links in the charging system to protect the alternator. As you learned in a previous chapter, a fuse link is a short length of wire built into the wiring harness. The fuse link is several gauges smaller than the wire used in the circuit that it protects. Thus, when excessive current flows in that circuit, the fuse link wire acts as a fuse and burns out. Fuse links in the charging system often are burned out when a booster battery is incorrectly connected.

On some cars, the fuse link that protects the alternator is wired as shown in Figure 10.33. If that fuse link burns out, current cannot flow to other electrical equipment. Some cars are wired as shown in Figure 10.34. When wired in that manner, a burned out fuse link in the charging circuit will have no effect on the other circuits. When a burned fuse link is found, you must determine the cause of the failure. You must also consult an appropriate manual for the specification of the replacement wire.

Replacing a Regulator   On most cars with externally mounted regulators, the regulator is mounted on an inner fender panel or on the firewall. They usually are secured with several self-tapping screws, and are easily replaced. The following steps outline a procedure for regulator replacement:

1 Disconnect the ground cable from the battery.

2 Disconnect the wiring plug from the regulator.

Note: Most regulators have a locking device that secures the plug. On some, the plug may be released by depressing the locking tab shown in Figure 10.35. On others, the plug may be released by using a screwdriver as shown in Figure 10.36.

2 Remove the mounting screws and the regulator.

3 Mount the replacement regulator.

Note: The base of the regulator and the mounting screws must be clean and free of dirt and corrosion so that the regulator is properly grounded.

4 Connect the wiring plug.

5 Connect the battery cable.

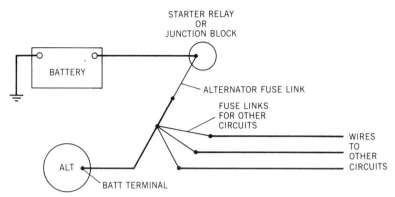

Figure 10.33 On some cars, the fuse link that protects the alternator is wired in a series with other fuse links that protect other circuits. If the alternator fuse link burns out, it will stop the flow of current to the other circuits.

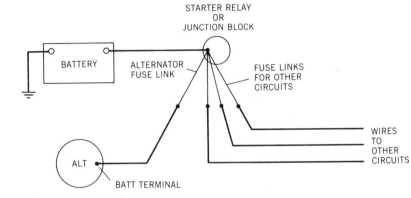

Figure 10.34 On some cars, the alternator fuse link is wired in parallel with the fuse links that protect other circuits. If the alternator fuse link burns out, the other circuits will still function.

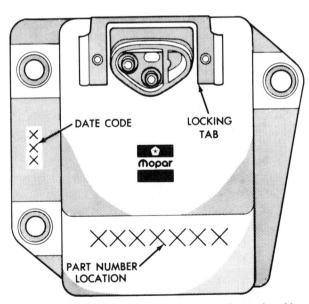

Figure 10.35 On voltage regulators of the type shown, the wiring plug is released by depressing the locking tab (courtesy of Chrysler Corporation).

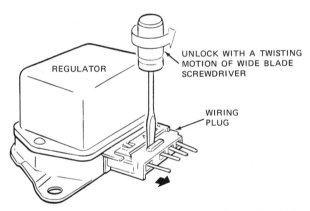

Figure 10.36 On voltage regulators of the type shown, the wiring plug is released by using a screwdriver as shown (courtesy of Ford Motor Company).

Replacing an Alternator On most cars, the alternator is mounted so that it is accessible from under the hood. On some cars, it must be removed from under the car. Figures 10.37, 10.38, and 10.39 show typical alternator installations and the attaching hardware and brackets. Cars equipped with power steering and air conditioning may have the alternator mounted in a different location than on similar cars without those options.

The following steps outline a general procedure for replacing an alternator. There are many variations in alternator mountings, and many different replacement procedures. Therefore, you should con-

Job 10e

PERFORM BASIC CHARGING SYSTEM CHECKS

SATISFACTORY PERFORMANCE
A satisfactory performance on this job requires that you do the following:

1 Perform basic charging system checks on the car assigned.
2 Following the steps in the "Performance Outline" and the procedures and specifications of the vehicle manufacturer, complete the job within 45 minutes.
3 Fill in the blanks under "Information."

PERFORMANCE OUTLINE

1 Inspect the drive belt(s) and check the belt tension.
2 Test the specific gravity of the battery electrolyte.
3 Load test the battery.
4 Check the battery terminals and cable clamps.
5 Clean and tighten the connections if necessary.
6 Test the voltage drop in the cables and connections.
7 Check the system wiring for loose connections and damaged wires.
8 Clean and tighten connections as required.
9 Check the condition of the fuse link (if present).

INFORMATION

Vehicle identification _____

Reference used _____ Page(s) _____

Alternator drive belt:

   Belt condition: _____ Good   _____ Glazed   _____ Damaged

   Belt tension specification _____

   Belt tension measurement _____

   Was belt adjusted? _____ Yes   _____ No

Specific gravity of battery electrolyte:

   Cell #1 _____   Cell #2 _____   Cell #3 _____

   Cell #4 _____   Cell #5 _____   Cell #6 _____

Does battery require charging? _____ Yes   _____ No

Battery load test:

   Battery ampere hour capacity   _____

Discharge rate (test load) _____

Amount of time load was maintained _____

Voltage indicated while under load _____

Approximate battery temperature _____

Recommendations: _____ Battery is serviceable

_____ Battery should be replaced

Battery terminals and connections:

_____ Were found clean and tight

_____ Were cleaned and tightened

Voltage drop test results:

Battery positive (+) cable: _____ volts

Battery negative (−) cable: _____ volts

Were cable services required? _____ Yes _____ No

Job 10f

REPLACE A REGULATOR

SATISFACTORY PERFORMANCE
A satisfactory performance on this job requires that you do the following:

1 Replace the regulator on the car assigned.
2 Following the steps in the "Performance Outline"
and the procedure of the car manufacturer, complete the job within 30 minutes.
3 Fill in the blanks under "Information."

PERFORMANCE OUTLINE
1 Disconnect the battery.
2 Disconnect the wiring plug.
3 Remove the regulator.
4 Install the replacement regulator.
5 Connect the wiring plug.
6 Connect the battery.

INFORMATION
Vehicle identification _____

Reference used _____ Page(s) _____

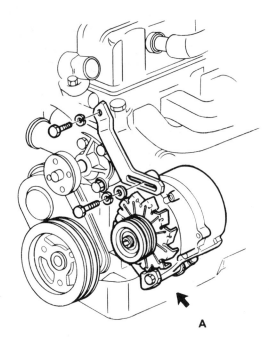

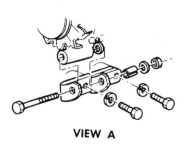

VIEW A

A

Figure 10.37 A typical alternator mounting on an in-line engine (courtesy of Chevrolet Motor Division).

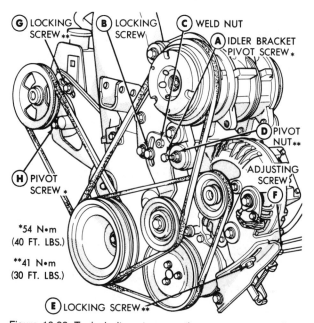

Figure 10.38 Typical alternator mounting on a transverse four cylinder engine (courtesy of Chrysler Corporation).

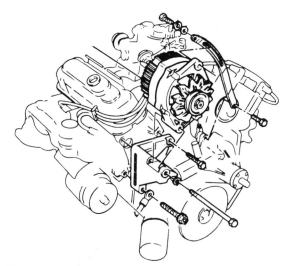

Figure 10.39 A typical alternator mounting on a V-type engine (courtesy of Chevrolet Motor Division).

sult an appropriate manual for the procedure required for the car on which you are working:

Removal
1 Disconnect the battery ground cable.
2 If the alternator must be removed from under the car, raise the car and support it with jack stands.
3 Loosen the alternator adjustment bolts and remove the belt from the alternator pulley.

4 Disconnect the wires from the alternator.
5 Remove any brackets, braces, or shields necessary.
6 Remove the alternator.

Preparation for Installation   Many replacement alternators are supplied without a pulley. The pulley must be removed from the old alternator and installed on the replacement alternator.

On most alternators, the pulley is held on the rotor

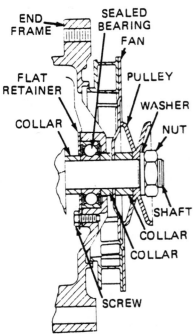

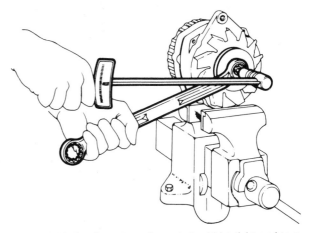

Figure 10.42 An alternator pulley nut should be tightened to a torque specification. A torque wrench can be used on the Allen wrench socket (courtesy of American Motors).

Figure 10.40 On most alternators, the pulley and fan are held on the shaft by a nut. Note the washers and collars also used on this assembly (courtesy of Chevrolet Motor Division).

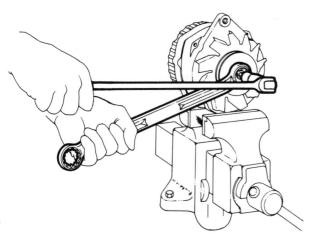

Figure 10.41 An alternator nut should be removed by holding the rotor shaft with an Allen wrench socket and turning the nut with a box wrench (courtesy of American Motors).

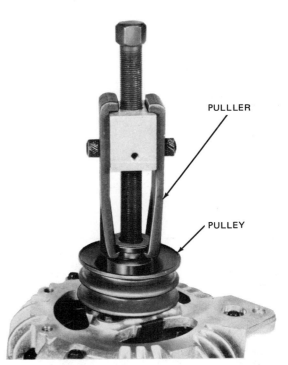

Figure 10.43 A puller installed on an interference fit alternator pulley (courtesy of Chrysler Corporation).

shaft by a nut as shown in Figure 10.40. The rotor shaft usually has a hex-shaped hole in its end. That hole enables you to hold the shaft with an *Allen,* or hex, wrench. As shown in Figure 10.41, the nut can then be removed with a box wrench. When the pulley is installed on the replacement alternator, it should be tightened to the manufacturer's torque specification. Figure 10.42 shows how a torque wrench can be used on an Allen socket.

Some alternators do not use a nut to retain the pul-

ley. The hole in the pulley is slightly smaller than the diameter of the rotor shaft. The pulley is forced on the shaft to obtain a very tight *interference fit.* Special pullers usually are required to remove interference fit pulleys. Some pulleys have a small groove machined on the end. Figure 10.43 shows how a special puller fits on a pulley of that type. As shown in Figure 10.44, other pulleys require a special puller that fits inside the pulley. Pulleys of that type are removed as shown in Figure 10.45.

Job 10g

REPLACE AN ALTERNATOR

SATISFACTORY PERFORMANCE

A satisfactory performance on this job requires that you do the following:

1 Replace the alternator on the car assigned.
2 Following the steps in the "Performance Outline" and the procedure and specifications of the manufacturer, complete the job within 200 percent of the manufacturer's suggested time.
3 Fill in the blanks under "Information."

PERFORMANCE OUTLINE

1 Disconnect the battery ground cable.
2 Raise and support the car if necessary.
3 Remove the alternator.
4 Install the replacement alternator.
5 Connect the battery ground cable.
6 Check the operation of the alternator.
7 Lower the car to the floor.

INFORMATION

Vehicle identification _____

Reference used _____ Page(s) _____

Was the pulley exchanged? _____ Yes   _____ No

Does the replacement alternator operate correctly?

_____ Yes   _____ No

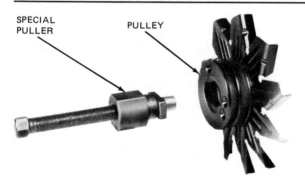

Figure 10.44 Some interference fit pulleys require a special puller that fits inside the pulley (courtesy of Chrysler Corporation).

The installation of interference fit pulleys usually requires the use of a press as shown in Figure 10.46. The alternator must be disassembled so that the rear of the rotor shaft can be supported.

The procedures required for pulley replacement are provided in the following chapter.

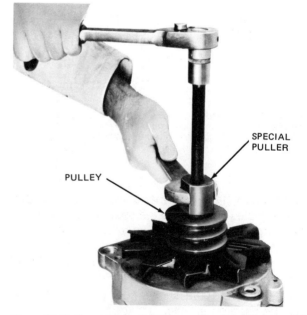

Figure 10.45 Removing an interference fit pulley with a special puller (courtesy of Chrysler Corporation).

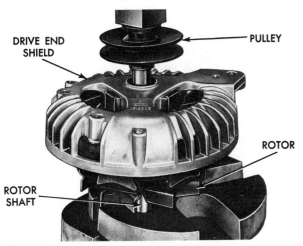

DRIVE END
SHIELD

PULLEY

ROTOR

ROTOR
SHAFT

Figure 10.46 Using a press to install an interference fit pulley. Notice that the alternator has been disassembled so that the rotor shaft can be supported on the press table (courtesy of Chrysler Corporation).

Installation

1 Hold the alternator in position and install the mounting bolts by hand.

Note: Do not attempt to tighten any of the bolts until all of the bolts have been installed.

2 Alternately tighten the bolts until they are all snug.
3 Install the belt on the pulley.
4 Adjust the belt tension.
5 Tighten all of the bolts to the torque specifications of the manufacturer.
6 Connect the wires to the alternator.
7 Connect the battery ground cable.
8 Check the operation of the alternator.
9 If the car is supported by jack stands, lower the car to the floor.

SUMMARY

In this chapter, you were introduced to the charging system. By studying the material presented, you became knowledgeable of the different types of alternators and regulators commonly used, and how they function in various systems. By performing the jobs provided, you developed diagnostic and repair skills that enable you to perform basic charging system checks and to replace many of the components in those systems.

SELF-TEST

Each incomplete statement or question in this test is followed by four suggested completions or answers. In each case select the *one* that best completes the sentence or answers the question.

1 In an alternator, current is induced in the
   A. diode
   B. rotor
   C. stator
   D. rectifier

2 Current is conducted to the field winding by
   A. lines of force
   B. the stator windings
   C. diodes and rectifiers
   D. brushes and slip rings

3 The strength of the magnetic field in an alternator is determined by the amount of current that flows in the
   A. solenoid
   B. rectifier
   C. field windings
   D. stator windings

4 In an alternator, alternating current is converted to direct current by the
   A. stator
   B. brushes
   C. rectifier
   D. regulator

5 Two mechanics are discussing alternator regulators.
   Mechanic A says that regulators control the current that flows through the field winding.
   Mechanic B says that regulators convert alternating current to direct current.
   Who is right?
   A. A only
   B. B only
   C. Both A and B
   D. Neither A nor B

6 Two mechanics are discussing alternator regulators.
   Mechanic A says that electro-mechanical regulators use magnetically operated switches to control voltage.
   Mechanic B says that electronic regulators use transistorized circuits to control voltage.
   Who is right?
   A. A only
   B. B only
   C. Both A and B
   D. Neither A nor B

7 Two mechanics are discussing drive belt adjustment.

Mechanic A says that an alternator belt should be adjusted while the engine is running at approximately 1,000 rpm.

Mechanic B says that the tension specification for a new belt is different than the specification for a used belt.

Who is right?

A. A only

B. B only

C. Both A and B

D. Neither A nor B

8 Two mechanics are discussing preliminary charging system checks.

Mechanic A says that the voltage drop in the battery cables should be checked.

Mechanic B says that the voltage drop in the stator windings should be checked.

Who is right?

A. A only

B. B only

C. Both A and B

D. Neither A nor B

9 Voltage drop in a battery cable should not exceed

A. 0.1 V

B. 0.2 V

C. 0.3 V

D. 0.4 V

10 Two mechanics are discussing the causes of excessive voltage drop.

Mechanic A says that excessive voltage drop can be caused by a burned-out fuse link.

Mechanic B says that excessive voltage drop can be caused by worn alternator brushes.

Who is right?

A. A only

B. B only

C. Both A and B

D. Neither A nor B

Chapter 11
The Charging System—Testing and Alternator Overhaul

In performing the inspections and checks outlined in Chapter 10, you may have found and corrected some problems. Tests should now be made to determine if the charging system is operating correctly. If the system is undercharging or overcharging, additional tests will allow you to determine if the problem is located in the alternator, in the regulator, or in the wiring. The results of those tests will enable you to isolate the cause of the problem so that you can perform the required repairs.

Based on the knowledge you gained in the previous chapters, you now will perform diagnostic and repair procedures that will help you to further develop your skills. Your specific objectives are to perform the following jobs:

A
Perform a three-stage alternator output test
B
Determine the cause of undercharging
C
Determine the cause of overcharging
D
Overhaul a Delco SI (Delcotron) alternator
E
Overhaul a Ford rear terminal alternator
F
Overhaul a Chrysler alternator

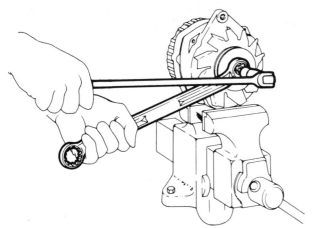

CHARGING SYSTEM TESTING All too often, a regulator or an alternator is replaced in the hope that a problem in the charging system will be eliminated. However, the problem frequently remains or returns within a short period of time. With a few simple tests, you can isolate the cause of a charging system problem. The correct repair procedures can then be performed and the problem can be eliminated.

The Three-Stage Charging Test A three-stage charging test will tell you if a charging system is operating correctly or if it is undercharging or overcharging. This test can be used not only for the diagnosis of problems, but for verifying that any repairs you made were effective.

Performing a Three-Stage Charging Test
The following steps outline a procedure for performing a three-stage charging test:

Stage 1 This part of the test provides you with a base voltage reading to which you can compare the readings that you obtain in stages 2 and 3.

1 Adjust a voltmeter to a scale exceeding battery voltage.

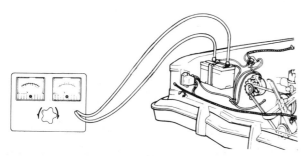

Figure 11.1 All the voltage readings taken during the three-stage charging test are taken with a voltmeter connected to the battery (courtesy of Ford Motor Company).

2 Connect the voltmeter to the battery terminals as shown in Figure 11.1.
3 Read the voltage and record the reading as the base voltage. (See Figure 11.2.)

Note: The engine should not be running and all accessories should be turned off.

Stage 2 This part of the test provides you with a no-load voltage reading with the engine running.

1 Apply the parking brake.
2 If the car has an automatic transmission, place the transmission selector lever in the Park position. If the car has a manual or standard transmission, place the shift lever in the Neutral position.
3 Start the engine and run it at approximately 2,000 rpm.

Note: Be sure that all accessories are turned off.

4 Read the voltmeter and record the reading as the no-load voltage.

Note: You may have to let the engine run for a short time if the voltmeter reading keeps rising. Read the voltmeter when the needle stops rising.

5 Compare the no-load voltage with the base voltage. See Figure 11.3.

Note: The no-load voltage should exceed the base voltage, but by no more than 2.0 V. A no-load voltage of more than 2.0 V over the base voltage may indicate an overcharging condition. The reading obtained in stage 3 will verify that condition.

Stage 3 This part of the test provides you with a load voltage reading while the engine is running and with a heavy load on the charging system.

1 Turn on the headlights in the high beam position.
2 Turn the heater blower switch to the High position.
3 Run the engine at approximately 2,000 rpm.

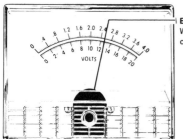

BASE VOLTAGE
With engine and all electrical loads turned off.

Figure 11.2 The base voltage reading is a measurement of battery voltage. It is taken with the engine and all electrical loads off (courtesy of Ford Motor Company).

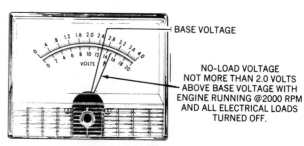

BASE VOLTAGE

NO-LOAD VOLTAGE
NOT MORE THAN 2.0 VOLTS
ABOVE BASE VOLTAGE WITH
ENGINE RUNNING @2000 RPM
AND ALL ELECTRICAL LOADS
TURNED OFF.

Figure 11.3 The no-load voltage reading is a measurement of the charging circuit voltage taken while the engine is running approximately 2,000 rpm with all electrical loads off (courtesy of Ford Motor Company).

Job 11a

PERFORM A THREE-STAGE CHARGING SYSTEM TEST

SATISFACTORY PERFORMANCE
A satisfactory performance on this job requires that you do the following:

1 Perform a three-stage test on the charging system of the vehicle assigned.
2 Following the steps in the "Performance Outline," complete the job within 15 minutes.
3 Fill in the blanks under "Information."

PERFORMANCE OUTLINE
1 Measure the base voltage.
2 Measure the no-load voltage.
3 Measure the load voltage.
4 Compare the readings.

INFORMATION
Vehicle identification _____

Reference used _____ Page(s) _____

Base voltage: _____

No-load voltage: _____

Load voltage: _____

Interpretation of test results _____

---

4 Read the voltmeter and record the reading as the load voltage.
5 Turn off the engine, the headlights, and the heater blower.
6 Disconnect the voltmeter.
7 Compare the load voltage with the base voltage. (See Figure 11.4.)

Note: The load voltage should exceed the base voltage by at least 0.5 V but not more than 2.0 V. If the load voltage is within that range, the charging system is operating correctly. If the difference is less than 0.5 V, the system is undercharging. If the difference is more than 2.0 V, the system is overcharging. Additional tests are required in both instances.

Testing for the Cause of Undercharging

The test procedures that follow are typical of those recommended by the various manufacturers. There are many different types of charging systems in use.

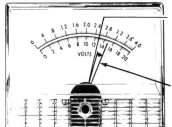

BASE VOLTAGE

LOAD VOLTAGE
AT LEAST 0.5 VOLT ABOVE
BASE VOLTAGE WITH ENGINE
RUNNING @2000 RPM AND
HEADLIGHTS AND BLOWER
TURNED ON.

Figure 11.4 The load voltage reading is a measurement of the charging circuit voltage taken while the engine is running approximately 2,000 rpm with a heavy load on the circuit (courtesy of Ford Motor Company).

And those systems use a variety of components. For those reasons, you should consult an appropriate manual for the specific test procedures recommended for the system on which you are working. In all instances, those test procedures should not be attempted until the preliminary checks and

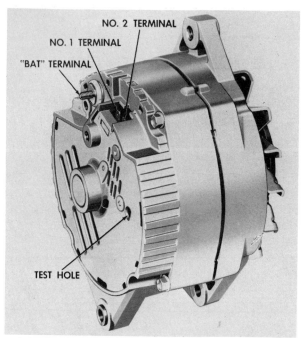

Figure 11.5 A Delco SI type (Delcotron) alternator with an internal regulator. A test hole is provided so that the regulator can be bypassed for testing (courtesy of Pontiac Motor Division, General Motors Corporation).

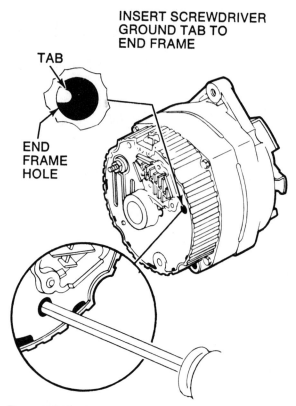

Figure 11.7 The internal regulator of a Delco type SI (Delcotron) alternator is bypassed by inserting a screwdriver into the test hole so that the internal tab is grounded (courtesy of Chevrolet Motor Division).

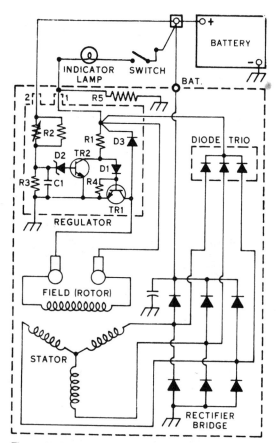

Figure 11.6 A typical wiring diagram for a Delco SI type (Delcotron) alternator used in a vehicle with a charge indicator lamp (courtesy of Chevrolet Motor Division).

the three-stage charging test have been performed:

Systems Using Alternators with Integral Regulators (Delco SI-type—Delcotron) Alternators of the type shown in Figure 11.5 have a D-shaped *test hole* in the slip ring end frame. The test hole allows you to bypass the integral regulator shown in the wiring diagram in Figure 11.6. The following procedures should be used:

1 Locate the test hole in the rear of the alternator.

Note: On some cars, access to the test hole is blocked. On those cars, the alternator must be removed for bench testing.

2 Insert a small screwdriver into the test hole as shown in Figure 11.7. Check to see that it will touch both the side of the hole and the *test tab* inside the alternator.

Note: The test tab is about ¾ inch (19 mm) inside the hole. Do not insert the screwdriver into the hole for a distance of more than 1 inch (25 mm). To do so may damage internal parts. If you cannot touch the test tab, the alternator must be removed for bench testing.

3 Adjust a voltmeter to a scale exceeding battery voltage.

4 Connect the voltmeter to the battery terminals. (Refer to Figure 11.1.)

5 Apply the parking brake.

6 If the car has an automatic transmission, place the transmission selector lever in the Park position. If the car has a standard or manual transmission, place the shift lever in the Neutral position.

7 Start the engine and allow it to idle.

8 Insert the small screwdriver into the test hole so that it contacts both the side of the hole and the test tab. (Refer to Figure 11.7.)

9 Gradually accelerate the engine and observe the voltmeter while holding the screwdriver in place. The reading should rise at least 2.0 V above the battery voltage.

10 Remove the screwdriver and turn off the engine.

Interpretation of Test Results   If the voltage rises considerably, the regulator in the alternator is defective. If the voltage does not rise, and the wiring is known to be in good condition, the alternator is defective. In both instances the alternator must be repaired or replaced.

Systems Using Alternators with Separate Regulators (Ford and Motorcraft)   Alternators of the type shown in Figures 11.8 and 11.9 are controlled by regulators that are mounted separately. Figure 11.10 shows how a typical system is wired. The following procedure explains how those regulators can be bypassed with jumper wires:

1 Determine the location of the battery (BAT) terminal and the field (FLD) terminal on the alternator. (Refer to Figures 11.8 and 11.9.)

2 Disconnect the regulator from the circuit.

Note: Most wiring harnesses have a plug that connects to the regulator. This plug can be unlocked by using a screwdriver as shown in Figure 11.11.

3 Adjust a voltmeter to a scale exceeding battery voltage.

4 Connect the voltmeter to the battery terminals. (Refer to Figure 11.1.)

5 Apply the parking brake.

6 If the car has an automatic transmission, place the transmission selector lever in the Park position. If the car has a standard or manual transmission, place the shift lever in the Neutral position.

7 Start the engine and allow it to idle.

8 Connect a jumper wire to the battery (BAT) and the field (FLD) terminals as shown in Figure 11.12.

9 Gradually accelerate the engine and observe

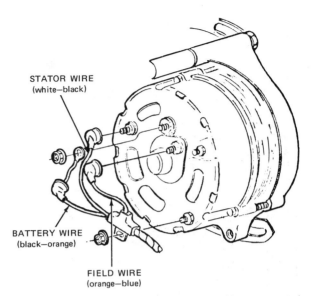

STATOR WIRE
(white—black)

BATTERY WIRE
(black—orange)

FIELD WIRE
(orange—blue)

Figure 11.8 An externally regulated alternator with rear-mounted terminals. Alternators of this type are used on some vehicles built by Ford Motor Company and by American Motors Corporation (courtesy of Ford Motor Company).

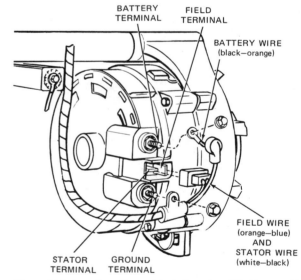

BATTERY TERMINAL

FIELD TERMINAL

BATTERY WIRE
(black—orange)

FIELD WIRE
(orange—blue)
AND
STATOR WIRE
(white—black)

STATOR TERMINAL

GROUND TERMINAL

Figure 11.9 An externally regulated alternator with side mounted terminals. Alternators of this type are used on some vehicles built by Ford Motor Company (courtesy of Ford Motor Company).

the voltmeter reading. The reading should rise at least 2.0 V above the battery voltage.

10 Remove the jumper wire and turn off the engine.

Interpretation of Test Results   If the voltage does not rise, the alternator is defective and must be repaired or replaced. If the voltage rises considerably, the alternator is functioning properly. The cause of undercharging lies either in the regulator or in the wires connecting the regulator to the alternator. To isolate the problem, continue with step 11.

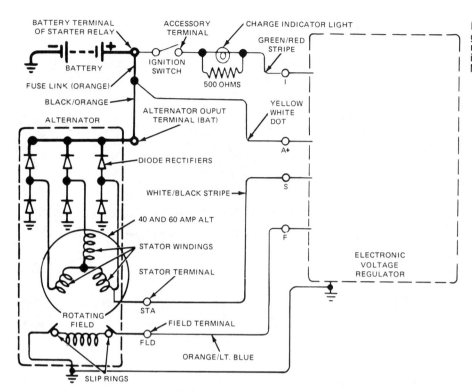

Figure 11.10 A typical wiring diagram for a Ford rear terminal alternator used in a vehicle with a charge indicator lamp (courtesy of Ford Motor Company).

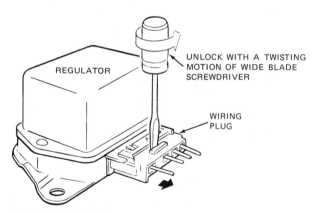

Figure 11.11 On some cars, the wiring plug is disconnected from the regulator by using a screwdriver as shown (courtesy of Ford Motor Company).

11 Connect a jumper wire between the "A" and "F" contacts in the plug that you disconnected from the regulator. (See Figure 11.13.)

12 Restart the engine.

13 Gradually accelerate the engine and observe the voltmeter reading.

14 Turn off the engine and remove the jumper wire.

Interpretation of Test Results   If the voltage rises considerably, the regulator is defective and should be replaced. If the voltage does not rise, the problem is caused by a break in the "A" wire or in the

"F" wire. To isolate the problem, continue with step 15.

15 Connect a jumper wire to the "F" contact in the plug and to the positive (+) battery terminal as shown in Figure 11.14.

16 Restart the engine.

17 Gradually accelerate the engine and observe the voltmeter reading.

18 Turn off the engine and remove the jumper wire.

Interpretation of Test Results   If the voltage does not rise, the problem is caused by a break in the "F" wire. If the voltage rises, the problem is caused by a break in the "A" wire. Perform the needed repair. Verify the repair by performing a three-stage charging test.

Systems Using Alternators with Separate Regulators (Chrysler)   Alternators of the type shown in Figure 11.15 are controlled by regulators that are mounted separately. Those regulators control the alternator field through its ground circuit as shown in Figure 11.16. The following procedure explains how those regulators can be bypassed with a jumper wire:

1 Determine the location of the field (FLD) terminal on the rear of the alternator. (Refer to Figure 11.15.)

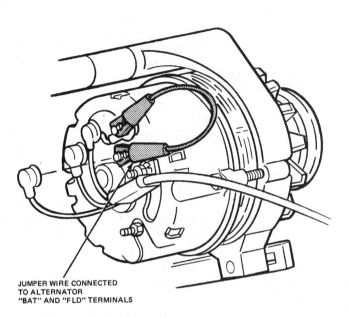

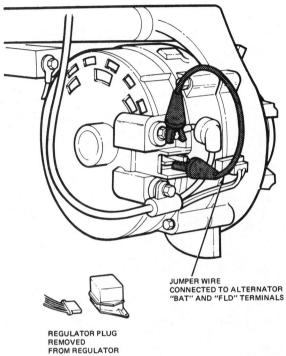

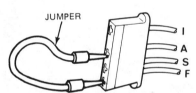

Figure 11.12 Jumper wire connections for checking alternator output (courtesy of Ford Motor Company).

JUMPER WIRE CONNECTED TO ALTERNATOR "BAT" AND "FLD" TERMINALS

JUMPER WIRE CONNECTED TO ALTERNATOR "BAT" AND "FLD" TERMINALS

REGULATOR PLUG REMOVED FROM REGULATOR

Figure 11.13 Jumper wire connections for bypassing the regulator at the wiring harness plug (courtesy of Ford Motor Company).

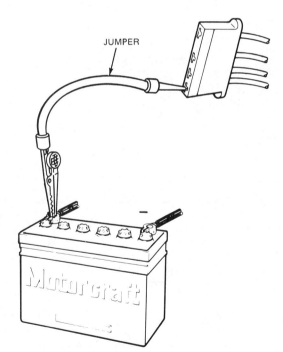

Figure 11.14 Jumper wire connections for testing the wires between the regulator and the alternator (courtesy of Ford Motor Company).

2 Disconnect (unplug) the wire from the field (FLD) terminal.

3 Adjust a voltmeter to a scale exceeding battery voltage.

4 Connect the voltmeter to the battery terminals. (Refer to Figure 11.1.)

5 Apply the parking brake.

6 If the car has an automatic transmission, place the transmission selector lever in the Park position. If the car has a standard transmission, place the shift lever in the Neutral position.

7 Start the engine and allow it to idle.

8 Connect a jumper wire to the field (FLD) terminal on the alternator and to a good ground.

9 Gradually accelerate the engine and observe the voltmeter. The reading should rise to at least 2.0 V above the battery voltage.

10 Remove the jumper wire and turn off the engine.

Interpretation of Test Results   If the voltage does not rise, the alternator is defective and must be repaired or replaced. If the voltage rises considerably, the alternator is functioning properly. The cause of the undercharging lies either in the regulator ground, in the regulator, or in the wire that connects the field terminal to the regulator. To isolate the problem, continue with step 11.

11 Connect the field wire to the field (FLD) terminal.
12 Connect a jumper wire between the base of the regulator and a good ground.
13 Start the engine.
14 Gradually accelerate the engine and observe the voltmeter. The reading should rise to at least 2.0 V above battery voltage.
15 Turn off the engine and remove the jumper wire.

Interpretation of Test Results   If the voltage rises considerably, the regulator has a poor ground. Remove the regulator and clean the regulator base to obtain a good connection with the body. The use of lock washers on the attaching screws is recommended. Verify the repair by performing a three-stage charging test. If the voltage does not rise, the cause of the undercharging lies in the regulator or in the wire connecting the regulator to the field (FLD) terminal. To isolate the cause of the problem, continue with step 16.

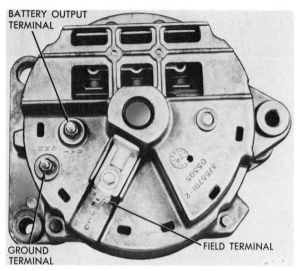

Figure 11.15 The rear view of an externally regulated alternator used on some vehicles built by Chrysler Corporation (courtesy of Chrysler Corporation).

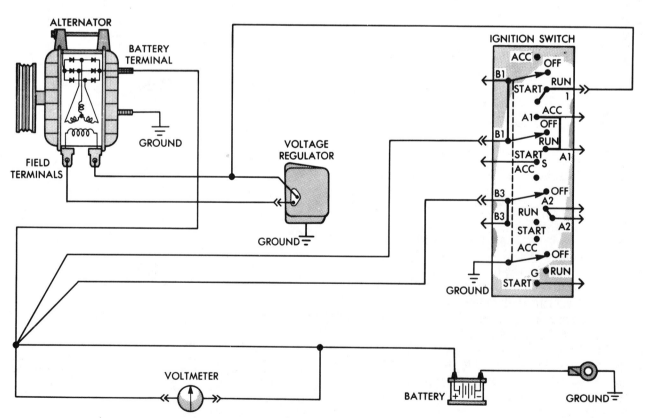

Figure 11.16 A typical wiring diagram for a Chrysler alternator used in a vehicle equipped with a voltmeter (courtesy of Chrysler Corporation).

16 Disconnect the field wire from the regulator.

Note: On vehicles with an electromechanical regulator, remove the green wire from the terminal. On vehicles with an electronic regulator, disconnect the wiring plug by releasing the locking tab. (See Figure 11.17).

17 Turn the ignition switch to the On position, but do not start the engine.

18 Remove the positive (+) voltmeter lead from the battery and touch it to the disconnected green wire.

19 Observe the voltmeter reading. Battery voltage should be indicated.

20 Turn the ignition switch to the Off position.

Interpretation of Test Results If battery voltage is indicated, the regulator is defective and should be replaced. If no voltage is indicated, the wire connecting the regulator to the field (FLD) terminal is broken and should be repaired. Verify the repair made by performing a three-stage charging test.

Testing for the Cause of Overcharging The following test procedures are typical of those recommended by the various manufacturers. You should consult an appropriate manual for the specific procedures recommended for the system on which you are working. In all instances, those procedures should be attempted only after the preliminary checks and the three-stage charging test have been performed:

Systems Using Alternators with Integral Regulators (Delco SI-type—Delcotron) Overcharging in systems using alternators of that type (refer to Figure 11.5) usually is caused by an internal failure in the alternator. The alternator should be repaired or replaced.

Systems Using Alternators with Separate Regulators (Ford and Motorcraft) The following steps outline a procedure for determining the cause of overcharging in systems using alternators of that type. (Refer to Figures 11.8 and 11.9.)

1 Adjust a voltmeter to a scale exceeding battery voltage.

2 Connect the voltmeter to the battery terminals. (Refer to Figure 11.1.)

3 Read the voltmeter and record the reading as the base voltage.

4 Connect a jumper wire to the base of the regulator and to a good ground. (See Figure 11.18.)

5 Apply the parking brake.

6 If the car has an automatic transmission, place the transmission selector lever in the Park position. If the car has a standard or manual transmission, place the shift lever in the Neutral position.

7 Start the engine and run it at approximately 2,000 rpm.

8 Read the voltmeter. The indicated voltage should not exceed 2.0 V above the base voltage.

9 Turn off the engine and remove the jumper wire.

Interpretation of Test Results If the voltage does not exceed 2.0 V over the base voltage, the regulator has a bad ground. Remove the regulator and clean the regulator base to obtain a good connection with the body. The use of lock washers on the attaching screws is recommended. Verify your re-

Figure 11.17 A typical electronic regulator. Note the locking tab that must be released to remove the wiring harness plug (courtesy of Chrysler Corporation).

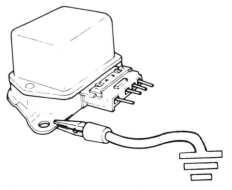

Figure 11.18 A poor regulator ground can be detected by grounding the regulator with a jumper wire (courtesy of Ford Motor Company).

Job 11b

DETERMINE THE CAUSE OF UNDERCHARGING

SATISFACTORY PERFORMANCE

A satisfactory performance on this job requires that you do the following:

1 Determine the cause of undercharging indicated by a three-stage charging test.

2 Following the steps in the "Performance Outline" and the procedure and specifications of the manufacturer, complete the tests required within 30 minutes.

3 Fill in the blanks under "Information."

PERFORMANCE OUTLINE

1 Bypass the regulator.

2 Check the system with the regulator bypassed.

INFORMATION

Vehicle identification _____

Reference used _____ Page(s) _____

Test Results: Alternator _____ OK _____ NG

Regulator _____ OK _____ NG

Wiring _____ OK _____ NG

Recommendations _____

_____

---

pair by repeating steps 7 and 8. If the voltage exceeds 2.0 V above the base voltage, continue with step 10.

10 Disconnect the wiring plug from the regulator. (Refer to Figure 11.11.)

11 Start the engine and run it at approximately 2,000 rpm.

12 Read the voltmeter. The voltage should not rise above the base voltage.

13 Turn off the engine.

Interpretation of Test Results   If the voltage rises above the base voltage, there is a short circuit in the wiring between the regulator and the alternator. Most likely, the "A" and "F" wires are in contact with each other. Check the wiring and perform the needed repairs. Verify your repair by repeating steps 11 through 13. If the voltage does not rise, the regulator is defective and should be replaced. Verify

your repairs by performing a three-stage charging test.

Systems Using Alternators with Separate Regulators (Chrysler)   The following steps outline a procedure for determining the cause of overcharging in systems using alternators of that type. (Refer to Figure 11.15.)

1 Adjust a voltmeter to a scale exceeding battery voltage.

2 Connect the voltmeter to the battery terminals. (Refer to Figure 11.1.)

3 Read the voltmeter and record the reading as the base voltage.

4 Disconnect the wire from the field (FLD) terminal on the alternator. (Refer to Figure 11.15.)

5 Apply the parking brake.

6 If the car has an automatic transmission, place the transmission selector lever in the Park position.

Job 11c

DETERMINE THE CAUSE OF OVERCHARGING

SATISFACTORY PERFORMANCE

A satisfactory performance on this job requires that you do the following:

1 Determine the cause of overcharging indicated by a three-stage charging test.
2 Following the steps in the "Performance Outline" and the procedure and specifications of the manufacturer, complete the tests required within 30 minutes.
3 Fill in the blanks under "Information."

PERFORMANCE OUTLINE

1 Bypass the regulator as required.
2 Check the system to isolate the defect.

INFORMATION

Vehicle identification _____

Reference used _____ Page(s) _____

Test results: Alternator _____ OK _____ NG

Regulator _____ OK _____ NG

Wiring _____ OK _____ NG

Grounds _____ OK _____ NG

Repairs required _____

If the car has a standard transmission, place the shift lever in the Neutral position.

7 Start the engine and allow it to idle.
8 Gradually accelerate the engine and observe the voltmeter. The reading should remain at the base voltage.
9 Turn off the engine.

Interpretation of Test Results   If the voltage rises above the base voltage, the alternator is defective and should be repaired or replaced. Verify your repair by performing a three-stage charging test. If the voltage does not rise above the base voltage, continue with step 10.

10 Disconnect the field wire from the regulator.

Note: On vehicles with an electromagnetic regulator, remove the green wire from the terminal. On vehicles with an electronic regulator, disconnect the wiring plug. (Refer to Figure 11.17.)

11 Disconnect the negative (−) voltmeter lead from the battery and connect it to the field wire.
12. Read the voltmeter. The voltmeter should indicate no voltage.
13 Disconnect the voltmeter lead.

Interpretation of Test Results   If voltage is indicated by the voltmeter, the field wire between the regulator and the alternator is grounded. Repair the wire. Verify your repair by repeating steps 11 through 13. If the voltmeter indicates no voltage, the regulator is defective and should be replaced. Connect the field wire to the field (FLD) terminal on the alternator and verify any repairs by performing a three-stage charging test.

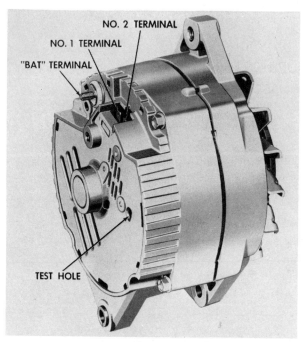

Figure 11.19 A Delco SI type alternator (courtesy of Pontiac Motor Division, General Motors Corporation).

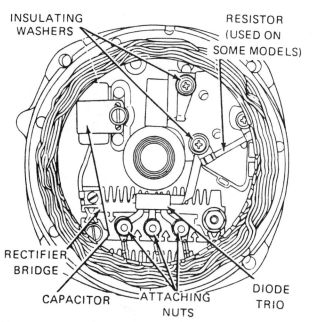

Figure 11.21 Internal view of the rear end frame assembly (courtesy of Chevrolet Motor Division).

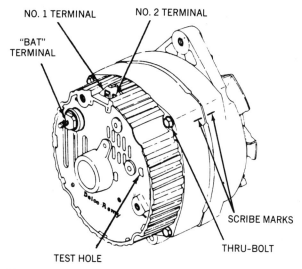

Figure 11.20 Before removing the thru-bolts, scribe marks should be made on both end frames so that they can be aligned during assembly (courtesy of Chevrolet Motor Division).

ALTERNATOR OVERHAUL

Overhauling a Delco SI-type (Delcotron) Alternator   The following steps outline a typical procedure for the disassembly, component testing, and reassembly of a Delco SI-type alternator. (See Figure 11.19.) You should consult an appropriate manual for the procedure and specifications that may be required for the particular alternator you overhaul.

Disassembly

1 Using a sharp tool or a chisel, scribe alignment marks on both end frames, as shown in Figure 11.20.

Note: The four thru-bolts are located 90° apart and the end frames may be assembled in four different positions. This allows the same alternator to be used on different engines and mounted in different locations. The scribe marks will enable you to assemble the alternator with the end frames properly aligned for the car from which it was removed.

2 Remove the four thru-bolts.

3 Separate the end frames so that the stator remains with the rear end frame assembly.

Note: Attempting to remove the stator with the front end frame (pulley end) may damage the stator leads.

4 Remove the three nuts that attach the diode trio. See Figure 11.21.

5 Remove the regulator attaching screws, noting the position of the insulating washers. (Refer to Figure 11.21.)

6 Remove the stator, the diode trio, the regulator, and the brush holder from the end frame.

7 Using a box wrench and an Allen socket as shown in Figure 11.22, remove the alternator pulley nut.

8 Remove the washer, pulley, fan, and fan collar from the rotor shaft.

9 Slide the rotor shaft from the bearing, noting the position of any collars or spacers.

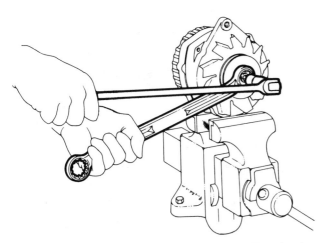

Figure 11.22 Removing an alternator pulley nut. Note that the nut is turned with a box wrench while the rotor is held by an Allen wrench socket (courtesy of American Motors).

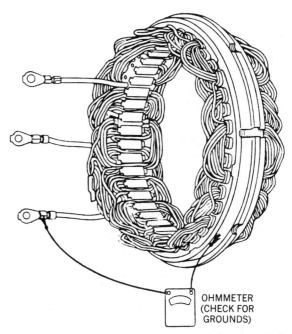

Figure 11.24 Checking a stator for grounds (courtesy of Chevrolet Motor Division).

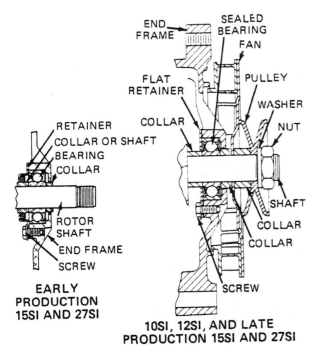

Figure 11.23 The stack-up of drive end parts differs on certain SI alternators (courtesy of Chevrolet Motor Division).

Note: The parts used and their order of assembly may differ on different versions of the SI alternator. (See Figure 11.23.)

10 Remove the screws holding the bearing retainer to the end frame. Then remove the bearing. (Refer to Figure 11.23.)

11 Push the bearing from the end frame.

Component Inspection and Testing

1 Using an ohmmeter as shown in Figure 11.24, check the stator for grounds. The meter should read infinity (∞). If low resistance is indicated, replace the stator. Repeat this test for each of the stator leads.

Note: The stator frame is usually coated with an insulating varnish. Be sure that the test probe has penetrated the varnish and is in contact with the metal laminations.

2 The stator used in the 10SI and 12SI alternators can be checked for opens, as shown in Figure 11.25. If either test indicates an open (a reading of infinity (∞), replace the stator.

Note: The 15SI and the 27SI alternators use delta wound stators that cannot be checked for opens by this method.

3 Check the bearing surfaces on the rotor shaft. If the surfaces are rough or damaged, replace the rotor.

4 Check the slip rings on the rotor. If they are dirty or rough, they may be cleaned and smoothed with polishing cloth of 400 grit or finer. The rotor should be turned against the polishing cloth held between your fingers to prevent the formation of flat spots.

5 Using an ohmmeter as shown in Figure 11.26, check the rotor to see that it is not grounded. When connected between a slip ring and the rotor shaft, the ohmmeter should indicate infinity (∞). If lower resistance is indicated, replace the rotor.

6 Check the rotor for an open circuit by connecting the ohmmeter between the two slip rings, as

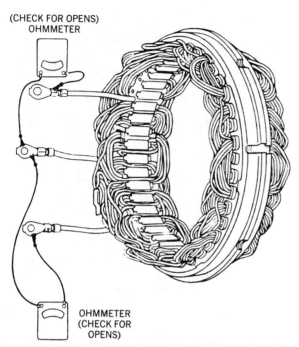

Figure 11.25 Checking a stator for opens. Note that two checks should be made (courtesy of Chevrolet Motor Division).

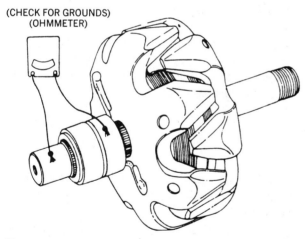

Figure 11.26 Checking a rotor for grounds. The ohmmeter is connected to a slip ring and to the rotor shaft (courtesy of Chevrolet Motor Division).

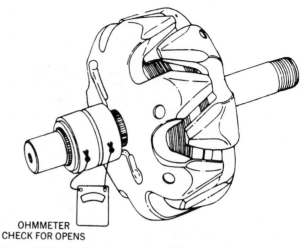

Figure 11.27 Checking a rotor for opens. The ohmmeter is connected to the two slip rings (courtesy of Chevrolet Motor Division).

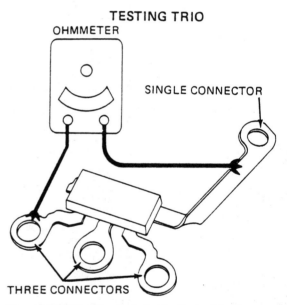

Figure 11.28 Checking a diode trio. Each of the three connections must be tested separately (courtesy of Chevrolet Motor Division).

shown in Figure 11.27. If the ohmmeter reading is not between 2.4 and 3.5 Ω, replace the rotor.

7 Insert the rotor shaft into the bearing in the rear end frame and spin the rotor to check the bearing. If the bearing feels rough, it should be replaced.

8 Check the drive end bearing. If the bearing feels loose or rough, replace the bearing.

9 Check the diode trio by connecting an ohmmeter to the single connector and to one of the three remaining connectors, as shown in Figure 11.28.

Alternately reverse the ohmmeter leads. The readings should alternate from infinity to low resistance. Repeat the test at each of the remaining connectors. The readings should be similar. If the readings are not similar or do not alternate during each of the three tests, replace the diode trio.

10 Check the rectifier bridge by connecting an ohmmeter between the grounded heat sink and one terminal as shown in Figure 11.29. Reverse the ohmmeter leads. If both readings are the same, replace the rectifier bridge. Repeat this test at the remaining two terminals.

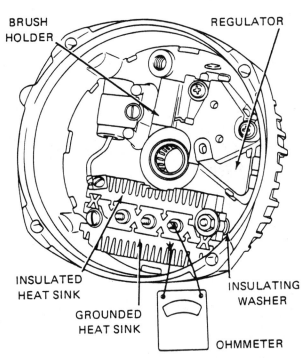

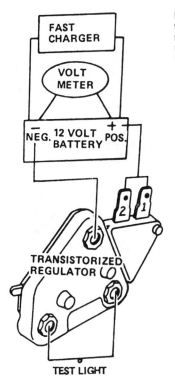

Figure 11.30 A transistorized voltage regulator may be bench tested in this manner (courtesy of Chevrolet Motor Division).

Figure 11.29 Checking the rectifier bridge (courtesy of Chevrolet Motor Division).

11 Repeat the tests in step 10 with the ohmmeter connected between the insulated heat sink and the terminals.

12 Clean and inspect the brushes. If the leads are frayed or if the brushes are worn to less than one half the length of a new brush, replace them.

13 Inspect the brush holder for cracks or damage. Replace the brush holder if it is damaged.

14 The transistorized voltage regulator may be tested by using a voltmeter and a test light connected as shown in Figure 11.30. The test procedure is as follows:

A. Connect a fast charger to a battery.

B. Connect a voltmeter to the battery.

C. Using jumper wires, connect the regulator to the battery. Be sure to observe the battery polarity shown.

D. Connect a test light to the regulator. The test light should go on.

E. Turn on the charger and gradually increase the charging rate while watching the voltmeter and the test light. The test light will go out when the voltage regulator setting is reached. The test light should go out when between 13.5 and 16.0 V is indicated.

Assembly

1 Install the bearing in the front end frame.

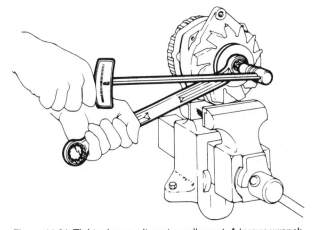

Figure 11.31 Tightening an alternator pulley nut. A torque wrench can be used on the Allen wrench socket (courtesy of American Motors).

Note: On some alternators, the bearing is not sealed and must be packed with lubricant before the bearing retainer is installed. Sealed bearings require no lubrication.

2 Install the bearing retainer.

3 Place the collar on the rotor shaft and push the shaft through the bearing.

4 Install the collar(s) fan, pulley, washer, and nut. (Refer to Figure 11.23.)

5 Using a torque wrench as shown in Figure 11.31, tighten the nut to 40 to 60 ft·lbs (54–82 N·m).

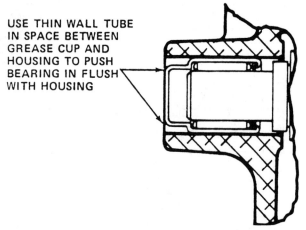

USE THIN WALL TUBE IN SPACE BETWEEN GREASE CUP AND HOUSING TO PUSH BEARING IN FLUSH WITH HOUSING

Figure 11.32 When replacing a stepped rear end frame bearing, a length of thin wall tubing should be used as an installation tool to avoid damage to the bearing (courtesy of Chevrolet Motor Division).

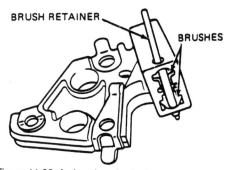

BRUSH RETAINER

BRUSHES

Figure 11.33 A short length of wire should be used to retain the brushes in the brush holder (courtesy of Chevrolet Motor Division).

6 If the bearing in the rear end frame must be replaced, push the defective bearing out from the outside toward the inside using a socket of the appropriate size.

Note: The inside of the end frame should be supported with a large socket or a piece of pipe to prevent distortion or breakage of the end frame.

7 Install the replacement bearing by pushing it in place from the outside toward the inside until it is flush with the outside of the end frame. Be sure to support the inside of the end frame.

Note: Some alternators use a variant bearing with a stepped shell as shown in Figure 11.32. Bearings of that type should be pushed in place using a length of thin wall tubing so that the force is transmitted to the step in the bearing.

8 Lubricate the bearing.

9 Install the brush springs and the brushes in the brush holder, making sure that they move freely. Push the brushes in against the spring pressure

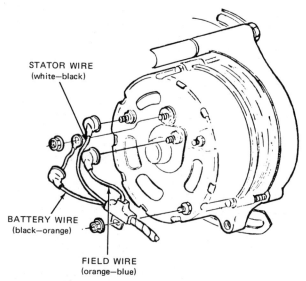

STATOR WIRE (white—black)

BATTERY WIRE (black—orange)

FIELD WIRE (orange—blue)

Figure 11.34 A Ford (Motorcraft) rear terminal alternator (courtesy of Ford Motor Company).

and hold them in place with a short length of stiff wire as shown in Figure 11.33 (The wire will protrude through a hole in the rear end frame and is removed after the alternator is assembled.)

10 Install the brush holder, regulator, and diode trio in the end frame.

Note: Check to see that the wire holding the brushes in place protrudes from the rear of the end frame.

11 Position the stator in the end frame and install the nuts that hold the stator leads. (Refer to Figure 11.21.)

12 Insert the rotor into the rear end frame and bring both the end frames together so that the marks made at the start of disassembly are aligned. (Refer to Figure 11.20.)

13 Install the four thru-bolts.

14 Tighten the thru-bolts alternately, checking to see that the end frames are aligned properly and that the rotor spins freely.

15 Remove the wire holding the brushes in position.

Overhauling a Ford Rear Terminal Alternator The following steps outline a typical procedure for the disassembly, component testing, and reassembly of a Ford rear terminal alternator. (See Figure 11.34.) You should consult an appropriate manual for the procedure and specifications that may be necessary for the specific alternator that you overhaul.

Disassembly

1 Remove the fan shield and shroud (if fitted). (See Figure 11.35.)

Job 11d

OVERHAUL A DELCO SI-TYPE (DELCOTRON) ALTERNATOR

SATISFACTORY PERFORMANCE

A satisfactory performance on this job requires that you do the following:

1 Overhaul the Delco SI-type (Delcotron) alternator assigned.

2 Following the steps in the "Performance Outline" and the procedure and specifications of the manufacturer, complete the job within 200 percent of the manufacturer's suggested time.

3 Fill in the blanks under "Information."

PERFORMANCE OUTLINE

1 Disassemble the alternator.

2 Clean and inspect the components.

3 Test the stator, rotor, diode trio, rectifier bridge, and regulator.

4 Assemble the alternator, replacing defective parts.

INFORMATION

Alternator identification _____

Reference used _____ Page(s) _____

Test results: Stator grounded _____ Yes _____ No

Stator open _____ Yes _____ No

Rotor grounded _____ Yes _____ No

Rotor winding resistance _____ ohms

Diode trio _____ OK _____ NG

Rectifier bridge _____ OK _____ NG

Regulator setting _____ volts

Parts replaced _____

---

2 Using a scriber or a chisel, scribe alignment marks on both end frames. (Refer to Figure 11.20.)

Note: The thru-bolts are located 120° apart and the end frames may be assembled in three different positions. The scribe marks will enable you to assemble the alternator with the end frames properly aligned for the vehicle from which it was removed.

3 Remove the three thru-bolts. (See Figure 11.36.)

4 Separate the end frames so that the stator remains with the rear end frame assembly.

Note: Attempting to remove the stator with the front end frame (pulley end) may damage the stator leads.

5 Remove the brush springs from the brush holder. (Refer to Figure 11.36.)

6 Remove the nuts, washers, and insulators from the terminals on the back of the rear end frame.

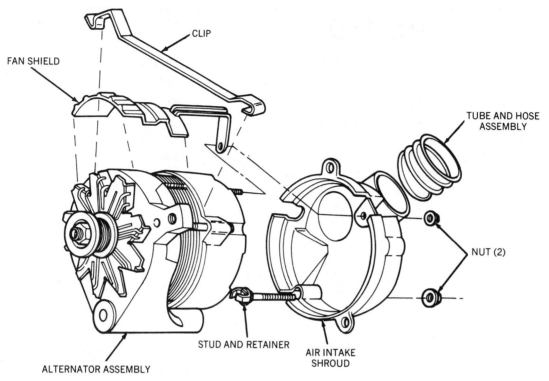

Figure 11.35 On some applications, the alternator is fitted with a fan shield or with both a fan shield and an air intake shroud (courtesy of Ford Motor Company).

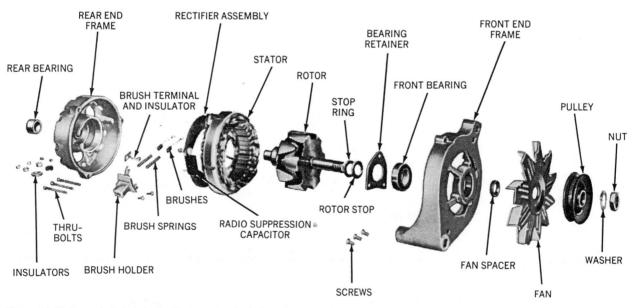

Figure 11.36 An exploded view of a Ford rear terminal alternator (courtesy of Ford Motor Company).

Note: Observe the color of the insulators and their location so that they will be correctly placed during reassembly.

7 Carefully remove the stator and rectifier assembly from the end frame.

8 Remove the screws securing the brush holder, and remove the brush holder, the brushes, and the brush terminal insulator.

9 Using a box wrench and an Allen wrench, remove the the nut that secures the pulley to the rotor

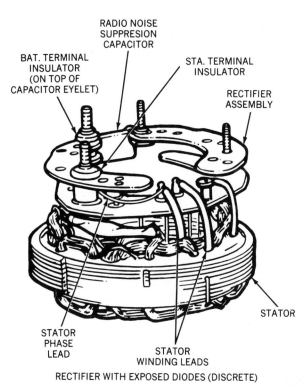

RADIO NOISE
SUPPRESION
CAPACITOR

BAT. TERMINAL
INSULATOR
(ON TOP OF
CAPACITOR EYELET)

STA. TERMINAL
INSULATOR

RECTIFIER
ASSEMBLY

STATOR

STATOR
PHASE
LEAD

STATOR
WINDING LEADS

RECTIFIER WITH EXPOSED DIODES (DISCRETE)

Figure 11.37 A rectifier with exposed diodes (discrete). Note the position of the stator winding leads, the insulators, and the radio noise suppression capacitor (courtesy of Ford Motor Company).

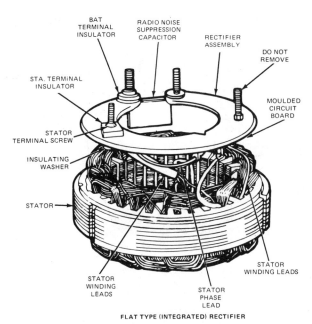

BAT
TERMINAL
INSULATOR

RADIO NOISE
SUPPRESSION
CAPACITOR

RECTIFIER
ASSEMBLY

DO NOT
REMOVE

STA. TERMINAL
INSULATOR

MOULDED
CIRCUIT
BOARD

STATOR
TERMINAL SCREW

INSULATING
WASHER

STATOR

STATOR
WINDING
LEADS

STATOR
WINDING LEADS

STATOR
PHASE
LEAD

FLAT TYPE (INTEGRATED) RECTIFIER

Figure 11.38 A flat type (integrated) of rectifier that contains the diodes in a moulded circuit board (courtesy of Ford Motor Company).

shaft. (Refer to Figure 11.22.)

10 Remove the lockwasher, pulley, fan, and fan spacer from the rotor shaft. (Refer to Figure 11.36.)

11 Slide the rotor shaft from the bearing.

12 Remove the rotor stop washer from the rotor shaft.

Note: Do not remove the stop ring from its groove in the rotor shaft. Spreading the ring to remove it will distort the ring and will require its replacement.

13 Remove the three screws holding the bearing retainer to the end frame. Remove the retainer.

14 Push the bearing from the end frame.

15 Remove the radio suppression capacitor and the battery terminal insulator from the rectifier assembly.

Note: Two types of rectifier assemblies are used. One, shown in Figure 11.37, has exposed diodes. Figure 11.38 shows a flat, integrated rectifier that contains the diodes in a molded circuit board.

16 Unsolder the stator leads from the rectifier assembly. As shown in Figure 11.39, needle nose pliers should be used to pull the tubular ends of the wires loose from the terminals when the solder is molten.

Note: It is recommended that a soldering iron or gun

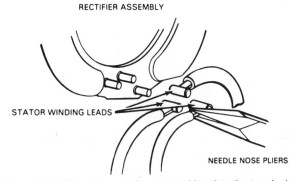

RECTIFIER ASSEMBLY

STATOR WINDING LEADS

NEEDLE NOSE PLIERS

Figure 11.39 The stator windings are soldered to the terminals on the rectifier assembly. They can be pulled off when the solder is molten (courtesy of Ford Motor Company).

of 100 W be used. Care must be taken so that the rectifier is not overheated during this step.

17 Disconnect the stator phase lead from the rectifier assembly.

Note: The method of removing the stator terminal screw holding the stator phase lead differs with the rectifier. On the rectifier with exposed diodes, the screw is turned one quarter of a turn (90°) to unlock it, as shown in Figure 11.40. On the flat integrated rectifier, shown in Figure 11.41, the terminal screw must be pressed straight out as it is fitted in a serrated hole. Any attempt to turn the screw may destroy the rectifier. (See Figure 11.42.)

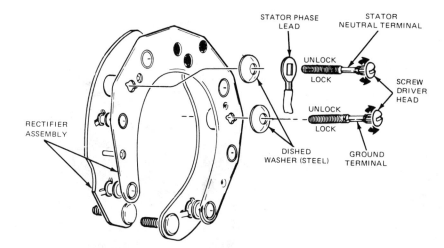

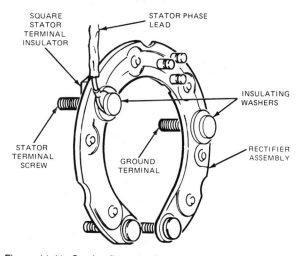

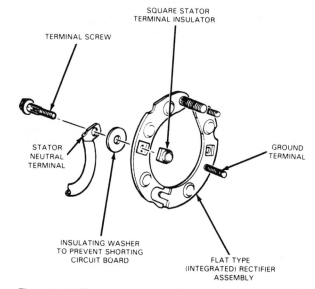

Figure 11.40 The stator phase lead and the ground terminal are disconnected from the exposed diode (discrete) rectifier by turning the screws ¼ turn to unlock them from the rectifier assembly (courtesy of Ford Motor Company).

Figure 11.41 On the flat type (integrated) rectifier, the stator phase lead is held by a serrated screw (courtesy of Ford Motor Company).

Figure 11.42 The stator phase lead is disconnected from the flat type rectifier by pressing the terminal screw straight out. Any attempt to turn the screw will damage the rectifier (courtesy of Ford Motor Company).

Component Inspection and Testing

1 Using an ohmmeter as shown in Figure 11.43, check the stator for grounds. The meter should read infinity (∞). If low resistance is indicated, replace the stator. Repeat this test for each of the stator leads.

Note: The stator frame is usually coated with an insulating varnish. Be sure that the test probe has penetrated the varnish and is in contact with the metal laminations.

2 Check the bearing surfaces on the rotor shaft. If the surfaces are rough or damaged, replace the rotor.

3 Check the slip rings on the rotor. If they are dirty or rough, they may be cleaned and smoothed with polishing cloth of 400 grit or finer. The rotor should

be turned against the polishing cloth held between your fingers to prevent the formation of flat spots.

4 Check the rotor to see that it is not grounded. Contact the ohmmeter probes to the rotor shaft and to one of the slip rings. (Refer to Figure 11.26.) The ohmmeter should indicate infinity (∞). If lower resistance is indicated, replace the rotor.

5 Check the rotor for an open circuit by connecting the ohmmeter between the two slip rings. (Refer to Figure 11.27.) If the ohmmeter reading is not between 2.0 and 3.5 Ω, replace the rotor.

6 Insert the rotor shaft into the bearing in the rear end frame and spin the rotor to check the bearing. If the bearing feels rough, it should be replaced.

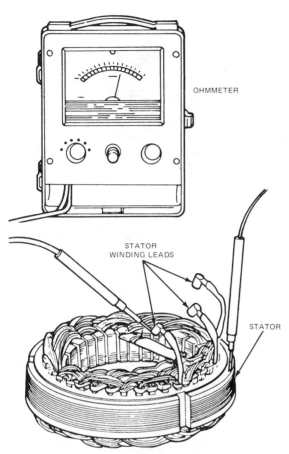

Figure 11.43 Checking a stator for grounds. This test should be repeated at each of the stator winding leads (courtesy of Ford Motor Company).

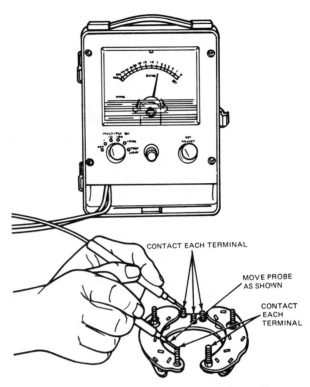

Figure 11.44 Testing a rectifier. All six diodes in the rectifier must be tested, reversing the probes during each test (courtesy of Ford Motor Company).

7 Check the drive end bearing. If the bearing feels loose or rough, replace the bearing.

8 Check the rectifier assembly as shown in Figure 11.44. One lead of the ohmmeter should be held on one of the two terminal bolts, while the other lead is held to one of the three stator lead terminals. The leads are then reversed. The ohmmeter should indicate a low reading (about 6 Ω) when the leads are held in one position and should indicate infinity (∞) when the leads are reversed. This test should be repeated at each of the three stator lead terminals.

9 Repeat step 8 at the remaining terminal bolt.

Note: It is necessary to repeat the test in step 9 so that all six diodes are checked. If the readings are not the same for all six diodes (about 6 Ω in one direction, infinity (∞) in the other direction) replace the rectifier assembly.

10 Clean and inspect the brushes. If the leads are frayed or if the brushes are worn to less than one half the length of a new brush, replace them.

11 Inspect the brush holder for cracks or damage. Replace the brush holder if it is damaged.

Assembly
1 Install the bearing in the front end frame.
2 Install the bearing retainer.
3 Place the rotor stop washer on the rotor shaft.

Note: The rotor stop washer has a recess on one side. The washer must be installed so that the recess is toward the stop ring.

4 Push the rotor shaft through the bearing and install the fan spacer, fan, pulley, washer, and nut. (Refer to Figure 11.36.)

5 Tighten the nut to 60 to 100 ft·lb (82 to 135N·m). (Refer to Figure 11.31.)

6 If the bearing in the rear end frame must be replaced, push the defective bearing out from the outside toward the inside using a socket of the appropriate size.

Note: The inside of the end frame must be supported with a large socket or a section of pipe to prevent distortion or breakage of the end frame.

7 Install the replacement bearing by pushing it in place from the outside toward the inside until it is

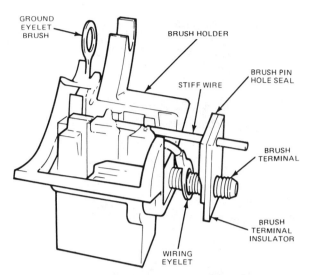

Figure 11.45 The brush holder with the brushes installed. Note that a piece of stiff wire is used to hold the brushes in place, and that the wiring eyelet is over the brush terminal (courtesy of Ford Motor Company).

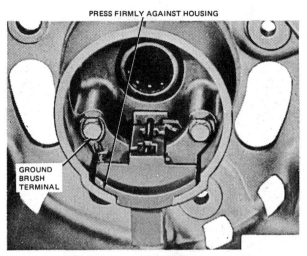

Figure 11.46 A correctly installed brush holder. Note that the ground brush terminal is held by one of the mounting screws. The brush holder should be held firmly in place while the mounting screws are tightened (courtesy of Ford Motor Company).

flush with the outside of the end frame. Be sure to support the inside of the end frame.

8 Lubricate the bearing.

9 Place the eyelet on the inner (insulated) brush over the brush terminal, as shown in Figure 11.45.

10 Install the brush springs and brushes in the brush holder, making sure that they move freely. Push the brushes in against the spring pressure and hold them in place with a short length of stiff wire. (Refer to Figure 11.45.) (The wire will protrude through a hole in the rear end frame and is removed after the alternator is assembled.)

11 Install the brush terminal insulator. (Refer to Figure 11.45.)

12 Install the brush holder assembly in the end frame and hold it firmly in place while tightening the attaching screws. (See Figure 11.46.)

Note: Be sure that the the eyelet on the ground brush is in place under one of the screws. Be sure that the wire holding the brushes in place protrudes from the rear of the end frame.

13 Connect the stator phase lead to the rectifier. (Refer to Figures 11.40, 11.41, and 11.42.)

14 Solder the stator leads to the rectifier, making sure that the terminals on the leads are pushed into place.

Note: Use rosin core solder and a 100-W soldering gun or iron. Care must be taken not to overheat the rectifier during this step.

15 Install the radio suppression capacitor and the insulators on the battery terminal and the stator ter-

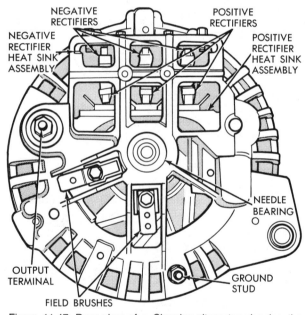

Figure 11.47 Rear view of a Chrysler alternator showing the location of the terminals (courtesy of Chrysler Corporation).

minal screws in the rectifier. (Refer to Figures 11.37 and 11.38.)

16 Position the rectifier and stator assembly in the end frame, aligning the terminal screws on the rectifier with the appropriate holes in the end frame.

17 Align the terminal screws so that they are centered in their holes and install the outer insulators.

Note: The insulators are color-coded. The black insulator should be installed on the stator (STA) terminal, the red on the battery (BAT) terminal, and the

Job 11e

OVERHAUL A FORD REAR TERMINAL ALTERNATOR

SATISFACTORY PERFORMANCE

A satisfactory performance on this job requires that you do the following:

1 Overhaul the Ford rear terminal alternator assigned.

2 Following the steps in the "Performance Outline" and the procedure and specifications of the manufacturer, complete the job within 200 percent of the manufacturer's suggested time.

3 Fill in the blanks under "Information."

PERFORMANCE OUTLINE

1 Disassemble the alternator.

2 Clean and inspect the components.

3 Test the stator, rotor, and rectifier.

4 Assemble the alternator, replacing defective parts.

INFORMATION

Alternator identification _____

Reference used _____ Page(s) _____

Test results: Stator grounded _____ Yes _____ No

Rotor grounded _____ Yes _____ No

Rotor open _____ Yes _____ No

Rotor winding resistance _____ ohms

Rectifier assembly _____ OK _____ NG

Parts replaced _____

---

orange on the field (FLD) terminal.

18 Install the washers and nuts on the terminals and tighten the nuts.

19 Insert the rotor into the rear end frame and bring both of the end frames together so that the marks made at the start of disassembly are aligned. (Refer to Figure 11.20.)

20 Install the three thru-bolts.

21 Tighten the thru-bolts alternately, checking to see that the end frames are properly aligned and that the rotor spins freely.

22 Remove the wire holding the brushes in position.

23 Install the fan shield and shroud (if fitted).

Overhauling a Chrysler Alternator The following steps outline a typical procedure for the disassembly, component testing, and reassembly of a Chrysler alternator of the type shown in Figure 11.47. You should consult an appropriate manual for the procedure and specifications that may be required for the specific alternator you overhaul:

Disassembly

1 Remove the screws holding the brushes to the rear end frame and remove the brushes and the brush holders. (See Figure 11.48.)

2 Remove the three thru-bolts. (Refer to Figure 11.48.)

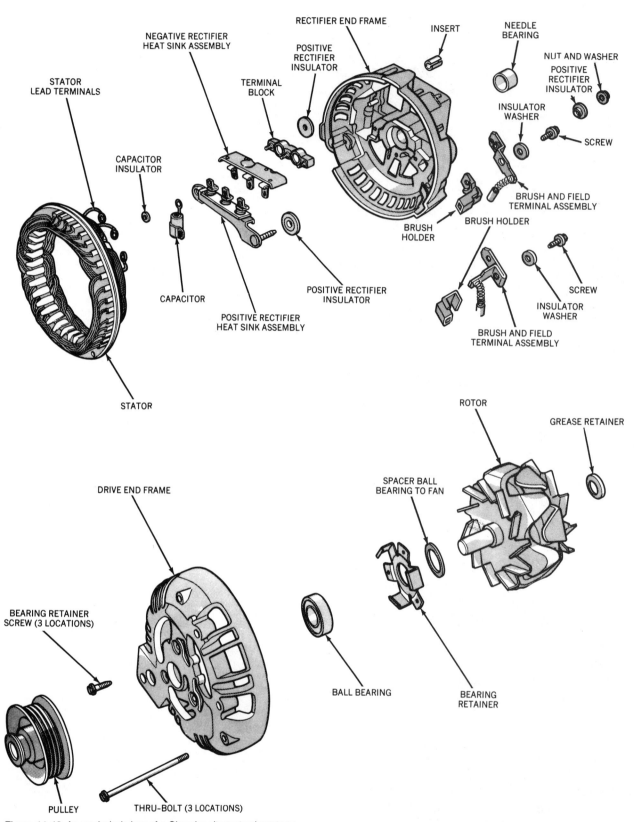

Figure 11.48 An exploded view of a Chrysler alternator (courtesy of Chrysler Corporation).

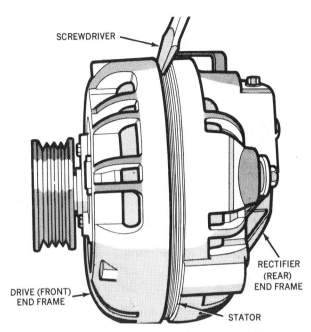

Figure 11.49 When separating the end frames, the stator should remain with the rear end frame (courtesy of Chrysler Corporation).

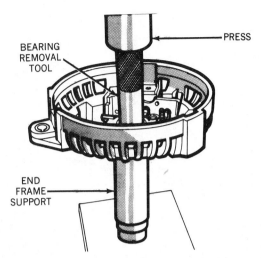

Figure 11.50 The needle bearing in the rear end frame should be pressed out from the inside toward the outside. Note that the end frame is supported at the bearing (courtesy of Chrysler Corporation).

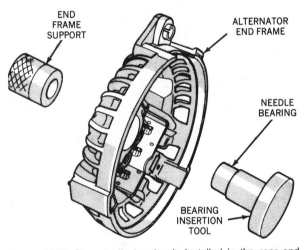

Figure 11.51 The needle bearing is installed in the rear end frame from the inside toward the outside. Lacking the special tools shown, the bearing can be installed by the careful use of appropriate size sockets (courtesy of Chrysler Corporation).

3 Separate the end frames so that the stator remains with the rear end frame assembly. A screwdriver may be used, as shown in Figure 11.49.

Note: Attempting to remove the stator with the front end frame (pulley end) may damage the stator leads.

4. Check the drive end bearing for noise, roughness, and looseness by holding the drive end frame and spinning the rotor. If the bearing is defective, it must be replaced.

Note: Both the pulley and the drive end bearing are pressed on the rotor shaft to obtain a very tight interference fit. Removal of the pulley and bearing is not recommended unless one of those parts must be replaced.

5 Check the bearing surface on the rear of the rotor shaft. If the surface is rough or damaged, the rotor should be replaced.

6 Insert the rotor shaft into the bearing in the rear end frame and spin the rotor to check the bearing. If the bearing feels rough, it should be replaced.

7 If the bearing in the rear end frame must be replaced, push the defective bearing out as shown in Figure 11.50.

Note: If you lack the special tools, the bearing can removed by the careful use of sockets of the appropriate size.

8 Install the replacement bearing by pressing it into place as shown in Figure 11.51.

Note: If you lack special tools, the bearing can be installed by the careful use of sockets of the appropriate size.

9 Check the slip rings on the rotor. If they are dirty or rough, they may be cleaned and smoothed with polishing cloth of 400 grit or finer. The rotor should be turned against the polishing cloth held between your fingers to prevent the formation of flat spots.

10 Using an ohmmeter, check the rotor to see that it is not grounded. When connected between a slip ring and the rotor shaft as shown in Figure 11.52, the ohmmeter should indicate infinity (∞). If lower resistance is indicated, replace the rotor.

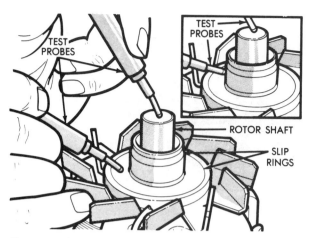

Figure 11.52 Checking a rotor for grounds. The ohmmeter is connected to a slip ring and the rotor shaft (courtesy of Chrysler Corporation).

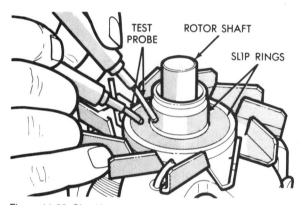

Figure 11.53 Checking a rotor for opens. The ohmmeter is connected to the two slip rings (courtesy of Chrysler Corporation).

11 Check the rotor for an open circuit by connecting an ohmmeter between the two slip rings, as shown in Figure 11.53. The ohmmeter should read between 1.5 and 2.0 Ω. An ohmmeter reading between 2.5 and 3.0 Ω indicates that the alternator has been operating at higher-than-normal temperatures. An ohmmeter reading over 3.5 Ω indicates excessive resistance, and the rotor should be replaced.

Pulley and Bearing Replacement Both the pulley and the drive end bearing are pressed on the rotor shaft to obtain a very tight interference fit. A special puller is required for removal, and a press is required for replacement. If the rotor, bearing, or pulley require replacement, the following procedure should be used:

Removal

1 Using a special puller as shown in Figure 11.54, remove the pulley from the rotor shaft.

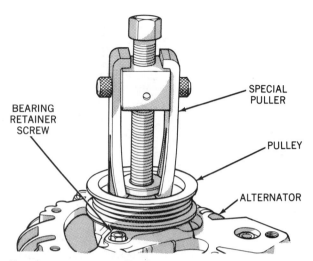

Figure 11.54 Removing the pulley with a special puller. Note that the puller is positioned on the small collar at the hub of the pulley (courtesy of Chrysler Corporation).

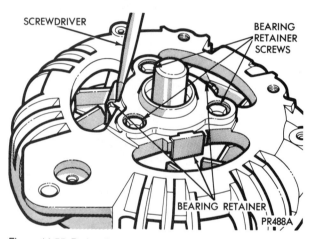

Figure 11.55 Prying the bearing retainer tabs from the front end frame (courtesy of Chrysler Corporation).

2 Remove the three screws that hold the bearing retainer. (Refer to Figure 11.54.)
3 Using a screwdriver as shown in Figure 11.55, disengage the three tabs on the bearing retainer.
4 Using a plastic hammer, drive the rotor shaft and bearing from the drive end frame.

Note: Do not strike the rotor shaft with a metal hammer as shaft damage will result.

5 Remove the bearing from the rotor shaft using the special puller as shown in Figure 11.56.
6 Remove the bearing spacer from the rotor shaft. (Refer to Figure 11.48.)

Installation

1 Insert the replacement bearing in the drive end frame.

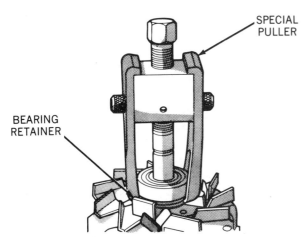

Figure 11.56 Removing the bearing from the rotor shaft (courtesy of Chrysler Corporation).

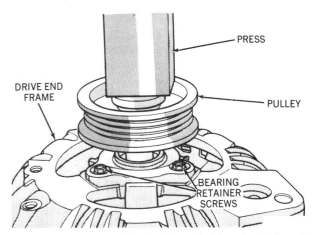

Figure 11.58 Installing the pulley. The end of the rotor shaft must be supported on the press during this operation (courtesy of Chrysler Corporation).

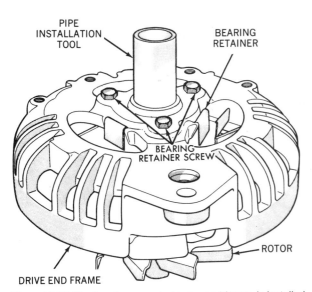

Figure 11.57 The bearing, mounted in the end frame, is installed on the rotor shaft by using a short length of pipe slipped over the shaft. The rotor must be supported in a press for this operation (courtesy of Chrysler Corporation).

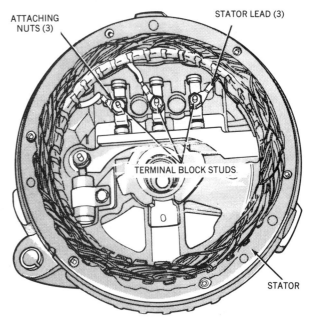

Figure 11.59 The stator leads are attached to the terminal block studs by three nuts. Those nuts must be removed before the stator is removed (courtesy of Chrysler Corporation).

2 Install the bearing retainer so that the three tabs snap over the drive end frame.

3 Install the three bearing retainer screws and tighten them.

4 Place the bearing spacer on the rotor shaft. (Refer to Figure 11.48.)

5 Position the drive end frame assembly over the end of the rotor shaft. Using a short length of pipe that contacts the inner bearing race as shown in Figure 11.57, press the bearing into place on the shaft.

Note: Be sure that the rotor and the bearing are in perfect alignment before any pressure is applied to the bearing. The bearing and the rotor shaft may be

damaged if the bearing is forced on at an angle.

6 Press the pulley on the shaft as shown in Figure 11.58.

Note: The pulley should be pressed on until it contacts the inner race of the bearing. Do not exceed 6,800 psi (47,000 kPa) installation pressure.

Stator and Rectifier Testing

1 Remove the attaching nuts from the terminal block studs on the rectifier. (See Figure 11.59.)

2 Remove the stator leads from the studs and carefully pry the stator from the end frame.

Job 11f

OVERHAUL A CHRYSLER ALTERNATOR

SATISFACTORY PERFORMANCE

A satisfactory performance on this job requires that you do the following:

1 Overhaul the Chrysler alternator assigned.
2 Following the steps in the "Performance Outline" and the procedure and specifications of the manufacturer, complete the job within 200 percent of the manufacturer's suggested time.
3 Fill in the blanks under "Information."

PERFORMANCE OUTLINE

1 Disassemble the alternator.
2 Clean and inspect the components.
3 Test the stator, rotor, and rectifiers.
4 Assemble the alternator, replacing defective parts.

INFORMATION

Alternator identification _____

Reference used _____ Page(s) _____

Test results:	Stator grounded	_____ Yes	_____ No
	Stator open	_____ Yes	_____ No
	Rotor grounded	_____ Yes	_____ No
	Rotor open	_____ Yes	_____ No
	Rotor winding resistance	_____ ohms	
	Positive rectifier	_____ OK	_____ NG
	Negative rectifier	_____ OK	_____ NG

Parts replaced _____

---

3 Using an ohmmeter, check the stator for grounds. The meter should read infinity (∞) when connected between the stator frame and all of the stator leads. (Refer to Figure 11.24.) Replace the stator if low resistance is indicated in any of the three windings.

Note: The stator frame is usually coated with an insulating varnish. Be sure that the test probe has penetrated the varnish and is in contact with the metal laminations.

4 Check the stator for opens. (Refer to Figure 11.25.) If either test indicates an open [a reading of infinity (∞)], replace the stator.

5 Check the positive rectifier by connecting an ohmmeter between the positive rectifier heat sink and one of the positive rectifier straps as shown in Figure 11.60. Reverse the ohmmeter leads. If both readings are the same, replace the rectifier. Repeat this test at the two remaining positive rectifier straps.

6 Test the negative rectifier by repeating the test in step 5 with the ohmmeter connected between the negative heat sink and the negative rectifier straps as shown in Figure 11.61.

7 Clean and inspect the brushes. If the leads are frayed or if the brushes are worn to less than one

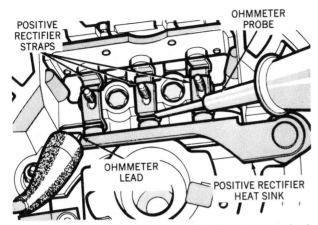

Figure 11.60 Testing the positive rectifier. One ohmmeter lead is connected to the positive heat sink while the other lead is touched to each of the three rectifier straps. The leads are then reversed and the test repeated (courtesy of Chrysler Corporation).

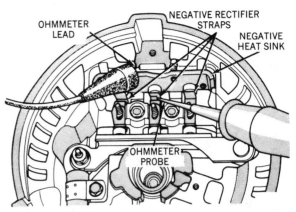

Figure 11.61 Testing the negative rectifier. One ohmmeter lead is connected to the negative heat sink while the other lead is touched to each of the three rectifier straps. The leads are then reversed and the test repeated (courtesy of Chrysler Corporation).

half the length of a new brush, replace them.

8 Inspect the brush holders for cracks and damage. Replace the brush holders if they are damaged.

Assembly

1 Insert the stator assembly into the rear end frame.

2 Position the stator leads on the studs of the terminal block. (Refer to Figure 11.59.)

3 Install and tighten the nuts securing the leads to the studs.

Note: Be sure that the leads are positioned so that they will not contact the rotor or the sharp edges of the negative heat sink. (Refer to Figure 11.59.)

4 Insert the rotor into the rear end frame and bring both frames together so that the mounting ears are aligned.

5 Holding the end frames together, install the three thru-bolts.

6 Tighten the thru-bolts alternately, checking to see that the end frames are properly aligned and that the rotor spins freely.

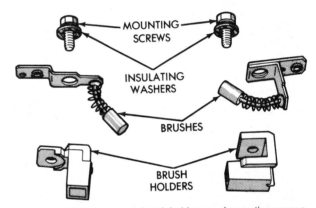

Figure 11.62 The brushes, brush holders, and mounting screws. Note that the mounting screws are fitted with insulating washers (courtesy of Chrysler Corporation).

7 Install the brushes in the brush holders and position them in the rear end frame. (See Figure 11.62.)

8 Secure the brushes and the brush holders with the mounting screws, making sure that the insulating washers are in place. (Refer to Figure 11.48.)

Note: Be sure that the terminals on the brushes are not grounded.

SUMMARY

By studying this chapter and completing the diagnostic and repair jobs provided, you have developed skills in locating the causes of problems in charging systems and performing the necessary repairs. You now can determine the cause of undercharging and overcharging and can overhaul the most commonly used alternators. The knowledge that you gained and the skills that you developed have advanced you further toward your goal.

SELF-TEST

Each incomplete statement or question in this test is followed by four suggested completions or answers. In each case select the *one* that best completes the sentence or answers the question.

1 When performing a three-stage charging test, a voltmeter should be connected to the
   A. field (FLD) terminal and ground
   B. stator (STA) and field (FLD) terminals
   C. field (FLD) and battery (BAT) terminals
   D. positive (+) and negative (−) battery terminals

2 Which of the following measurements are NOT taken as a part of a three-stage charging test?
   A. Base voltage
   B. Load voltage
   C. Field voltage
   D. No-load voltage

3 Two mechanics are discussing alternators.
   Mechanic A says that a Delco SI-type (Delcotron) alternator has a built-in regulator.
   Mechanic B says that a Ford rear terminal alternator has an external regulator.
   Who is right?
   A. A only
   B. B only
   C. Both A and B
   D. Neither A nor B

4 When bypassing a regulator on a Ford charging system, a jumper wire should be connected between the alternator
   A. BAT and FLD terminals
   B. BAT and GND terminals
   C. FLD and GND terminals
   C. STA and FLD terminals

5 Some rotor shafts have a hex-shaped hole in the drive end to aid in
   A. removing the pulley
   B. adjusting the diodes
   C. aligning the brushes
   D. installing the stator

6 An ohmmeter connected to one stator lead and to the laminated iron frame of the stator reads zero (0). This indicates that the stator winding is
   A. open
   B. not open
   C. grounded
   D. not grounded

7 An ohmmeter connected to the slip rings of an alternator reads infinity (∞). This indicates that the rotor winding is
   A. open
   B. not open
   C. grounded
   D. not grounded

8 An ohmmeter connected to one of the slip rings and the shaft of a rotor reads infinity (∞). This indicates that the rotor is
   A. open
   B. not open
   C. grounded
   D. not grounded

9 The internal regulator used in Delco SI-type (Delcotron) alternators can be bench tested by using a battery and a battery charger while observing
   A. an ohmmeter and an ammeter
   B. an ohmmeter and a test light
   C. a voltmeter and an ammeter
   D. a voltmeter and a test light

10 An ohmmeter connected to a diode in a rectifier reads infinity (∞). When the test probes are reversed, the ohmmeter reads zero (0). This indicates that the diode is
   A. OK
   B. open
   C. shorted
   D. grounded

Chapter 12
Instrumen-tation– Warning Light and Gauge Circuit Service

In some cars, a few warning lights and a couple of gauges provide the absolute minimum amount of information thought to be necessary. In other cars, the instrument panels contain an array of lights, instruments, and displays that monitor more conditions and functions than most drivers care to know about.

In this chapter, you will be introduced to the basic circuitry used to operate most warning lights and gauges. You will learn about various switches, senders, and gauges, and how they function. You also will perform diagnostic and repair procedures in warning light and gauge circuits.

Your specific objectives in this chapter are to perform the following jobs:

A
Test the operation of warning light circuits
B
Replace a mechanical switch in a warning light circuit
C
Replace a pressure-sensitive switch in a warning light circuit
D
Replace a temperature-sensitive switch in a warning light circuit
E
Identify the function of warning light and gauge circuit components
F
Determine the cause of problems in gauge circuits

WARNING LIGHTS   Most automotive instrument panels include various warning lights. (See Figure 12.1.) Those lights may warn of serious mechanical or electrical problems, or they may inform the driver that a seat belt is unfastened or that a door is not fully closed.

Most warning light circuits are easily checked. Testing begins with placing the ignition switch in the On or Ign position without starting the engine. This should cause some of the lights, including the oil and charge lights, to glow. Fastening and unfastening the seat belts should cause the seat belt light to operate, and opening and closing the doors should operate the door ajar light. Applying and releasing the parking brake should cycle the brake light if the car is so wired.

Some warning lights, including the light that warns of excessive engine temperature and the light that signals a failure in the brake system, cannot be checked in this manner. On many cars, an extra set of contacts in the ignition switch completes the circuit to those lights so that they will glow when the ignition switch is turned to the Start position.

Some warning lights on some cars are controlled by the car's computer or another electronic device. Problems in circuits of this type should be diagnosed by following the specific procedures established by the car manufacturer. In most warning light circuits, the lights are activated by switches. The types of switches used and their locations usually are determined by their function.

A light that indicates that the parking brake is not released is controlled most often by a mechanical switch operated by the parking brake lever or pedal. A light that warns of low engine oil pressure usually is activated by a pressure-sensitive switch threaded into an engine oil passage. A light that indicates excessive engine temperature usually is turned on by a temperature-sensitive switch mounted so that it is immersed in the engine coolant.

In most instances, those switches can be tested in the same manner that you would test any other switch. In circuits where the switch completes the ground circuit, the wire can be disconnected from the switch and grounded to bypass the switch, as shown in Figure 12.2. When the switch is used in the "hot" side of the circuit, it usually can be bridged with a jumper wire, as shown in Figure 12.3.

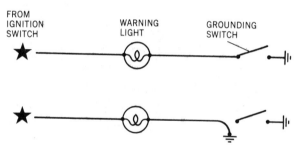

Figure 12.2 In circuits where the switch supplies the ground, the circuits and the switch may be tested by disconnecting the wire from the switch and grounding it.

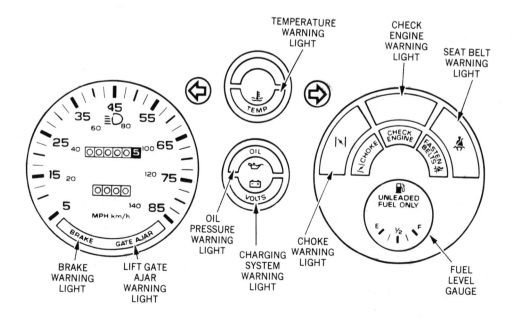

Figure 12.1 An instrument panel fitted with eight warning lights and only one electric gauge (courtesy of Chevrolet Motor Division).

Mechanical Switches Some cars have a door ajar light on the instrument panel to warn the driver if any of the doors are not fully closed. *Door jamb switches* similar to those shown in Figures 12.4 and 12.5 usually are mounted in the door posts. The switches are spring-loaded so that the plunger protrudes, and are normally in the On position. When a door is fully closed, the switch plunger is pushed in and the switch contacts are opened.

On some cars, the door jamb switches are used in the ground side of the circuit, as shown in Figure 12.6. Those switches usually are of the single contact, self-grounding type. (Refer to Figure 12.4.) On other cars, the door jamb switches are wired in the "hot" side of the circuit, as shown in Figure 12.7. The switches used in those circuits have two or more terminals. (Refer to Figure 12.5.) On many cars, the door jamb switches also are a part of the courtesy light circuit.

Door jamb switches are easily removed for testing or replacement. Threaded switches usually have a hex head and can be unscrewed with a wrench. (See Figure 12.8.) Flush mounted or "snap in" switches usually are friction fitted and can be pried out easily.

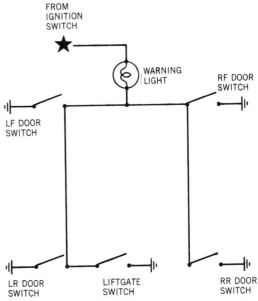

Figure 12.6 A typical door ajar circuit wired with switches on the ground side of the circuit. If any of the switches are closed, the ground circuit will be completed and the warning light will go on.

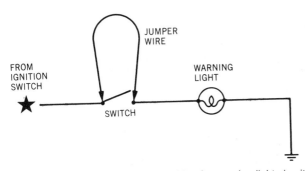

Figure 12.3 A switch on the "hot" side of a warning light circuit can be by-passed with a jumper wire.

Figure 12.4 Typical self-grounding door jamb switches. Switches of this type complete the ground side of a circuit. They are threaded into a hole in a door post so that the switch plunger is pushed in when the door is fully closed.

Figure 12.5 Insulated door jamb switches of this type are usually used in circuits where the switch is wired in the "hot" side of the circuit. Switches of this type have two or more terminals.

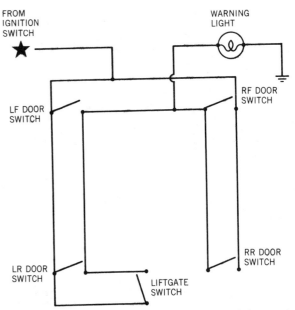

Figure 12.7 A door ajar circuit wired so that the switches are in the "hot" side of the circuit. If any of the switches are closed, current will flow to the warning light.

Job 12a

TEST THE OPERATION OF WARNING LIGHT CIRCUITS

SATISFACTORY PERFORMANCE
A satisfactory performance on this job requires that you do the following:

1 Test the operation of the warning light circuits on the car assigned.
2 Following the steps in the "Performance Outline," complete the job within 15 minutes.
3 Fill in the blanks under "Information."

PERFORMANCE OUTLINE
1 If the car has an automatic transmission, place the transmission selector lever in the Park position. If the car has a manual transmission, place the selector in the Neutral position.
2 Place the ignition switch in the On or Ign position. Do not start the engine.
3 Check the operation of the oil and charge lights.
4 Operate the parking brake, buckle and unbuckle the seat belts, and open and close the doors. Check the operation of the appropriate warning lights.
5 Turn the ignition switch to the Start position and immediately check the operation of the brake and temperature warning lights.
6 With the engine running, check that the oil and charge warning lights go out.
7 Place the ignition switch in the Off position.
8 Check the operation of any other warning lights.

INFORMATION
Vehicle identification _____

Reference used _____

Warning Light Checked	Working	Not Working
Charge (Alternator)	_____	_____
Oil (Oil pressure)	_____	_____
Seat Belts	_____	_____
Brake	_____	_____
Temperature	_____	_____
Door Ajar	_____	_____
_____	_____	_____
_____	_____	_____
_____	_____	_____
_____	_____	_____

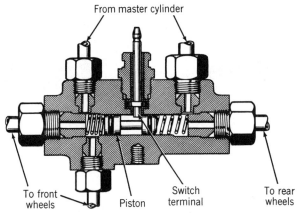

Figure 12.10 A pressure differential switch in the "failed" position. A loss of pressure in the front brake system has caused the pressure in the rear brake system to move the piston into contact with the switch terminal, grounding the warning light circuit (courtesy of Chevrolet Motor Division).

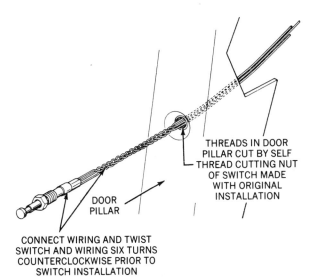

Figure 12.8 A typical threaded door jamb switch that serves several functions. Note that the wires should be twisted prior to installation so that they will be untwisted when the switch is threaded into place (courtesy of Ford Motor Company).

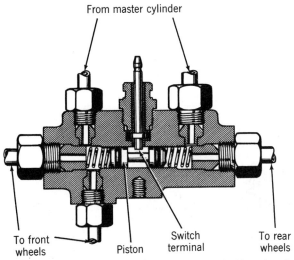

Figure 12.9 A typical pressure differential switch. Note that the piston is centered and is not in contact with the switch terminal as long as the pressure on both sides of the piston is the same (courtesy of Chevrolet Motor Division).

Most instrument panels contain a brake warning light to notify the driver of a failure in the service brake system. When an imbalance in the hydraulic system occurs, a hydraulically operated *pressure differential switch,* shown in Figures 12.9 and 12.10, completes the circuit and turns on the light. If the brake warning light stays on, the circuit can be checked by disconnecting the wire from the pressure differential switch. If the light goes out when the wire is disconnected, there is a problem in the service brake system. The failure should be located

and repaired. If the light remains on with the wire disconnected from the pressure differential switch, the wire is grounded or the parking brake switch is in the On position.

On some cars, the brake warning light also is used to notify the driver that the parking brake is not fully released. A mechanical switch similar to a door jamb switch usually is mounted near the parking brake lever or pedal, as shown in Figure 12.11. When the lever or pedal is completely released, it holds in the switch plunger. When the parking brake is applied, the switch plunger moves out, closing the switch contacts and turning on the light. Figure 12.12 shows how both switches usually are wired in the circuit.

The switch at the parking brake is easily tested and replaced, but the replacement of a pressure differential switch may require special brake bleeding procedures that are beyond the scope of this text. You should consult an appropriate manufacturer's manual before attempting to replace any part in a hydraulic system.

On some cars, the brake warning light also serves to notify the driver that the brake shoes require replacement. On those cars, a wire is imbedded at a predetermined depth in a brake shoe at each wheel. When the brake lining wears down far enough to expose the wire, the wire contacts the brake rotor, completing the circuit to ground.

Pressure-Sensitive Switches Many cars are equipped with a warning light that goes on if engine oil pressure drops below a certain limit. The oil warning light is controlled by a switch similar to those shown in Figure 12.13. Threaded into an oil

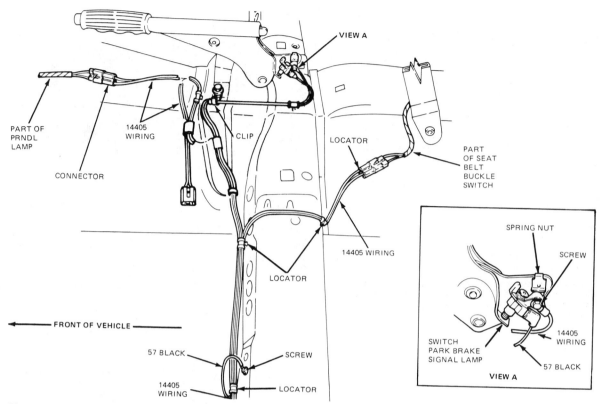

Figure 12.11 A typical parking brake warning light switch mounting. When the brake lever is completely released, the switch plunger is held in, breaking the circuit (courtesy of Ford Motor Company).

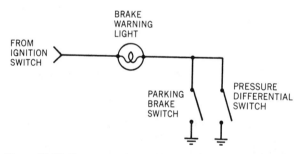

Figure 12.12 On some cars, the brake warning light is wired so that both the pressure differential switch and the parking brake switch can complete its circuit to ground.

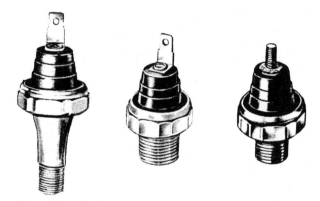

Figure 12.13 Typical pressure-sensitive switches. Switches of this type are used to activate a low oil pressure warning light.

passage in the engine, as shown in Figure 12.14, the switch contacts are held open by the oil pressure in the engine's lubrication system. If the oil pressure drops too low, the switch contacts close, allowing current to flow to the light. Figure 12.15 shows a typical oil pressure warning light circuit. Figure 12.16 shows another type of pressure-sensitive switch that is operated by oil pressure. Switches of this type contain two sets of contacts. One set of contacts are held open by oil pressure to break the ground side of the warning light circuit. Another set of contacts are held closed by oil pressure. On

some cars, those contacts allow current to flow to an electrically heated choke. A similar switch often is used on cars with electric fuel pumps. The switch will open the fuel pump circuit if the oil pressure drops too low. This will cause the engine to stop running, hopefully before extensive engine damage occurs.

Oil pressure switches are easily replaced, although a special socket usually is required. As long as the

Job 12b

REPLACE A MECHANICAL SWITCH IN A WARNING LIGHT CIRCUIT

SATISFACTORY PERFORMANCE

A satisfactory performance on this job requires that you do the following:

1 Replace the designated switch in the car assigned.
2 Following the steps in the "Performance Outline" and the procedure and specifications of the car manufacturer, complete the job within 200 percent of the manufacturer's suggested time.
3 Fill in the blanks under "Information."

PERFORMANCE OUTLINE

1 Locate the designated switch.
2 Remove the switch.
3 Install the replacement switch.
4 Adjust the switch if necessary.
5 Check the operation of the circuit.

INFORMATION

Vehicle identification _____

Reference used _____ Page(s) _____

Switch replaced _____

Switch type: _____ Threaded   _____ Snap in

_____ Single terminal   _____ Multi terminal

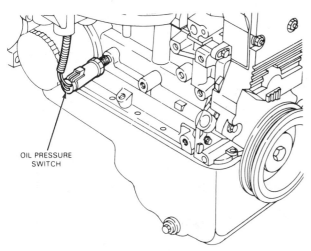

Figure 12.14 An oil pressure switch is usually threaded into an oil passage in the engine block (courtesy of Ford Motor Company).

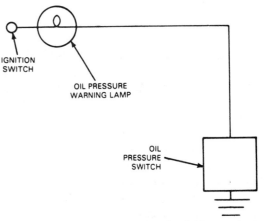

Figure 12.15 Most oil pressure warning lights are wired so that the switch completes the circuit to ground (courtesy of Ford Motor Company).

engine is not running, the switch may be removed without the loss of oil. After replacing an oil pres-

sure switch, you should always check for the possibility of an oil leak in addition to checking the operation of the light.

Job 12c

REPLACE A PRESSURE-SENSITIVE SWITCH IN A WARNING LIGHT CIRCUIT

SATISFACTORY PERFORMANCE
A satisfactory performance on this job requires that you do the following:

1 Replace the designated pressure-sensitive switch in the car assigned.
2 Following the steps in the "Performance Outline" and the procedure and specifications of the car manufacturer, complete the job within 200 percent of the manufacturer's suggested time.
3 Fill in the blanks under "Information."

PERFORMANCE OUTLINE
1 Locate the designated switch.
2 Remove the switch.
3 Install the replacement switch.
4 Check the operation of the circuit.
5 Check for leaks.

INFORMATION
Vehicle identification _____

Reference used _____ Page(s) _____

Switch replaced _____

Was a special socket required? _____ Yes _____ No

Figure 12.16 Pressure-sensitive switches with multiple contacts are often used to control electrically heated chokes and electric fuel pumps in addition to activating a low oil pressure warning light.

TEMPERATURE-SENSITIVE SWITCHES

Temperature-sensitive switches, sometimes called *thermal switches,* will open and close a circuit with changes in temperature. Most warning light circuits that indicate excessive engine temperature are triggered by a temperature-sensitive switch that is mounted in the cylinder head of the engine, as shown in Figure 12.17. The switch is positioned so that its sensing bulb or element is immersed in the coolant. When the temperature of the coolant exceeds a certain specification, the contacts in the switch close, completing the circuit for the light.

Most thermal switches used in warning light circuits are built to close when the coolant reaches a temperature of approximately 5 to 10°F below the boiling point of the coolant mixture. If plain water without permanent antifreeze is used in the cooling system, the warning light usually will not operate, even though the water may be boiling. Proper operation of the cooling system and the warning light require that a mixture of approximately 50 percent water and 50 percent permanent antifreeze be used in the system. Your diagnosis of any problem thought to be in the cooling system should include a check of the coolant. A coolant tester should be used to determine that the freezing point of the mixture is at least 0°F (−18°C).

Most switches used to control the temperature warning light complete the circuit to ground and have but one terminal, as shown in Figure 12.18. In some cars, however, the switch performs additional duties. Those switches will usually be found with two or more terminals. (See Figure 12.19.)

One variant system sometimes found operates two

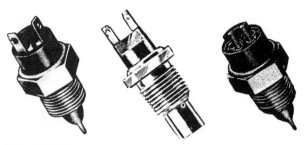

Figure 12.19 Temperature-sensitive switches with two or more terminals are often used in other circuits as well as in warning light circuits.

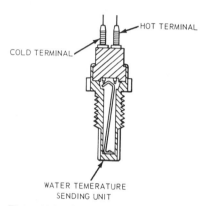

HOT TERMINAL

COLD TERMINAL

WATER TEMERATURE SENDING UNIT

Figure 12.20 A typical cold/hot water temperature sending switch. Note that the bimetallic strip is in contact with the cold terminal. When overheated, the strip will move to the hot terminal (courtesy of Ford Motor Company).

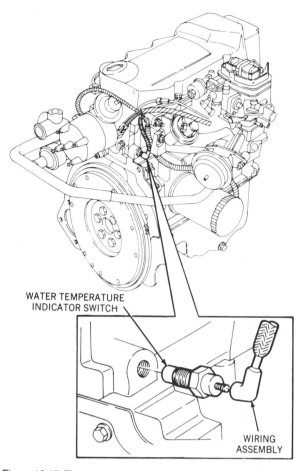

WATER TEMPERATURE INDICATOR SWITCH

WIRING ASSEMBLY

Figure 12.17 The water temperature warning light switch is usually fitted in a threaded hole in the cylinder head (courtesy of Ford Motor Company).

Figure 12.18 Typical temperature-sensitive switches used to operate warning lights.

warning lights. One light indicates a cold engine condition, and the other light signals a hot engine. Switches used in those systems usually use a bimetallic strip that can contact two terminals, as shown in Figure 12.20. When the engine is cold, the bimetallic strip is in contact with the "cold" terminal, completing the circuit of the "cold" light. If the coolant temperature rises too high, the strip moves into contact with the "hot" terminal, energizing the "hot"

light. As long as the engine temperature remains in the normal operating range, the bimetallic strip will remain between the two terminals and neither light will glow. Figure 12.21 shows how such a system is wired.

Temperature-sensitive switches also are used to control the operation of electric cooling fans, and to signal computer-controlled engine systems of changes in engine temperature.

Some cars do not have separate warning lights for oil and temperature. The switch operated by oil pressure and the switch operated by engine temperature are wired as shown in Figure 12.22. In this manner, they provide alternate ground paths for a single light. On those cars, the light is usually marked "Engine."

Temperature-sensitive switches are easily changed. Since most are provided with a hex head, they can be removed from the engine with a socket or a box wrench. Since the switch enters the engine water jacket and is in contact with the coolant, the cooling system must be drained to prevent leakage and loss of coolant. The following steps outline a pro-

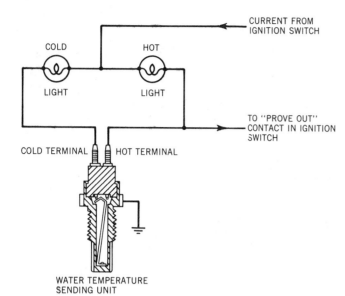

Figure 12.21 A typical circuit for a cold/hot water temperature indicating system. When the switch is cold, the COLD light circuit is completed through ground. When the switch is heated above a certain temperature, the ground is switched to the HOT light circuit. The HOT light is tested by means of "prove out" contacts that are grounded when the ignition switch is turned to the START position (courtesy of Ford Motor Company).

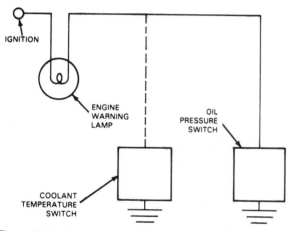

Figure 12.22 On some cars, the oil pressure and engine temperature are monitored by a single light. If either switch is closed, the light will glow (courtesy of Ford Motor Company).

cedure that you should follow to perform the job safely.

WARNING: Never remove a radiator cap while the engine is running. Failure to follow this advice could cause serious burns and possible damage to the cooling system. Since scalding hot coolant or steam may erupt from the radiator filler neck, use extreme care when removing a radiator cap. It is best to wait until the engine has cooled.

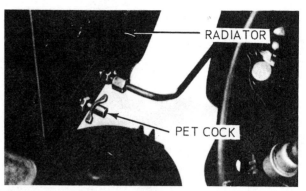

Figure 12.23 Most cooling systems can be drained by means of a petcock located in the lower radiator tank (courtesy of Ford Motor Company).

1 Wrap a heavy rag or wiper around the radiator cap.

2 Turn the radiator cap counterclockwise until you feel it stop at its first "stop" or detent. This position releases the pressure in the system.

Note: If pressure is released, stand away from the radiator until you are sure that all pressure is released. Do not attempt to remove the cap in one motion. The pressure in the system could force hot coolant or steam to erupt from the filler neck.

3 Wiggle the cap to be sure that it is loose and has released all the pressure in the system.

4 Push down on the cap and slowly turn it counterclockwise to its second "stop" and lift the cap from the filler neck.

5 Place a clean drain pan under the radiator and open the radiator petcock. (See Figure 12.23.)

Note: The entire cooling system need not be drained, but the coolant level must be lowered to below the location of the switch.

6 Disconnect the wire from the switch.

7 Using an appropriate socket or box wrench, remove the switch.

8 Install and tighten the replacement switch.

Note: Some manufacturers recommend coating the switch threads with a sealing compound or Teflon tape.

9 Connect the wire to the switch.
10 Close and tighten the radiator petcock.
11 Install the coolant drained from the system.
12 Check the coolant level and add coolant if necessary.
13 Install the radiator cap.
14 Start the engine and check for coolant leaks at the switch.
15 Check the operation of the circuit.
16 Turn off the engine.

Job 12d

REPLACE A TEMPERATURE-SENSITIVE SWITCH IN A WARNING LIGHT CIRCUIT

SATISFACTORY PERFORMANCE

A satisfactory performance on this job requires that you do the following:

1 Replace the designated temperature-sensitive switch in the assigned car.
2 Following the steps in the "Performance Outline" and the procedure and specifications of the manufacturer, complete the job within 200 percent of the manufacturer's suggested time.
3 Fill in the blanks under "Information."

PERFORMANCE OUTLINE

1 Drain the coolant.
2 Locate the designated switch.
3 Install the replacement switch.
4 Install the coolant.
5 Check for leaks.
6 Check the operation of the circuit.

INFORMATION

Vehicle identification _____

Reference used _____ Page(s) _____

Switch replaced _____

"Switchless" Warning Light Circuits Some warning light circuits are operated by means other than switches. The light that indicates if the charging system is functioning is in one of those circuits. On most cars, the warning light for the charging system is grounded through the alternator. Figure 12.24 shows a typical wiring diagram for a system of that type. When the alternator output voltage equals battery voltage, the voltage drop, or difference in voltage, across the warning light bulb ceases and the bulb goes out. On some cars, the warning light is operated by a solid state electronic voltage monitor. Wired as shown in Figure 12.25, that unit turns on the light when system voltage drops below a certain value.

Some fuel gauge systems contain a warning light that glows when the fuel in the tank drops below one quarter full. Some systems of that type are controlled by a *thermistor,* or thermal resistor. A thermistor is a special resistor whose resistance decreases as its temperature rises. The thermistor is attached to the pick-up tube of the fuel sender unit

in the tank, as shown in Figure 12.26. When the thermistor is immersed in gasoline, it is kept cool and has a high resistance. When the fuel level drops and exposes the thermistor, it heats up and its resistance decreases, allowing current to flow through the winding of a relay. When the relay points close, current flows to the warning light.

Many warning light circuits are activated by electronic sensors and control modules. Some are turned on as a function of a computer. Since those systems vary, even among cars built by the same manufacturer, the diagnostic and repair procedures cannot be covered in this text. Those procedures can be found only by reference to the car maker's service manuals.

GAUGES In addition to warning lights, most instrument panels include gauges. (See Figure 12.27.) Those gauges are not as accurate as the instruments that you use for diagnosis, but they are sufficiently accurate to provide the driver with a means of monitoring certain conditions.

While some mechanical gauges will be found, most

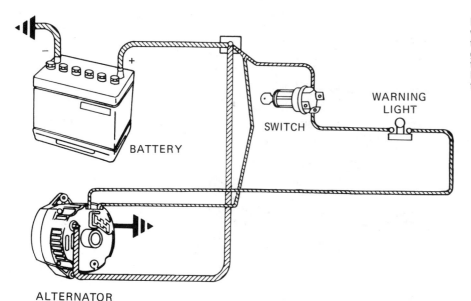

Figure 12.24 On many cars, the warning light for the charging system is wired so that its circuit to ground is completed through the alternator (courtesy of Pontiac Motor Division, General Motors Corporation).

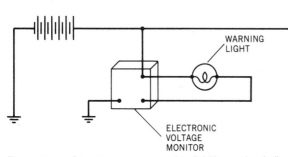

Figure 12.25 On some cars, a warning light is used to indicate low system voltage. An electronic unit monitors system voltage and sends current to the light when the system voltage drops below a predetermined value.

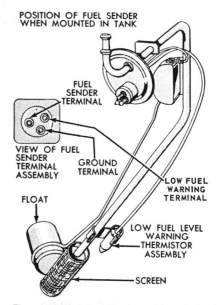

Figure 12.26 A fuel level sending unit that incorporates a thermistor as a low fuel level warning device (courtesy of Ford Motor Company).

are electrically operated. Ammeters and voltmeters used on automotive instrument panels usually contain simplified d'Arsonval movements. You are familiar with the construction and operation of such gauges from your work in Chapter 3. Oil pressure, engine temperature, and fuel level gauges usually contain either thermal or magnetic movements controlled by variable resistors called *senders* or *sending units*.

Thermal Gauges   Thermal gauges are operated by heat. Since they contain a bimetallic strip, they often are referred to as bimetallic gauges. As shown in Figures 12.28 and 12.29, the current that flows through the gauge heats the bimetallic strip, causing the pointer or needle to move across the gauge face. The amount of current that flows through the gauge determines the amount of pointer movement.

Properly operating thermal gauges should return to

zero when the ignition switch is turned off. Since they react rather slowly, thermal gauges are self-dampening and tend to give steady readings. Since the gauges are sensitive to current flow, any change in system voltage would result in a change of gauge reading. For thermal gauges to operate with any degree of accuracy, the voltage applied to the gauge circuit must be constant.

Gauge Voltage Regulation   To eliminate any fluctuations or changes in system voltage, an *instru-*

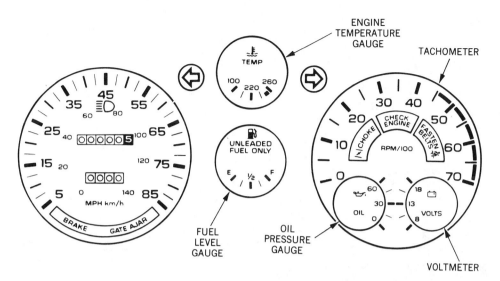

Figure 12.27 An instrument panel fitted with five electric gauges supplemented with five warning lights (courtesy of Chevrolet Motor Division).

ENGINE TEMPERATURE GAUGE

TACHOMETER

FUEL LEVEL GAUGE

OIL PRESSURE GAUGE

VOLTMETER

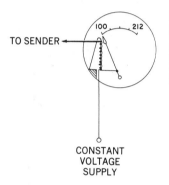

Figure 12.28 A typical thermal or bimetallic gauge. When no current is flowing through the gauge, the bimetallic strip is cold and the pointer reads zero (courtesy of Chrysler Corporation).

TO SENDER

CONSTANT VOLTAGE SUPPLY

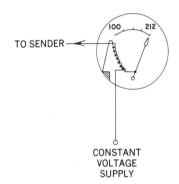

Figure 12.29 When current flows through a thermal gauge, the bimetallic strip bends, causing the pointer to move across the face of the gauge (courtesy of Chrysler Corporation).

TO SENDER

CONSTANT VOLTAGE SUPPLY

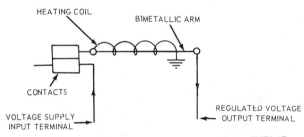

HEATING COIL

BIMETALLIC ARM

CONTACTS

VOLTAGE SUPPLY INPUT TERMINAL

REGULATED VOLTAGE OUTPUT TERMINAL

Figure 12.30 A typical instrument voltage regulator (IVR). Because the contacts open and close very rapidly, the average output voltage is regulated to about 5 V. (courtesy of Ford Motor Company).

opens the contacts. When the contacts open, the circuit is broken and the arm cools, closing the contacts and repeating the cycle. The cycles are repeated very rapidly, providing a pulsating voltage at the output terminal. Since the output voltage cycles from 0 V to system voltage, the actual output voltage is a regulated average voltage. Most IVRs provide an output of approximately 5 V, and the gauges that they are matched with are designed to operate accurately at that voltage.

On most cars that use thermal gauges, the IVR is located on the back of the instrument panel cluster. On some cars, it is combined with one of the gauges in a single assembly. The exact location of the IVR must be determined by reference to a manufacturer's shop manual.

Magnetic Gauges In a magnetic gauge, a permanent magnet is mounted and balanced so that it can move in a magnetic field created by two or three electromagnets. A pointer is attached to the permanent magnet so that the movement is indi-

ment voltage regulator (IVR), sometimes called a *constant voltage regulator (CVR)*, is used in thermal gauge circuits.

An IVR, another thermal device, operates in a manner similar to a circuit breaker. As shown in Figure 12.30, system voltage at the input terminal pushes current through a set of contacts and through a heating coil wrapped around a bimetallic arm. When the arm is heated by current flow, it bends and

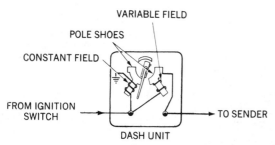

Figure 12.31 A typical magnetic gauge. The constant field tends to hold the pointer in the zero position. The variable field offsets the constant field and tends to pull the pointer away from zero (courtesy of Chrysler Corporation).

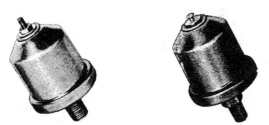

Figure 12.32 Typical pressure-sensitive senders used in oil pressure gauge circuits.

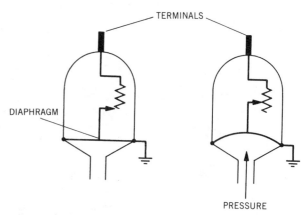

Figure 12.33 The operation of a typical pressure-sensitive sending unit. When no pressure is applied to the sender, the unit has an infinite resistance. When pressure is applied, the diaphragm flexes, decreasing the resistance of the unit.

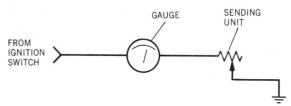

Figure 12.34 The variable resistance in the sender controls the current flow through the gauge and determines the position of the pointer on the gauge face.

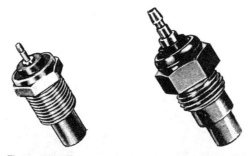

Figure 12.35 Temperature-sensitive senders of this type are used in most temperature gauge circuits. They contain a thermistor that changes its resistance with temperature changes.

cated on the face of the gauge. As shown in Figure 12.31, one electromagnet provides a constant field that tends to hold the pointer in the zero position. The remaining electromagnet(s) create a variable magnetic field that tends to offset the constant field. Since current flow in the gauge circuit determines the strength of the variable field, the amount of current flow through the gauge determines the amount of pointer movement.

Since system voltage is applied to both the constant field and the variable field(s), a magnetic gauge is not sensitive to changes in system voltage and an IVR is not required.

When the ignition switch in a car with magnetic gauges is turned off, the gauges usually do not return to zero, but the readings that they maintain should not be considered accurate.

Pressure-Sensitive Senders   Most gauges that measure oil pressure are operated by pressure-sensitive senders of the type shown in Figure 12.32. Threaded into an oil passage in the engine, oil pressure causes a flexible diaphragm to operate a variable resistor. As shown in Figure 12.33, when no oil pressure is present, the resistance of the sender is infinite (∞). As oil pressure increases, the diaphragm moves, decreasing the resistance of the sender. Connected as shown in Figure 12.34, the

sender can change the current flow in the circuit and thus change the position of the pointer on the gauge face.

The procedure for changing an oil pressure sender is the same that you used in changing an oil pressure switch.

Temperature-Sensitive Senders   Gauges that measure temperature usually are operated by temperature-sensitive senders of the type shown in Figure 12.35. They are threaded into the cylinder head

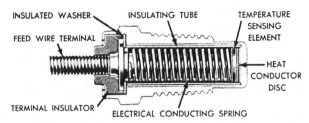

Figure 12.36 A cutaway view of a typical temperature sender. A thermistor is used for the temperature sensing element (courtesy of Chrysler Corporation).

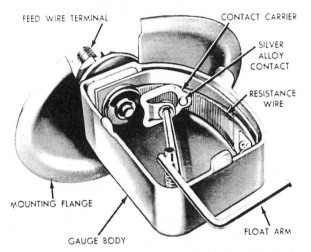

Figure 12.37 The construction of a typical fuel sender. As the float arm moves, a contact slides across a coiled resistance wire, changing the resistance of the unit (courtesy of Chrysler Corporation).

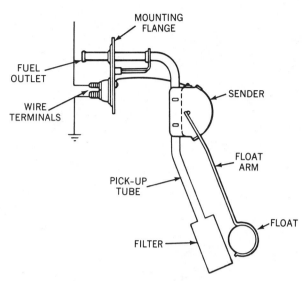

Figure 12.38 A typical fuel sender is combined in an assembly that is mounted inside the fuel tank (courtesy of Ford Motor Company).

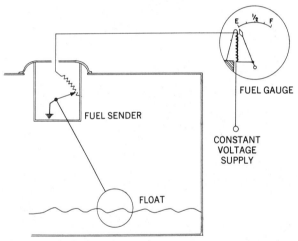

Figure 12.39 The operation of the tank sending unit and gauge when the tank is near empty (courtesy of Chrysler Corporation).

of the engine, where they are in contact with the coolant. Figure 12.36 shows the construction of a typical temperature-sensitive sender. The temperature sensing element is a thermistor. As you may recall, a thermistor is a resistor with a negative temperature coefficient. When heated, the resistance of a thermistor decreases. The changing resistance of the sender controls the operation of the gauge. (Refer to Figure 12.34.)

The replacement of a temperature sender is similar to the replacement of a temperature switch. The same procedure should be followed and the same safety precautions should be observed.

Mechanical Senders   In a mechanical sender, a contact is physically moved along a resistance wire to increase or decrease the resistance of the sender. (See Figure 12.37.) Mechanical senders most often are used to measure the level of fuel in a fuel tank. As shown in Figure 12.38, the sender usually is incorporated in a unit that includes the fuel pick-up tube and a filter. Figures 12.39 and

12.40 show how the sender operates the fuel gauge.

Since the sender is mounted in the fuel tank, any attempt to replace a fuel sender should be preceded by reference to an appropriate service manual. Failure to observe safety precautions and to follow the correct procedure could result in fuel spillage and a serious fire.

On some cars, the fuel sender is located in the front or in the side of the fuel tank. (See Figure 12.41.) If the fuel level is above the location of the sender, any attempt to remove the sender will result in fuel spillage. You must be sure that the fuel level is well below the sender opening.

The following steps are provided only to outline a typical procedure for the replacement of a fuel tank sender. Before you attempt to perform the job, you should consult the manufacturer's manual for the car on which you are working.

1 Determine that the fuel level is below the sender opening in the tank.
2 Raise and support the car.

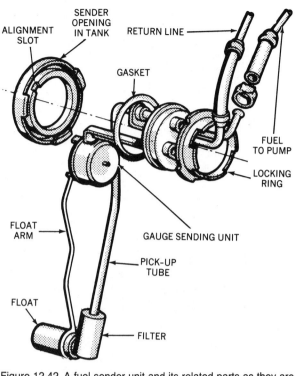

Figure 12.42 A fuel sender unit and its related parts as they are installed in a fuel tank (courtesy of Ford Motor Company).

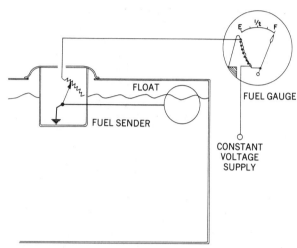

Figure 12.40 The operation of the tank sending unit and gauge when the tank is near full (courtesy of Chrysler Corporation).

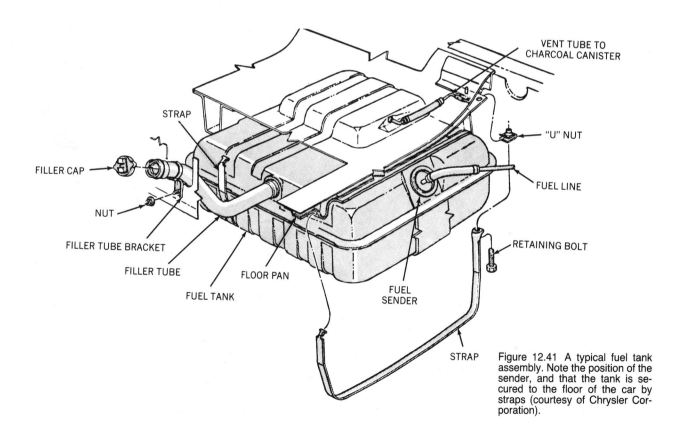

Figure 12.41 A typical fuel tank assembly. Note the position of the sender, and that the tank is secured to the floor of the car by straps (courtesy of Chrysler Corporation).

Job 12e

IDENTIFY THE FUNCTION OF WARNING LIGHT AND GAUGE CIRCUIT COMPONENTS

SATISFACTORY PERFORMANCE
A satisfactory performance on this job requires that you do the following:

1 Identify the function of warning light and gauge circuit components.
2 Place the number of each component listed in the "Performance Situation" in front of the phrase that best describes its function. Complete the job within 15 minutes.

PERFORMANCE SITUATION

1 Pressure-sensitive switch
2 Temperature-sensitive sender
3 Thermistor
4 Warning light
5 Magnetic gauge

6 Temperature-sensitive switch
7 Pressure-sensitive sender
8 IVR
9 Thermal gauge
10 Mechanical sender

_____ Operates a fuel gauge

_____ Regulates voltage for thermal gauges

_____ Operates parking brake warning light

_____ Does not require an IVR

_____ Operates an oil pressure warning light

_____ Operates an oil pressure gauge

_____ Used in temperature-sensitive senders

_____ Requires an IVR for accuracy

_____ Usually operated by a switch

_____ Operates an engine temperature warning light

_____ Operates an engine temperature gauge

---

3 Disconnect the wire(s) from the sender.

Note: Access to the sender may be blocked by the body floor pan. On some cars, the fuel tank must be lowered or removed before the sender can be removed.

4 Disconnect the hose(s) from the sender.
5 Remove the sending unit locking ring. (See Figure 12.42.)

Note: Lacking a special locking ring wrench, the ring may be removed by carefully turning it counterclockwise with a blunt punch on the locking ring lugs. (Refer to Figure 12.42.)

6 Remove the locking ring, the sender, and the sender gasket. (Refer to Figure 12.42.)
7 Install a new gasket in the gasket groove.

Note: A slight coating of grease on the gasket may aid in keeping the gasket in place during assembly.

8 Install the replacement sending unit, taking care

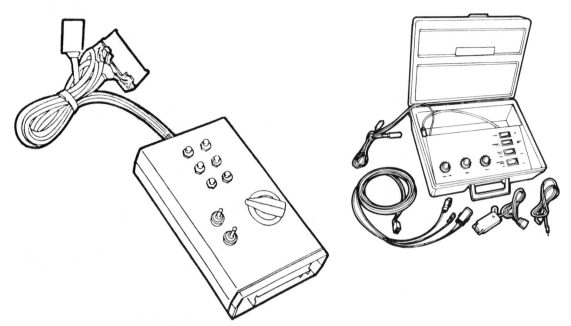

Figure 12.43 The instruments shown are specified by one car manufacturer for the diagnosis of problems in gauge circuits. The instruments are connected to the gauges and to the senders and used in a prescribed series of tests (courtesy of Chevrolet Motor Division).

that it is centered and that the alignment tab is positioned in the slot in the tank. (Refer to Figure 12.42.)

9 Holding the sender in position, install the locking ring.

10 Tighten the locking ring with the special wrench or by carefully turning it with a blunt punch.

11 Install the hose(s) and secure them with new clamps.

12 Install the wire(s).

13 Lower the car to the floor.

14 Check the operation of the gauge circuit.

Diagnosis of Gauge Circuit Problems Although most gauge circuits use similar components and function in the manner explained earlier in this chapter, many differences exist in those systems, even in cars built by the same manufacturer. Most car makers provide specific test procedures for each variation in the manuals they publish for their cars. In most instances, those procedures require the use of special test equipment similar to the instruments shown in Figure 12.43. Without access to those instruments, you must rely on your knowledge of basic circuitry to diagnose any problems that you may encounter.

After verifying that a problem exists, the diagnosis of a gauge circuit problem should start with the isolation and inspection of the circuit involved. The following information should aid you in your diagnostic procedure.

Inoperative Gauges The most common causes of an inoperative gauge are a disconnected terminal, a corroded connector, and a broken wire. A gauge wire often is accidentally disconnected from a sender while work is being performed under the hood. As most senders are subjected to extremes of heat and pressure, and may be exposed to road splash, a faulty sender also should be considered.

As previously mentioned, jumper wires can be used to bridge or bypass the switches in warning light circuits. Although some car manufacturers warn against this practice, in most instances the senders in gauge circuits also can be bridged. To ensure against gauge damage, the sender should be bridged only for the amount of time that it takes for the pointer to move across the gauge face. Most mechanics use a long jumper wire that enables them to ground the wire removed from the sender while they observe the gauge. With the ignition switch turned on, the pointer should move completely across the gauge face when the wire is grounded.

This quick check determines if continuity exists in the gauge and in the wiring. If the gauge operates with the jumper wire and does not operate when connected to the sender, the sender probably is

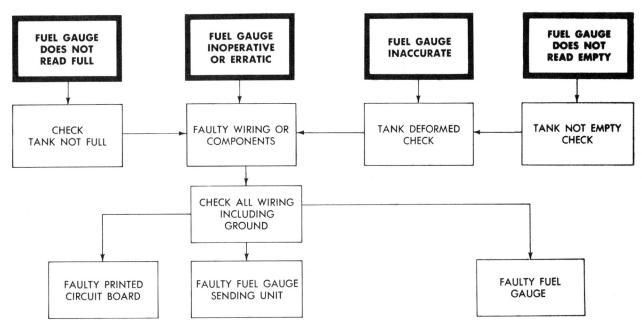

Figure 12.44 A diagnostic flow chart for determining the cause of problems in a fuel gauge circuit as provided by one car manufacturer (courtesy of Chrysler Corporation).

faulty. Replacing the sender will verify your diagnosis. If the gauge does not operate with the jumper wire in place, the sender probably is in good condition, and the problem is in the wiring or the gauge. The wiring should be checked for an open before any attempt is made to replace the gauge.

Gauges That Read Full Scale You occasionally will find a gauge that reads full scale when the ignition switch is turned on. The problem may be that the wire to the sender is grounded. A visual inspection usually will determine if this is the case. A thermal gauge will read full scale if its IVR is faulty or improperly grounded. If the instrument panel contains two or more thermal gauges, and they all read full scale, you can be almost certain that the problem is related to the IVR. A jumper wire can be held to the IVR to establish a temporary ground for testing. If the jumper corrects the condition, the IVR ground should be cleaned and tightened. If the jumper does not correct the condition, the IVR should be replaced. A faulty sender may also cause a gauge to read full scale. Removing the wire from the sender should cause the pointer to return to zero if this is the problem. A faulty sender should be replaced.

Inaccurate Gauges A gauge thought to be inaccurate should not be replaced until it is determined that the problem actually is the fault of the gauge.

The gauge may be accurately reading a faulty condition.

An inaccurate fuel level reading may be caused by a damaged or collapsed fuel tank. The tank should be inspected carefully. Indicated low oil pressure may be a true reading of the pressure in the lubricating system. The sender can be removed from the engine and a test gauge inserted to measure the actual system pressure. An engine temperature reading thought to be inaccurate can be checked by the use of a coolant thermometer.

Some car manufacturers' service manuals contain diagnostic flowcharts of the type shown in Figure 12.44 to help you to determine the cause of problems in gauge circuits. Some manuals also provide resistance specifications for gauges and senders so that you can use an ohmmeter for further testing.

Gauge Replacement On some cars, the face of the instrument cluster can be removed to gain access to the gauges. On others, the entire cluster must be removed if a gauge must be replaced. Each model car, even among those built by the same manufacturer, requires different procedures. If, through your diagnosis of a gauge circuit problem, you have found that a gauge must be replaced, you should consult the manufacturer's manual for the car on which you are working to find the precautions to take and the correct procedure to follow.

Job 12f

DETERMINE THE CAUSE OF PROBLEMS IN GAUGE CIRCUITS

SATISFACTORY PERFORMANCE

A satisfactory performance on this job requires that you do the following:

1 Determine the cause of problems in the gauge circuits of the car(s) assigned.
2 Following the steps in the "Performance Outline" and the manufacturer's procedure and specifications, determine the cause of each problem within 15 minutes.
3 Fill in the blanks under "Information."

PERFORMANCE OUTLINE

1 Determine each problem.
2 Isolate the circuit.
3 Inspect and test the circuit.
4 Determine the cause of the problem.

INFORMATION

Vehicle #1 identification _____

Reference used _____ Page(s) _____

Existing problem(s) _____

_____

Cause(s) of problem(s) _____

_____

Recommended repair(s) _____

_____

Vehicle #2 identification _____

Reference used _____ Page(s) _____

Existing problem(s) _____

_____

Cause(s) of problem(s) _____

_____

Recommended repair(s) _____

_____

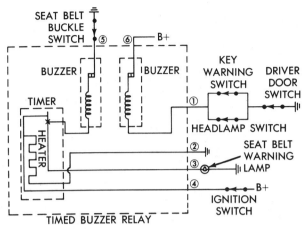

Figure 12.46 The wiring diagram for a seat belt and key warning system used by one car manufacturer (courtesy of Chrysler Corporation).

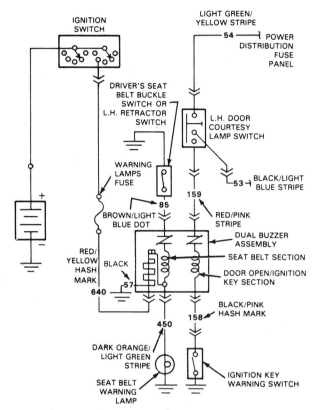

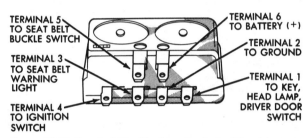

Figure 12.45 A dual warning buzzer system that combines the seat belt warning buzzer with the key-in-switch warning buzzer (courtesy of Ford Motor Company).

Figure 12.47 Terminal identification of a typical timed buzzer-relay (courtesy of Chrysler Corporation).

Warning Buzzers and Chimes On many cars, the warning lights are supplemented with buzzers or chime signals to remind the driver that the seat belts are unbuckled, that the key was left in the ignition switch, or that the headlights were left on. As shown in Figures 12.45 and 12.46, some of the switches used in those circuits also are a part of the warning light circuits.

Buzzer systems usually contain a timed buzzer-relay of the type shown in Figure 12.47. The timer is a thermal switch that opens the circuit after 5 to 10 seconds of buzzer operation.

Tone or chime signals usually use an audible tone generator with varied timing to provide different signals. Figure 12.48 shows a pictorial diagram of a chime system that monitors three functions.

As each manufacturer uses different circuitry and different components in those systems, the diagnostic and repair procedures required should be obtained from manufacturers' service manuals. Fig-ure 12.49 shows a typical diagnostic flowchart for a seat belt buzzer system.

Electronic Monitor System Some cars are equipped with electronic monitor systems that convey a visual message through a graphic display, and sometimes supplement that display with an audible message from a speech synthesizer. Some circuits in those systems are controlled by the same switches and senders that operate warning lights and gauges. Others are controlled by float-activated switches and additional switches that are sensitive to temperature and pressure. Still others are controlled by electronic modules and as a function of a computer. Since the systems used by different manufacturers use different components and circuitry, the diagnosis and repair of problems must be performed by following specific procedures outlined in appropriate manufacturers' manuals.

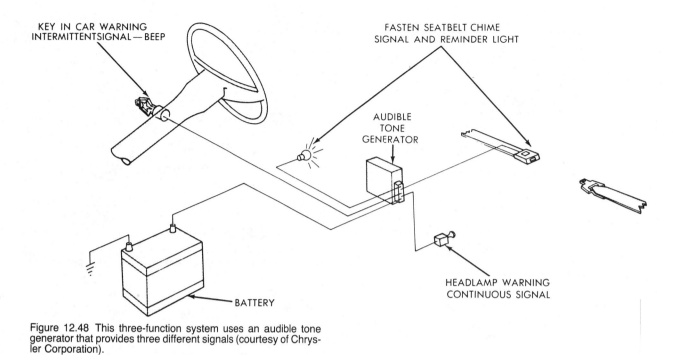

Figure 12.48 This three-function system uses an audible tone generator that provides three different signals (courtesy of Chrysler Corporation).

SEAT BELT WARNING SYSTEM DIAGNOSIS

Figure 12.49 A diagnostic flow chart for a seat belt buzzer system (courtesy of Chrysler Corporation).

SUMMARY

In this chapter, you were introduced to basic warning light circuits and are now aware of the various switches used and how they operate to control those lights. You also learned about the different types of senders and gauges used in gauge circuits, and how those components operate. Based on that knowledge, you can perform basic diagnostic checks and tests to determine the causes of many circuit problems. You also can replace defective components and perform necessary repairs.

SELF-TEST

Each incomplete statement or question in this test is followed by four suggested completions or answers. In each case select the *one* that best completes the sentence or answers the question.

1 The warning light that indicates an imbalance in the hydraulic service brake system is operated by a
   A. mechanical switch
   B. pressure-sensitive switch
   C. temperature-sensitive switch
   D. pressure differential switch

2 Two mechanics are discussing pressure-sensitive switches.
   Mechanic A says, that on some cars, engine oil pressure operates a switch that turns off the oil pressure warning light.
   Mechanic B says, that on some cars, engine oil pressure operates a switch that turns on an electric choke heater.
   Who is right?
   A. A only
   B. B only
   C. Both A and B
   D. Neither A nor B

3 A gauge whose pointer is moved by a bimetallic strip is usually referred to as a
   A. thermal gauge
   B. magnetic gauge
   C. d'Arsonval gauge
   D. mechanical gauge

4 An instrument voltage regulator (IVR) is required for a
   A. thermal gauge
   B. magnetic gauge
   C. d'Arsonval gauge
   D. mechanical gauge

5 An instrument voltage regulator (IVR) operates through the action of
   A. a thermistor
   B. a bimetallic strip
   C. a variable resistor
   D. opposing magnetic fields

6 A pressure-sensitive sender usually contains a
   A. thermistor
   B. bimetallic strip
   C. permanent magnet
   D. variable resistor

7 A temperature-sensitive sender usually contains a
   A. thermistor
   B. bimetallic strip
   C. permanent magnet
   D. variable resistor

8 A mechanically operated sender usually is used in a gauge circuit that indicates
   A. fuel level
   B. engine temperature
   C. engine oil pressure
   D. alternator charging rate

9 The pointer on an oil pressure gauge moves completely across the gauge face when the ignition switch is turned on, even though the engine is not running. What is the most probable cause of this problem?
   A. The wire from the gauge to the sender is broken.
   B. The wire from the gauge to the sender is grounded.
   C. The wire from the ignition switch to the gauge is broken.
   D. The wire from the ignition switch to the gauge is grounded.

10 Both the oil pressure gauge and the engine-temperature gauge read full scale when the ignition switch is turned on, even though the engine is cold and not running. What is the most probable cause of this problem?
   A. The wire from the ignition switch to the gauges is grounded.
   B. The instrument voltage regulator (IVR) is not functioning.
   C. Both the oil pressure sender and the engine-temperature sender are defective.
   D. The wires on the oil pressure sender and on the engine-temperature sender have been reversed.

Chapter 13 Electrical Accessory Service

Introduced as options, many accessories have contributed so much to safety and convenience that they are now considered standard equipment. Many accessories now incorporate electronic components, and the circuits used by different car makers require specific diagnostic and repair procedures. Service in those circuits requires both reference to manufacturers' manuals and the knowledge and skills required to follow the procedures.

In this chapter, you will be introduced to some basic accessory circuits. You will learn about their components and how they function. You also will perform certain diagnostic and repair procedures that will help you develop additional skills.

Your specific objectives are to perform the following jobs:

A
Test and replace horn circuit parts
B
Test and replace radiator fan circuit parts
C
Test and repair a rear window defogger
D
Test and repair a heater blower
E
Test and repair windshield wiper and washer systems
F
Diagnose externally caused radio problems

HORNS Most cars have one or two horns mounted behind the grill or under the hood as shown in Figure 13.1. On all but a few cars, the horns are grounded through their attachment to the car body. Most horn circuits incorporate a relay, as shown in Figure 13.2. The relay allows the use of a horn switch that completes the ground side of the circuit for the relay coil winding. This simplifies the wiring to the horn switch and reduces the current that must flow through its contacts. The relay may be mounted under the hood on the radiator support or fender panel, or may be located on the fuse block, as shown in Figure 13.3.

Some cars do not use a horn relay, but wire the horn switch as shown in Figure 13.4. Circuits of this type often are used when the horn switch is included in an assembly of switches mounted on the steering column, as shown in Figure 13.5.

Troubleshooting If a horn does not operate, its wire should be disconnected and a jumper wire should be connected between the horn terminal and the positive (+) battery terminal. If the horn oper-

ates with the jumper wire in place, it can be considered to be in good condition and properly grounded. If the horn does not operate, it is either defective or improperly grounded.

A second jumper wire can be used to provide a temporary ground between the body of the horn and the car body. If the horn operates with the temporary ground, the ground connection is faulty and

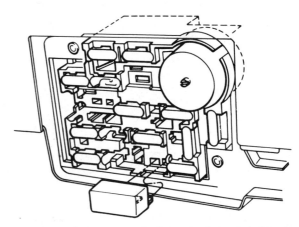

Figure 13.3 On some cars, the horn relay is located at the fuse block (courtesy of Ford Motor Company).

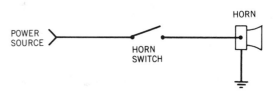

Figure 13.4 Some cars do not use a horn relay, but use a simple switch between the power source and the horn.

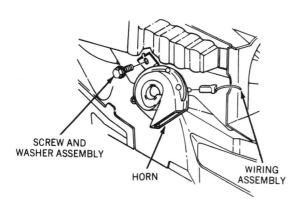

SCREW AND WASHER ASSEMBLY

HORN

WIRING ASSEMBLY

Figure 13.1 On most cars, the horns are mounted on or near the radiator support and are accessible from under the hood (courtesy of Ford Motor Company).

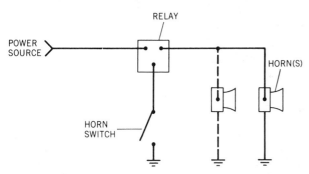

Figure 13.2 Many horn circuits incorporate a relay so that a simple grounding switch can be used on the steering wheel.

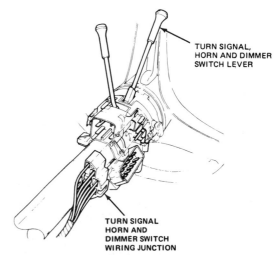

TURN SIGNAL, HORN AND DIMMER SWITCH LEVER

TURN SIGNAL HORN AND DIMMER SWITCH WIRING JUNCTION

Figure 13.5 On some cars, the horn switch is built into an assembly of switches mounted on the steering column under the steering wheel (courtesy of Ford Motor Company).

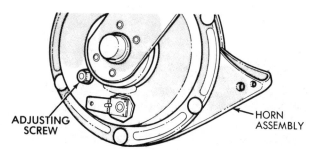

Figure 13.6 Some horns are provided with an adjusting screw so that the points within the horn may be adjusted. While an adjustment may clear up the sound or cause an inoperative horn to work, it cannot change the pitch or frequency of the horn (courtesy of Chrysler Corporation).

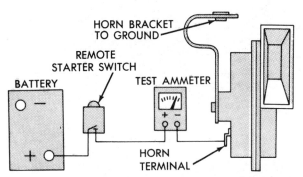

Figure 13.7 Connected as shown, an ammeter and a remote starter switch can be used to adjust a horn to the current draw specified by the manufacturer (courtesy of Chrysler Corporation).

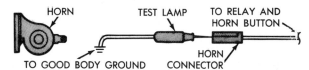

Figure 13.8 A test lamp can be used to check for the presence of voltage when the horn button is depressed (courtesy of Chrysler Corporation).

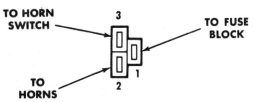

Figure 13.9 Typical terminal arrangement in a horn relay connector or socket (courtesy of Chrysler Corporation).

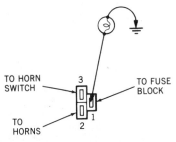

Figure 13.10 Connected between terminal #1 and ground, the test lamp should light. If the lamp does not light, the fuse may be blown or there may be an open in the wire from the power source. If the lamp lights, continue with Test Two (courtesy of Chrysler Corporation).

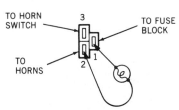

Figure 13.11 Connected between terminal #1 and terminal #2, the test lamp should light. If the lamp does not light, there is an open in the wire between the horn and the terminal in the socket. If the lamp lights, continue with Test Three (courtesy of Chrysler Corporation).

should be cleaned and tightened. If the horn does not operate, it can be considered defective or in need of adjustment.

Some horns are fitted with an adjustment screw, as shown in Figure 13.6, and some manufacturers recommend that an adjustment be attempted before a horn is replaced. With the jumper wires attached, the adjustment screw should be turned slightly, first one way and then the other, until the horn operates. If movement of the adjustment screw does not result in horn operation, the horn should be replaced. If the horn operates, a final adjustment should be made with an ammeter connected as shown in Figure 13.7. Average current flow will range from 4 to 7 A, depending on the design of the horn. The specification for the horn you are testing must be

obtained from an appropriate manual. If the horn will not operate at the specified current flow, or if the specified current flow cannot be obtained, the horn should be replaced.

If the horn is in good condition but will not operate when connected in its circuit, the wire should be disconnected from the horn and a test lamp should be used, as shown in Figure 13.8, to determine if current is available when the horn button is pressed. If the test lamp does not light, there is an open in the horn circuit.

On cars that use a horn relay, the relay often can be tested by substitution. As shown in Figure 13.9, the relay connector or socket usually has three terminals. If you can identify those terminals by testing or by reference to an appropriate manual, they can provide test points by which you can determine where the problem in the horn circuit exists. Figures 13.10, 13.11, and 13.12 illustrate those tests and provide the interpretation of their results.

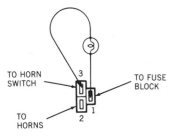

Figure 13.12 Connected between terminal #1 and terminal #3, the test lamp should light when the horn button is pressed. If the lamp does not light, the horn switch is defective, the wire between the horn switch and the terminal in the socket is open, or the steering column is not grounded. If the lamp lights, the switch, its wire, and ground are OK (courtesy of Chrysler Corporation).

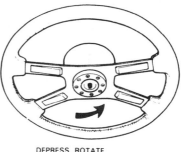

DEPRESS, ROTATE COUNTERCLOCKWISE AND REMOVE.

Figure 13.15 On some cars, the horn button is removed by depressing it and turning it counterclockwise (courtesy of Ford Motor Company).

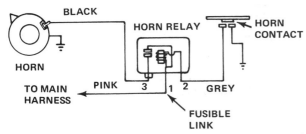

Figure 13.13 A typical horn circuit wiring diagram (courtesy of American Motors).

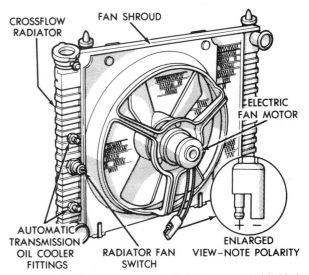

Figure 13.16 Many cars have an electric fan mounted behind the radiator. Note that in this installation, the fan switch is mounted in the radiator tank (courtesy of Chrysler Corporation).

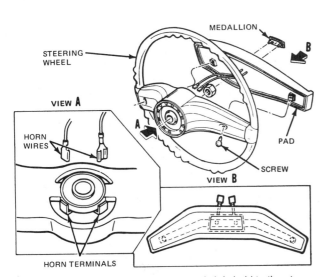

Figure 13.14 On many cars, the horn switch is held to the steering wheel by screws, and is connected to the horn terminals by short wires (courtesy of Ford Motor Company).

On cars that do not use a horn relay, a test lamp can be used to check through the circuit. As shown in Figure 13.13, the test points will differ with different cars and should be located by reference to an appropriate manual.

If your diagnosis indicates that the horn switch could be defective, it can be removed for further testing. As shown in Figures 13.14 and 13.15, all horn but-

tons are not mounted in the same manner. An appropriate manual should be consulted if the horn button cannot be removed easily. An ohmmeter or a continuity tester can be used to check the operation of the horn switch after it has been removed. A defective switch should be replaced. On some cars, the horn switch is connected to its wire by means of a small brush (similar to an alternator brush) that is held in contact with a ring beneath the steering wheel. The brush should be replaced if it is worn or if its wire is broken.

RADIATOR FANS   As shown in Figure 13.16, many vehicles are equipped with a radiator fan driven by an electric motor. The motor is controlled by a temperature-sensitive switch located in the radiator tank or in the cylinder head of the engine. (See Figure 13.17.) When the coolant in contact with the

Job 13a

TEST AND REPLACE HORN CIRCUIT PARTS

SATISFACTORY PERFORMANCE

A satisfactory performance on this job requires that you do the following:

1 Test the parts and wires in a horn circuit and repair or replace the parts required.
2 Following the steps in the "Performance Outline" and the procedure and specifications of the car manufacturer, complete the job within 200 percent of the manufacturer's suggested time.
3 Fill in the blanks under "Information."

PERFORMANCE OUTLINE

1 Check the horn and its ground.
2 Check for voltage to the horn.
3 Check the relay, fuse, and switch.
4 Perform the required repair or part replacement.

INFORMATION

Vehicle identification _____

Reference used _____ Page(s) _____

Defect(s) found _____

_____

Repairs performed _____

_____

Parts replaced _____

_____

switch exceeds a predetermined temperature, the switch contacts close, energizing the motor. When the coolant temperature drops below the operating range of the switch, the contacts open, breaking the circuit. On some cars, the switch merely completes the ground side of the fan motor circuit as shown in Figure 13.18. On others, it operates a relay that sends current to the fan motor. (See Figure 13.19).

On air conditioned cars that have the condenser mounted in front of the radiator, additional air flow through the condenser and the radiator often is required when the air conditioner is operating. On some cars, the fan motor is energized at all times when the air conditioner is turned on. On others, the fan is energized by both a temperature-sensi-

tive switch and by a pressure-sensitive switch in the air conditioning system. Some cars have a two-speed fan motor to provide the correct air flow for different operating conditions. (See Figure 13.20).

On some cars, the fan will operate at any time that the contacts in the temperature-sensitive switch close, even if the ignition switch is turned off. To avoid the possibility of injury, disconnect the fan wire connector when working near the fan blades. (Refer to Figure 13.16.)

Testing   There are several different fan operating systems in common use, each requiring a slightly different test procedure. The following steps outline a typical procedure, but reference to an appropriate manual is suggested:

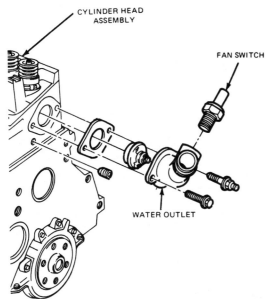

Figure 13.17 In this application, the fan switch is installed in the water outlet connection on the cylinder head (courtesy of Ford Motor Company).

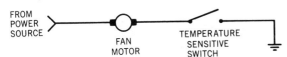

Figure 13.18 On some cars, the temperature-sensitive switch is placed in the ground side of the motor circuit.

1 Check the coolant level in the radiator.

WARNING: Never remove a radiator cap while the engine is running. Failure to follow this advice could cause serious burns and possible damage to the cooling system. Since scalding hot coolant or steam may erupt from the radiator filler neck, use extreme care when removing a radiator cap. It is best to wait until the engine has cooled.

2 Adjust the coolant level if necessary.

Note: The coolant should consist of a mixture of approximately 50 percent water and 50 percent permanent antifreeze (ethylene glycol). The addition of water will dilute the coolant and will adversely affect the operation of the cooling system.

3 Start the engine and allow it to run for several minutes until the coolant temperature rises to above the specified operating temperature of the temperature-sensitive switch. [The specified temperature may range from 195°F (91°C) to 225°F (107°C).]

If the fan operates, the circuit can be considered to be working correctly.

If the fan does not operate, turn off the engine and continue with step 4.

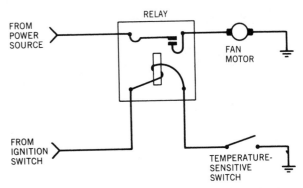

Figure 13.19 Some fan motor circuits incorporate a relay to activate the fan motor.

4 Disconnect the wire connector from the temperature-sensitive switch in the radiator tank or in the cylinder head.

5 If the connector has a single wire, ground the wire with a jumper. If the connector has two wires, use a jumper as shown in Figure 13.21 to bridge the two wires. Turn on the ignition switch.

If the fan operates, the temperature-sensitive switch should be replaced.

If the fan does not operate, continue with step 6.

6 Turn off the ignition switch.

7 Connect the wire connector to the temperature-sensitive switch.

8 Unplug the wire(s) at the fan. (Refer to Figure 13.16.)

9 Using jumper wires from the battery, energize the fan motor at its connector. (Refer to Figure 13.16.)

If the fan does not operate, the fan motor should be replaced.

If the fan operates, refer to an appropriate manual or wiring diagram and check the wires and any fuses, fuse links, and relays that may be used in the circuit.

Fan and Motor Replacement   On most cars, the fan motor is mounted on a *shroud* or frame that is bolted to the radiator. (Refer to Figure 13.16.) As shown in Figures 13.22 and 13.23, the entire assembly must be removed from the radiator for service. While the replacement procedure will vary slightly with different systems, the following steps outline a typical replacement procedure:

1 Unplug the wire connector at the fan and free the wires from any clip or bracket that may be present.

2 Remove the bolts or nuts holding the shroud or fan frame to the radiator.

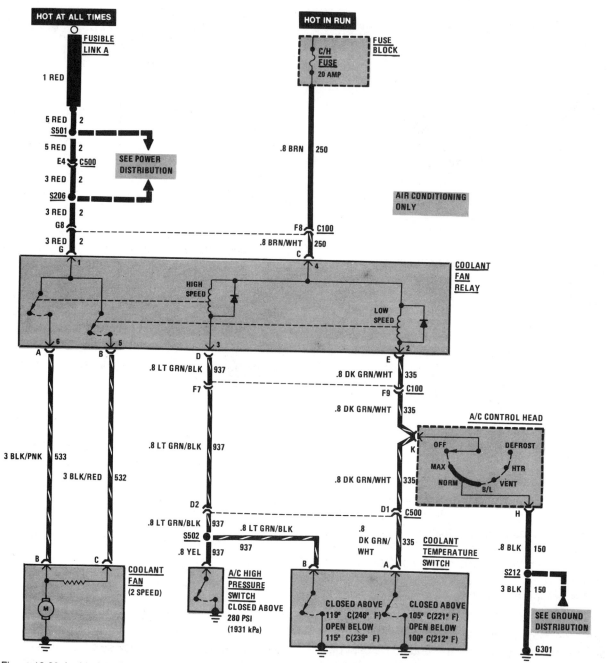

Figure 13.20 In this fan circuit, a two-speed motor is controlled by a relay activated by a dual temperature-sensitive switch and a pressure switch (courtesy of Pontiac Motor Division, General Motors Corporation).

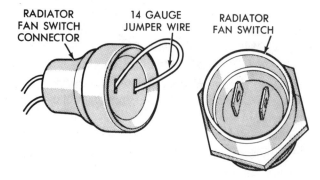

Figure 13.21 A jumper wire can be used to bridge the terminals in the fan switch connector (courtesy of Chrysler Corporation).

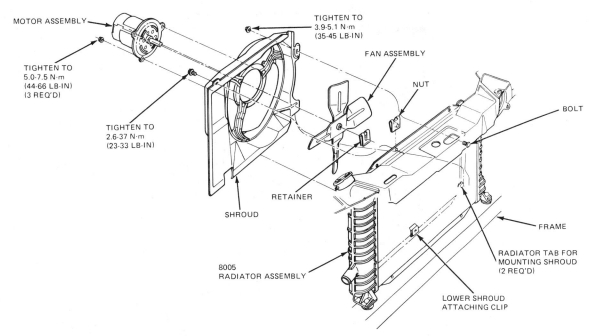

Figure 13.22 An exploded view of a typical fan assembly. In this application, the fan motor is mounted in a shroud (courtesy of Ford Motor Company).

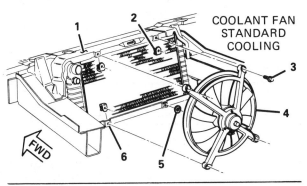

COOLANT FAN
STANDARD
COOLING

1 – PANEL (UPPER)   4 – FAN ASSEMBLY

2 – NUT   5 – WASHER

3 – BOLT   6 – PANEL (LOWER)

Figure 13.23 In this application, the fan motor is mounted in a simple frame and a shroud is not used (courtesy of Pontiac Motor Division, General Motors Corporation).

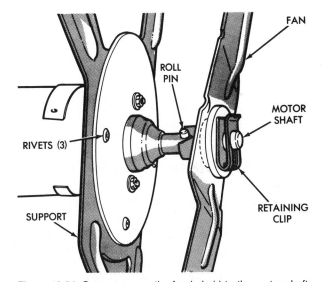

Figure 13.24 On some cars, the fan is held to the motor shaft by a retaining clip (courtesy of Chrysler Corporation).

3 Remove the assembly from the vehicle.

4 Remove the nut or clip that holds the fan to the motor shaft. (See Figure 13.24.)

5 Remove the fan from the motor shaft.

Note: If the fan is cracked, bent, or damaged in any way, it should be replaced. Never attempt to straighten, weld, or otherwise repair a damaged fan.

6 Remove the nuts holding the fan motor to the shroud or frame and remove the fan motor.

Note: On some cars, the fan motor is not considered

a separate part, and the entire assembly must be replaced.

7 Install the replacement motor in the shroud or frame.

8 Install the fan and the retaining nut or clip.

9 Position the assembly on the radiator and install the mounting nuts or bolts.

10 Connect the fan wires.

11 Check the operation of the fan.

Job 13b

TEST AND REPLACE RADIATOR FAN CIRCUIT PARTS

SATISFACTORY PERFORMANCE

A satisfactory performance on this job requires that you do the following:

1 Test the parts and wires in a radiator fan circuit and repair or replace the parts required.
2 Following the steps in the "Performance Outline" and the procedure and specifications of the car manufacturer, complete the job within 200 percent of the manufacturer's suggested time.
3 Fill in the blanks under "Information."

PERFORMANCE OUTLINE

1 Check the coolant level.
2 Check the fan operation.
3 Check the switch.
4 Check the motor.
5 Check remaining circuit parts.
6 Perform the required repairs.

INFORMATION

Vehicle identification _____

Reference used _____ Page(s) _____

Defect(s) found _____

Repairs performed _____

Parts replaced _____

GRID-TYPE REAR WINDOW DEFOGGERS

On many cars, the rear window contains horizontal element lines of a silver-bearing conductive material bonded to the inner surface of the glass. (See Figure 13.25.) Those *grid lines* are connected at their ends to heavy vertical lines or *bus bars*. One bus bar is grounded, and the other is wired to a switch through a relay. When the switch is closed, current flows through the grid lines, raising their temperature and heating the glass.

Depending on the number of grid lines and their length, current flow in a 12-V system will range from 15 to 25 A. For this reason, extended operation of the defogger is not advisable. Most defogger circuits incorporate a *relay timer*. The relay timer allows the use of a small switch and, in case the driver neglects to do so, automatically turns off the circuit after about 10 minutes of operation. Most circuits also contain an indicator lamp on the instrument panel to inform the driver when the system

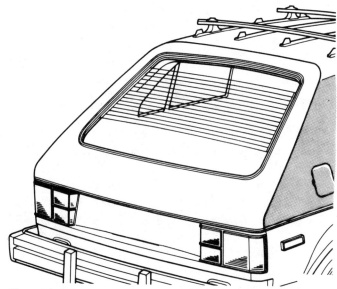

Figure 13.25 A typical grid-type rear window defogger (courtesy of Chrysler Corporation).

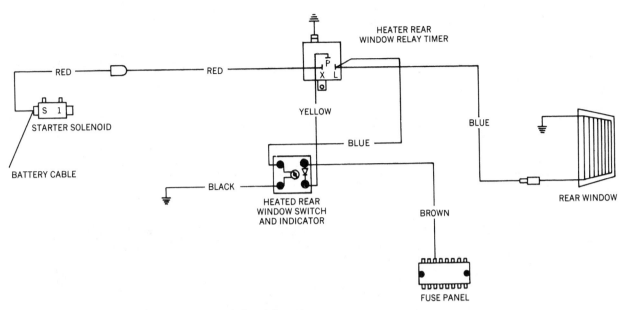

Figure 13.26 A wiring diagram for a typical rear window defogger. Note that the switch contains an indicator lamp (courtesy of American Motors).

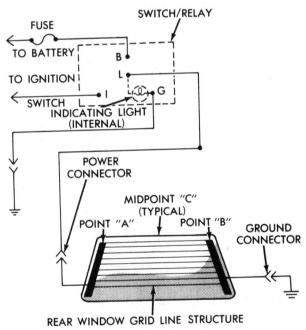

Figure 13.27 On some cars, the switch, indicator lamp, and relay are combined in one unit (courtesy of Chrysler Corporation).

is turned on. Figure 13.26 shows a typical rear window defogger circuit.

Troubleshooting An inoperative defogger may be caused by a failed component or by an open circuit. On most cars, the indicator lamp is actuated by the relay timer. (Refer to Figure 13.26.) If the indicator lamp does not light when the defogger is turned on, a test lamp or a voltmeter should be used to determine if a fuse is blown, if the switch or the relay timer is at fault, or if an open circuit exists in the wires connecting and feeding those parts. Since many different components and circuits are used by different car makers, reference to an appropriate wiring diagram is recommended to determine the test points. (See Figure 13.27.)

An operative indicator lamp usually indicates that the switch and the relay timer are functioning, and that an open circuit exists between the relay timer and the grid. A test lamp or a voltmeter connected between the "hot" bus bar terminal and ground can be used to determine if voltage is available to the grid. If no voltage is detected, there is a break in the wire between the relay timer and the bus bar. If voltage is available at the "hot" bus bar, the test lamp or voltmeter lead should be touched to the grounded bus bar. Any indication of voltage at this point is caused by a bad ground.

At times, the terminals or the wires attached to the bus bars will be found to be loose. Those connections should be checked. In most instances, a loose terminal or wire can be reattached.

Reinstalling Bus Bar Terminals and Wires
A loose terminal or wire usually can be reattached to a bus bar by soldering. Best results will be obtained by using a solder with a 3 percent silver content and a rosin-base flux. If the conductive material of the bus bar is torn or broken, it must be

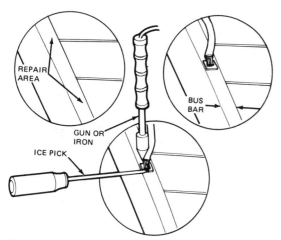

Figure 13.28 To obtain a good connection, the terminal or wire end must be held firmly in place while it is being soldered (courtesy of Ford Motor Company).

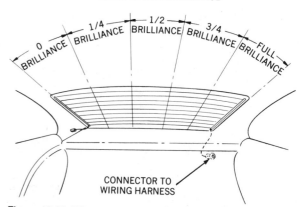

ZONES OF BULB BRILLIANCE

Figure 13.29 When checked with a test lamp, a properly operating rear window defogger will have the zones of bulb brilliance shown above (courtesy of Pontiac Motor Division, General Motors Corporation).

repaired before soldering is attempted. Most car manufacturers make available a grid repair compound for that purpose. The compound contains powdered silver and is painted across any break or tear in the conductive material. The procedure for its use is presented later in this chapter. The following steps outline a typical procedure for reattaching terminals and wires:

1 Carefully clean the bus bar at the repair area.
2 Using fine steel wool (3/0 or 4/0 grade), buff the repair area to remove any oxide coating that may be present.
3 Clean and buff the terminal or wire end to be attached.
4 Apply a thin coating of rosin-base flux to the bus bar and to the terminal or wire end.
5 Using a clean, tinned soldering iron, tin the terminal or wire end.
6 Tin the section of the bus to which the terminal or wire is to be attached.

Note: The glass must be at room temperature or higher. The use of a heat gun or a hair dryer to preheat the glass is recommended.

7 Position the terminal or wire on the tinned section of the bus bar and hold it in place with an ice pick or a thin screwdriver as shown in Figure 13.28.
8 Hold the soldering iron to the terminal or wire end until the solder flows.

Note: Hold the soldering iron in place only long enough for the solder to flow. Excessive heat may damage the bus bar.

9 Start the engine, turn on the defogger, and test its operation. Allow the defogger to operate through

its cycle (about 10 minutes).
10 Inspect the repair.

Note: Some manufacturers recommend coating the repair with repair compound to improve its conductivity and appearance.

Testing Grid Lines　A common problem with grid-type defoggers is partial operation. Since the grid material is relatively fragile, the grid lines often are broken through contact with articles carried in the car or by improper cleaning procedures. In most instances, a broken grid line can be repaired.

At times, a break in a grid line is obvious, but in many cases a slight break cannot easily be seen. Individual grid lines can be tested with a test lamp, and the exact location of the break can be found. The following steps outline a typical procedure for testing grid lines:

1 Carefully clean the inside of the window with a mild glass cleaner and a soft cloth.
2 Start the engine and turn on the defogger.
3 Connect a test lamp to a good ground near the window.
4 Touch the probe of the test lamp to the "hot" bus bar to check the operation of the test lamp. The lamp should light at normal brilliance.
5 Carefully touch the probe of the test lamp to the center of each grid line. As shown in Figure 13.29, the test lamp should light at approximately one half normal brilliance.

If the test lamp does not light, there is a break in the grid line between the probe and the "hot" bus bar. If the probe is slowly moved toward the "hot"

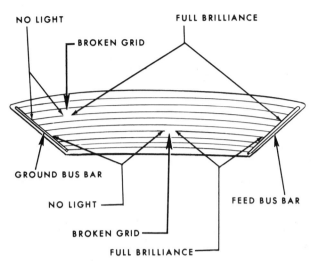

Figure 13.30 Grid lines can be tested and the exact location of breaks can be found by the use of a test lamp. The breaks in the grid lines shown are exaggerated for illustration (courtesy of Pontiac Motor Division, General Motors Corporation).

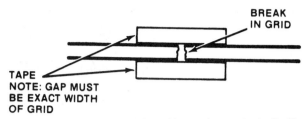

Figure 13.31 The edges of the grid must be masked off with celophane tape so that the repair matches the width of the grid line (courtesy of American Motors).

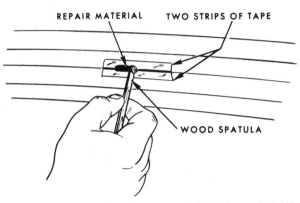

Figure 13.32 The grid repair compound should be applied with a small brush or spatula and should extend at least ¼ inch on both sides of the break (courtesy of Pontiac Motor Division, General Motors Corporation).

bus bar, the lamp will light when the break is passed. (See Figure 13.30.)

If the lamp lights at or near full brilliance, the break is between the probe and the grounded bus bar. By moving the probe toward the grounded bus bar, the break will be located when the light dims. (Refer to Figure 13.30.)

6 Using a crayon, place small marks on the glass above and below any breaks found.

Note: Do not make a crayon mark across the grid lines.

7 Repeat steps 5 and 6 at all the grid lines.

8 When all the breaks have been located, turn off the engine.

9 Working from the outside of the car, place vertical crayon marks on the outside of the window across the breaks in the grid lines.

Repairing Grids   A broken or torn grid line or bus bar usually can be repaired by the careful application of a paint-like grid repair compound. The compound contains powdered silver and will restore continuity between separated grid parts. While different manufacturers provide different types of compound and recommend slightly different procedures, the following procedure can be considered typical:

1 Allow the window to reach room temperature.

2 Carefully clean the grid line or bus bar at the repair area.

3 Using fine steel wool (3/0 or 4/0 grade), buff the

repair area to remove any oxide coating that may be present.

Note: Buff the repair area at least ¼ inch beyond the break.

4 Using cellophane tape, mask off the repair area above and below the break as shown in Figure 13.31. The gap between the pieces of tape should be the exact width of the grid line.

5 Mix the grid repair compound thoroughly until the silver powder is evenly distributed and the compound reaches a uniform consistency.

6 Using a small brush or a spatula, as shown in Figure 13.32, apply a heavy coating of the compound to the repair area, extending the coating to at least ¼ inch beyond each side of the break.

Note: Some compounds require additional applications.

7 After waiting about 5 minutes for the compound to dry, remove the tape used to mask the repair.

Note: Some compounds will cure properly only with the application of heat. A heat gun should be used as shown in Figure 13.33 to raise the temperature of the repair to about 300°F (150°C) for 1 to 2 min-

Job 13c

TEST AND REPAIR A REAR WINDOW DEFOGGER
SATISFACTORY PERFORMANCE
A satisfactory performance on this job requires that you do the following:

1 Test the parts and grid lines in a rear window defogger and perform the necessary repairs.
2 Following the steps in the "Performance Outline" and the procedure and specifications of the car manufacturer, complete the job within 200 percent of the manufacturer's suggested time.
3 Fill in the blanks under "Information."
PERFORMANCE OUTLINE
1 Check the operation of the defogger.
2 Check for the presence of current at the defogger.
3 Repair or replace any circuit parts necessary to obtain circuit operation.
4 Check the bus bars and grids.
5 Repair bus bars and grids as necessary.
INFORMATION
Vehicle identification _____

Reference used _____ Page(s) _____

Defect(s) found _____

_____

Parts replaced and repairs performed _____

_____

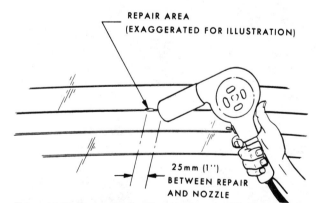

Figure 13.33 Some compounds must be cured at high temperatures, and some manufacturers recommend the use of a heat gun as shown (courtesy of Pontiac Motor Division, General Motors Corporation).

utes. Use the heat gun carefully so that nearby trim and upholstery are not damaged.

8 Start the engine, turn on the defogger, and check the repair.
9 Turn off the engine.
10 Check the appearance of the repair.

Note: The repaired area can be colored to match the grid lines by touching up the repair with tincture of iodine.

11 Clean the crayon marks from the window.

Note: Do not attempt to clean the repair area for at least 24 hours.

HEATER BLOWERS   The forced air flow through the heater, defroster, and air conditioning ducts is provided by a fan motor or blower motor similar to the one shown in Figure 13.34. On most cars, the speed of the motor is controlled by a multiple contact switch that routes current through various resistors. Typical blower motor circuits are shown in Figures 13.35 and 13.36.

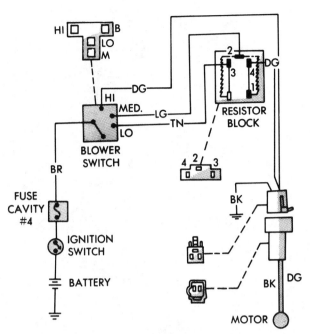

Figure 13.35 In this heater blower circuit, two resistors are used to provide three motor speeds (courtesy of Chrysler Corporation).

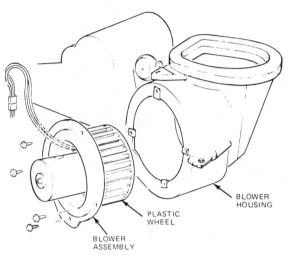

Figure 13.34 A typical blower motor removed from its housing (courtesy of Ford Motor Company).

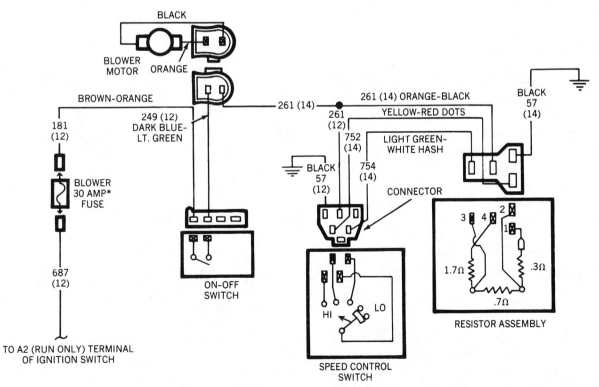

Figure 13.36 In this blower motor circuit, a switch is located in the "hot" side to turn the system on and off. A separate switch controls the speed of the blower through resistors placed in the ground side of the circuit (courtesy of Ford Motor Company).

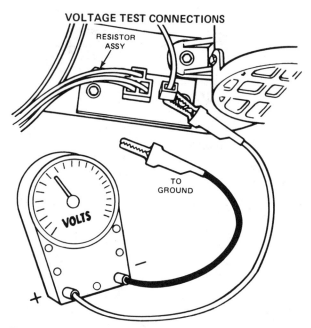

Figure 13.37 The resistor assembly provides a good test point to determine if voltage is available to the motor (courtesy of Ford Motor Company).

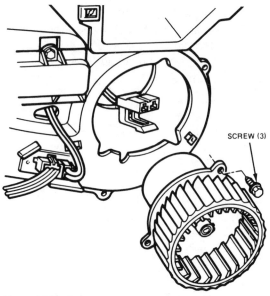

Figure 13.38 On some cars, the blower motor and fan assembly can be removed from the inside of the car after removing a housing cover (courtesy of Ford Motor Company).

Troubleshooting   A noisy blower may indicate that leaves or other debris is trapped in the blower housing. You may find that the blower assembly must be removed to gain access to the foreign matter. A squeal or squeak that varies with the speed of the motor, especially when it is cold, usually indicates that the motor bearings are dry or worn. In most instances, a noisy motor should be replaced.

If the motor operates only when the switch is in the high speed position, the resistor assembly usually is at fault. A voltmeter or a test lamp can be used, as shown in Figure 13.37, to check for voltage from the switch and through the resistor assembly.

An inoperative blower motor can be caused by a blown fuse, an open circuit, or a faulty switch or motor. A voltmeter or a test lamp can be used to check for voltage at various points in the circuit. Reference to a wiring diagram should be made to determine the test points.

Parts Replacement   The location of the parts in a heating system and the means of access to those parts varies greatly among different cars and are best determined by reference to an appropriate manual. While some motors are positioned so they can easily be replaced, in some installations the entire heater assembly must be removed to gain access to the motor. Figures 13.38 and 13.39 show

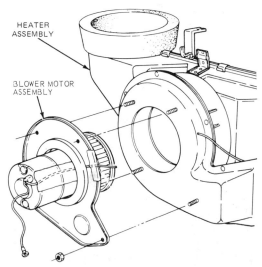

Figure 13.39 In this installation, the blower motor is mounted in the front of the heater assembly. The entire assembly must be removed from under the instrument panel to gain access to the motor (courtesy of Ford Motor Company).

typical motor installations. On some cars, the switch can be readily removed from the instrument panel. On others, the switch is accessible only after the removal of other parts. Figures 13.40 and 13.41 show typical switch mounting systems. While the location of the resistor assembly will vary, it usually is mounted on or near the blower housing and is removed easily. (See Figure 13.42.)

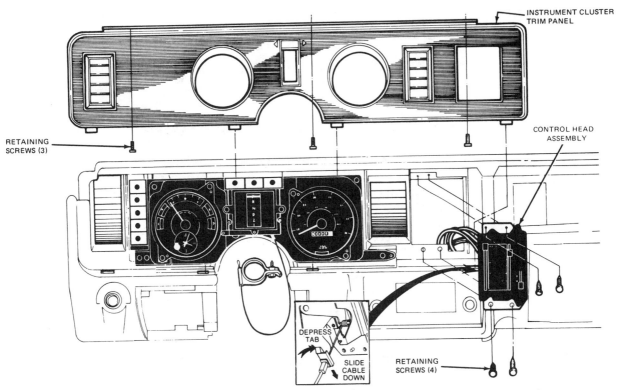

INSTRUMENT CLUSTER
TRIM PANEL

CONTROL HEAD
ASSEMBLY

RETAINING
SCREWS (3)

DEPRESS
TAB

SLIDE
CABLE
DOWN

RETAINING
SCREWS (4)

Figure 13.40 To gain access to the heater control assembly in this installation, it is necessary to remove the instrument cluster trim panel (courtesy of Ford Motor Company).

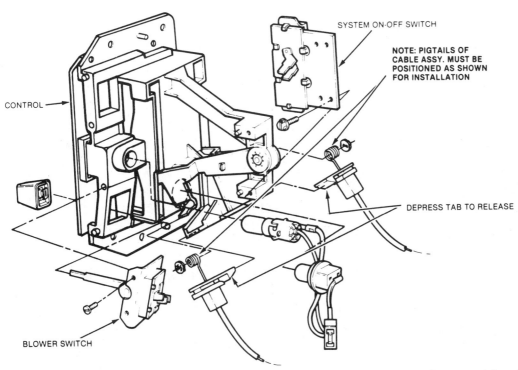

SYSTEM ON-OFF SWITCH

NOTE: PIGTAILS OF
CABLE ASSY. MUST BE
POSITIONED AS SHOWN
FOR INSTALLATION

CONTROL

DEPRESS TAB TO RELEASE

BLOWER SWITCH

Figure 13.41 In this installation, the on-off switch and the speed control switch are both mounted on the heater control assembly. The assembly must be removed from the instrument panel to gain access to the switches (courtesy of Ford Motor Company).

Job 13d

TEST AND REPAIR A HEATER BLOWER

SATISFACTORY PERFORMANCE

A satisfactory performance on this job requires that you do the following:

1 Test the parts in a blower motor circuit and perform the necessary repairs.
2 Following the steps in the "Performance Outline" and the procedure and specifications of the car manufacturer, complete the job within 200 percent of the manufacturer's suggested time.
3 Fill in the blanks under "Information."

PERFORMANCE OUTLINE

1 Check the operation of the heater blower.
2 Check for the presence of current at the blower motor.
3 Repair or replace any circuit parts necessary to obtain blower motor operation.

INFORMATION

Vehicle identification _____

Reference used _____ Page(s) _____

Defect(s) found _____

_____

Parts replaced and repairs performed _____

_____

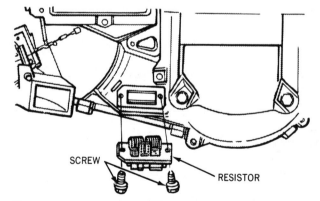

SCREW

RESISTOR

Figure 13.42 The resistor assembly is usually mounted on or near the motor housing. The coiled resistance wires are kept cool by the flow of air through the housing (courtesy of Ford Motor Company).

WINDSHIELD WIPERS AND WASHERS

On most cars, the windshield wipers are powered by a multispeed motor. The speed of some motors is controlled by changing current flow through the field windings by means of a resistor. Other motors have permanent magnet field poles. The speed of those motors is controlled by switching current through different brushes. Typical wiper motor control circuits are shown in Figures 13.43 and 13.44. Through a set of reduction gears, the motor turns a drive crank that pushes and pulls the wiper arm pivots by means of links. Figures 13.45 and 13.46 show those parts and their assembly.

On some cars, the windshield washer pump is mounted on the wiper motor gear box and is driven by the wiper motor. An assembly of that type is shown in Figure 13.47. When the washer circuit is activated, a relay coil moves an armature to provide a mechanical connection between a small piston pump and a cam driven by the motor. Figure 13.48 shows the mechanism used.

Many cars are equipped with a separate motor-driven washer pump. In some systems, the pump

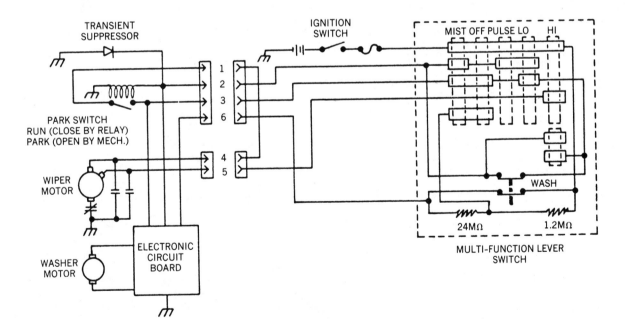

HOT IN ACCY OR RUN

12
6A   FUSE
     BLOCK

63 R

WINDSHIELD
WIPER/WASHER
SWITCH

LO   WIPER   LO
     SWITCH
OFF   HI   OFF   HI   WASHER
     HI   SWITCH

B   H   L   P   W

63 R   56 DB/O   28 BK/PKD   941 BK/W

          58 W

                    C201   C237

56 DB/O   28 BK/PKD   941 BK/W
63 R   58 W

                    C311

56 BLUE   28 BLACK
         58 WHITE   WASHER
63 RED   PUMP

M   RUN   WIPER MOTOR   57 BK
         PARK   AND SWITCH

                                     57 BK

GROUND IS MOTOR
ATTACHING BOLT

TRANSIENT
SUPPRESSOR

IGNITION
SWITCH

MIST OFF PULSE LO   HI

1
2
3
6

PARK SWITCH
RUN (CLOSE BY RELAY)
PARK (OPEN BY MECH.)

WIPER
MOTOR

4
5

WASH

WASHER
MOTOR

ELECTRONIC
CIRCUIT
BOARD

24MΩ   1.2MΩ

MULTI-FUNCTION LEVER
SWITCH

Figure 13.44 A windshield wiper and washer circuit that incorporates an electronic circuit board to program the washer operation (courtesy of Chevrolet Motor Division).

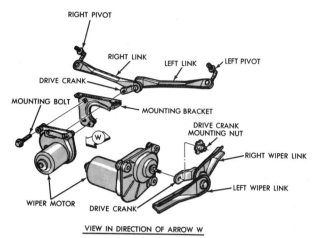

Figure 13.45 Typical wiper motor and linkage parts. The drive crank converts the rotating motion of the motor to reciprocating motion which is transmitted to the wiper arm pivots by links (courtesy of Chrysler Corporation).

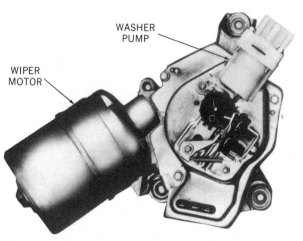

Figure 13.47 A windshield wiper motor and windshield washer pump assembly (courtesy of Chevrolet Motor Division).

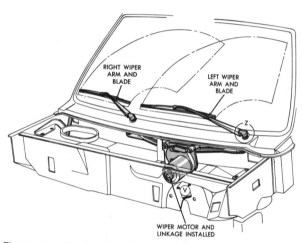

Figure 13.46 The location of the wiper motor and linkage assembly in a typical installation (courtesy of Chrysler Corporation).

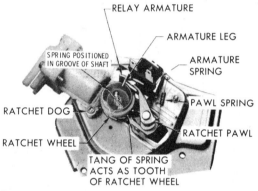

Figure 13.48 A windshield washer pump mechanism that is driven by the wiper motor. The armature of a relay provides the mechanical connection between the pump and the motor (courtesy of Chevrolet Motor Division).

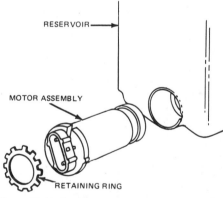

Figure 13.49 On some cars, the windshield washer pump motor assembly is mounted in the bottom of the reservoir (courtesy of Ford Motor Company).

is mounted in the bottom of the fluid reservoir as shown in Figure 13.49. In other systems, the pump motor is mounted in an assembly with the wiper motor. In systems of this type, the operation and timing of the pump motor is controlled by an electronic printed circuit board located under the washer pump cover. Figure 13.50 illustrates such an assembly.

The windshield washer fluid flows from the reservoir through a hose as shown in Figure 13.51. On some cars, the fluid is discharged through a pair of nozzles mounted under the rear edge of the hood or in the cowl. On other cars, the nozzles are mounted on the windshield wiper arms. On still other cars, a single *fluidic* nozzle is used. A fluidic nozzle, shown

in Figure 13.52, discharges the fluid in a rapidly oscillating stream that appears as a solid fan of fluid.

Windshield Wiper Troubleshooting An inoperative wiper motor may be caused by a blown fuse, a poor ground connection, or an open in the circuit. A test lamp or a voltmeter used with reference to an appropriate wiring diagram will enable you to check through the circuit and determine if voltage is available at the motor. If voltage is available, the motor may be defective. As shown in Figure 13.53, the test points for checking motor operation are best obtained from the car maker's service manual. The wiper systems used on some cars are so complex that their makers suggest that a special tester be used to simplify diagnosis. Lacking such a tester, reference to a shop manual usually is necessary so that a specific test procedure can be followed.

Windshield Washer Troubleshooting In many instances, an inoperative washer is caused by an empty reservoir, collapsed or plugged hoses, or dirt-clogged nozzles. Operating the washer with the discharge hose disconnected from the pump is the easiest way to check pump operation. If the pump delivers fluid, the hoses and nozzles should be checked and cleared with air pressure. Collapsed or dried hoses, and nozzles that cannot be blown clear should be replaced.

An inoperative pump motor may be caused by a blown fuse, a poor ground connection, or an open

1 PUMP AND CIRCUIT BOARD COVER
2 WINDSHIELD WASHER PUMP
3 PERMANENT MAGNET WIPER MOTOR

Figure 13.50 A windshield wiper and washer assembly. The washer pump has a separate motor which is controlled by an electronic circuit board located beneath the cover (courtesy of Chevrolet Motor Division).

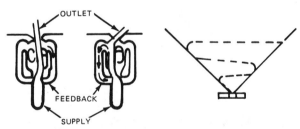

Figure 13.52 A fluidic system washer nozzle. The feedback in the flow tube causes the water stream to oscillate rapidly, appearing as a solid fan of fluid to the naked eye (courtesy of Ford Motor Company).

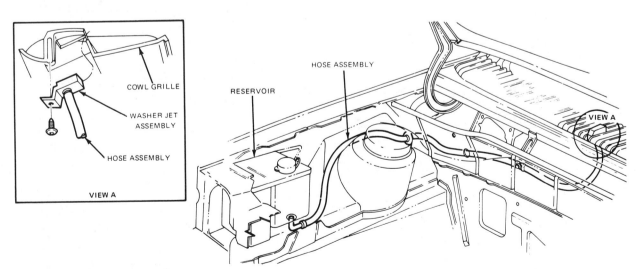

Figure 13.51 A typical windshield washer system. In this system, the pump is located in the base of the reservoir. Fluid is pushed through a hose to a single fluidic nozzle located in the center of the cowl (courtesy of Ford Motor Company).

in the circuit. When the pump motor is mounted in the reservoir (refer to Figure 13.49), the wiring harness can be unplugged from the motor so that a test lamp or a voltmeter can be used to check for voltage when the washer is activated. If voltage is available and the ground connection is good, the motor probably is defective. Jumper wires can be used to energize the motor to verify your diagnosis. When the motor is mounted with the wiper motor, a jumper wire can be used as shown in Figure 13.54 to check the operation of the washer pump motor.

When the washer pump is driven by the wiper motor (refer to Figure 13.47), the cover can be removed to observe the action of the mechanical components. An inoperative pump of that type usually can be repaired by the replacement of certain parts provided in a kit available for that purpose.

Windshield Wiper Repairs   While the service manuals of some manufacturers provide detailed procedures for the overhaul of wiper motors and drive assemblies, parts for those assemblies are usually available only on special order. If a wiper motor or drive is found to be defective, common practice is to replace the assembly with a new or rebuilt unit.

In most instances, the wiper arm linkage must be disconnected from the drive assembly before the assembly can be removed. Access to the motor and linkage may require that you work under the hood, under the instrument panel, inside the cowl, or a combination of those locations. Because of the many different systems and mounting locations in use, the specific procedure should be obtained from a manual for the car on which you are working.

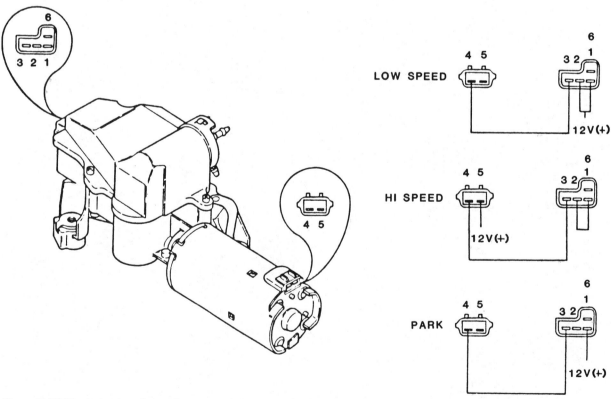

Figure 13.53 The test points shown above are specified by one manufacturer for on-car testing of a wiper motor (courtesy of Chevrolet Motor Division).

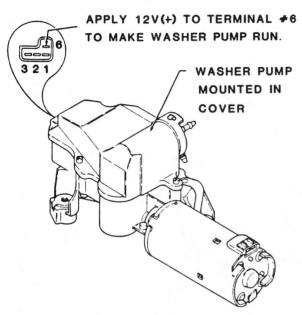

APPLY 12V(+) TO TERMINAL #6
TO MAKE WASHER PUMP RUN.

WASHER PUMP
MOUNTED IN
COVER

Figure 13.54 Test points specified by one manufacturer for activating a windshield washer pump motor (courtesy of Chevrolet Motor Division).

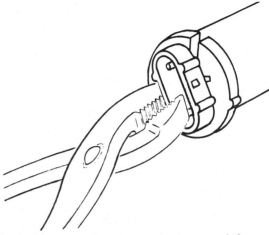

Figure 13.55 After the retaining ring is removed, the pump assembly can be pulled from the reservoir (courtesy of Ford Motor Company).

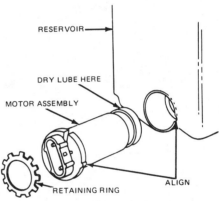

RESERVOIR

DRY LUBE HERE

MOTOR ASSEMBLY

RETAINING RING

ALIGN

HAND PRESS COMPONENTS TOGETHER

Figure 13.56 The seal on the pump motor assembly should be lubricated with graphite before assembly. Note the alignment tab and notch (courtesy of Ford Motor Company).

Windshield Washer Repairs   In those systems where the pump is located in the reservoir, the pump and motor assembly usually can be replaced in the following manner:

1 Remove the discharge hose from the reservoir and allow the fluid to drain into a pan.

2 Disconnect the wiring harness plug from the motor.

3 Remove the screws holding the reservoir and remove it from the car.

4 Using a small screwdriver, pry out the retaining ring that secures the motor assembly. (Refer to Figure 13.49.)

5 Grasp the assembly with a pair of pliers as shown in Figure 13.55 and pull the assembly from the reservoir.

6 Carefully clean the pump chamber in the reservoir.

Note: This will make replacement easier and insure against leakage.

7 Lubricate the seal on the replacement assembly with powdered graphite. (See Figure 13.56.)

8 Align the tab on the motor assembly with the notch in the reservoir (refer to Figure 13.56) and push the assembly into its chamber so that it bottoms.

9 Install the retaining ring.

Note: A 12-point, 1-inch socket can be used as an installation tool. The ring should be pushed in by hand pressure only.

10 Install the reservoir, discharge hose, and wiring plug.

11 Fill the reservoir with clean fluid.

12 Check the operation of the washer.

Note: Since the pump is lubricated by the washer

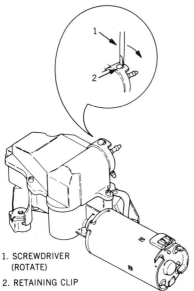

1. SCREWDRIVER
   (ROTATE)

2. RETAINING CLIP

Figure 13.57 The windshield washer pump can be removed after unlocking the retaining clip with a screwdriver (courtesy of Chevrolet Motor Division).

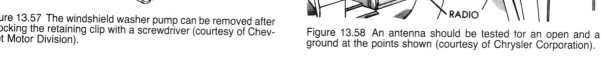

Figure 13.58 An antenna should be tested for an open and a ground at the points shown (courtesy of Chrysler Corporation).

fluid, avoid operating the pump when the reservoir is empty.

In those systems where the pump and motor assembly is mounted in an assembly with the wiper motor, the pump can be replaced as follows:

1 Using a small screwdriver as shown in Figure 13.57, release the clip retaining the pump.

2 Pull the pump assembly from the cover.

3 Clean the pump chamber in the cover.

4 Insert the replacement pump in the chamber.

Note: Be sure that the pump is fully seated so that the internal terminals make proper contact.

5 Secure the retaining clip.

6 Check the operation of the pump.

RADIOS   Radio repair is not usually considered part of the work of an automotive electrician, but many problems blamed on the radio have external causes. An inoperative radio, intermittent operation, distorted sound, poor sensitivity, and excessive noise or static often are caused by external problems that can be diagnosed and repaired without removing the radio.

Troubleshooting and Repair   Before a radio is removed for repair, several checks should be made to determine if the radio is at fault.

Inoperative Radio   An inoperative radio is often caused by a blown fuse. If the radio fuse is found

to be blown, or defective, it should be replaced after both the ignition switch and the radio switch are turned off. If the replacement fuse blows when the ignition switch is turned on and before the radio is turned on, the wire from the fuse to the radio is probably grounded and should be repaired or replaced. If the fuse blows when both the ignition switch and the radio is turned on, the problem probably is in the radio. If the original fuse is not blown, or if the replacement fuse does not blow and the radio is still inoperative, the problem may be in the antenna.

An open or a grounded antenna lead-in wire also will cause a radio to be inoperative. Intermittent operation may be caused by a poor antenna connection or a break in the wire. Connecting a substitute antenna known to be good provides a quick method of checking for antenna problems.

A mast-type antenna and its lead-in can be checked with a ohmmeter or a continuity tester. Continuity should exist between the tip of the antenna mast and the pin connector on the end of the lead-in cable. Very high resistance or no continuity should be found between the mast and ground. Figure 13.58 shows the test points. A defective antenna should be replaced.

Some cars do not use a mast antenna, but have antenna wires in the windshield. Windshield antennas can be tested with an antenna tester of the type shown in Figure 13.59. The tester is a small signal generator and is used in the following manner:

Job 13e

TEST AND REPAIR WINDSHIELD WIPER AND WASHER SYSTEMS

SATISFACTORY PERFORMANCE

A satisfactory performance on this job requires that you do the following:

1 Test the parts in wiper and washer systems and perform the necessary repairs.

2 Following the steps in the "Performance Outline" and the procedures and specifications of the manufacturer, complete the job within 200 percent of the manufacturer's suggested time.

3 Fill in the blanks under "Information."

PERFORMANCE OUTLINE

1 Check the operation of the wiper and washer circuits.

2 Locate the causes of any circuit problems.

3 Repair or replace any parts necessary to obtain circuit operation.

INFORMATION

Vehicle identification _____

Reference used _____ Page(s) _____

Defect(s) found _____

_____

Parts replaced and repairs performed _____

_____

---

1 Check that the antenna lead-in is connected to the radio.

2 Turn on the radio and tune the radio to a dead area between stations on the AM band.

3 Turn on the tester and hold it over the end of the antenna wire at one corner of the windshield. The speaker will emit a whistle if the wire has continuity.

4 Repeat step 3 at the opposite corner of the windshield.

The speaker should whistle when the tester is used on both antenna wires. The location of a break or open in the wires can be detected by sliding the tester along the wires until the speaker whistles. A windshield antenna cannot be repaired, and the windshield should be replaced. To avoid the high cost of windshield replacement, many car owners elect to have a mast-type antenna installed.

A defective speaker, a defect in the speaker wiring, or a disconnected speaker can also result in a silent radio. If a speaker is found to be disconnected, the radio may have to be removed for repair. Operating a radio with the speaker disconnected may cause damage to internal parts. A speaker known to be good can be connected to the radio for another test by substitution. If the radio operates with the substitute speaker, the original speaker should be replaced.

Distorted or "Fuzzy" Sound The quality of the sound can also be adversely affected by a defective speaker. Substitution of a temporary speaker will enable you to eliminate that possibility.

Poor Sensitivity Many radios have an externally adjustable *antenna trimmer*. An antenna trimmer is a variable capacitor used to "trim" or "peak" the radio and antenna combination for the greatest

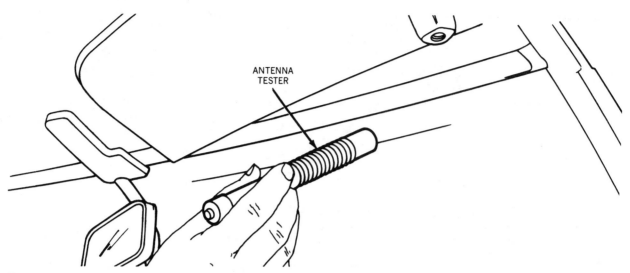

Figure 13.59 A windshield antenna can be tested with a small signal generator made for that purpose (courtesy of Chevrolet Motor Division).

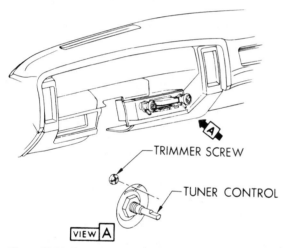

Figure 13.60 On some cars, the antenna trimmer access hole is located behind the tuning knob (courtesy of Chevrolet Motor Division).

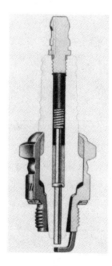

Figure 13.61 Resistor type spark plugs are often used to reduce radio interference. The resistor forms a part of the center electrode (courtesy of Champion Spark Plug Company).

sensitivity. An improperly adjusted trimmer may result in poor reception of distant stations. In most instances, the trimmer is accessible through a small hole in the radio case, and can be turned with a small screwdriver. On some radios, the access hole is located near the antenna wire socket. On other radios, the access hole is located behind the tuning knob, as shown in Figure 13.60.

An antenna trimmer can be adjusted or "peaked" in the following manner:

1 Raise the antenna to a height of approximately 30 inches.
2 Turn on the radio and turn up the volume. Tune in a weak station near 1400 KC on the AM band.
3 Using a small screwdriver, turn the trimmer screw slowly back and forth until the loudest volume is obtained.

Excessive Noise or Static An improperly adjusted antenna trimmer may also allow noise and static to become more obvious. Any check for the cause of excessive noise or static should start with an antenna trimmer adjustment.

In most instances, noise or static is caused by external interference. Many electrical circuits or components in a car transmit signals that are received by the radio through its wiring or antenna system. Those signals must be suppressed if static is to be minimized.

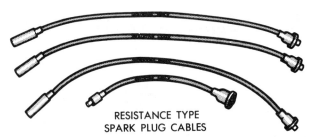

RESISTANCE TYPE
SPARK PLUG CABLES

Figure 13.62 Resistance type spark plug wires are used to suppress the static created by the ignition system. Those wires are usually marked TVRS to indicate Television Radio Suppression (courtesy of Chrysler Corporation).

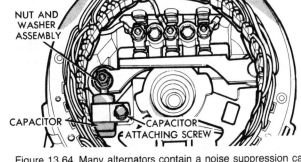

Figure 13.64 Many alternators contain a noise suppression capacitor connected internally between the output stud and ground (courtesy of Chrysler Corporation).

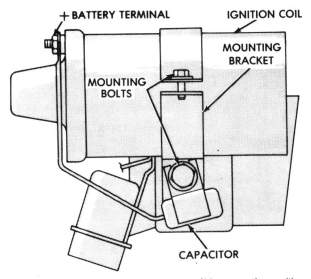

Figure 13.63 A capacitor is often mounted between the positive coil terminal and ground to aid the elimination of static caused by the ignition system (courtesy of Chrysler Corporation).

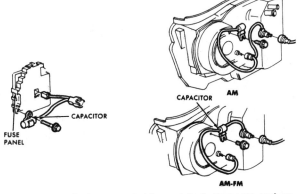

Figure 13.65 Static generated by certain instruments and accessories is often suppressed by capacitors (courtesy of Chevrolet Motor Division).

Each time a spark plug fires, a radio frequency signal is generated and transmitted. Those signals are received by the radio and are heard as static. To suppress those signals, all car makers install resistance type spark plugs and spark plug wires in the secondary circuit of their ignition systems. (See Figures 13.61 and 13.62.) If the plugs or wires are defective, if the ends of the wires are not properly installed, or if those parts have been replaced with aftermarket nonresistance types, excessive static usually will be experienced.

Diagnosis consists of checking for static both with the engine running and with the engine turned off. Secondary circuit static will be present only when the engine is running and increases in frequency as the engine is accelerated. If the static is found to be caused by the secondary circuit, the spark plugs and wires should be inspected and tested. Procedures for testing and replacing those parts are provided in Chapter 14.

The primary circuit of an ignition system also can produce static. This interference usually is suppressed by a capacitor or condenser connected between the positive (+) coil terminal and ground, as shown in Figure 13.63.

An alternator also will emit a signal that often is heard through the radio as a whine or whistle that increases in pitch as the engine is accelerated. This static often is suppressed by an externally mounted capacitor connected between the output terminal and ground. Many alternators are now fitted with an internally mounted capacitor to minimize the possibility of radio noise. (See Figure 13.64.)

Static also is generated by the windshield wiper motor, by the make-and-break action of switches, and even by the operation of the instrument voltage regulator. Capacitors often are installed across the terminals of those devices to minimize radio noise. Figures 13.65 and 13.66 show some typical installations.

In many instances, static generated by engine accessories can be minimized by establishing a good

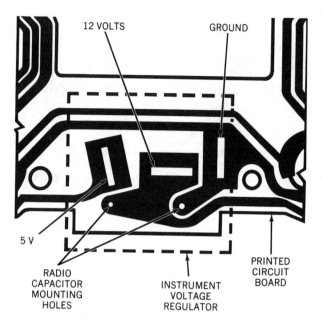

12 VOLTS   GROUND

5 V

RADIO
CAPACITOR
MOUNTING
HOLES

INSTRUMENT
VOLTAGE
REGULATOR

PRINTED
CIRCUIT
BOARD

Figure 13.66 The rear view of a section of an instrument panel printed circuit board showing how one manufacturer provides for mounting a capacitor on the IVR (courtesy of Chrysler Corporation).

HOOD GROUND
CLIP

Figure 13.68 On many cars, a ground clip is mounted on the cowl to insure that the hood is grounded when closed (courtesy of Chevrolet Motor Division).

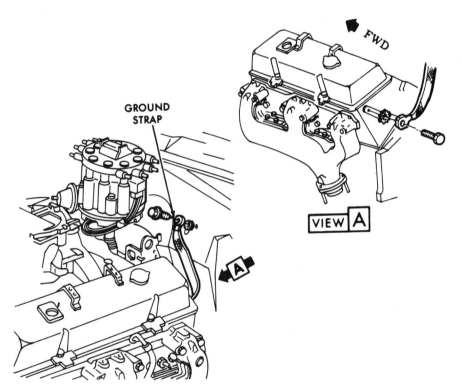

GROUND
STRAP

FWD

VIEW A

Figure 13.67 A ground strap connected between the engine and the firewall often eliminates certain types of static (courtesy of Chevrolet Motor Division).

ground connection between the engine and the car body. To ensure a good ground, some car makers install a ground strap between the engine and the firewall, as shown in Figure 13.67. If that ground strap is disconnected or missing, it should be repaired or replaced.

The metal hood of a car provides a shield against

the transmission of certain signals, but only if the hood is grounded to the body. As the hood hinges do not always provide a good ground, a ground clip or spring often is mounted on the cowl, as shown in Figure 13.68. The sharp edges of the ground clip contact the bottom of the hood when it is closed, providing a positive ground connection.

Job 13f

DIAGNOSE EXTERNALLY CAUSED RADIO PROBLEMS

SATISFACTORY PERFORMANCE
A satisfactory performance on this job requires that you do the following:

1 Determine problems in a car radio system, identify the cause(s) of the problems, and indicate the repairs required.
2 Following the steps in the "Performance Outline," complete the job within 30 minutes.
3 Fill in the blanks under "Information."

PERFORMANCE OUTLINE
1 Identify the problem(s).
2 Perform the checks and tests necessary to determine the cause(s) of the problem(s).

INFORMATION
Vehicle identification _____

Reference used _____ Page(s) _____

Problem(s) found _____

_____

Cause(s) _____

_____

Suggested repairs _____

_____

SUMMARY

In this chapter, you were introduced to a few of the more commonly used accessory circuits. By studying the material presented, you became aware of how some of those circuits operate. By completing the jobs provided, you further developed your diagnostic and repair skills.

SELF-TEST

Each incomplete statement or question in this test is followed by four suggested completions or answers. In each case select the *one* that best completes the sentence or answers the question.

1 Some car manufacturers specify that their horns be adjusted while
   A. an ammeter is connected in series with the horn
   B. a voltmeter is connected in series with the horn
   C. an ammeter is connected in parallel with the horn

D. a voltmeter is connected in parallel with the horn

2 Many vehicles are equipped with an electric radiator fan controlled by a relay. The relay is controlled by
   A. a time-delay switch
   B. the heater control switch
   C. a switch sensitive to coolant pressure
   D. a switch sensitive to coolant temperature

3 The coolant used in an engine cooling system should consist of approximately
   A. 30 percent water and 70 percent permanent antifreeze

B. 40 percent water and 60 percent permanent antifreeze

C. 50 percent water and 50 percent permanent antifreeze

D. 60 percent water and 40 percent permanent antifreeze

4 Current flow through the grid lines of a rear window defogger is usually controlled by a
A. bus bar
B. relay timer
C. thermal switch
D. variable resistor

5 A squeal or a squeak in a heater blower motor that varies with motor speed is usually caused by
A. worn brushes
B. loose field poles
C. dry or worn bearings
D. excessive current flow

6 A heater blower motor operates only when the switch is placed in the high speed position. The most probable cause of this problem is a defective
A. fuse
B. switch
C. blower motor
D. resistor assembly

7 The resistor assembly used in the heater blower circuit usually is mounted on the
A. firewall
B. field frame
C. motor housing
D. instrument panel

8 The speed of a wiper motor with permanent magnet field poles usually is controlled by routing current through
A. a relay
B. resistors
C. a magnetic clutch
D. different brushes

9 Poor reception by a car radio of distant stations can be caused by
A. a defective IVR
B. defective spark plug wires
C. an improperly grounded engine
D. an improperly adjusted antenna trimmer

10 Coarse static that is present only when the engine is running and that changes with engine speed usually is caused by
A. the alternator
B. a defective antenna
C. a defective speaker
D. the ignition system

Chapter 14
The Ignition System— The Secondary Circuit

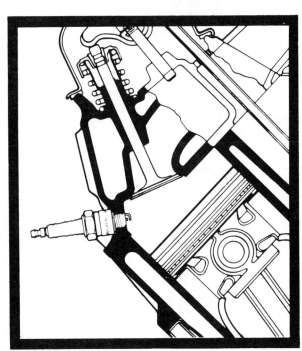

The ignition system supplies the spark that ignites the fuel-air mixture in the engine's cylinders. Most ignition systems consist of two main circuits. The primary circuit boosts battery voltage to the high voltage required and controls the timing of the delivery of that high voltage. The secondary circuit distributes the high voltage to the spark plugs.

In cars of current production, the primary circuit consists of electronic circuitry and contains electronic components that differ even among cars built by the same manufacturer. On most cars, however, the secondary circuit components are similar.

This chapter provides an introduction to basic ignition systems and covers the most commonly performed secondary circuit services. Your specific objectives are to perform the following jobs:

A
Identify the parts in a basic ignition system
B
Identify the function of ignition system parts
C
Identify spark plug operating conditions
D
Replace spark plugs
E
Recondition used spark plugs
F
Inspect and test spark plug wires
G
Replace a distributor cap

THE IGNITION SYSTEM   There are many different ignition systems in use. So many, in fact, that it is beyond the scope of this book to attempt to include all of them. All of those systems, however, operate on the same principle. Battery voltage is boosted to high voltage that may at times exceed 30,000 V. That high voltage is conducted, at the proper time, to a spark plug in each of the engine's cylinders.

A spark plug, shown in Figure 14.1, contains a pair of electrodes that are spaced or *gapped* so that the high voltage will push current across the gap, causing a spark. Most ignition systems contain two circuits. A *primary circuit* handles battery voltage. A *secondary circuit* handles the high voltage produced to jump the gap at the spark plugs.

The Primary Circuit   The primary circuit actually is the control circuit of the ignition system. Two basic types of primary circuits are used in automotive ignition systems. Most older cars have a primary circuit that uses a set of mechanically

Figure 14.1 A typical spark plug. A spark occurs when high voltage jumps the gap between the two electrodes (courtesy of Chevrolet Motor Division).

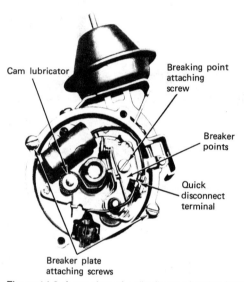

Figure 14.2 A top view of a distributor showing how the breaker points are installed (courtesy of Chevrolet Motor Division).

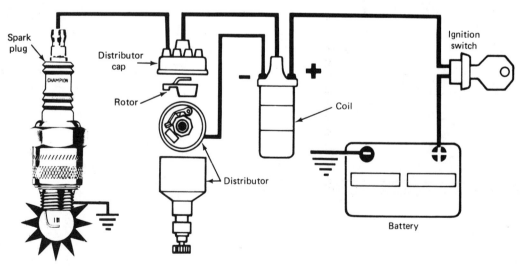

Figure 14.3 A pictorial diagram of a basic breaker point ignition system (courtesy of Champion Spark Plug Company).

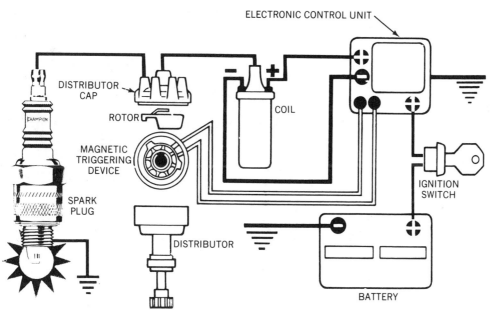

Figure 14.4 A pictorial diagram of a basic electronic ignition system (courtesy of Champion Spark Plug Company).

operated *breaker points.* (See Figure 14.2.) Cars in current production do not use breaker points. Electronic components eliminate the need for those mechanically operated parts.

A primary circuit using breaker points consists of the battery, the ignition switch, the *ignition coil,* a set of breaker points, and a *condenser.* Those parts are connected as shown in Figure 14.3.

A primary circuit using electronic components consists of the battery, the ignition switch, the ignition coil, the *electronic control unit,* or *module,* and some type of *magnetic triggering device.* Those parts are connected as shown in Figure 14.4.

Primary Circuit Components

The Battery   The battery supplies the energy to operate the system during starting and during those times when the charging system is not producing sufficient current.

The Ignition Switch   The ignition switch enables the driver to turn the system, and thus the engine, on and off.

The Ignition Coil   The ignition coil boosts the battery voltage to the high voltage needed to jump the gap at the spark plugs. Since the coil handles both low and high voltage, it is actually a part of both the primary and secondary circuits.

A typical coil, shown in Figure 14.5, consists of two separate windings of insulated wire. Those wind-

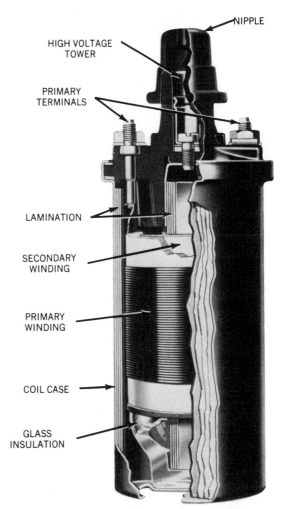

Figure 14.5 A typical ignition coil. Note the inner, secondary winding, the outer, primary winding, and the laminated core and case (courtesy of Chevrolet Motor Division).

TO THE
DISTRIBUTOR
CAP

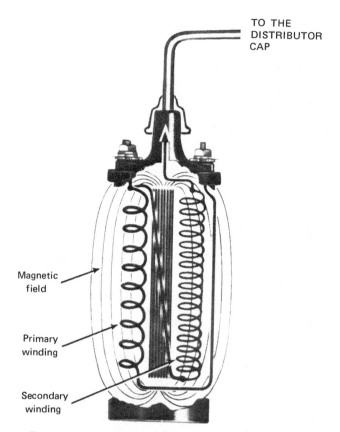

Magnetic field

Primary winding

Secondary winding

Figure 14.6 When current flows in the primary winding of a coil, a magnetic field is created. When the current flow is cut off, the magnetic field collapses, inducing high voltage in the secondary winding (courtesy of Chevrolet Motor Division).

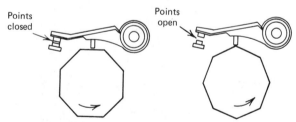

Points closed

Points open

Figure 14.7 The breaker points in a distributor are closed and opened by the action of a rotating cam. The cam is driven at one half engine speed.

ings are wrapped around a laminated iron core. The *primary winding* is part of the primary circuit and consists of about 100 turns of heavy gauge wire. The *secondary winding* is part of the secondary circuit and consists of several thousand turns of very fine wire. When current from the battery flows through the primary winding, a strong magnetic field is produced. The magnetic field is aligned and reinforced by the iron core and the iron case surrounding the coil. (See Figure 14.6.) When the flow of current through the primary winding is interrupted,

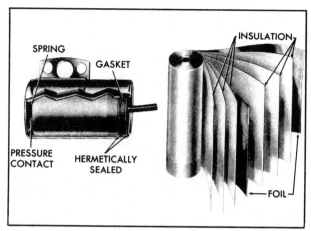

SPRING

GASKET

INSULATION

PRESSURE CONTACT

HERMETICALLY SEALED

FOIL

Figure 14.8 A typical condenser. A condenser is connected across the points to minimize arcing when the points open (courtesy of Chevrolet Motor Division).

the magnetic field collapses. When the magnetic field collapses, a surge of voltage is *induced,* or generated in the secondary winding. Due to the difference in the number of turns in the two windings, the induced voltage is very high.

The Breaker Points The breaker points, commonly called *points,* are a pair of switch contacts located in the distributor. By means of a rotating cam, those contacts are brought together and then separated, as shown in Figure 14.7. This action takes place when the engine is running and when it is being cranked by the starter motor. The alternate closing and opening of the points turns the primary circuit on and off, repeatedly building and collapsing the magnetic field in the coil.

The Condenser The condenser, shown in Figure 14.8, is usually located in the distributor or attached to its side. The condenser acts as an electrical "shock absorber" in the primary circuit. As shown in Figure 14.9, it provides an alternate path for the flow of current when the points start to open. In performing that function, the condenser reduces arcing at the points, extending their life. In addition, the action of the condenser helps to induce a higher voltage in the secondary coil winding.

The Electronic Control Unit In a breakerless system, the electronic control unit, or *module,* takes the place of the points. That unit, shown in Figures 14.10, 14.11, and 14.12, uses a transistorized circuit to turn the primary circuit on and off. Since the module has no moving parts, it does not require

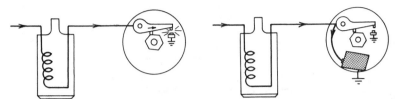

Figure 14.9 The action of a condenser. When the points open, the condenser acts to provide an alternate path for current flowing in the primary circuit. This minimizes arcing at the points.

Figure 14.10 A typical electronic control unit used in some electronic ignition systems. Units of this design are usually mounted on an inner fender panel or on the firewall and are connected to the distributor by a wire harness (courtesy of American Motors).

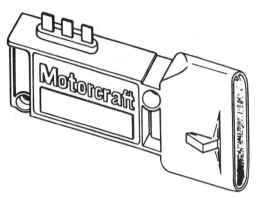

Figure 14.12 Some electronic ignition systems use a module of this type attached to the outside of the distributor (courtesy of Ford Motor Company).

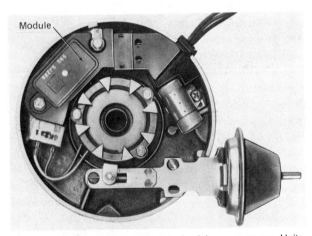

Module

Figure 14.11 A small electronic control unit in common use. Units of this type are usually called modules and are mounted inside the distributor (courtesy of Pontiac Motor Division, General Motors Corporation).

Figure 14.13 A typical sensor mounted on a distributor breaker plate (courtesy of American Motors).

adjustment or replacement because of wear. The electronic control unit is operated by a signal from a magnetic triggering device.

The Magnetic Triggering Device   The magnetic triggering device usually consists of some type of *pickup assembly,* or *sensor,* and an *armature,* or *trigger wheel.* A typical sensor and trigger wheel are shown in Figures 14.13 and 14.14. The sensor is mounted in the distributor and detects changes

in a magnetic field caused by the movement of the trigger wheel. The trigger wheel is rotated by the distributor shaft as is the cam in a breaker point system. The operation of one type of magnetic triggering device is shown in Figures 14.15, 14.16, and 14.17.

The Secondary Circuit   The secondary circuit in both breaker point systems and electronic systems is the same. The function of the secondary circuit is to distribute the high voltage surges produced by the coil to each cylinder, where it can jump the gap at the spark plug. The secondary cir-

Figure 14.14 A typical armature or trigger wheel. The trigger wheel is mounted on the distributor shaft (courtesy of American Motors).

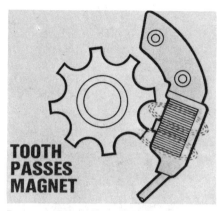

Figure 14.17 The third stage in the operation of a magnetic triggering device. As the tooth passes the magnet in the sensor, the magnetic field moves back to its normal position (courtesy of American Motors).

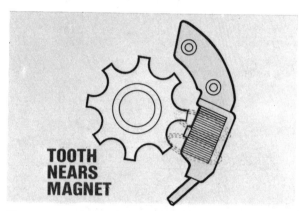

Figure 14.15 The first stage in the operation of a magnetic triggering device. As a trigger wheel tooth approaches the magnet in the sensor, the magnetic field shifts toward the tooth (courtesy of American Motors).

Figure 14.18 Secondary circuit components of a six-cylinder engine. Shown is the distributor cap, rotor, spark plugs, and high tension wires (courtesy of American Motors).

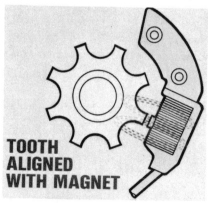

Figure 14.16 The second stage in the operation of a magnetic triggering device. When the tooth is aligned with the magnet in the sensor, the magnetic field has shifted to its maximum (courtesy of American Motors).

cuit consists of the ignition coil, the *distributor cap,* the *rotor,* the spark plugs, and the secondary wires or *high tension wires* connecting those parts. Figure 14.18 shows those components.

Secondary Circuit Components

The Ignition Coil   As mentioned previously, the coil is shared by both the primary and the secondary circuits. The high voltage induced in the coil secondary winding flows out through the tower at the top of the coil. (Refer to Figures 14.5 and 14.6.) The high voltage surges are carried from the coil to the center tower of the distributor cap by a high tension wire.

The Distributor Cap   The distributor cap is made of plastic or other insulating material. In addition to the center tower, a distributor cap has other towers or terminals spaced around its circumference. One tower or terminal is provided for each of the engine's cylinders. Inside the cap, at the base of each of those towers, is a metal post as shown in Figure 14.19. Those posts, one at a time, receive a surge of high voltage from the center tower by means of a rotating contact called a rotor.

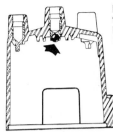

Figure 14.19 A cutaway view of a distributor cap showing the terminals at the base of the side towers and the carbon button at the base of the center tower (courtesy of AC-Delco, General Motors Corporation).

Figure 14.20 Typical distributor rotors used in breaker point ignition systems (courtesy of AC-Delco, General Motors Corporation).

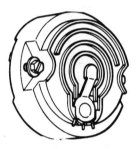

Figure 14.21 A multi-point rotor used in an electronic distributor (courtesy of Ford Motor Company).

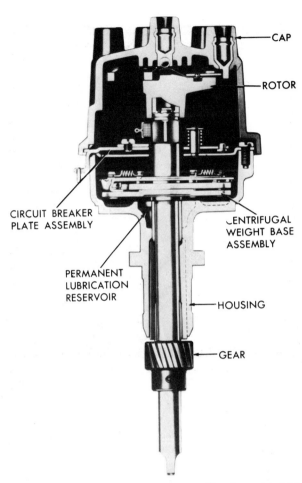

Figure 14.22 A sectioned view of a typical breaker point distributor. Note the position of the rotor and the terminals in the distributor cap (courtesy of Chevrolet Motor Division).

The Rotor The rotor, usually made of plastic, is mounted on the distributor shaft and turns with it. Some typical rotors are shown in Figures 14.20 and 14.21. A metal strip or spring on the rotor contacts a *button* or *brush* at the base of the center tower of the distributor cap. (Refer to Figure 14.19.) As the distriburor shaft rotates, the tip of the rotor passes from one outer terminal to the next. This allows the high voltage surge to flow to each terminal in turn. Thus, in one turn of the distributor shaft, high voltage is distributed to all the outer terminals. The relationship of the rotor and the distributor cap is shown in Figure 14.22.

The Spark Plugs While there are many types of spark plugs, they all serve the same purpose. A spark plug provides the gap, or air space, across which the high voltage produced by the coil can jump. A spark occurs when the high voltage jumps the gap. Spark plugs are threaded so that they can be screwed into the cylinder head(s) of an engine. This places the gap inside the combustion chamber as shown in Figure 14.23.

The High Tension Wires Since the secondary circuit may handle in excess of 30,000 V, special high tension wires must be used. Those wires, commonly called *spark plug wires*, must have extra insulation to prevent any leakage of that high voltage. Factory-installed wires have a nonmetallic conductor. Some replacement spark plug wires have a metal conductor. The construction of those two types of spark-plug wires is shown in Figure 14.24. To provide good connections at the distributor cap and at the spark plugs, special terminals are used at the ends of the wires. Some of those terminals are shown in Figure 14.25.

SPARK PLUGS Of all the parts in an ignition system, the spark plugs usually require the most service and the most frequent replacement. This is understandable when you consider the conditions under which they operate.

When the average car has traveled 10,000 mi (16,000 km), each spark plug in the engine has fired about 15 to 20 million times. Each time a plug fires, the

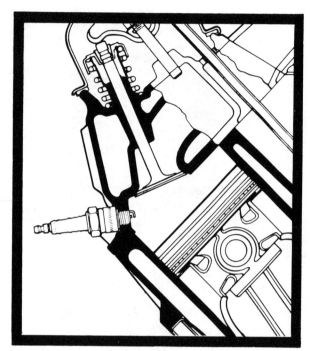

Figure 14.23 The position of the spark plug inside a typical combustion chamber (courtesy of Champion Spark Plug Company).

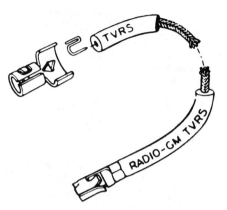

Figure 14.25 Typical terminals used with high tension wire. Note that a wire insert is used to obtain a good electrical connection between the wire core and the terminal (courtesy of AC-Delco, General Motors Corporation).

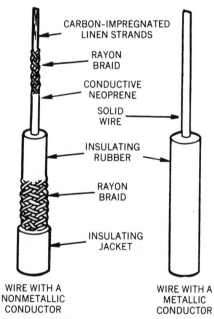

CARBON-IMPREGNATED
LINEN STRANDS

RAYON
BRAID

CONDUCTIVE
NEOPRENE

SOLID
WIRE

INSULATING
RUBBER

RAYON
BRAID

INSULATING
JACKET

WIRE WITH A
NONMETALLIC
CONDUCTOR

WIRE WITH A
METALLIC
CONDUCTOR

Figure 14.24 Construction of the two types of high tension wire used in the secondary circuit. Note that the conductor has a very small cross-section compared to the cross-section of the insulation (courtesy of AC-Delco, General Motors Corporation).

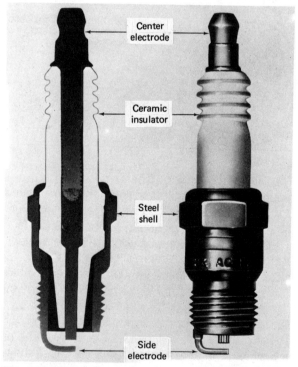

Center
electrode

Ceramic
insulator

Steel
shell

Side
electrode

Figure 14.26 The construction of a typical spark plug (courtesy of Chevrolet Motor Division).

spark is pushed across the gap by a voltage surge that can exceed 30,000 V. During the combustion of the fuel-air mixture, the plug is exposed to temperatures of over 5,000°F (2760°C) and subjected to pressures that can exceed 700 psi (4226 kPa).

The skills required for spark plug service are easily developed, but the application of those skills must be based on an understanding of spark plug construction and operation.

Spark Plug Construction   As shown in Figure 14.26, a spark plug consists of a *center electrode* enclosed in a ceramic *insulator* held by a steel *shell*. The center electrode provides one side of the gap that the spark must jump. The remaining side of the

Job 14a

IDENTIFY THE PARTS IN A BASIC IGNITION SYSTEM

SATISFACTORY PERFORMANCE
A satisfactory performance on this job requires that you do the following:

1 Identify the numbered parts on the drawing by placing the number of each part in front of the correct part name listed below.
2 Correctly identify all of the parts within 10 minutes.

PERFORMANCE SITUATION

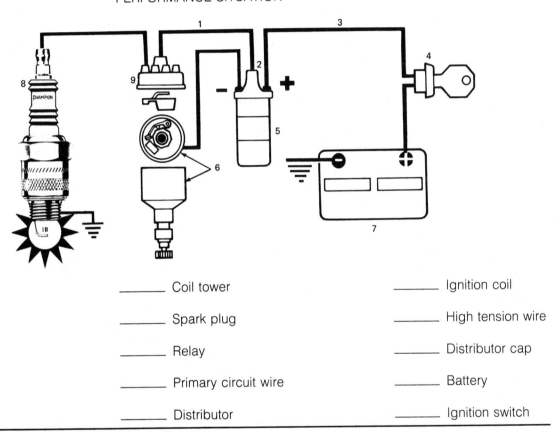

_____ Coil tower	_____ Ignition coil
_____ Spark plug	_____ High tension wire
_____ Relay	_____ Distributor cap
_____ Primary circuit wire	_____ Battery
_____ Distributor	_____ Ignition switch

gap is formed by a *side electrode,* or ground electrode, that is welded to the shell.

The lower part of the shell is threaded to fit a threaded hole in the cylinder head. The upper part of the shell is hexagonal, allowing you to use a socket to remove and install the plug.

Spark-Plug Design Differences While most spark plugs appear similar, they have many design differences. Some of those differences are not apparent. Although a certain spark plug may appear to fit a particular engine, its use may cause poor engine performance and even serious engine damage.

Thread Size Thread size, or thread diameter, is the most obvious design difference. Most car manufacturers have adopted two sizes as standard. As shown in Figure 14.27, those sizes are 14 mm and 18 mm. Spark plugs with other thread sizes are available, but they rarely are used in automobile engines.

Seat Types When a spark plug is installed in an engine, it must provide a perfect seal against the pressures of compression and combustion. This seal also must allow heat to flow from the plug to the cylinder head. A tight, heat-conducting seal is provided between the spark plug seat and a matching

Job 14b

IDENTIFY THE FUNCTION OF IGNITION SYSTEM PARTS

SATISFACTORY PERFORMANCE
A satisfactory performance on this job requires that you do the following:

1 Identify the function of the ignition system parts listed below by placing the part number in the space provided in front of the correct part function.
2 Correctly identify the function of all the parts within 15 minutes.

PERFORMANCE SITUATION

1 Ignition coil	6 Spark plug
2 Electronic control unit	7 Primary wires
3 Breaker points	8 Armature and sensor
4 High tension wires	9 Ignition switch
5 Condenser	10 Battery

_____ Provides current for the ignition system while the engine is cranking

_____ Conducts high voltage from the distributor cap to the spark plugs

_____ Generates high voltage through induction

_____ Minimizes arcing at the points

_____ Conducts battery voltage through the system

_____ Allows the driver to turn the system on and off

_____ Makes and breaks the primary circuit in a breakerless ignition system

_____ Generates high voltage in the primary circuit

_____ Makes and breaks the primary circuit in an ignition system using breaker points

_____ Provides a gap for the high voltage to jump

_____ Triggers the electronic control unit

seat that is machined in the cylinder head. In some engines, gaskets made of soft metal are used between the seats. In other engines, no gaskets are used.

Two commonly used seat designs are shown in Figure 14.28. On plugs that use gaskets, the seat is flat. The gasket is compressed between the seat on the plug and a matching flat seat in the cylinder head. On plugs that do not use gaskets, the seat is tapered. A tapered seat spark plug is designed to wedge into a similar taper formed in the cylinder head. Since the seats in all cylinder heads are not the same, the selection of plugs with the wrong seat will allow leakage around the spark plug threads and will restrict heat flow.

Reach   Reach, or thread length, is the distance between the seat and the end of the thread. Figure 14.29 shows some of the thread lengths in common use. If the reach of a plug is too short for a particular

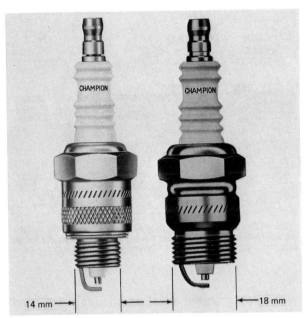

Figure 14.27 The two most commonly used spark plug thread sizes are 14 mm and 18 mm (courtesy of Champion Spark Plug Company).

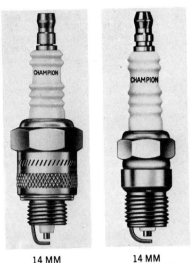

14 MM
FLAT SEAT

14 MM
TAPERED SEAT

Figure 14.28 Examples of flat and tapered seat plug designs. Notice that the plug with the flat seat is fitted with a gasket (courtesy of Champion Spark Plug Company).

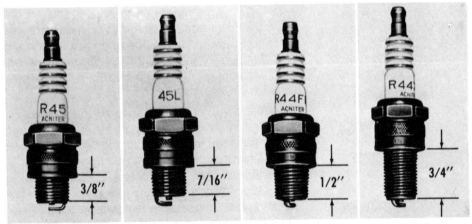

Figure 14.29 Different engines require plugs with different thread length or reach (courtesy of AC-Delco, General Motors Corporation).

engine, the electrodes will not be properly positioned in the combustion chamber. This condition is shown in Figure 14.30. Moreover, the exposed threads in the cylinder head will become filled with carbon and other deposits of combustion. Those deposits will make it difficult for you to install the correct plug at a later date.

If the reach of a spark plug is too long, as shown in Figure 14.31, the plug will extend too far into the combustion chamber. This can cause trouble in several ways:

1 The threads of the plug will become filled with combustion deposits. This will make it difficult to remove the plug.

COMBUSTION CHAMBER

Figure 14.30 A spark plug with a short reach installed in a cylinder head designed for a plug with a long reach (courtesy of Autolite Division, Allied Automotive).

Figure 14.31 A spark plug with a long reach installed in a cylinder head designed for a plug with a shorter reach (courtesy of Autolite Division, Allied Automotive).

COMBUSTION CHAMBER

Figure 14.32 A fouled spark plug (courtesy of Ford Motor Company).

Figure 14.33 Pre-ignition damage caused by the installation of a spark plug whose heat range was too high (courtesy of Ford Motor Company).

2 The exposed threads may glow when heated, causing the fuel-air mixture to ignite before the spark jumps the gap. This early firing is called *pre-ignition,* and can cause damage to the spark plug and to the engine.

3 The end of the plug may contact the valves or the piston when the engine is running. That contact can cause serious engine damage.

Heat Range Spark plug heat range is the range of temperature within which a spark plug normally operates. Heat range is determined by the plug's ability to dissipate heat. To provide good performance, the spark plugs in a particular engine must operate within a certain temperature range. If a plug's operating temperature is too low, the plug will become *fouled.* As shown in Figure 14.32, fouling is a build-up of oil, carbon, and other combus-

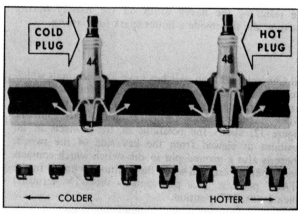

Figure 14.34 Examples of spark plugs with different heat ranges. The ability of a plug to dissipate heat is determined by the length of its insulator tip (courtesy of Chevrolet Motor Division).

tion deposits on the insulator tip and in the shell. Those deposits conduct electricity and allow secondary voltage to leak to ground. Since the spark does not jump the gap, the plug misfires and combustion does not occur.

If the operating temperature of a spark plug is too high, the electrodes wear rapidly. Under extreme conditions, pre-ignition, spark plug failure, and engine damage will result. Figure 14.33 shows a spark plug whose heat range was too hot for the engine in which it was installed.

The heat range of a plug is usually determined by the length of its lower insulator. As shown in Figure 14.34, the short insulator tip in the "cold" plug provides a short path for heat to flow from the insulator tip to the cylinder head, where it is dissipated in the coolant. Since the plug can pass off its heat rather rapidly, its operating temperature remains low. The "hot" plug has a long insulator that requires heat to travel a greater distance before it is dissipated. Thus, its operating temperature remains high.

Extended Tips On some spark plugs, the electrodes and the tip of the insulator are extended so that they project farther beyond the end of the shell. A spark plug of that design is shown in Figure 14.35. In some engines, that design extends the heat range of the plug. When the engine is running at low speed, the extended tip retains more heat and acts as a hotter plug. This prevents fouling, which is common when an engine is operated at low speeds. At higher speeds, when a colder plug is desirable, the extended tip is cooled by the flow of the incoming fuel-air mixture. Due to engine design differences, extended tip plugs cannot be used in all engines.

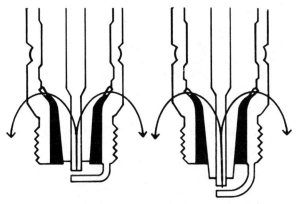

Figure 14.35 Conventional and extended tip design spark plugs. Note the heat flow paths (courtesy of Autolite Division, Allied Automotive).

CONVENTIONAL DESIGN   RESISTOR DESIGN

Figure 14.36 A comparison of a conventional spark plug and a resistor spark plug. The resistor forms a part of the center electrode (courtesy of Champion Spark Plug Company).

Resistor Types Some spark plugs have a built-in resistor. The resistor forms a part of the center electrode as shown in Figure 14.36. The resistor serves two purposes: (1) it reduces radio and television interference caused by the ignition system, and (2) it reduces electrode wear and thus extends spark-plug life.

Spark Plug Selection The correct spark plugs for a particular engine are determined by the engine manufacturer. Spark plugs are identified by code numbers and letters. The identifying code is a specification and can be found in owner's manuals, shop manuals, and the catalogs of spark plug manufacturers. Examples of some of those listings are shown in Figures 14.37 and 14.38.

MAKE-MODEL-YEAR-ENGINE	PLUG NUMBER		GAP
	RESISTOR	NON-RESISTOR	

CHEVROLET - Passenger Cars
3 CYLINDER

1985	1.0L	63	53	.035

4 CYLINDER

1985	1.5L	64	54	.035
	1.6L (C)	23	13	.035
	2.0L F/inj.	23	13	.035
	2.5L F/inj.	664	–	.060
1984	1.6L 2 bbl.	23	13	.035
	2.0L 2 bbl. or F/inj.	23	13	.035
	2.5L F/inj.	664	–	.060
1983	1.6L (2 bbl.)	23	13	.035
	2.0L (F/inj.)	23	13	.035
	2.5L (F/inj.)	665	–	.060
1982	1.6L (2 bbl.)	23	13	.035
	1.8L (2 bbl.)	23	13	.035
	2.5L (F/inj.)	665	–	.060
1981	98 C.I.D. (1.6L)	23	13	.035
	151 C.I.D. (2.5L)	665	–	.060
1980	98 C.I.D. (1.6L)	23	13	.035
	151 C.I.D. (2.5L)	665	–	.060

NOTE: Original equipment or equivalent plug shown in bold print.

Figure 14.37 Spark plug specifications for certain 3- and 4-cylinder engines used in Chevrolet passenger cars (courtesy of Autolite Division, Allied Automotive).

MAKE-MODEL-YEAR-ENGINE	PLUG NUMBER		GAP
	RESISTOR	NON-RESISTOR	

FORD (Includes Fiesta) - Passenger Cars
4 CYLINDER

1985	1.6L 2 bbl. and H.O.	2544	–	.044
	1.6L F/inj. (exc. Turbo)	2543	–	.044
	1.6L F/inj. (Turbo)	763	–	.044
	2.3L (O.H.C.)	2545	–	.044
	2.3L F/inj. H.S.C. (exc. H.O.)	766	–	.044
	2.3L F/inj. H.S.C. (H.O.)	764 or 104	–	.044
	2.3L F/inj. (Turbo)	764 or 104	–	.044
1984	1.6L 2 bbl. and H.O.	2544	–	.044
	1.6L F/inj.	2543	–	.044
	1.6L Turbo	763	–	.044
	2.3L (O.H.C.)	2545	–	.044
	2.3L (H.S.C.)	766	–	.044
	2.3L (Turbo)	764 or 104	–	.034
1983	1.6L (2 bbl. or H.O.) After 9/1/82	2544	–	.044
	Before 9/1/82	764 or 104	–	.044
	1.6L (F/inj.)	2543	–	.044
	2.3L (1 bbl.)	2545	–	.044
	2.3L (Turbo)	764 or 104	–	.034
	2.3L (H.S.C.)	766	–	.044
	2.3L (Propane)	764 or 104	–	.034
1982	1.6L (Escort)	764 or 104	–	.044
	1.6L (Escort) H.O.	764 or 104	–	.044
	1.6L (EXP) Early Prod.	3924	–	.044
	Late Prod.	764 or 104	–	.044
	H.O.	764 or 104	–	.044
	2.3L (Fairmont, Granada)	765/ 865[1]	–	.034
	(Mustang)	765/ 865[1]	–	.034
1981	98 C.I.D. (1.6L)	3924	–	.044
	140 C.I.D. (2.3L)	765/ 865[1]	–	.034
1980	97.6 C.I.D. (1.6L)	764 or 104	–	.034
	140 C.I.D. (2.3L) (exc. Turbo)	765/ 865[1]	–	.034
	Turbo	764 or 104	–	.034

NOTE: Original equipment or equivalent plug shown in bold print.

[1] Resistor equivalent to the original equipment suppressor spark plug.

Figure 14.38 Spark plug specifications for certain 4-cylinder engines used in Ford passenger cars (courtesy of Autolite Division, Allied Automotive).

PREFIX	SUFFIX

PREFIX

B —Series Gap
C —Commercial
CS—Chain Saw
G —Gas Engine
H —High altitude or weatherproof (shield
  connector, ¾-20 thread)
M —Marine
LM—Lawn mower type
R —Resistor (ACniter)
S —Shielded (⅝-24 thread)
S —Sport vehicle type
V —Surface gap
W —Waterproof (shield connector, ⅝-24
  thread)

SUFFIX

E —Engineer Corps. (Not an Aircraft type) Shielded.
F —½-inch reach with pilot design (14mm.) (Foreign)
FF —½-inch reach fully threaded (14mm.)
G —Pin gap
I —Iridium Electrode
K —Hi-Perf. Marine
L —Long reach (⁷⁄₁₆-inch for 14mm., ¾-inch for
  18mm.)
LTS—Long reach, taper seat, extended tip
M —Special center electrode
XL —Extra long reach (¾-inch for 14mm.)
N —Extra long reach (14mm.)(¾-inch reach with
  ⅜-inch thread length)
R —Resistor
P —Platinum Electrodes
S —Extended tip
S —(⅞-inch) Moderate long reach (²³⁄₃₂-inch)
T —Tapered engine seat
TS —Taper seat with extended tip
W —Recessed termination
X —Special gap
Y —3 prong cloverleaf electrode

**THREAD SIZE—FIRST DIGITS
IN THE AC TYPE NUMBER**

TYPE NUMBER STARTS WITH	THREAD SIZE
2	½-inch
4	14mm
5	½-inch
6	¾-inch
7	⅞-inch
8	18mm
10	10mm
12	12mm

**HEAT RANGE—LAST DIGIT
IN THE AC TYPE NUMBER**

The last digit of the AC type number is the heat range of the spark plug. Read these numbers like a thermometer . . . the higher the last number, the "hotter" the spark plug will operate, the lower the last number, the "cooler" the spark plug will operate. For example, an 8 indicates a hot spark plug, 6 medium hot, 4 medium cold, and 2 cold, etc.

Figure 14.39 The code system used to identify AC spark plugs (courtesy of AC-Delco, General Motors Corporation).

Spark-Plug Catalogs The catalogs distributed by the manufacturers of spark plugs are as important as the shop manuals that you use. Different plug manufacturers use different code systems and explain their system in their catalogs. As an example, one of those code systems is explained in Figure 14.39.

At times, the replacement plugs that you have available may be of a different brand than those specified by the engine manufacturer. In those instances, you must be able to convert from one code system to another. The catalogs contain conversion charts so that you can easily locate the correct identification code for the replacement plugs.

"Reading" Spark Plugs You occasionally will find that an engine requires spark plugs different than those specified by the manufacturer. The problem usually is one of incorrect heat range. The condition of the electrodes and of the insulators of the plugs that you remove from an engine provide a fairly accurate means of determining (1) the condition of the engine and (2) the type of service to which the engine is subjected. With a little practice, you can learn to "read" a spark plug and select the correct replacement.

Normal Appearance The spark plug shown in Figure 14.40 was removed from an engine that is in

Figure 14.40 A spark plug with normal deposits. The insulator tip is dry and is light tan or gray in color (courtesy of Ford Motor Company).

Figure 14.41 An oil fouled spark plug. Wet, oily carbon deposits cover the firing tip (courtesy of Ford Motor Company).

Figure 14.42 A carbon fouled spark plug. The firing tip is coated with dry, fluffy carbon deposits (courtesy of Ford Motor Company).

Figure 14.43 A spark plug that has been overheated. The insulator is white or light gray with random colored specks (courtesy of Ford Motor Company).

good condition and is in average service. The insulator tip has a light tan or gray color and the electrodes are not excessively burned. There are no heavy deposits present. The heat range of this plug obviously is correct for the engine from which it was removed. The replacement plug should be of the same heat range.

Oil Fouled Plugs Worn piston rings, cylinder walls, bearings, and valve guides may allow excessive amounts of oil to enter the combustion chambers

of an engine. The oil is not completely burned during combustion, and oily carbon deposits build up on the spark plug tip. In a short time, the plug becomes oil fouled and it misfires. A typical oil fouled plug is shown in Figure 14.41. An oil fouled plug usually indicates the need for major engine repairs. When it is not practical for those repairs to be made, the use of a hotter plug may help to burn off the oil fouling deposits before the plug misfires.

Carbon Fouled Plugs As shown in Figure 14.42, a carbon fouled plug has dry, fluffy carbon deposits on the insulator, shell, and electrodes. Such a condition may be caused by excessive idling or low speed operation. If that is the case, a hotter spark plug may be appropriate. Since the dry, fluffy carbon is caused by unburned fuel, you also should check for a plugged air filter and other faults that may cause the fuel-air mixture to be too rich.

Overheating A plug that is operating at too high a temperature will exhibit a white insulator tip containing random colored specks. The electrodes may appear burned. A plug that has been overheated is shown in Figure 14.43. An overheated plug usually indicates excessive combustion chamber temperatures and may indicate that the plug is too hot. A plug that shows overheating should be replaced with a colder plug, but the engine and its accessories should be checked for other causes of high combustion temperatures.

Figure 14.44 provides additional illustrations of spark plugs that have been removed from various engines. Each illustration depicts a different engine operating condition. By studying the illustrations you can become skilled at reading spark plugs.

SPARK PLUG SERVICES As an automotive electrician, you may be required to perform spark plug services. Those services are (1) the adjustment of spark plug gap, (2) the installation of new spark plugs, and (3) the reconditioning of used spark plugs.

Adjusting Spark Plug Gap The distance between the spark plug electrodes must be adjusted to a specification. Although the gap on all new plugs is preset by the plug manufacturer, the gap may not be correct for the engine on which you are working. The plugs also may have been dropped during handling and shipping, and the gap may have changed. Different gap measurements are required for different engines. The correct measurement may

CARBON FOULED

IDENTIFIED BY BLACK, DRY FLUFFY CARBON DEPOSITS ON INSULATOR TIPS, EXPOSED SHELL SURFACES AND ELECTRODES.
CAUSED BY TOO COLD A PLUG, WEAK IGNITION, DIRTY AIR CLEANER, DEFECTIVE FUEL PUMP, TOO RICH A FUEL MIXTURE, IMPROPERLY OPERATING HEAT RISER OR EXCESS IDLING.
CAN BE CLEANED.

GAP BRIDGED

IDENTIFIED BY DEPOSIT BUILD-UP CLOSING GAP BETWEEN ELECTRODES.
CAUSED BY OIL, OR CARBON FOULING. IF DEPOSITS ARE NOT EXCESSIVE, THE PLUG CAN BE CLEANED.

OIL FOULED

IDENTIFIED BY WET BLACK DEP-OSITS ON THE INSULATOR SHELL BORE ELECTRODES. CAUSED BY EXCESSIVE OIL ENTERING COMBUS-TION CHAMBER THROUGH WORN RINGS AND PISTONS, EXCESSIVE CLEARANCE BETWEEN VALVE GUIDES AND STEMS, OR WORN OR LOOSE BEARINGS, CAN BE CLEANED. IF ENGINE IS NOT REPAIRED, USE A HOTTER PLUG.

LEAD FOULED

IDENTIFIED BY DARK GRAY, BLACK, YELLOW OR TAN DEPOSITS OR A FUSED GLAZE COATING ON THE INSULATOR TIP.
CAUSED BY HIGHLY LEADED GASOLINE, CAN BE CLEANED.

NORMAL

IDENTIFIED BY LIGHT TAN OR GRAY DEPOSITS ON THE FIRING TIP.
CAN BE CLEANED.

WORN

IDENTIFIED BY SEVERELY ERODED OR WORN ELECTRODES.
CAUSED BY NORMAL WEAR.
SHOULD BE REPLACED.

FUSED SPOT DEPOSIT

IDENTIFIED BY MELTED OR SPOTTY DEPOSITS RESEMBLING BUBBLES OR BLISTERS.
CAUSED BY SUDDEN ACCELERATION. CAN BE CLEANED.

OVERHEATING

IDENTIFIED BY A WHITE OR LIGHT GRAY INSULATOR WITH SMALL BLACK OR GRAY BROWN SPOTS AND WITH BLUISH-BURNT APPEAR-ANCE OF ELECTRODES, CAUSED BY ENGINE OVERHEATING, WRONG TYPE OF FUEL, LOOSE SPARK PLUGS, TOO HOT A PLUG, LOW FUEL PUMP PRESSURE OR INCOR-RECT IGNITION TIMING. REPLACE THE PLUG.

PRE-IGNITION

IDENTIFIED BY MELTED ELECTRODES AND POSSIBLY BLISTERED INSULATOR. METALLIC DEPOSITS ON INSULATOR INDICATE ENGINE DAMAGE.
CAUSED BY WRONG TYPE OF FUEL, INCORRECT IGNITION TIMING OR ADVANCE, TOO HOT A PLUG, BURNT VALVES OR ENGINE OVERHEATING.
REPLACE THE PLUG.

Figure 14.44 A diagnostic chart for "reading" spark plugs (cour-tesy of Ford Motor Company).

Job 14c

IDENTIFY SPARK PLUG OPERATING CONDITIONS
SATISFACTORY PERFORMANCE
A satisfactory performance on this job requires that you do the following:

1 Identify the operating conditions of the spark plugs shown below by placing the number of each plug in front of the description of its condition.
2 Identify correctly four of the five conditions indicated within 5 minutes.
PERFORMANCE SITUATION

 Plug #1   Wet, black deposits on the insulator, shell interior, and electrodes

 Plug #2   Black, dry, fluffy deposits on the insulator, shell interior, and electrodes

 Plug #3   Melted electrodes, blistered insulator

 Plug #4   White or light gray insulator with small black or gray-brown spots

 Plug #5   Light tan or gray deposits on the insulator tip

_____ Overheating

_____ Ceramic fusing

_____ Carbon fouled

_____ Pre-ignition

_____ Normal appearance

_____ Oil fouled

Figure 14.45 Checking the gap on a spark plug (courtesy of American Motors).

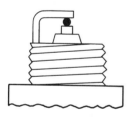

Figure 14.46 A corrrectly adjusted gap, the gauge wire slides between the electrodes and touches both of them.

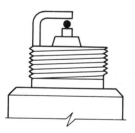

Figure 14.47 A gap that is too wide. The gauge wire does not contact both electrodes.

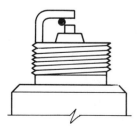

Figure 14.48 A gap that is too narrow. The gauge wire does not fit between the electrodes.

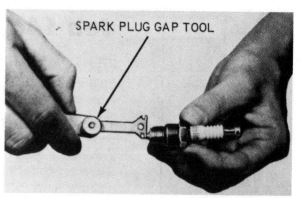

Figure 14.49 Adjusting the gap of a spark plug by bending the side electrode (courtesy of Ford Motor Company).

range from 0.025 inch (0.635 mm) to 0.080 inch (2.03 mm). It is important that you refer to an appropriate shop manual or a spark plug catalog for the gap measurement specified for the engine on which you are working. To ensure the quality of your work, you always should check the gap of a new plug and adjust it if adjustment is required.

Spark plug gap is easily measured by using a wire *feeler gauge* or a *spark plug gap gauge* of the type shown in Figure 14.45. A wire of the correct gap measurement is passed between the electrodes. If the wire passes through with a slight drag, as shown in Figure 14.46, the gap is correct.

If, as shown in Figure 14.47, the wire passes through loosely, without touching both electrodes, the gap is too large. If the wire will not pass between the electrodes, as shown in Figure 14.48, the gap is too small.

The gap of a spark plug is adjusted by bending the side electrode. This is easily done with the bending tool attached to the gauge. The use of that tool is shown in Figure 14.49.

Replacing Spark Plugs   Spark plugs probably are the most frequently replaced parts in an engine. The accessibility of the spark plugs will vary from car to car, but the procedure that you follow will be the same. The following steps outline the procedure, but you should refer to an appropriate manual for the specifications required for the car on which you are working.

Removal

1 Grasp the rubber boot at the end of the spark plug wire and twist it as shown in Figure 14.50. This will loosen the boot on the plug. Continue twisting the boot and pull it off the spark plug.

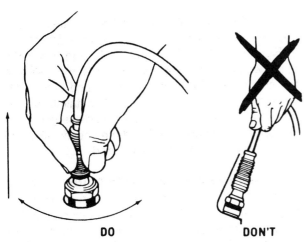

DO   **DON'T**

Figure 14.50 A spark plug wire should be removed from a spark plug by twisting and pulling on the boot. Never attempt to remove a wire by pulling on the wire (courtesy of AC-Delco, General Motors Corporation).

Note: Special pliers such as those shown in Figure 14.51 can be used for this step. If you attempt to remove the boot from the plug by pulling on the wire, you may pull the wire from the boot or cause the wire to break within the insulation.

2 Label the wire with a piece of masking tape numbered to the cylinder number, or position the wire so that you can reinstall it on the correct plug.

3 Remove and label or position the remaining wires in the same manner.

4 Using a spark plug socket, loosen all of the spark plugs about one full turn.

5 Using a compressed air blowgun, blow away all dirt from around the base of the plug. This will minimize the possibility of dirt entering the engine when the plugs are removed.

Note: Be sure to wear your safety glasses during this step.

6 Using a spark plug socket, remove all the plugs and arrange them in the order of their position in the engine. The use of a rack similar to the one shown in Figure 14.52 is suggested.

Note: By arranging the plugs in order, you can determine which cylinder(s) may be causing problems if one or more of the plugs have an abnormal reading.

7 Read the spark plugs to determine if they are of the correct heat range and if engine defects are indicated.

Installation The replacement spark plugs should be of the type specified by the engine manufacturer. If, by reading the old plugs, you have deter-

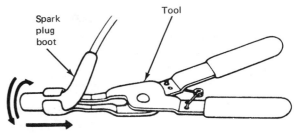

Spark plug boot   Tool

TWIST & PULL

Figure 14.51 Spark plug boots can be removed from spark plugs without damage to the wire by the use of special pliers (courtesy of Ford Motor Company).

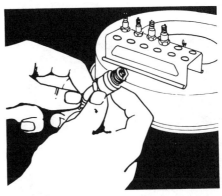

Figure 14.52 When removing spark plugs from an engine, you should place them in a tray or rack so that you can identify any cylinders that may have trouble (courtesy of AC-Delco, General Motors Corporation).

mined that hotter or colder plugs are required, consult an appropriate spark plug catalog to find the identification code of the correct plug. It is suggested that you change heat range by only one or two steps or numbers. Too great a change may cause poor performance or possible engine damage.

1 Check to be sure that the gap of all of the replacement plugs meets the engine manufacturer's specifications. Adjust the gaps as necessary. (Refer to Figure 14.49.)

2 Thread all of the spark plugs into place using finger pressure only.

Note: Do not use a wrench for this step. If access to the spark plug holes is difficult, a boot from a discarded spark plug wire can be used as an installation tool. The use of a wrench or ratchet handle to install spark plugs may result in your stripping the threads on the plugs and in the spark plug holes.

3 After all of the plugs have been threaded into their holes, tighten the plugs to the torque specification of the engine manufacturer.

Note: When the use of a torque wrench proves difficult, you can apply the proper torque by an alternate method shown in Figure 14.53.

4 Install the wires on the plugs, making sure that they are installed in their original locations.

Note: Spark plug wires must be installed to conform to the *firing order* of the engine. The firing order specifies the sequence in which the plugs fire. This specification can be found in an appropriate manual. Typical firing orders and wire locations are shown in Figures 14.54, 14.55, and 14.56. On some engines with electronic ignition systems, the inside of the spark plug boots should be coated with a silicone dielectric compound prior to installation. The engine manufacturer's manual will advise you as to the need for this operation and the specifications of the compound that may be required.

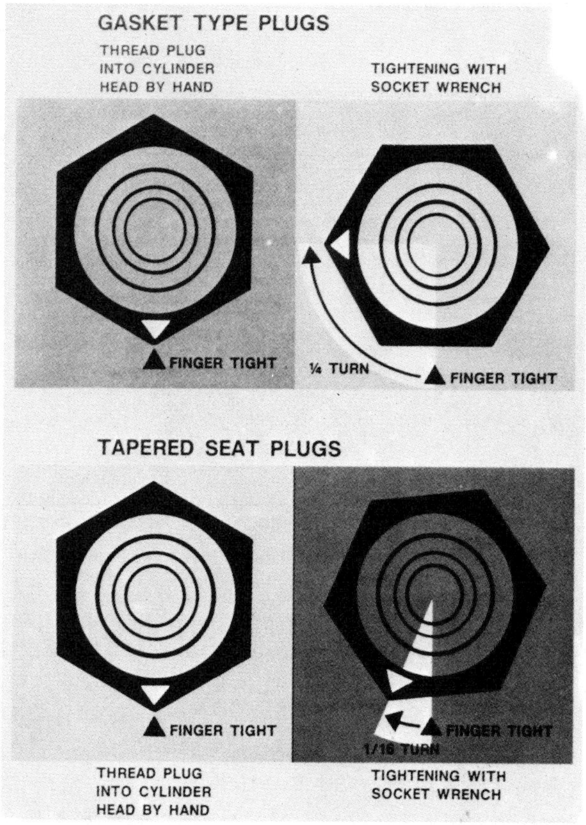

Reconditioning Used Spark Plugs   In some instances, used spark plugs can be reconditioned and reinstalled for further use. Spark plugs used in engines with breaker point ignition systems have an average life of 10,000 miles (16,000 km). When used in engines with certain electronic ignition systems, the efficient life of a spark plug may exceed 20,000 miles (32,000 km). Within those mileage limits you may find plugs that are dirty or fouled, especially in engines using leaded fuel. It usually is possible to recondition those plugs and return them to service.

Regardless of the mileage, the condition of the electrodes usually determines if a plug can be reconditioned successfully. As shown in Figure 14.57,

GASKET TYPE PLUGS

THREAD PLUG
INTO CYLINDER
HEAD BY HAND

TIGHTENING WITH
SOCKET WRENCH

▲ FINGER TIGHT   ¼ TURN   ▲ FINGER TIGHT

TAPERED SEAT PLUGS

▲ FINGER TIGHT   1/16 TURN   ▲ FINGER TIGHT

THREAD PLUG
INTO CYLINDER
HEAD BY HAND

TIGHTENING WITH
SOCKET WRENCH

Figure 14.53 Suggested methods of tightening spark plugs when a torque wrench cannot be used (courtesy of Champion Spark Plug Company).

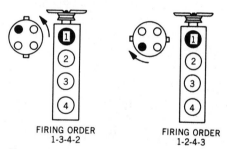

FIRING ORDER
1-3-4-2

FIRING ORDER
1-2-4-3

Figure 14.54 While these two four cylinder engines may appear similar, the firing order, the direction of distributor rotation, and the location of #1 wire is different.

Figure 14.55 While all these in-line six cylinder engines have the same firing order, the location of the #1 wire, the direction of distributor rotation, and even the numbering of the cylinders are different.

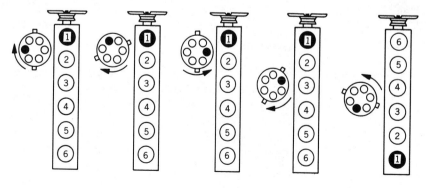

FIRING ORDER
1-5-3-6-2-4

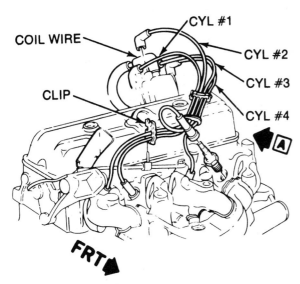

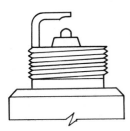

VIEW A

Figure 14.56 Spark plug wire routing on a typical four cylinder engine (courtesy of Chevrolet Motor Division).

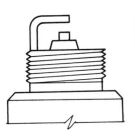

Figure 14.57 The electrodes of a spark plug should be square and have sharp corners.

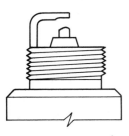

Figure 14.58 Normal electrode wear. In most instances, electrodes with this degree of wear can be restored by filing.

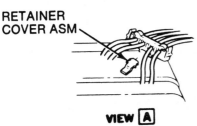

Figure 14.59 Excessive electrode wear. Electrodes worn to this degree cannot be restored by filing. The plug should be discarded.

Figure 14.60 A typical spark plug cleaner (courtesy of Champion Spark Plug Company).

the edges of the electrodes should be sharp and square. Electrode edges that are slightly rounded can be corrected by filing, but a plug that shows excessive electrode wear should be discarded. Those conditions are shown in Figures 14.58 and 14.59.

Fouling and other combustion deposits on the insulator tip, inside the shell, and on the electrodes can be removed by the use of a spark plug cleaner. A spark plug cleaner, shown in Figure 14.60, is a machine that directs a blast of abrasive sand or grit into the open end of a spark plug. The abrasive blast chips away the deposits on the plug surfaces.

Spark plugs that do not exhibit excessive electrode wear can be reconditioned by performing the following operations:

1 Clean the exterior surfaces of the plug.
2 Clean the interior surfaces of the plug.
3 File the electrodes.
4 Adjust the gap.
5 Clean the threads.

The procedures for performing those operations are outlined in the following steps:

1 Clean the exterior surface of the plug with a rag or a wiper dipped in solvent. All deposits should be removed from the insulator.

2 Examine the insulator for cracks. Discard any plug that has a cracked insulator.

Job 14d

REPLACE SPARK PLUGS

SATISFACTORY PERFORMANCE

A satisfactory performance on this job requires that you do the following:

1 Replace the spark plugs in the assigned engine.
2 Following the steps in the "Performance Outline" and the procedure and specifications of the manufacturer, complete the job within 200 percent of the manufacturer's suggested time.
3 Fill in the blanks under "Information."

PERFORMANCE OUTLINE

1 Disconnect the wires from the plugs.
2 Loosen the plugs and blow away the dirt from around the base of the plugs.
3 Remove the plugs.
4 Read the plugs.
5 Select the correct replacement plugs.
6 Check the gap on the replacement plugs and adjust to specifications if necessary.
7 Install the plugs.
8 Tighten the plugs to the manufacturer's torque specifications.
9 Install the wires.

INFORMATION

Vehicle identification _____

Engine identification _____

Reference used _____ Page(s) _____

Spark plug brand and identification code specified _____

Spark plug gap specification _____

Spark plug torque specification _____

Brand and identification of spark plugs removed _____

Spark plugs removed were

_____ Normal   _____ Too hot   _____ Too cold

Brand and identification code of spark plugs installed

---

3 Determine the thread size of the plug and select the correct adapter for the spark plug cleaner. (See Figure 14.61.)
4 Install the adapter in the spark plug cleaner.

5 Position the plug in the adapter as shown in Figure 14.62.

Note: Be sure to wear your safety glasses during the following operations. Most spark plug cleaners

Figure 14.61 Rubber adapters allow plugs with different thread sizes to be cleaned (courtesy of Champion Spark Plug Company).

Figure 14.62 Installing a spark plug in the adapter (courtesy of Champion Spark Plug Company).

Figure 14.64 Cleaning a spark plug. Tipping and rotating the plug allows the abrasive blast to reach all parts of the insulator and shell (courtesy of Champion Spark Plug Company).

Figure 14.63 A spark plug cleaner that has the control valve built into the safety cover (courtesy of AC-Delco, General Motors Corporation).

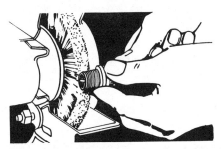

Figure 14.65 Cleaning the threads of a spark plug on a wire wheel (courtesy of AC-Delco, General Motors Corporation).

operate as sand blasters, and if the plug or the adapter should slip, the abrasive sand could be blown into your eyes.

6 Holding the plug in place in the adapter, move the cleaner control valve to the "Cleaning Blast" or "Abrasive Blast" position.

Note: Some spark plug cleaners of the type shown in Figure 14.63 have the control valve built into a safety cover.

7 Rotate the plug in the adapter and rock it from side to side while holding the valve in the "Cleaning Blast" position. (See Figure 14.64.)

8 After about 3 or 4 seconds of cleaning, move the control lever to the "Air Blast" position.

Note: This step is necessary to blow out any abrasive sand that may be lodged between the insulator and the shell.

9 Continue to rotate and rock the plug while holding the valve in the "Air Blast" position for about 5 seconds.

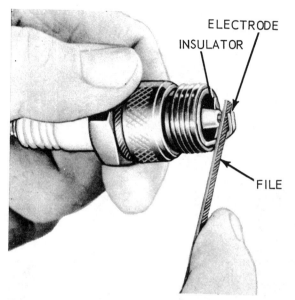

Figure 14.66 Cleaning and squaring spark plug electrodes with a flat file (courtesy of Ford Motor Company).

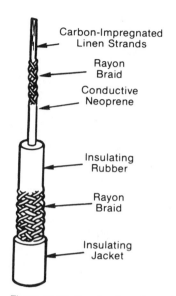

Figure 14.67 Construction of a typical resistance wire. Wires of this type are usually marked TVRS to indicate Television Radio Suppression (courtesy of AC-Delco, General Motors Corporation).

WIRE LENGTH	RESISTANCE (Ω)
6 to 15 in. (15 to 38 cm)	3000 to 10000
15 to 25 in. (38 to 63 cm)	4000 to 15000
25 to 35 in. (63 to 88 cm)	6000 to 20000
Over 35 in. (Over 88 cm)	8000 to 25000

Figure 14.68 Typical spark plug wire resistance specifications.

10 Release the control valve.

11 Remove the plug from the adapter.

12 Examine the nose of the plug. If all deposits have not been removed, repeat steps 5 through 12 until the nose of the plug is clean.

13 Inspect the insulator tip. Discard the plug if the tip of the insulator is cracked or chipped.

14 Clean the threads of the plug with a stiff wire brush or with a wire wheel as shown in Figure 14.65.

15 Using a small file, clean the firing surfaces of the electrodes as shown in Figure 14.66. It may be necessary to open the gap slightly to provide clearance for the file.

Note: This operation is necessary to flatten and square the electrode surfaces.

16 Adjust the gap to the manufacturer's specifications.

SPARK PLUG WIRES, DISTRIBUTOR CAPS, AND ROTORS

The spark plug wires, the distributor cap, and the rotor are responsible for delivering the high voltage to the spark plugs. Excessive resistance in those parts may prevent a spark from jumping the plug gap. Defects in the insulation of those parts may allow voltage to leak to ground before it reaches the plug. Problems in the secondary circuit may require that you inspect and test those parts and replace those found defective.

Spark Plug Wires

Most spark plug wires used as original equipment have a nonmetallic conduc-

tor. As shown in Figure 14.67, the core of those wires is usually made of carbon-impregnated linen strands. The carbon conducts electricity, but has a relatively high resistance. Those *resistance wires:* (1) *suppress,* or reduce, radio and television interference caused by the ignition system, and (2) reduce spark plug electrode wear. The core of a spark plug wire is easily broken, and can be pulled loose from its terminal. Therefore, spark plug wires should be handled carefully and should be removed from spark plugs only by twisting and pulling on the boots. (Refer to Figures 14.50 and 14.51.)

Spark plug wires must be able to handle up to 30,000 V while being subjected to almost continuous vibration. The wires also are exposed to extremes in temperature and are many times wet with fuel, oil, and water. Because of those severe service conditions, spark plug wires deteriorate and occasionally will require replacement. A spark plug wire may fail in several ways:

Job 14e

RECONDITION USED SPARK PLUGS

SATISFACTORY PERFORMANCE

A satisfactory performance on this job requires that you do the following:

1 Recondition the spark plugs assigned.
2 Following the steps in the "Performance Outline" and the specifications of the manufacturer, complete the job within 60 minutes.
3 Fill in the blanks under "Information."

PERFORMANCE OUTLINE

1 Remove the plugs from the engine.
2 Inspect the plugs.
3 Clean the plugs.
4 File the electrodes.
5 Adjust the gaps.
6 Install the plugs.

INFORMATION

Vehicle identification _____

Engine identification _____

Reference used _____ Page(s) _____

Spark plug brand and identification code _____

Spark plug gap specification _____

Spark plug torque specification _____

1 *The insulation may fail.* Heat may cause the wire insulation and the boots and nipples to dry and crack. Oil and fuel may cause the insulation to become porous. Any failure of the insulation can allow voltage to leak to ground and cause the spark plug to misfire.

2 *The resistance of the wire may increase.* Vibration and stress may cause the carbon particles in the core to separate slightly. That separation increases the resistance of the conductor and thus increases the voltage required to fire the plug. When the required voltage exceeds the voltage available from the coil, the plug will misfire.

3 *The continuity of the wire may be lost.* The conductive core may have an internal break or may have been pulled loose from one of its terminals. Current must then jump the gap in the wire as well as the gap of the plug. The break in the wire will burn increasingly larger until the available voltage

can no longer jump both gaps. The plug will then misfire.

The condition of the insulation can be checked visually. The resistance and continuity should be checked with an ohmmeter.

Inspecting and Testing Spark Plug Wires

The following steps outline a procedure for inspecting and testing the spark plug wires used on most cars. The specifications for wire resistance should be obtained from an appropriate manual, but usually fall into the ranges shown in Figure 14.68. When two or more wires on an engine fail to pass the inspection or tests, it is advisable to replace the entire set. The remaining wires have been subjected to equal service, and they most likely will fail within a short time.

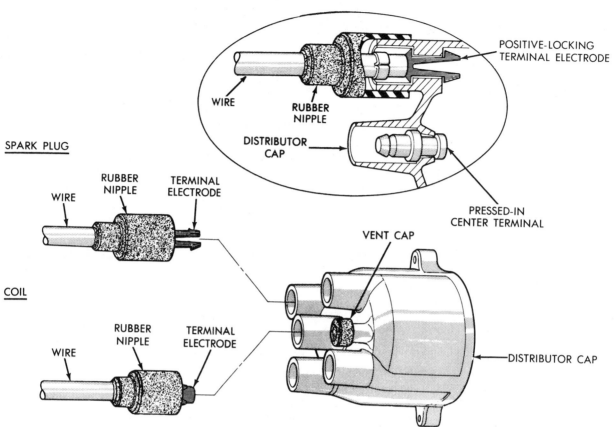

Figure 14.69 In some electronic ignition systems, the terminals on the spark plug wire lock into the cap and take the place of the terminals at the base of the towers (courtesy of Chrysler Corporation).

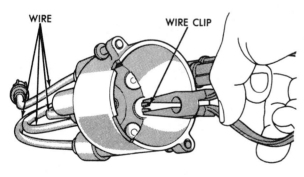

DISTRIBUTOR CAP

Figure 14.70 Spark plug wires that lock into the distributor cap can be released by squeezing the ends of the terminal electrodes with a pair of needle nose pliers (courtesy of Chrysler Corporation).

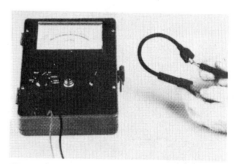

Figure 14.71 Using an ohmmeter to check the resistance of a spark plug wire (courtesy of American Motors).

1 Carefully remove one spark plug wire from a spark plug.

2 Remove the remaining end of the wire from its distributor cap tower.

Note: The manufacturers of some cars with elec-tronic ignition systems do not recommend removing the wires from the cap for testing. On those cars, the distributor cap can be removed to gain access to the inner tower terminals. (See Figures 14.69 and 14.70.) Be sure to check an appropriate manual for the procedure recommended for the car on which you are working.

3 If the wire is dirty or oil soaked, clean the in-

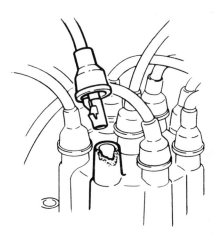

Figure 14.72 A distributor cap with an eroded or cracked tower should be replaced (courtesy of AC-Delco, General Motors Corporation).

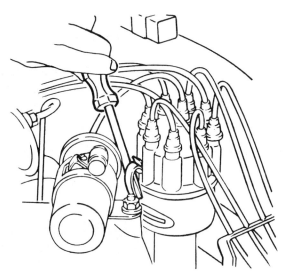

Figure 14.73 A distributor cap retained by spring clips. The clips can be released by prying with a screwdriver as shown (courtesy of AC-Delco, General Motors Corporation).

sulation with a rag or a wiper wet with cleaning solvent and dry the wire.

4 Examine the wire insulation, the boot, and the nipple for cracks, burned areas, abrasions, porosity, and other damage. Check the metal terminals at the ends of the wire. They should be clean and bright and show no corrosion.

5 Turn the selector switch on the ohmmeter to the "X 1000" scale and adjust the zero.

6 Connect the ohmmeter leads to the terminals on the wire as shown in Figure 14.71.

Note: It may be necessary to insert an adapter or a short piece of wire into the spark plug boot to contact the terminal at that end.

7 Read and record the resistance indicated by the meter.

8 Compare the resistance of the wire to the manufacturer's specifications.

Note: A reading of zero (0) indicates that the wire is not a resistance wire, but contains a metallic conductor. A reading of infinity (∞) indicates that the wire lacks continuity—that there is a break in the conductor.

9 If the wire was removed from the distributor cap, inspect the cap tower. The tower should not be cracked or eroded, as shown in Figure 14.72. The metal insert in the tower should be clean and have no corrosion.

Note: The use of a mirror and a droplight or flashlight will enable you to inspect the inside of a distributor cap tower when it is otherwise impossible to do so.

10 Slide the nipple up on the wire.

11 Insert the wire into the cap tower, pushing it

firmly down into the place until you feel the terminal bottom in the tower.

12 Slide the nipple down over the tower, squeezing it to expel any air that may be trapped inside.

Note: Some manufacturers recommend coating the inside of the nipple with a silicone dielectric compound prior to installation. Be sure to check the manual for this recommendation and the specification of the compound that may be required.

13 Install the remaining end of the wire on the spark plug, pushing the boot firmly over the insulator.

Note: Some manufacturers recommend coating the inside of the boot with a silicone dielectric compound.

14 Position the wire so that it will not contact the exhaust manifold and secure it in any guides or wire holders that may be present.

15 Repeat the previous steps on the remaining wires.

Note: It is advisable to remove, test, and install each wire one at a time. By following that procedure, you will maintain the correct firing order and wire positions.

Distributor Caps and Rotors   Distributor caps and rotors should be checked for cracks, erosion, corrosion, and damage. Cracks and erosion on those parts provide a path for voltage to leak to ground. Corrosion increases the resistance to current flow. Those defects usually can be located by a visual inspection.

Job 14f

INSPECT AND TEST SPARK PLUG WIRES

SATISFACTORY PERFORMANCE

A satisfactory performance on this job requires that you do the following:

1 Inspect and test the spark plug wires on the car assigned.
2 Following the steps in the "Performance Outline" and the procedure and specifications of the manufacturer, complete the job within 60 minutes.
3 Fill in the blanks under "Information."

PERFORMANCE OUTLINE

1 Disconnect one wire.
2 Clean and inspect the wire.
3 Test the wire for continuity and resistance.
4 Connect the wire.
5 Repeat the above steps on the remaining wires.

INFORMATION

Vehicle identification _____

Reference used _____ Page(s) _____

Wire resistance specifications _____

Test results:

Wire #	Insulation	Length	Continuity	Resistance
1	____OK ____NG	_____	____Yes ____No	_____
2	____OK ____NG	_____	____Yes ____No	_____
3	____OK ____NG	_____	____Yes ____No	_____
4	____OK ____NG	_____	____Yes ____No	_____
5	____OK ____NG	_____	____Yes ____No	_____
6	____OK ____NG	_____	____Yes ____No	_____
7	____OK ____NG	_____	____Yes ____No	_____
8	____OK ____NG	_____	____Yes ____No	_____

Is the resistance of all the wires within specifications?

_____ Yes _____ No

Which wires, if any, should be replaced? _____

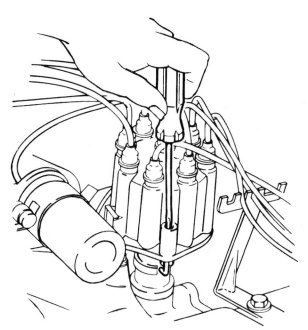

Figure 14.74 Some distributor caps are retained by spring loaded hooks. They are released by using a screwdriver (courtesy of AC-Delco, General Motors Corporation).

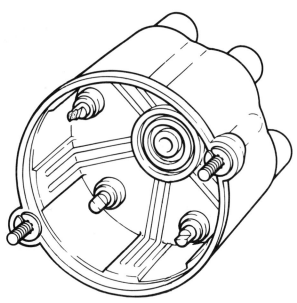

Figure 14.75 A distributor cap that is retained by screws (courtesy of American Motors).

Inspecting Distributor Caps A distributor cap may be cracked and eroded internally without any external sign of those defects. Therefore, you must remove the distributor cap from the distributor for a thorough inspection.

The following three methods are commonly used to hold a distributor cap in place on a distributor.

Spring Clips The most commonly used method of retaining a distributor cap is by spring clips, as shown in Figure 14.73. Hinged at the distributor, spring clips hook into projections on the side of the cap. They are easily removed by prying them loose with a screwdriver. When you wish to install the cap, the spring clips can be snapped back into position.

Spring Loaded Hooks As shown in Figure 14.74, some caps are fitted with spring loaded hooks. Those hooks engage in notches on the bottom of the distributor. Caps retained in that manner are removed by turning the slotted head of the hooks 180° with a screwdriver. When installing these caps, the hooks are pushed down with a screwdriver and turned so that they fit into their notches under the distributor.

Screws Some distributor caps are held to the distributor by screws. In most instances, screws are built into the cap so that they cannot be dropped and lost. A cap of that type is shown in Figure 14.75.

Figure 14.76 The outside of a distributor cap should be cleaned and checked for cracks and carbon paths (courtesy of Chevrolet Motor Division).

All distributor caps are made with a tab or key that aligns with a notch in the distributor. When installing a distributor cap, you should check the position of the cap before attempting to secure it. A properly positioned cap cannot be rotated on the distributor.

The following steps outline a procedure for inspecting a distributor cap that has been removed from a distributor. A cap that exhibits cracks, erosion, carbon paths, corrosion, or damage should be replaced.

1 Clean the outside surface of the cap, carefully examining it for cracks and erosion.

Note: Cracks and erosion usually are indicated by the presence of a carbon path, as shown in Figure 14.76. A carbon path is formed when high voltage leaks along a crack or flaw in the cap and erodes the cap surface.

2 Clean the inside of the cap with a wiper or with

Job 14g

REPLACE A DISTRIBUTOR CAP

SATISFACTORY PERFORMANCE
A satisfactory performance on this job requires that you do the following:

1 Replace the distributor cap on the car assigned.
2 Following the steps in the "Performance Outline" and the procedure and specifications of the manufacturer, complete the job within 200 percent of the manufacturer's suggested time.
3 Fill in the blanks under "Information."

PERFORMANCE OUTLINE
1 Remove and inspect the distributor cap.
2 Remove and inspect the rotor.
3 Install the rotor.
4 Install the distributor cap.
5 Check the firing order and the wire locations.

INFORMATION
Vehicle identification _____

Reference used _____ Page(s) _____

Results of distributor cap inspection:

_____ Defective   _____ Suitable for reuse

Defect found _____

Cap was secured by:

_____ Spring clips   _____ Spring loaded hooks   _____ Screws

Figure 14.77 The inside of a distributor cap should be cleaned and checked for cracks, carbon paths, and burnt or eroded terminals (courtesy of Chevrolet Motor Division).

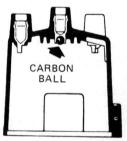 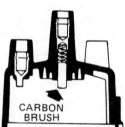

Figure 14.78 A distributor cap inspection should include a check of the carbon ball or brush under the center tower (courtesy of AC-Delco, General Motors Corporation).

a compressed air blowgun, as shown in Figure 14.77. Check for cracks, carbon paths, and erosion of the terminals.

Note: Be sure to wear your safety glasses when cleaning parts with compressed air.

3 Check the carbon ball or brush under the center tower.

Note: Where a carbon brush is used, it is backed by a small spring that holds the brush out in contact with the rotor. In caps of that type, the brush should

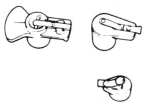

Figure 14.79 Examples of distributor rotors that slide over the distributor shaft. They are positioned by a flat or keyway on the shaft (courtesy of AC-Delco, General Motors Corporation).

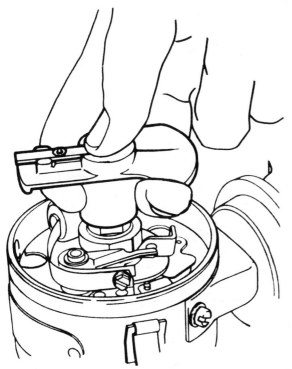

Figure 14.80 Removing a distributor rotor. If the rotor resists removal, a screwdriver can be used to carefully pry it off the distributor shaft (courtesy of AC-Delco, General Motors Corporation).

be checked for freedom of movement. (See Figure 14.78.)

Inspecting Rotors While the distributor cap is off, the rotor should be removed and inspected. Many different types of rotors are used. The rotors shown in Figure 14.79 are merely pushed on over the distributor shaft. They are positioned by a projection or key that fits into a matching flat or keyway on the shaft. Rotors of that type are removed by pulling them off the shaft as shown in Figure 14.80.

Rotors of the type shown in Figure 14.81 are held by two screws. Correct positioning is obtained by two protrusions molded into the bottom of the rotor.

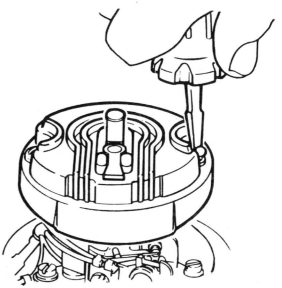

Figure 14.81 A rotor that is held to the distributor shaft by screws (courtesy of AC-Delco, General Motors Corporation).

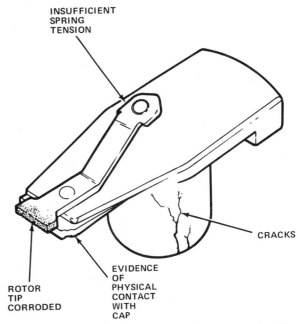

Figure 14.82 Possible rotor defects (courtesy of American Motors).

One protrusion is round; the other is square. The two protrusions fit into matching holes in a plate on the distributor shaft. After the screws are removed, the rotor can be lifted from the distributor.

When installing a rotor, be sure that it is correctly positioned and securely in place. Failure to do so may result in damage to the rotor and to the cap.

The rotor should be cleaned and inspected for

cracks, corrosion, and insufficient spring tension, as shown in Figure 14.82. The rotors used in electronic systems should be checked carefully for carbon tracks and carbon smudges on their underside. This indicates that current has been leaking through the rotor to the distributor shaft. Rotors that exhibit those faults should be replaced.

Replacing Distributor Caps   Replacing a distributor cap is an easy job, but each wire must be installed in its correct tower. To avoid mixing the wires, each wire should be removed from the old cap and immediately installed in the replacement cap. The old cap and the replacement cap should be aligned and held side-by-side as shown in Figure 14.83. If space permits, the new cap should be installed on the distributor. Then the wires should be transferred one at a time.

Figure 14.83 Replacing a distributor cap. The new cap should be aligned with the old cap and the wires transfered one at a time (courtesy of Chevrolet Motor Division).

SUMMARY

By completing this chapter, you learned some of the fundamentals of ignition systems. You are knowledgeable of the functions of the primary and secondary circuits. You learned to identify the parts in those circuits, and are aware of their operation. You have gained diagnostic skills in that you now can read spark plugs to determine engine operating conditions, and you can inspect and test certain secondary circuit components. You also have gained repair skills. You now can replace spark plugs, recondition used plugs, and replace spark plug wires, distributor caps, and rotors.

SELF-TEST

Each incomplete statement or question in this test is followed by four suggested completions or answers. In each case select the *one* that best completes the sentence or answers the question.

1 Two mechanics are discussing the operation of a breaker point ignition system.
   Mechanic A says that high voltage is induced in the coil secondary winding when the distributor points open.
   Mechanic B says that the condenser reduces arcing at the distributor points.
   Who is right?
   A. A only
   B. B only
   C. Both A and B
   D. Neither A nor B

2 Which of the following components is NOT a part of the primary circuit in a breaker point ignition system?
   A. Rotor
   B. Condenser
   C. Breaker points
   D. Ignition switch

3 Two mechanics are discussing ignition systems.
   Mechanic A says that the ignition coil is part of the primary circuit.
   Mechanic B says that the ignition coil is part of the secondary circuit.
   Who is right?
   A. A only
   B. B only
   C. Both A and B
   D. Neither A nor B

4 Two mechanics are discussing ignition systems.
   Mechanic A says that the breaker points in a distributor are opened by the rotor.
   Mechanic B says that the rotor is turned by the distributor shaft.
   Who is right?
   A. A only
   B. B only
   C. Both A and B
   D. Neither A nor B

5 The most commonly used spark plug thread sizes are
   A. 10 mm and 14 mm
   B. 14 mm and 18 mm
   C. 18 mm and 22 mm
   D. 22 mm and 26 mm

6 Spark plugs removed from an engine have dry, fluffy carbon deposits on the insulator tip, inside the shell, and on the electrodes. This condition could be caused by
   A. worn piston rings
   B. a plugged air cleaner
   C. excessive wire resistance
   D. excessive combustion chamber temperatures

7 Two mechanics are discussing resistance-type spark plug wires.
   Mechanic A says that the resistance wires are used to suppress radio and television interference caused by the ignition system.
   Mechanic B says that the resistance wires are used to reduce spark plug electrode wear.
   Who is right?
   A. A only
   B. B only
   C. Both A and B
   D. Neither A nor B

8 When tested with an ohmmeter, a spark plug wire is found to have a resistance of 0 Ω. The wire
   A. lacks continuity
   B. has a metallic core
   C. has excessive resistance
   D. has a carbon-impregnated linen core

9 When tested with an ohmmeter, a spark plug wire is found to have a resistance of infinity (∞). The wire
   A. lacks continuity
   B. has a metallic core
   C. has insufficient resistance
   D. has a carbon-impregnated linen core

10 Two mechanics are discussing distributor caps.
   Mechanic A says that some distributor caps are held to the distributor by spring loaded hooks.
   Mechanic B says that a carbon path on the surface of a distributor cap indicates that high voltage has been leaking.
   Who is right?
   A. A only
   B. B only
   C. Both A and B
   D. Neither A nor B

Chapter 15
The Ignition System— The Primary Circuit

In Chapter 14, you were introduced to the ignition system and worked with some of the parts that make up basic secondary circuits. In this chapter, you will learn about the primary circuit. While all of the different types of primary circuits used in electronic ignition systems cannot be covered in this book, the operation of a basic primary circuit and how spark timing is provided will be covered. To help you further develop your diagnostic and repair skills, some of the most commonly performed primary circuit services will be outlined.

Your specific objectives in this chapter are to perform the following jobs:

A
Identify terms relating to the primary circuit
B
Perform a primary circuit inspection
C
Measure dwell
D
Adjust dwell—external adjustment
E
Adjust dwell—internal adjustment
F
Identify the four strokes of the four-stroke cycle and related engine parts
G
Identify terms relating to spark timing
H
Check and adjust timing

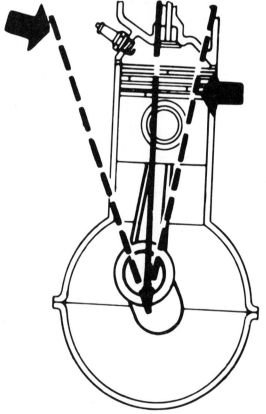

THE PRIMARY CIRCUIT As mentioned in the previous chapter, the primary circuit is actually the control circuit of the ignition system. As such, it performs the following functions:

1 It provides current to the primary winding in the coil so that a strong magnetic field is created.
2 It interrupts the flow of that current so that the magnetic field collapses.
3 It times the occurrence of the first two functions so that a high voltage surge is produced in the secondary circuit at the exact instant it is required.

To appreciate the performance of the primary circuit, you must consider the speed at which it operates. At cruising speed in a car with an eight-cylinder engine, the primary circuit must perform its functions about 200 times per second.

Primary Circuit Components and Their Function The primary circuits of the various electronic ignition systems are designed to require practically no maintenance. Each of those systems were designed to meet the requirements of the car manufacturer for a particular application. Thus, the components and circuitry differ greatly, even in cars built by the same maker. When service is required in the primary circuits of those systems, a detailed procedure established by the manufacturer must be followed. Those procedures are listed in appropriate factory manuals and are not within the scope of this book.

Systems using breaker points require the frequent inspection and adjustment of certain components. All of those systems use similar components and circuitry. Therefore, common procedures are followed in the maintenance of most of those systems. If you are to follow those procedures correctly, a knowledge of the components used in the primary circuit of breaker point systems is necessary.

The Breaker Points The breaker points, commonly referred to as points, form the contacts of a mechanically operated switch. A typical set of points is shown in Figure 15.1. As mentioned in Chapter 14, when the points are closed, or together, current flows in the primary circuit. When the points are open, or separated, the current flow is stopped. This action is shown in Figure 15.2. It is not enough that the points merely open and close. Three things must be considered if the points are to function properly: (1) point resistance, (2) dwell, and (3) point gap.

Point Resistance As you know, resistance in an electrical circuit will limit the amount of current that flows in the circuit. If the surfaces of the points are dirty, burned, or misaligned, they may provide a resistance to current flow. That resistance could limit the current available to the primary winding in the coil. If insufficient current flows in the primary winding, a weak magnetic field will be produced. The collapse of a weak magnetic field may not induce sufficient voltage in the secondary winding. Point resistance is easily measured. In most instances, points with excessive resistance should be replaced.

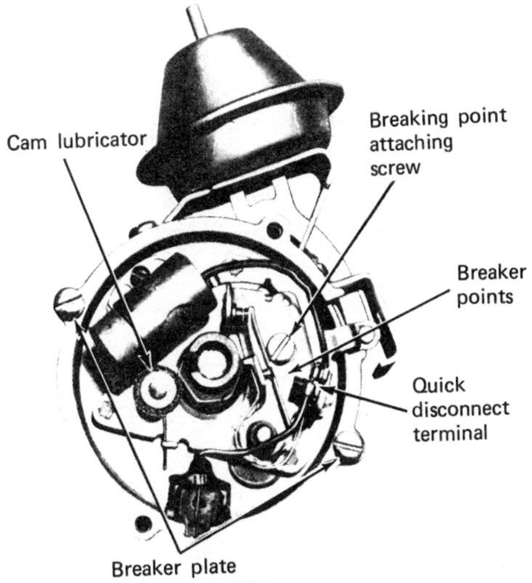

Cam lubricator

Breaking point attaching screw

Breaker points

Quick disconnect terminal

Breaker plate attaching screws

Figure 15.1 A typical set of points in a distributor. Note the location of the condenser (courtesy of Chevrolet Motor Division).

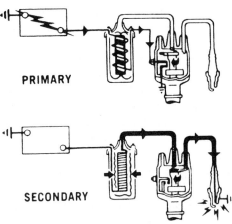

PRIMARY

SECONDARY

Figure 15.2 The operation of the breaker points in an ignition system. Current flows in the primary circuit when the points are closed. The current flow stops when the points are opened (courtesy of AC-Delco, General Motors Corporation).

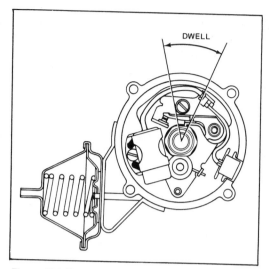

Figure 15.3 Dwell is the amount of time the points remain closed measured in degrees of cam rotation (courtesy of Chevrolet Motor Division).

Dwell   Dwell refers to the amount of time that the points remain closed. The points must remain closed for a certain amount of time to allow the primary winding to build a strong magnetic field. When the strongest possible magnetic field is produced in a coil, the coil is said to be *saturated*. The collapse of a saturated magnetic field induces the highest voltage in the secondary winding.

If the points do not remain closed for a sufficient amount of time, the secondary voltage will be low. If the points remain closed for too long a period of time, saturation of the coil will not be increased. However, excessive arcing and burning of the points usually will occur. This problem occurs because of the relationship between dwell and point gap.

Dwell, sometimes referred to as *cam angle*, is measured in degrees of distributor cam rotation as shown in Figure 15.3. An ignition system will operate properly only when the dwell is maintained within specifications. Dwell is easily measured and adjusted.

Point Gap   Point gap is the distance between the points when they are held open by the highest part of a cam lobe. (See Figure 15.4.) Point gap is measured with a feeler gauge in much the same manner that you measured the gap of spark plugs. Figure 15.5 illustrates the measurement of point gap. Point gap is adjustable. In most instances, the stationary point can be moved closer to, or farther away from, the movable point. As shown in Figure 15.6, point gap and dwell are related. Closing the gap of a set of points increases the dwell. Opening the gap decreases the dwell.

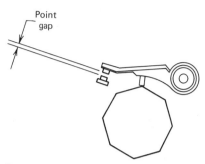

Figure 15.4 Point gap is the distance between the point contact surfaces when the points are held open by the highest point of a cam lobe.

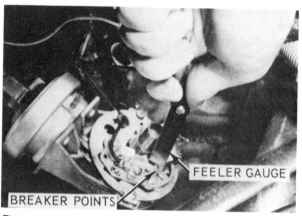

Figure 15.5 Measuring point gap with a flat feeler gauge (courtesy of Ford Motor Company).

In most instances, a set of points adjusted to the correct dwell specification will have the correct gap. This does not mean that point gap can be ignored. Both the dwell and the gap should be checked. If the gap is too small, current may jump the gap after the points open. Such current flow interferes with the collapse of the magnetic field in the coil, resulting in low secondary voltage. Current jumping the gap also causes arcing and burning of the points.

The Condenser   The flow of current in the primary circuit is sufficient to jump the small gap formed by the points as they start to open. Connected across the points as shown in Figure 15.7, the condenser provides an alternate path for the flow of that current.

A typical condenser, shown in Figure 15.8, is made of two strips of conductive foil separated by strips of insulation. The strips of foil form a large conductive area that will temporarily store electricity. The foil and the insulation are rolled and sealed in a small cylinder. The cylinder body contacts one foil strip and is grounded to the distributor. The other foil strip is connected to a wire that is attached to

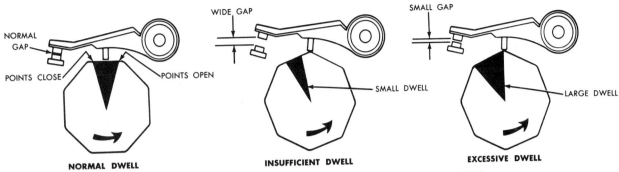

▼ = **DWELL ANGLE** (POINTS ARE CLOSED DURING THIS PERIOD OF CAM ROTATION)

Figure 15.6 Dwell and point gap are related. A wide gap results in insufficient dwell. A small gap results in excessive dwell (courtesy of Ford Motor Company).

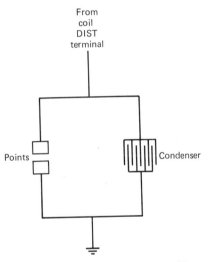

Figure 15.7 A condenser is connected in parallel with the points.

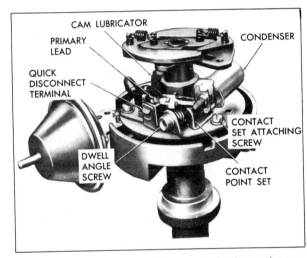

Figure 15.9 A distributor containing a set of points and a condenser combined in one assembly (courtesy of Chevrolet Motor Division).

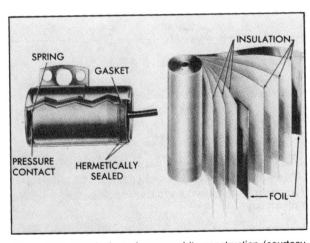

Figure 15.8 A typical condenser and its construction (courtesy of Chevrolet Motor Division).

the terminal for the movable point. Some condensers are not made as a separate part. They are combined with the points in one assembly as shown in Figure 15.9.

As mentioned in Chapter 14, the condenser in an ignition system performs two functions:

1 It reduces arcing at the points and thereby extends their life. It does this by providing an alternate path for the flow of current at the instant of point opening.

2 It aids in the collapse of the magnetic field in the coil, thereby increasing the voltage induced in the secondary winding.

PRIMARY CIRCUIT INSPECTION Primary circuit inspection includes those checks and tests that enable you to determine if the circuit is operating within the manufacturer's specifications. Based on the results of that inspection, you can

Job 15a

IDENTIFY TERMS RELATING TO THE PRIMARY CIRCUIT
SATISFACTORY PERFORMANCE
A satisfactory performance on this job requires that you do the following:

1 Identify terms relating to the primary circuit by placing the number of each listed term in front of its correct definition.
2 Correctly identify all the terms within 5 minutes.
PERFORMANCE SITUATION

1 Points
2 Condenser
3 Saturation
4 Induction

5 Point gap
6 Dwell
7 Arcing
8 Resistance

———— Sparks that occur when an electrical connection is broken

———— The amount of time that points remain open

———— The point where a magnetic field is at full strength

———— Contacts that control the flow of current in the primary circuit

———— The creation of current by means of magnetism

———— The amount of time that points remain closed

———— The distance between the points when they are at their widest opening

———— A device that minimizes arcing at the points

———— The opposition to the flow of current

———— The creation of magnetism by electricity

make the necessary adjustments and replace the required parts.

Visual Inspection A visual inspection may reveal obvious problems and should be performed before any electrical tests are made. The following steps outline a procedure that can be used on any breaker point system:

1 Check for dirty, corroded, and loose connections at the battery terminals. Clean the terminals and clamps and tighten the clamps if necessary.
2 Check for dirty and loose connections at the battery ground cable, the starter solenoid, and the coil. Clean and tighten those connections as required.
3 Check all of the connecting wires for broken or frayed insulation. Replace any damaged wires.
4 Remove the distributor cap and rotor.
5 Using a small screwdriver or a hook made from stiff wire, separate the points. Check the condition of the point surfaces.

Note: The surfaces of used points normally appear gray or "frosted." They also may exhibit slight pitting, or metal transfer. Points that are burned or exhibit excessive pitting should be replaced. (See Figures 15.10 and 15.11.)
6 Check the connection of the primary lead and the condenser lead at the points. Tighten the connection if it is loose.
7 Check the mounting of the condenser. Tighten the mounting screw if it is loose.

Figure 15.10 Burned points. Burned points can be caused by oil or dirt on the point surfaces, or by excessive current flow in the primary circuit (courtesy of Ford Motor Company).

Figure 15.11 Points with excessive pitting. Pitting, or metal transfer, can be caused by a condenser of the wrong capacity (courtesy of Ford Motor Company).

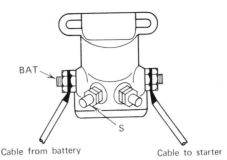

Cable from battery   Cable to starter

Figure 15.12 A starter solenoid of the type used by Ford Motor Company and by American Motors Corporation. Remote starter switch connections should be made at the BAT and S terminals.

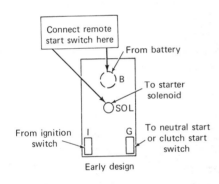

Early design

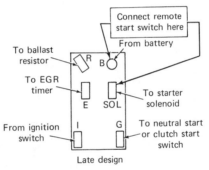

Late design

Figure 15.13 Relays of two different designs will be found on cars built by Chrysler Corporation. Remote starter switch connections should be made at the battery and solenoid terminals (courtesy of Chrysler Corporation).

Point Resistance   A test of point resistance will give you an accurate indication of the electrical condition of the point contact surfaces. Point resistance is checked while the points are closed. While the distributor cap and rotor are off, you should check the position of the points. If the points are not closed, you can "tap" or "bump" the engine over with the starter motor.

A remote starter switch will make it easier for you to crank the engine while you watch the points. Figures 15.12, 15.13, and 15.14 show how a remote starter switch may be connected.

As an antitheft feature, some cars are wired so that the use of a remote starter switch could damage certain parts of the ignition system. To eliminate the possibility of damage, the ignition switch should be placed in the On position before using a remote starter switch. As you have the distributor cap off, the engine will not start. If you use a remote starter switch with the cap on, the coil secondary wire should be removed from the center tower of the cap and grounded with a jumper wire.

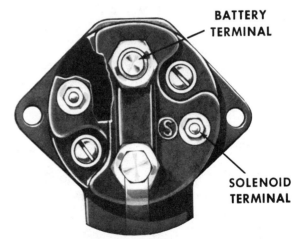

Figure 15.14 A front view of the starter solenoid used on most cars built by General Motors Corporation. When using a remote starter switch, the switch should be connected to the battery terminal and to the solenoid (S) terminal (courtesy of Chevrolet Motor Division).

Testing Point Resistance with a Dwell Meter   As its name implies, a dwell meter is an instrument that measures dwell. Most shops use an instrument that

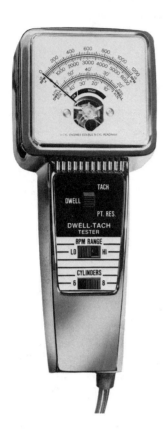

Figure 15.15 A typical hand-held tach-dwell meter (courtesy of Kal-Equip Company).

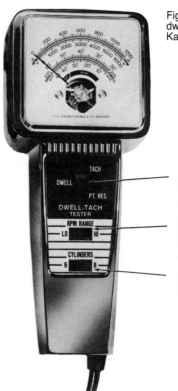

Figure 15.16 Typical tach-dwell controls (courtesy of Kal-Equip Company).

The function switch enables you to select the test you wish to perform

The tachometer scale switch enables you to select a low (under 1000 rpm) or high (over 1000 rpm) scale

The cylinder selector switch enables you to match the meter to the number of cylinders in the engine

combines a dwell meter with a tachometer. Such an instrument, commonly called a tach-dwell meter, is shown in Figure 15.15. Most tach-dwell meters have a function switch that enables you to match the meter to the test that you wish to perform. Figure 15.16 shows the function switch and other switches on a typical tach-dwell meter. The following steps outline a procedure for testing point resistance with a tach-dwell meter:

1 Apply the parking brake.
2 If the car has an automatic transmission, place the transmission selector lever in the Park position. If the car has a standard or manual transmission, place the shift lever in the Neutral position.
3 Check the position of the points. If the points are not closed, "tap" or "bump" the engine over with the starter motor until the distributor cam is not holding the points open.

Note: Where space permits, you can use a wrench on the crankshaft pulley retaining bolt to turn the engine as shown in Figure 15.17.

4 Place the function switch on the tach-dwell meter in the Point Resistance position.
5 Connect the tach-dwell meter to the DIST (−) terminal on the coil and to a good ground as shown in Figure 15.18.

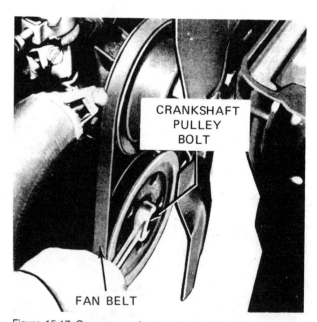

CRANKSHAFT PULLEY BOLT

FAN BELT

Figure 15.17 On some engines, you can turn the crankshaft by turning the pulley retaining bolt with a wrench (courtesy of Ford Motor Company).

6 Turn the ignition switch to the On position.
7 Read the point resistance on the meter scale.

Note: Most meter scales have a small box or band that indicates acceptable point resistance. If the me-

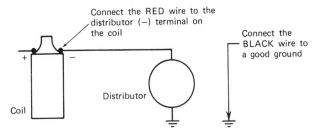

Figure 15.18 Meter connections for measuring point resistance.

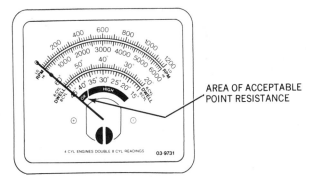

Figure 15.19 On most tach-dwell meters, acceptable point resistance is indicated when the pointer remains inside a small box or band (courtesy of Kal-Equip Company).

ter pointer falls outside that box or band, the point resistance is excessive. (See Figure 15.19.)

8 Place the ignition switch in the Off position.
9 Disconnect and remove the tach-dwell meter.

DWELL AND POINT GAP
As previously mentioned, dwell and point gap are related. Dwell is measured electrically with a dwell meter. Point gap is measured manually with a feeler gauge. The use of a dwell meter can, in most instances, eliminate the mechanical and human error that can affect the measurement of point gap. A dwell meter provides another advantage. It can be used while an engine is running or while an engine is being cranked by the starter motor. The use of a feeler gauge to measure point gap requires that you remove the distributor cap. It also requires that you "tap" or "bump" the engine over with the starter motor or turn it over by hand until the points are held open by the highest point of one of the cam lobes. Most mechanics use a dwell meter to save time and to obtain a more accurate measurement.

Measuring Dwell
Dwell should be checked as a part of routine maintenance of the ignition system and as a part of any diagnostic procedure in de-

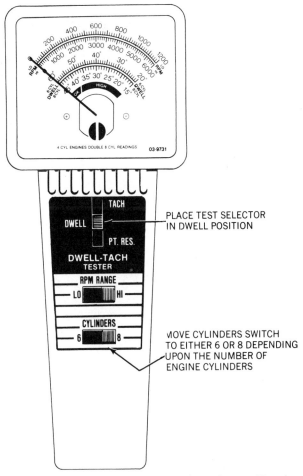

Figure 15.20 Tach-dwell meter function switches positioned so that the meter will measure the dwell in a four-cylinder and an eight-cylinder engine (courtesy of Kal-Equip Company).

termining the cause of ignition system problems. The meter connections for measuring dwell are the same as those you made when you measured point resistance. (Refer to Figure 15.18.) To save time, the measurement of dwell normally should follow the point resistance test. Dwell measurement then requires only that you change the position of the function switch. (See Figure 15.20.)

Most dwell meters require that you adjust the meter for the number of cylinders in the engine. Some meters allow you to select only six or eight cylinders. When using those meters, the dwell on four-cylinder engines is measured as follows:

1 Place the switch in the eight-cylinder position.
2 Double the dwell indicated by the meter. As an example, an indicated dwell of 25° would be doubled to obtain the actual dwell of 50°.

Job 15b

PERFORM A PRIMARY CIRCUIT INSPECTION

SATISFACTORY PERFORMANCE

A satisfactory performance on this job requires that you do the following:

1 Perform a primary circuit inspection on the car assigned.
2 Following the steps in the "Performance Outline" and the procedures and specifications of the car manufacturer, complete the job within 20 minutes.
3 Fill in the blanks under "Information."

PERFORMANCE OUTLINE

1 Check the battery terminals and cables. Clean and tighten the connections where necessary.
2 Check the primary wiring and connections. Clean and tighten the connections where necessary.
3 Check the condition of the point surfaces.
4 Check the condenser connection and mounting. Clean and tighten the connection and mounting if necessary.
5 Check the point resistance.

INFORMATION

Vehicle identification _____

Reference used _____ Page(s) _____

Condition of parts and services performed:

Battery cables and connections _____ OK

_____ Cleaned and tightened

Primary wiring and connections _____ OK

_____ Cleaned and tightened

Point surfaces   _____ OK (normal wear)

_____ Burned

_____ Excessively pitted

Condenser connections   _____ Wire lead tight

_____ Mounting screw tight

Point resistance   _____ Acceptable

_____ Excessive

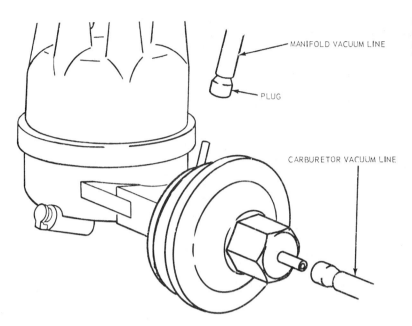

Figure 15.21 When checking dwell, the hose(s) at the distributor diaphragm should be disconnected and plugged (courtesy of Ford Motor Company).

MANIFOLD VACUUM LINE

PLUG

CARBURETOR VACUUM LINE

The dwell specification for any engine is determined by the engine manufacturer. For four-cylinder engines, dwell specifications range from 40° to 60°. Specifications for dwell on six-cylinder engines range from 31° to 47°. Eight-cylinder engines usually require from 24° to 32°. Because of the wide range of dwell specifications, the correct dwell for the engine on which you are working should be obtained from an appropriate manual. On many cars, the dwell specification is provided on the decal on the engine or in the engine compartment.

The following steps outline a procedure for measuring dwell:

1 Determine the dwell specification for the engine on which you are working.

2 Place the function switch on the meter in the Dwell position. (Refer to Figure 15.20.)

3 Position the cylinder selector switch to match the number of cylinders in the engine.

4 Connect the meter to the DIST (–) terminal on the coil and to a good ground. (Refer to Figure 15.18.)

5 Disconnect and plug the vacuum hose(s) that are connected to the distributor diaphragm. (See Figure 15.21.)

Note: This step should be performed as some advance mechanisms change the dwell and thus may cause a false dwell measurement.

6 Apply the parking brake.

7 If the car has an automatic transmission, place the transmission selector lever in the Park position. If the car has a standard or manual transmission,

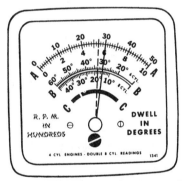

Figure 15.22 On most cars, the dwell should not change more than 3° as the engine is accelerated from idle speed to 2,000 rpm (courtesy Kal-Equip Company).

place the shift lever in the Neutral position.

8 Start the engine and allow it to run at idle speed.

9 Read the dwell indicated on the meter and compare it to the specification.

10 Slowly accelerate the engine to approximately 2,000 rpm while observing the meter. The dwell should not change more than 3°. (See Figure 15.22.)

Note: A change of more than 3° in dwell may indicate wear in the distributor shaft bushings or in the breaker plate. Such wear may require the overhaul or replacement of the distributor.

11 Place the ignition switch in the Off position.

12 Connect the vacuum hose(s) to the distributor diaphragm.

13 Disconnect and remove the meter.

Job 15c

MEASURE DWELL

SATISFACTORY PERFORMANCE

A satisfactory performance on this job requires that you do the following:

1 Measure the dwell on the car assigned.
2 Following the steps in the "Performance Outline" and the procedure and specifications of the car manufacturer, complete the job within 15 minutes.
3 Fill in the blanks under "Information."

PERFORMANCE OUTLINE

1 Adjust the meter and connect it to the engine.
2 Read the indicated dwell at idle and while accelerating.
3 Compare the indicated dwell with the specification.

INFORMATION

Vehicle identification _____

Reference used _____ Page(s) _____

Dwell specification _____

Indicated dwell at idle: _____

Dwell change during acceleration: _____

Indicated dwell is: _____ Within specification

                               _____ Too high

                               _____ Too low

Based on the indicated dwell reading, the point gap is:

                               _____ Within specifications

                               _____ Too wide

                               _____ Too narrow

Based on the indicated dwell reading while accelerating, the distributor:

                               _____ Appears to be in good condition

                               _____ Requires further inspection

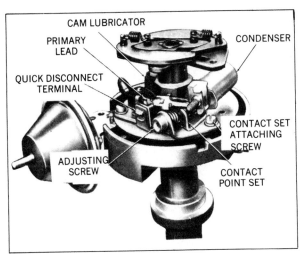

Figure 15.23 A distributor that uses points that are adjusted externally. The adjusting screw is accessible through a window in the distributor cap (courtesy of Chevrolet Motor Division).

Figure 15.24 Adjusting dwell with an Allen wrench, or hex-type wrench. Note that the window in the distributor cap has been raised for access to the adjusting screw (courtesy of Chevrolet Motor Division).

Adjusting Dwell Dwell is not adjustable in electronic ignition systems. The correct dwell is maintained by the electronic control unit. In breaker point systems, dwell is adjusted by changing the point gap. Closing the gap increases the amount of time that the points remain closed, increasing the dwell. Opening the point gap decreases the amount of time the points remain closed, decreasing the dwell. (Refer to Figure 15.6.)

External Adjustment Some distributors are designed so that you can adjust the points externally, without having to remove the distributor cap. Such a design allows you to adjust the dwell while the engine is running. Distributors of that design use a set of points similar to those shown in Figure 15.23. Those points have an adjusting screw that can be turned with a ⅛ inch Allen wrench, or hex-type wrench. Access to the points is provided by a window in the distributor cap. When the window is opened, an Allen wrench can be inserted in the adjusting screw, as shown in Figure 15.24. Turning the screw clockwise closes the point gap, increasing the dwell. Turning the screw counterclockwise opens the gap, decreasing the dwell.

The following steps outline a procedure for adjusting dwell where an external adjustment is provided:

1 Determine the dwell specification for the engine.

2 Adjust a dwell meter and connect it to the DIST (−) terminal on the coil and to a good ground. (Refer to Figure 15.18.)

3 Disconnect and plug the vacuum hose(s) that

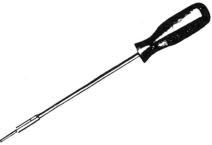

Figure 15.25 A distributor adjusting tool. A ⅛-in. Allen wrench is mounted at the end of a flexible shaft, allowing the tool to be used where access to the distributor cap window is limited (courtesy of Snap-on Tools Corporation).

are connected to the distributor diaphragm. (Refer to Figure 15.21.)

4 Raise the window on the distributor cap and insert a ⅛ inch Allen wrench into the head of the adjusting screw. (Refer to Figure 15.24.)

Note: A special point adjusting tool similar to the one shown in Figure 15.25 can be used when it is difficult to gain access to the window.

5 Apply the parking brake.

6 If the car has an automatic transmission, place the transmission selector lever in the Park position. If the car has a standard or manual transmission, place the shift lever in the Neutral position.

7 Start the engine and allow it to run at idle speed.

8 Read the dwell indicated on the meter and compare it to the specification.

9 Adjust the dwell to the specification by turning the adjusting screw.

Note: On some engines, the distributor is mounted

Job 15d

ADJUST DWELL—EXTERNAL ADJUSTMENT

SATISFACTORY PERFORMANCE

A satisfactory performance on this job requires that you do the following:

1 Adjust the dwell on the car assigned.
2 Following the steps in the "Performance Outline" and the procedure and specifications of the manufacturer, complete the job within 15 minutes.
3 Fill in the blanks under "Information."

PERFORMANCE OUTLINE

1 Determine the dwell specification.
2 Measure the dwell and determine the correction needed.
3 Adjust the dwell to the manufacturer's specifications.

INFORMATION

Vehicle identification _____

Engine identification _____

Reference used _____ Page(s) _____

Dwell specification _____

Actual dwell indicated _____

Direction adjusting screw was turned:

_____Clockwise   _____Counterclockwise

Dwell at completion of adjustment _____

---

close to the fan and to the fan belts. If your hand comes in contact with those parts while they are moving, you could be seriously injured. On those engines, it is best to turn off the engine and make an approximate adjustment while the engine is not running. The engine can then be restarted to check the adjustment. This procedure should be repeated until the adjustment is correct.

10 Turn the ignition switch to the Off position.
11 Remove the Allen wrench from the adjusting screw and close the window.
12 Disconnect and remove the dwell meter.
13 Connect the vacuum hose(s) to the distributor diaphragm.

Internal Adjustment   On most distributors, the dwell can be adjusted only after removing the distributor cap. A distributor of that type is shown in Figure 15.26. Dwell is adjusted by moving the stationary point closer to, or farther away from, the movable point. In order to move the stationary point, the attaching screw(s) must be loosened. A screwdriver can then be used as shown in Figure 15.27 to pry the stationary point in the required direction. Since the distributor cap must be removed, the engine cannot be running when the adjustment is being made.

One method of adjusting dwell in a distributor of that type requires that you adjust the point gap with a feeler gauge. You must then install the cap and measure the dwell with the engine running. If the dwell is not as specified, the point gap must be readjusted and the dwell rechecked. That procedure must be repeated until the dwell is correct.

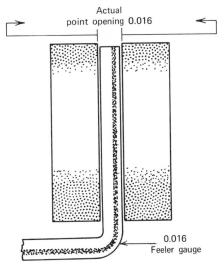

Figure 15.28 A feeler gauge will provide accurate measurement of point gap only when the points have smooth, parallel surfaces.

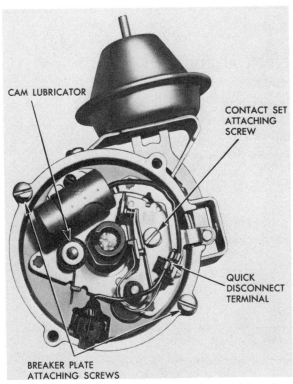

Figure 15.26 A top view of a distributor that has internally adjusted points (courtesy of Chevrolet Motor Division).

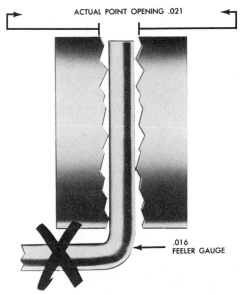

Figure 15.29 The gap between points with rough or pitted surfaces cannot be accurately measured with a feeler gauge (courtesy of Chevrolet Motor Division).

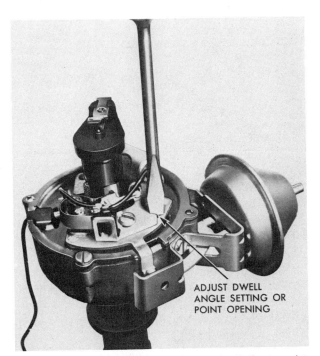

Figure 15.27 Using a screwdriver to move the stationary point (courtesy of Chevrolet Motor Division).

That method works fairly well when the points are new. The gap between new point surfaces can be accurately measured with a feeler gauge as shown

in Figure 15.28. When the points are used, the slight roughness or pitting on their surfaces does not allow you to make an accurate measurement. This condition is shown in Figure 15.29.

Even though the engine cannot be started, a dwell meter can be used to obtain an accurate dwell adjustment. Dwell will be indicated on the meter while the engine is being cranked by the starter motor. The use of a dwell meter in this manner will enable you to adjust dwell accurately even when the points are rough or pitted. The steps that follow outline a procedure for adjusting dwell while the distributor cap is off:

Job 15e

ADJUST DWELL—INTERNAL ADJUSTMENT

SATISFACTORY PERFORMANCE

A satisfactory performance on this job requires that you do the following:

1 Adjust the dwell on the car assigned.
2 Following the steps in the "Performance Outline" and the procedure and specifications of the car manufacturer, complete the job within 30 minutes.
3 Fill in the blanks under "Information."

PERFORMANCE OUTLINE

1 Determine the dwell specification.
2 Measure the dwell and determine the correction needed.
3 Adjust the dwell to the manufacturer's specification.

INFORMATION

Vehicle identification _____

Engine identification _____

Reference used _____ Page(s) _____

Dwell specification _____

Actual dwell indicated _____

Direction stationary point was moved:

_____ Toward movable point   _____ Away from movable point

Dwell at completion of adjustment _____

---

1 Determine the dwell specification for the engine.

2 Adjust a dwell meter and connect it to the DIST (−) terminal on the coil and to a good ground. (Refer to Figure 15.18.)

3 Remove the coil wire from the center tower of the distributor cap and ground it with a jumper wire.

4 Remove the distributor cap and the rotor.

5 Apply the parking brake.

6 If the car has an automatic transmission, place the transmission selector lever in the Park position. If the car has a standard or manual transmission, place the shift lever in the Neutral position.

7 Turn the ignition switch to the On position.

8 Using a remote starter switch, crank the engine with the starter motor and observe the dwell indicated on the meter.

Note: Dwell indicated while an engine is being cranked by the starter motor will usually be about 2° higher than the actual dwell. For example, an actual dwell of 32° will be indicated on the meter as 34°.

9 Compare the dwell with the specification.

10 Determine the direction in which you must move the stationary point.

11 Slightly loosen the point attaching screw(s). (Refer to Figure 15.26.)

Note: Loosen the screws only enough to permit movement of the stationary point by the screwdriver. This helps to keep the adjustment from slipping before you tighten the screw(s) to lock your adjustment.

12 Place a screwdriver in the notch at the base of

the stationary point. (Refer to Figure 15.27.)

13 While cranking the engine with the starter motor, move the stationary point in the desired direction until the correct dwell reading is obtained.

14 Tighten the point attaching screw(s).

15 Crank the engine and observe the dwell meter. If the dwell is not correct, repeat steps 9 through 15. If the dwell is correct, continue with step 16.

16 Install the rotor and the distributor cap.

17 Remove the jumper wire from the coil wire and install the coil wire in the center tower of the cap.

18 Start the engine and check the dwell adjustment. If the dwell is incorrect, repeat steps 3 through 18. If the dwell is correct, continue with step 19.

19 Turn the ignition switch to the Off position.

20 Disconnect and remove the dwell meter.

21 Disconnect and remove the remote starter switch.

SPARK TIMING Through your study of this and the previous chapter, and your completion of the jobs provided, you have gained some basic knowledge of how the ignition system produces high voltage and delivers it to the spark plugs. That voltage must be delivered at the right time if the engine is to run efficiently. Before you can appreciate the importance of spark timing, you must have an understanding of the four-stroke cycle.

The Four-Stroke Cycle Most automotive engines are of the four-stroke cycle type. A four-stroke cycle engine utilizes a piston that moves up and down in a cylinder. By means of a connecting rod, the piston is connected to a crankshaft. Through that arrangement, the *reciprocating*, or up-and-down, motion of the piston is converted to rotary motion. Figure 15.30 shows a four-stroke cycle engine and identifies the major parts.

During each stroke of the piston, the crankshaft rotates one-half turn (180°). Therefore, a complete four-stroke cycle requires that the crankshaft rotate two full turns (720°). The fuel-air mixture is burned and power is delivered to the crankshaft during only one stroke. The remaining three strokes are required to fill the cylinder with a fuel-air mixture, to compress the mixture, and to exhaust the burned gases.

The Intake Stroke The intake stroke starts with the piston at the extreme top of the cylinder. This position is called *Top Dead Center*, or *TDC*. As the crankshaft turns, the piston is pulled down in the cylinder as shown in Figure 15.31. The intake valve opens, allowing atmospheric pressure to push a mixture of gasoline vapor and air into the cylinder.

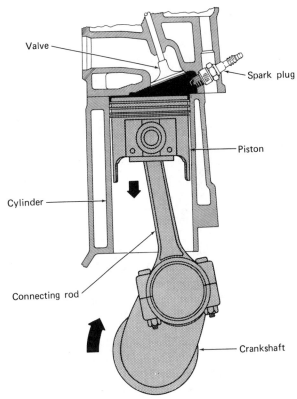

Figure 15.30 The major parts of a four-stroke cycle engine (courtesy of Ford Motor Company).

The intake stroke is completed when the piston reaches the limit of its travel at the bottom of the cylinder. This position is called *Bottom Dead Center* or *BDC*.

The Compression Stroke The fuel-air mixture in the cylinder must be squeezed, or *compressed*, for efficient burning. The intake valve closes and the rotating crankshaft forces the piston back up into the cylinder as shown in Figure 15.32. Since the fuel-air mixture cannot escape, it is compressed into a small space between the cylinder head and the top of the piston. That space is called the *combustion chamber*. (Refer to Figure 15.30.) The compression stroke is completed when the piston reaches TDC.

The Power Stroke The compressed fuel-air mixture is now ignited by the spark that jumps the gap at the spark plug. Both valves remain closed, and the expansion of the burning mixture forces the piston down as shown in Figure 15.33. The fuel mixture is completely burned and the power stroke completed when the piston reaches BDC.

The Exhaust Stroke The cylinder is now filled with

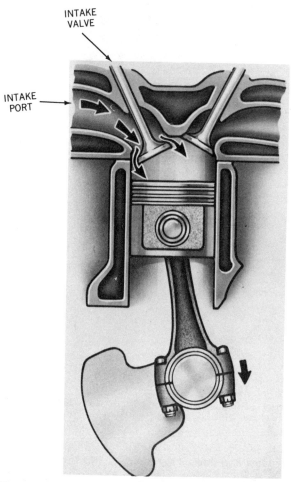

INTAKE
VALVE

INTAKE
PORT

Figure 15.31 The intake stroke. The intake valve opens and the piston is pulled down in the cylinder, creating a partial vacuum. Atmospheric pressure forces the fuel-air mixture into the cylinder through the intake port (courtesy of Chevrolet Motor Division).

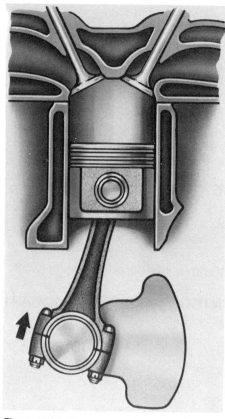

Figure 15.32 The compression stroke. The piston is pushed up in the cylinder while both valves remain closed. The fuel-air mixture is compressed into the combustion chamber (courtesy of Chevrolet Motor Division).

burned gases that must be expelled. As shown in Figure 15.34, the exhaust valve opens. The rotating crankshaft pushes the piston back up into the cylinder, forcing the burned gases out through the exhaust port. When the piston reaches TDC, the exhaust stroke is completed and the cycle is repeated.

Multicylinder Engines   A single cylinder engine delivers power to the crankshaft only once during two revolutions. Therefore, single cylinder engines run roughly, especially at low speed. The use of multiple cylinders provides a smoother running engine. Most automobile engines are made with four, six, or eight cylinders. The power strokes of all the cylinders are timed so that they occur within the same two revolutions of the crankshaft. In a four-cylinder engine, a power stroke occurs during each 180° of crankshaft rotation. In most six-cylinder en-

gines, a power stroke occurs every 120°. In eight-cylinder engines, power is delivered to the crankshaft during each 90° of crankshaft rotation.

Timing the Spark   For a four-stroke cycle engine to operate, the spark must occur at the end of each compression stroke. If the spark is timed to occur when each piston reaches TDC, the engine will run. However, it will not run well. Actually, the spark must occur at different times before each piston reaches TDC. The exact time the spark must occur depends on many factors. The speed at which the engine is running, the load under which it is operating, the operating temperature, the octane rating of the fuel used, and many other variables must be considered.

Many electronic ignition systems monitor engine operating conditions and automatically adjust spark timing to obtain optimum efficiency while minimizing objectionable exhaust emissions. Those systems adjust spark timing by means of computerized circuitry. The diagnosis and repair of those systems

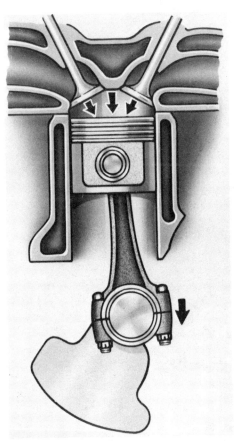

Figure 15.33 The power stroke. The fuel-air mixture is ignited by the spark jumping the gap at the plug and the pressure of the burning gases pushes the piston down in the cylinder (courtesy of Chevrolet Motor Division).

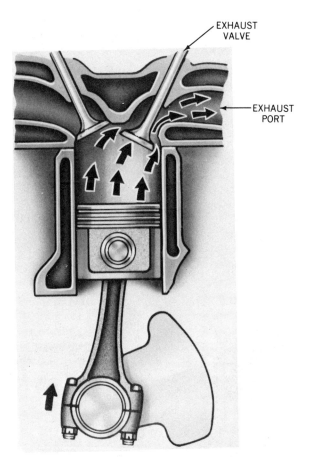

Figure 15.34 The exhaust stroke. The exhaust valve opens and the piston is pushed up in the cylinder. The burnt gases are forced out of the cylinder through the exhaust port (courtesy of Chevrolet Motor Division).

are beyond the scope of this book. Ignition systems using breaker points and some electronic systems use mechanical devices to adjust spark timing. On most engines, those devices adjust the spark timing for engine speed and load.

Engine speed and engine load are subject to constant change while driving. Two methods of providing for those changes are commonly used: *centrifugal spark advance* and *vacuum spark advance*.

Centrifugal Spark Advance The centrifugal spark advance mechanism changes the spark timing to match engine speed. When an engine is running at idle speed, a spark timed to occur at, or slightly before, TDC is satisfactory. When the speed of an engine is increased, the timing must be moved ahead, or *advanced*, so that the spark occurs earlier. This is because there is less time available for the fuel-air mixture to burn. The mixture does not explode, but burns rapidly. The average burning time of the fuel-air mixture in a cylinder is 0.003 of 1 second. That means that the maximum force is

exerted on the piston 0.003 of 1 second after the mixture is ignited.

An engine runs most efficiently when the maximum force is exerted on a piston while the crankshaft is located approximately 10° after TDC. During the 0.003 of 1 second burning time, the crankshaft of an engine idling at 550 rpm rotates about 10°. If the spark is timed to occur when the piston is at TDC (0°), the crankshaft will be 10° past TDC when the maximum force is applied to the piston. Given that situation, illustrated in Figure 15.35, you may assume that a spark timed to occur at TDC will be correct for an engine operating at 550 rpm. However, when the engine speed is increased, the spark timing will be incorrect.

When an engine is operated at 1,000 rpm, its crankshaft turns 18° in 0.003 of 1 second. If the spark occurs at TDC, the crankshaft will be 8° past the point where the piston should receive the maximum force. To achieve the greatest efficiency, the spark should be timed to ignite the mixture 8° before TDC.

Job 15f

IDENTIFY THE FOUR STROKES OF THE FOUR-STROKE CYCLE AND RELATED ENGINE PARTS

SATISFACTORY PERFORMANCE

A satisfactory performance on this job requires that you do the following:

1 Identify the four strokes of the four-stroke cycle shown by placing the letter of each stroke in front of the correct stroke name.

2 Identify the numbered engine parts shown by placing the number of each part in front of the correct part name.

3 Correctly identify all the strokes and all the parts within 10 minutes.

PERFORMANCE SITUATION

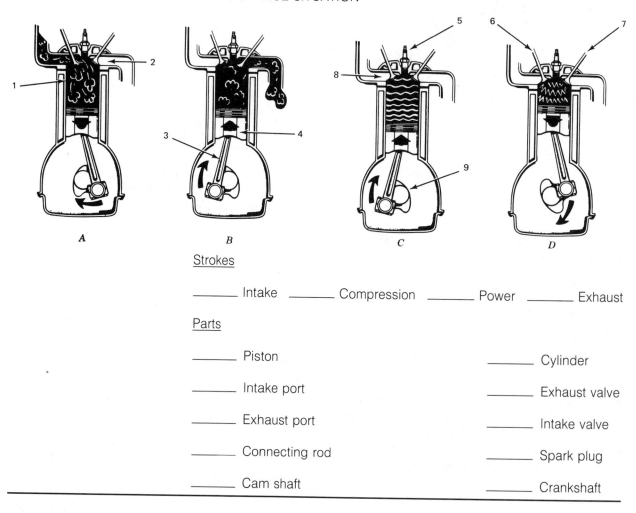

A   B   C   D

Strokes

_____ Intake   _____ Compression   _____ Power   _____ Exhaust

Parts

_____ Piston   _____ Cylinder

_____ Intake port   _____ Exhaust valve

_____ Exhaust port   _____ Intake valve

_____ Connecting rod   _____ Spark plug

_____ Cam shaft   _____ Crankshaft

This need for spark advance is shown in Figure 15.36.

If the engine speed is increased to 2,000 rpm, the crankshaft now turns 36° while the fuel is burning.

This requires that the spark occur 26° before TDC if the mixture is to be burned completely by the time the crankshaft reaches 10° after TDC (see Figure 15.37).

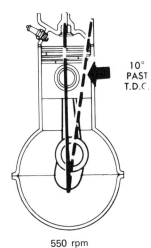

Figure 15.35 When an engine is operating at 550 rpm, the fuel-air mixture must be ignited at about TDC to obtain the maximum force on the piston when the crankshaft is 10° past TDC (courtesy of Chevrolet Motor Division).

10° PAST T.D.C.

550 rpm

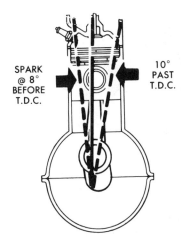

Figure 15.36 When an engine is operating at 1000 rpm, the fuel-air mixture must be ignited at about 8° before TDC to obtain the maximum force on the piston when the crankshaft is 10° past TDC (courtesy of Chevrolet Motor Division).

SPARK @ 8° BEFORE T.D.C.

10° PAST T.D.C.

1000 rpm

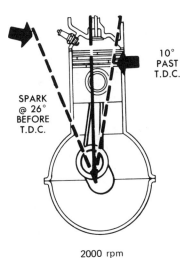

Figure 15.37 When an engine is operating at 2000 rpm, the fuel-air mixture must be ignited at about 26° before TDC to obtain the maximum force on the piston when the crankshaft is 10° past TDC (courtesy of Chevrolet Motor Division).

SPARK @ 26° BEFORE T.D.C.

10° PAST T.D.C.

2000 rpm

Since automobile engines operate at constantly varying speeds, the spark timing must change with the speed. This is accomplished by the use of cen-

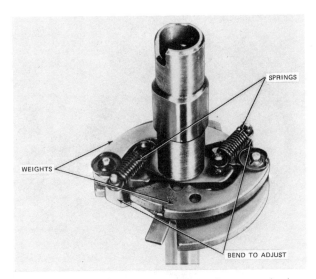

Figure 15.38 A typical centrifugal spark advance mechanism. Note that the springs hold the weights in toward the shaft when the unit is at rest (courtesy of Ford Motor Company).

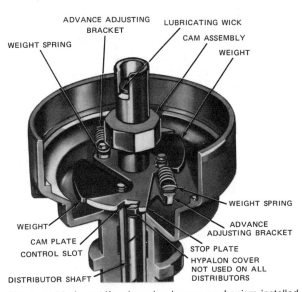

Figure 15.39 A centrifugal spark advance mechanism installed in the base of a distributor (courtesy of Ford Motor Company).

trifugal force. As mentioned in Chapter 1, centrifugal force is a force that tends to cause a body to move away from its center of rotation. You may have seen the effect of centrifugal force when you placed a small article on a spinning record. Centrifugal force caused the article to move to the edge of the record and slide off.

Most distributors use two weights, two springs, and a cam to harness centrifugal force (see Figure 15.38). In some distributors, those parts are housed under the breaker plate as shown in Figure 15.39. In other distributors they are mounted on top of the distrib-

ADVANCE WEIGHT (2)

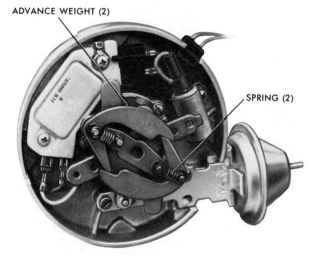

SPRING (2)

Figure 15.40 A centrifugal spark advance mechanism installed on top of a distributor shaft (courtesy of Chevrolet Motor Division).

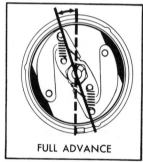

NO ADVANCE   FULL ADVANCE

Figure 15.41 The operation of a centrifugal spark advance mechanism. Note that as the wieghts move outward, they "advance" the position of the cam on the distributor shaft (courtesy of Chevrolet Motor Division).

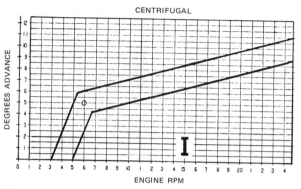

CENTRIFUGAL

Figure 15.42 Typical centrifugal spark advance specifications (courtesy of Cadillac Motor Car Division).

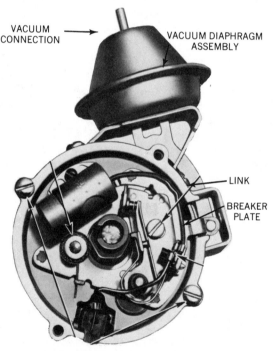

VACUUM CONNECTION   VACUUM DIAPHRAGM ASSEMBLY

LINK

BREAKER PLATE

Figure 15.43 A typical vacuum spark advance mechanism. When vacuum pulls the diaphragm, the link rotates the breaker plate, causing the points to open earlier (courtesy of Chevrolet Motor Division).

utor shaft as shown in Figure 15.40. The weights are held close to the shaft by the springs. As engine speed increases, the weights tend to move outward, against the tension of the springs. The movement of the weights turns the cam. In breaker point distributors, that cam turns the multi-lobed cam that opens the points so that the points open earlier. In the distributors of electronic systems, that cam turns the trigger wheel so that it passes the sensor earlier. The action of the weights is shown in Figure 15.41. As engine speed decreases, the springs return the weights toward their original position. As the weights move inward, they move the cam toward its original position.

The actual centrifugal advance required varies with different engines. Each engine manufacturer selects the weights, springs, and cam best suited for a particular engine design. Those parts are then adjusted so that the spark advance provides the best performance at all engine speeds. Figure 15.42 shows a centrifugal spark advance curve established by one manufacturer for a particular engine.

Vacuum Spark Advance Automobile engines operate under constantly changing loads. An engine running at 2,000 rpm in a car going uphill is subjected to a far greater load than if it were running at 2,000 rpm in a car going downhill. Some driving conditions such as acceleration and hill climbing place heavy loads on an engine. But during most driving, an engine is subjected to light or

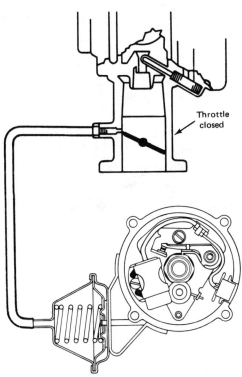

Figure 15.44 The vacuum spark advance is operated by vacuum from the carburetor. Note that when the throttle is closed, no vacuum is available and no additional advance is provided (courtesy of Chevrolet Motor Division).

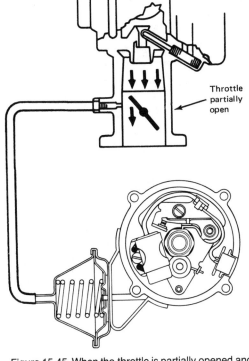

Figure 15.45 When the throttle is partially opened and the engine is operating under a light load, vacuum is present at the vacuum advance port (courtesy of Chevrolet Motor Division).

moderate loads. An engine that is running under light or moderate load conditions can be given additional spark advance. That additional spark advance usually results in increased fuel economy.

The centrifugal spark advance mechanism is sensitive only to speed, and cannot detect engine load conditions. Therefore, a different method of providing additional advance must be used. Most manufacturers use a vacuum advance mechanism of the type shown in Figure 15.43. That mechanism consists of a spring-loaded diaphragm that is connected to the breaker plate that holds the points or the sensor. When vacuum is applied to the diaphragm, the breaker plate is pulled in the direction opposite that of cam rotation. This advances the spark timing by causing the points to open earlier or, in electronic systems, by causing the trigger wheel to pass the sensor earlier. The diaphragm is connected to an opening, or *port,* in the carburetor. As shown in Figure 15.44, that port is located so that it is above the throttle plate when the throttle is in the idle or closed position. When the throttle is in the idle position, there is no vacuum available to pull the diaphragm. Thus, there is no additional spark advance at that speed.

When the engine is operated under a light load with the throttle partially open, manifold vacuum is relatively high. As shown in Figure 15.45, vacuum is present at the port under those conditions. When the throttle is wide open, as it is when the engine is heavily loaded, manifold vacuum is very low. No vacuum is available to provide additional spark advance (see Figure 15.46.)

At any speed above idle, both the centrifugal and the vacuum systems may function. The centrifugal system will advance the spark timing in relation to engine speed. The vacuum system will further advance the spark timing in relation to throttle opening and engine load.

Checking Spark Timing All engines are provided with a means by which spark timing can be checked. Most engines have timing marks on the crankshaft pulley and on the front of the engine. Figure 15.47 shows how those marks may appear. On some engines, the pulley is marked in degrees of crankshaft rotation. Those degree markings usually extend both before TDC (BTDC) and after TDC (ATDC). The desired degree marking can be aligned with a pointer attached to the front of the engine.

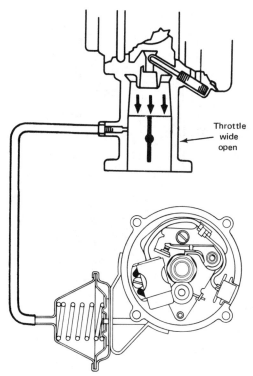

Figure 15.46 When the throttle is wide open, manifold vacuum is so low that none is present at the vacuum advance port (courtesy of Chevrolet Motor Division).

Figure 15.47 Typical timing marks as they appear on an engine (courtesy of Pontiac Motor Division, General Motors Corporation).

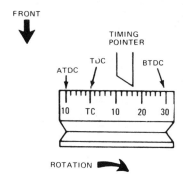

Figure 15.48 Typical timing marks as found on a crankshaft pulley. Those marks can be aligned with a pointer attached to the engine. Note that in this drawing the crankshaft is at 16° before TDC (courtesy of Ford Motor Company).

Timing marks of that type are shown in Figure 15.48. Other engines have a single mark or notch on the pulley as shown in Figure 15.49. That mark can be aligned with the desired marking on a degree scale attached to the engine.

Still other engines have the timing marks on the flywheel or torque converter and on the bell housing. On some engines, the degree markings are on the flywheel and can be aligned with a pointer on the bell housing as shown in Figure 15.50. You also will find engines with degree markings on the bell housing. On those engines, the flywheel or torque converter is notched or marked. (See Figure 15.51.)

Timing marks are used to check the *initial spark timing*. Initial spark timing is the starting point for all spark timing. It indicates when the spark occurs while an engine is idling and before the centrifugal and vacuum advance mechanisms begin to function. The specifications for initial spark timing vary with different engines and with different engine applications. The specifications for a particular car should be obtained from the tune-up decal or from an appropriate manual.

Initial spark timing can be checked and adjusted while the engine is not running. This method is known as *static timing*. Static timing may be necessary

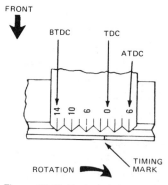

Figure 15.49 Typical timing marks as found on the front of an engine. The mark or notch on the crankshaft pulley can be aligned with those marks. Note that in this drawing the crankshaft is at 1° before TDC (courtesy of Ford Motor Company).

after a distributor has been replaced, but it is rarely used during routine service.

Most mechanics check spark timing *dynamically*, or while the engine is running. Since the marks can-

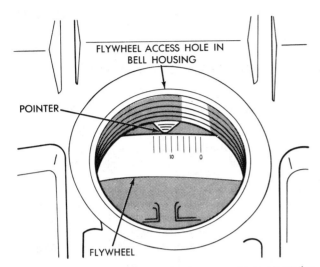

Figure 15.50 On some engines, the degree markings are on the flywheel. Those marks can be aligned with a pointer by looking through an access hole in the bell housing (courtesy of Chrysler Corporation).

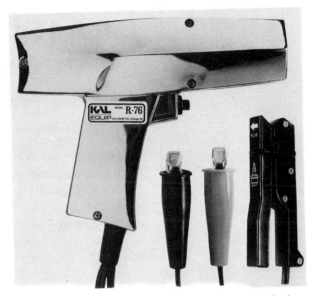

Figure 15.52 A typical power timing light. The two small wires are attached to the car battery. The remaining wire is connected to the spark plug wire for #1 cylinder. The induction pick-up shown merely clamps over the wire (courtesy of Kal-Equip Company).

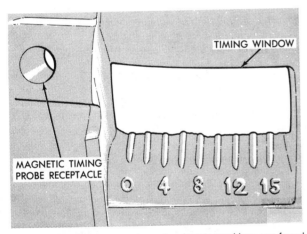

Figure 15.51 On some engines, the degree markings are found on the bell housing. A timing mark or notch on the flywheel or torque converter can be aligned with those markings by observing the mark through a window or opening in the bell housing (courtesy of Chrysler Corporation).

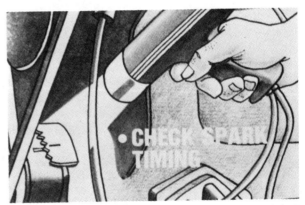

Figure 15.53 A timing light "stops" the motion of the crankshaft pulley so that the timing marks can be observed (courtesy of American Motors).

not be aligned while the crankshaft is turning, a *timing light,* or *strobe light,* is used to "stop" the pulley or flywheel. A typical timing light is shown in Figure 15.52. A timing light provides brilliant flashes of light. When that light is directed at moving parts, those parts appear to be at rest.

A timing light is connected to the car battery and to the spark plug wire that leads to #1 cylinder. When the high voltage surge flows to the #1 plug, it triggers the light. Thus, the light flashes at exactly the same time that the plug fires. When the light flashes, the timing marks can easily be seen. The alignment of the marks indicate when the plug fires

in relation to the position of the crankshaft. If the spark occurs when the piston is at TDC, the timing marks will be illuminated at the instant the pointer is aligned with the TDC mark. Figure 15.53 shows a timing light in use.

A timing light also can be used to check the operation of the centrifugal and vacuum advance mechanisms. As the engine is accelerated, the change in spark timing causes the timing marks to move. This movement can be observed and, with some timing lights, measured. Figure 15.54 shows a timing light combined with an advance meter.

The following steps outline a procedure for check-

Job 15g

IDENTIFY TERMS RELATING TO SPARK TIMING

SATISFACTORY PERFORMANCE

A satisfactory performance on this job requires that you do the following:

1 Identify the terms relating to spark timing by placing the number of each listed term in front of its correct definition.
2 Correctly identify all the terms within 10 minutes.

PERFORMANCE SITUATION

1 TDC
2 Distributor diaphragm
3 Centrifugal advance
4 0.003 of 1 second

5 Vacuum advance
6 Zero degrees
7 Centrifugal weights and cam
8 Spark advance

_____ Timing the spark to occur before the piston reaches TDC

_____ Moves the breaker plate to advance the spark timing

_____ A method of advancing spark timing in relation to engine speed

_____ The approximate time required for the fuel-air mixture to burn

_____ Top dead center

_____ The approximate time required for the piston to move from TDC to BDC

_____ The position of the crankshaft when the piston is at TDC

_____ Moves the breaker point cam or trigger wheel to advance the spark timing

_____ A method of providing additional spark advance during periods of light engine load

---

ing spark timing with a timing light. Timing specifications and the location and type of timing marks vary with different engines. Such information should be obtained from an appropriate manual:

1 Connect a tach-dwell meter to the engine and check the dwell. Any change in dwell will cause the timing to change. Before the timing can be accurately checked, the dwell must be checked and, if necessary, adjusted to specifications.

Note: Leave the dwell meter connected after this step since it will be used later in this procedure.

2 Locate the initial timing specification and the information giving the type of timing marks used and their location. (Refer to Figures 15.47 through 15.51.)

3 Using a droplight or a flashlight, locate the timing marks on the engine and on the pulley.

Note: On some cars, it is easier to locate the pulley markings from under the car. A remote starter switch will be of help in "tapping" or "bumping" the crankshaft over while looking for the pulley markings. If you use a remote starter switch, don't forget to disconnect the coil wire from the distributor cap and to ground it with a jumper wire. That will prevent the engine from starting. If the engine is equipped with an electronic ignition system using a distributor of the type shown in Figure 15.55, you can disable the ignition system by unplugging the connector.

4 Thoroughly clean the markings on the engine and on the pulley.

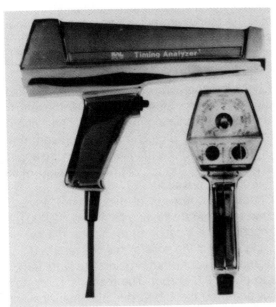

Figure 15.54 A timing light combined with a spark advance meter. This instrument will check initial spark timing and accurately measure both centrifugal and vacuum spark advance (courtesy of Kal-Equip Company).

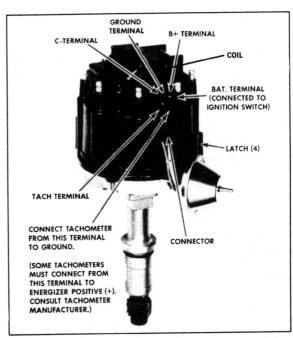

Figure 15.55 In an electronic ignition system using a distributor of this type, the coil is housed inside the distributor cap. This type system is disabled by unplugging the connector (courtesy of Pontiac Motor Division, General Motors Corporation).

Note: It is suggested that you accent the pointer and the degree mark specified. A sharp piece of chalk or a piece of stiff wire dipped in light-colored paint can be used. The accented marks are easier to see when the engine is running.

5 Connect the timing light battery leads to the battery terminals.

6 Connect the remaining timing light lead to the wire that runs to the spark plug in #1 cylinder.

Note: If the timing light is equipped with an induction pickup, merely clamp the pickup over the wire. (Refer to Figure 15.52.) If the timing light does not have an induction pickup, you must disconnect the spark plug wire from the plug or from the distributor cap. An adapter can then be used as shown in Figure 15.56. Never pierce a spark plug wire or attempt to insert a probe between the boot and the wire. To do so will damage the wire.

7 Position the timing light so that it cannot fall when the engine is started. Make sure that the wires are clear of the fan and fan belts.

8 Install the coil wire or connect the connector if you disabled the ignition system while you were looking for the timing marks.

9. Disconnect and plug the vacuum hose(s) that are connected to the distributor diaphragm. (Refer to Figure 15.21.)

Note: Failure to disconnect and plug the hose(s) may result in a false timing indication.

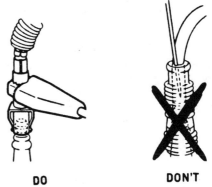

Figure 15.56 When you connect test equipment to a spark plug wire, use an adapter between the wire and the plug or between the wire and the distributor cap tower. Never pierce a wire or insert a probe between the wire and a boot or nipple (courtesy of AC-Delco, General Motors Corporation).

10 Set the function switch on the tach-dwell meter to TACH and the cylinder selector switch to the number of cylinders in the engine. (Refer to Figure 15.16.)

11 Apply the parking brake.

12 If the car has an automatic transmission, place the transmission selector lever in the Park position. If the car has a standard or manual transmission, place the shift lever in the Neutral position.

13 Start the engine and allow it to idle.

14 Check the idle speed and, if necessary, adjust it to specifications.

Note: If the idle speed is too high, the centrifugal spark advance mechanism may advance the spark timing, causing a false reading when you check the initial timing.

15 Aim the timing light at the timing marks and observe their position. The pointer and the specified degree mark should be aligned.

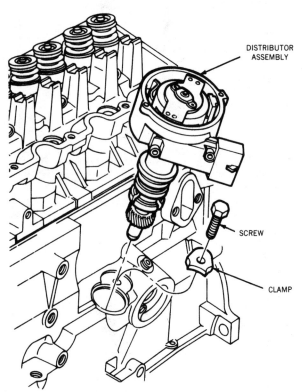

Figure 15.57 Most distributors are held in place by a clamp. After loosening the clamp bolt, the distributor can be rotated to adjust the initial timing (courtesy of Ford Motor Company).

Note: If the marks are not aligned, the initial spark timing must be adjusted. The procedure for that adjustment follows the completion of the procedure for checking timing.

16 Check the operation of the centrifugal spark advance mechanism. Slowly accelerate the engine while watching the timing marks. The marks should move in the BTDC direction and may even move beyond the range of the markings. When the engine speed is returned to idle, the marks should return to their original position.

17 Check the operation of the vacuum advance mechanism. Operate the engine at about 2,000 rpm and observe the location of the timing marks. Unplug and connect the hose(s) that you removed from the distributor diaphragm. Again accelerate the engine to about 2,000 rpm. The marks should move farther ahead in the BTDC direction than they did when the hose(s) were disconnected. Allow the engine to return to idle. The marks should return to their original position.

18 Turn the ignition switch to the OFF position.

19 If the initial spark timing is correct, disconnect and remove the timing light and the tach-dwell meter. If the initial spark timing requires adjustment, continue with the following procedure.

Adjusting Spark Timing   Initial spark timing is adjusted on most engines by rotating the distributor. The distributor is held in position by a clamp as shown in Figure 15.57. After the bolt holding the clamp is loosened, the distributor can be rotated in either direction. On some engines, access to that bolt is extremely limited. A special distributor wrench similar to one of those shown in Figure 15.58 may be required. On some engines, the distributor is secured with a special bolt as shown in Figure 15.59. Those bolts require the use of a special socket.

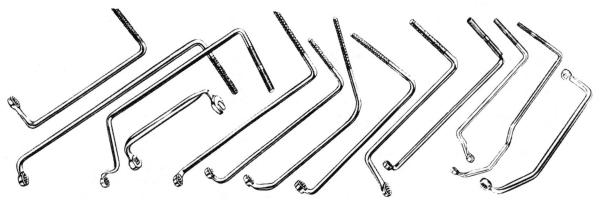

Figure 15.58 Special distributor wrenches are needed on some engines (courtesy of Snap-on Tools Corporation).

Job 15h

CHECK AND ADJUST TIMING

SATISFACTORY PERFORMANCE

A satisfactory performance on this job requires that you do the following:

1 Check and, if necessary, adjust the timing on the car assigned.
2 Following the steps in the "Performance Outline" and the procedure and specifications of the manufacturer, complete the job within 30 minutes.
3 Fill in the blanks under "Information."

PERFORMANCE OUTLINE

1 Check and, if necessary, adjust the dwell and the engine idle speed.
2 Check the initial spark timing and compare it to the specification.
3 Check the operation of the centrifugal and vacuum spark advance mechanisms.
4 Adjust the initial spark timing if necessary.

INFORMATION

Vehicle identification _____

Engine identification _____

Reference used _____ Page(s) _____

Specifications: Dwell _____

Initial timing _____

Idle speed _____

Measurements taken: Dwell _____

Initial timing _____

Idle speed _____

Adjustment made: Dwell _____

Initial timing _____

Idle speed _____

Operation of advance
mechanisms: Centrifugal _____

Vacuum _____

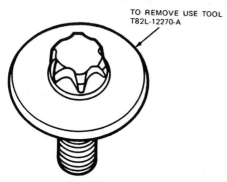

TO REMOVE USE TOOL
T82L-12270-A

Figure 15.59 On some engines, the distributor is locked by a special "security-type" hold down bolt. This prevents tampering with the timing adjustments by those who may not be qualified to make the proper adjustment (courtesy of Ford Motor Company).

The following steps outline a procedure for adjusting spark timing. As those steps are a continuation of the previously listed steps for checking initial spark timing, they are numbered in the same sequence:

20 Disconnect and plug the vacuum hose(s) that are connected to the distributor diaphragm. (Refer to Figure 15.21.)

Note: Failure to disconnect and plug the hose(s) may result in an incorrect adjustment.

21 Loosen the bolt or nut that secures the distributor clamp.

22 Start the engine and allow it to idle.

Note: Check the tachometer to be sure that the engine is idling at the specified idle speed.

23 While watching the timing marks, carefully rotate the distributor until the specified degree mark and the pointer are aligned.

Note: To advance timing, the distributor should be rotated in the direction opposite that of shaft rotation. Rotating the distributor in the same direction as shaft rotation will retard the timing.

24 Tighten the bolt or nut that holds the distributor clamp.

25 Recheck the alignment of the timing marks.

26 Recheck the idle speed and, if necessary, adjust it to specifications.

27 Turn the ignition switch to the off position.

28 Connect the vacuum hose(s) to the distributor diaphragm.

29 Disconnect and remove the timing light and the tach-dwell meter.

SUMMARY

By completing this chapter, you have gained additional knowledge of basic ignition systems and their operation. You have learned about some basic primary circuit components and their functions. You also are now aware of some of the different means of mechanically advancing spark timing. You also have gained additional diagnostic and repair skills including the measurement of dwell, the adjustment of dwell, and the adjustment of timing.

SELF-TEST

Each incomplete statement or question in this test is followed by four suggested completions or answers. In each case select the *one* that best completes the sentence or answers the question.

1 Two mechanics are discussing dwell.
  Mechanic A says that excessive dwell will cause the magnetic field in the coil to become oversaturated.
  Mechanic B says that insufficient dwell will cause the points to arc and burn.
  Who is right?
  A. A only
  B. B only
  C. Both A and B
  D. Neither A nor B

2 Cam angle refers to the number of degrees that the
  A. distributor shaft rotates while the points remain open
  B. crankshaft rotates while a piston moves from TDC to BDC
  C. distributor shaft rotates while the points remain closed
  D. crankshaft rotates while the fuel-air mixture is burning

3 Two mechanics are discussing the function of the condenser in a breaker point ignition system.
  Mechanic A says that the condenser reduces the arcing at the points.
  Mechanic B says that the condenser aids in the collapse of the magnetic field in the coil.
  Who is right?
  A. A only
  B. B only
  C. Both A and B
  D. Neither A nor B

4 When measuring dwell in a breaker point igni-

tion system, the meter should be connected to the

A. BAT (+) terminal on the coil and to the good ground

B. BAT (+) terminal on the coil and to the S terminal on the solenoid

C. DIST (−) terminal on the coil and to a good ground

D. DIST (−) terminal on the coil and to the S terminal on the solenoid

5 Two mechanics are discussing the adjustment of dwell.

Mechanic A says that closing the gap of the points will increase the dwell.

Mechanic B says that on distributors that have a window in the cap for dwell adjustment, the dwell is increased by turning the adjustment screw clockwise.

Who is right?

A. A only

B. B only

C. Both A and B

D. Neither A nor B

6 Two mechanics are discussing spark advance.

Mechanic A says that when the engine is operating under a heavy load, it requires more spark advance.

Mechanic B says that when an engine is operating at high speed, it requires more spark advance.

Who is right?

A. A only

B. B only

C. Both A and B

D. Neither A nor B

7 An engine is operating under a heavy load at 2,000 rpm with the throttle wide open. Under those conditions, additional spark advance is provided

A. only by the vacuum advance mechanism

B. only by the centrifugal advance mechanism

C. by both the vacuum and the centrifugal advance mechanisms

D. by neither the vacuum nor the centrifugal advance mechanisms

8 Two mechanics are discussing the procedure for checking initial spark timing.

Mechanic A says that the dwell must be checked and, if necessary, adjusted before checking the timing.

Mechanic B says that the idle speed must be checked and, if necessary, adjusted before checking the timing.

Who is right?

A. A only

B. B only

C. Both A and B

D. Neither A nor B

9 Two mechanics are discussing methods of checking spark advance mechanisms by using a timing light.

Mechanic A says that when an engine is accelerated, the timing marks should move in the ATDC direction.

Mechanic B says that the vacuum hose(s) at the distributor diaphragm should be disconnected while checking the centrifugal advance.

Who is right?

A. A only

B. B only

C. Both A and B

D. Neither A nor B

10 Initial spark timing is adjusted by rotating the

A. cap on the distributor

B. distributor in the engine

C. cam on the distributor shaft

D. breaker point in the distributor.

Appendix— Reference Material

Throughout this text you are constantly advised to consult various manuals. As an automotive electrician, you should have these manuals available so that you can refer to them for the information you may need on a particular job. You should start now to build your reference library so that you will be well equipped when you start to work.

MANUFACTURER'S SERVICE MANUALS Most car manufacturers publish complete service manuals for each model year of the cars they build. Information on the availability and price of these manuals can be obtained by contacting the manufacturers at the following addresses:

American Motors	American Motors Corporation 14250 Plymouth Road Detroit, MI 48232
Buick	Buick Motor Division General Motors Corporation Flint, MI 48550
Cadillac	Cadillac Motor Car Division General Motors Corporation 2860 Clark Avenue Detroit, MI 48232
Chevrolet	Chevrolet Motor Division General Motors Corporation General Motors Building Detroit, MI 48202
Chrysler	Chrysler Corporation P.O. Box 1919 Detroit, MI 48231
Dodge	Dodge Division Chrysler Corporation P.O. Box 857 Detroit, MI 48231
Ford	Ford Motor Company The American Road Dearborn, MI 48121
Jeep	Jeep Corporation American Motors Corporation Toledo, OH 43600

Lincoln and Mercury

Lincoln-Mercury Division
Ford Motor Company
3000 Schaefer Road
Dearborn, MI 48216

Oldsmobile

Oldsmobile Division
General Motors Corporation
920 Townsend Street
Lansing, MI 48921

Plymouth

Plymouth Division
Chrysler Corporation
P.O. Box 857
Detroit, MI 48231

Pontiac

Pontiac Motor Division
General Motors Corporation
1 Pontiac Plaza
Pontiac, MI 48053

COMPREHENSIVE SHOP MANUALS The most commonly used specifications and the procedures for the most frequently performed service operations for recent model cars are compiled in these manuals. Published yearly, they provide the best single source of reference material you will need in your daily work. The most widely used manuals are as follows:

Motor Auto Repair Manual

Motor
250 West 55th Street
New York, NY 10019

Chilton's Automotive Service Manual

Chilton Book Company
Chilton Way
Radnor, PA 19089

Mitchell Manuals

Mitchell Manuals, Inc.
9889 Willow Creek Road
San Diego, CA 92126

MANUALS AND CATALOGS PUBLISHED BY MANUFACTURERS OF AUTOMOTIVE PARTS, TOOLS, AND EQUIPMENT The makers of the parts, tools, and equipment you use in servicing automobiles offer a variety of reference materials. Many of these materials are available free of charge upon request from the following companies:

AC-Delco Division
General Motors Corporation
400 Renaissance Center
Detroit, MI 48243

Electrical test equipment, batteries, alternators, starting motors, spark plugs

Allen Electric and Equipment
Company
2101 North Pitcher Street
Kalamazoo, MI 49007

Electrical test equipment

Alltest, Incorporated
715 South Vermont Street
Palatine, IL 60067

Electrical test equipment

Auto Specialties Manufacturing
Company
St. Joseph, MI 49085

Jacks, stands, lifts, cranes, presses, hydraulic equipment

Auto-Test Incorporated 411 West 8th Street Neillsville, WI 54456	Automotive test equipment
Belden Wire and Cable Company Transportation Division 2625 West Butterfield Road Oak Brook, IL 60521	Wire and cable
Bendix Automotive Aftermarket 1217 South Walnut Street South Bend, IN 46620	Tune-up parts
Black and Decker Manufacturing Company Towson, MD 21204	Air and electric power tools and accessories
Blackhawk Manufacturing Company Applied Power Industries Incorporated P.O. Box 8720 Milwaukee, WI 53227	Hand and power tools, jacks and car stands
Borroughs Tool and Equipment Corporation 2429 North Burdick Street Kalamazoo, MI 49007	Tools and test instruments
Brookstone Company 127 Vose Farm Road Peterborough, NH 03458	Specialty hand tools
Champion Spark Plug Company P.O. Box 910 Toledo, OH 43601	Spark plugs, spark plug cleaners, hand tools
Coats Diagnostic Equipment Division 608 Country Club Drive Bensenville, IL 60106	Electrical test equipment
Dresser Industries Hand Tool Division 3201 North Wolf Road Franklin Park, IL 60131	Hand tools
Easco/K-D Tools Easco Hand Tools, Incorporated 3575 Hempland Road Lancaster, PA 17604	Hand tools
Echlin Manufacturing Company Branford, CN 06405	Electrical system parts
E. Edelmann and Company Chicago, IL 60647	Hydrometers

ESB Brands Incorporated
P.O. Box 6949
Cleveland, OH 44101

Batteries

The Gates Rubber Company
999 South Broadway
Denver, CO 80217

Drive belts

Hein-Werner Corporation
Waukesha, WI 53186

Jacks, car stands, presses

Hopkins Manufacturing
  Corporation
428 Peyton Street
Emporia, KS 66801

Headlight aimers

Ideal Corporation
10000 Pennsylvania Avenue
Brooklyn, NY 11207

Flashers

Ingersol Rand/Proto Tool
  Company
2309 Santa Fe Avenue
Los Angeles, CA 90058

Hand and power tools

Interstate Battery System of
  America, Incorporated
9304 Forest Lane
Suite 200
Dallas, TX 75243

Batteries

Kal-Equip Company
411 Washington Street
Otsego, MI 49078

Automotive test equipment

Kem Manufacturing Company
  Incorporated
Fair Lawn, NJ 07410

Electrical parts, ignition parts, starter
  drives

Lisle Corporation
Clarinda, IA 51632

Hand tools

Mac Tools, Incorporated
Washington Court House, OH
  43160

Hand tools and equipment

Marquette Corporation
307 East Hennepin Avenue
Minneapolis, MN 55414

Electrical test equipment, battery
  chargers

Milwaukee Hydraulic Products
  Corporation
Milwaukee, WI 53214

Jacks, lifts, presses

Neward Enterprises,
  Incorporated
9251 Archibald Avenue
Cucamonga, CA 91730

Vacuum pumps and testers

NGK Spark Plugs (U.S.A.)
  Incorporated

Spark plugs

20608 Madrona Avenue
Torrance, CA 90503

Owatonna Tool Company   Test equipment, hand tools, pullers
655 Eisenhower Drive
Owatonna, MN 55060

P & C Hand Tool Company   Hand tools
P.O. Box 22066
Portland, OR 93222

Rockwell International   Air and electric power tools
Power Tool Division
6263 Poplar Avenue
Memphis, TN 37901

Snap-on Tools Corporation   Hand and power tools, automotive test
8028 28th Avenue   equipment
Kenosha, WI 53140

Standard Motor Products   Tune-up parts
  Incorporated
37-18 Northern Boulevard
Long Island City
New York, NY 11101

Stewart-Warner Corporation   Instruments
Instrument Division
1826 Diversey Parkway
Chicago, IL 60614

Sun Electric Corporation   Automotive test equipment
Harlem and Avondale
Chicago, IL 60631

Thexton Manufacturing   Hand tools and test equipment
  Company
7685 Parklawn Avenue
Minneapolis, MN 55435

Walker Manufacturing   Jacks, car stands, presses
  Company
1201 Michigan Boulevard
Racine, WI 53402

Weaver Corporation   Jacks, car stands, lifts
Fords Mill Road
Paris, KY 40361

Metric System—SI   The International System of Units (Système International d'Unités) officially abbreviated "SI" in all languages—the modern metric system

QUANTITY	EXAMPLES OF APPLICATIONS	METRIC UNIT	SYMBOL
Length	Dimensions Tire rolling circumference Turning circle/radius Braking distance	meter	m
	Greater than 999 meter	kilometer	km
	Dimensions	millimeter	mm
	Depth of surface finish	micrometer	μm
Area	Glass & Fabrics Brake & Clutch linings Radiator area etc.	square centimeter	cm^2
	Small areas	square millimeter	mm^2
Volume	Car Luggage Capacity	cubic meter	m^3
	Vehicle fluid capacity	liter	l
	Engine Capacity	cubic centimeter	cm^3
Volume Flow	Gas & Liquid	liter per second	l/s
Time Interval	Measurement of elapsed time	second	s
		minute	min
		hour	h
		day	d
Velocity	General use	meter per second	m/s
	Road speed	kilometer per hour	km/h
Acceleration & Deceleration	General use	meter per second squared	m/s^2
Frequency	Electronics	hertz	Hz
		kilohertz	kHz
		megahertz	mHz
Rotational Speed	General use	revolution per minute	rpm
		revolution per second	rps
Mass	Vehicle mass Legal load rating	megagram	t
	General use	kilogram	kg
	Small masses	gram	g
		milligram	mg
Density	General use	kilogram per cubic meter	kg/m^3
		gram per cubic centimeter	g/cm^3
		kilogram per liter	kg/l
Force	Pedal effort Clutch spring force Handbrake lever effort etc.	newton	N
Moment of Force (Torque)	Torque	newton meter	N-m
Power, Heat Flow Rate	General use	watt	W
	Bulbs	kilowatt	kW
	Alternator output Engine performance Starter performance		
Celsius Temperature	General use	degree Celsius	°C
Thermodynamic Temperature	General use	kelvin	k

Metric System—SI *continued on p. 432.*

Metric System—SI (*continued*)   The International System of Units (Système International d'Unités) officially abbreviated "SI" in all languages—the modern metric system

QUANTITY	EXAMPLES OF APPLICATIONS	METRIC UNIT	SYMBOL
Electric Current	General use	ampere	A
		milliampere	mA
		microampere	μA
Potential Difference (Electromotive Force)	General use	kilovolt	kV
		volt	V
		millivolt	mV
		microvolt	μV
Electric Resistance	General use	megohm	MΩ
		kilohm	kΩ
		ohm	Ω
Electric Capacitance	General use	farad	F
		microfarad	μF
		picofarad	pF
Fuel Consumption	Vehicle performance	liter per 100 kilometer	1/100 km
Oil Consumption	Vehicle performance	liter per 1000 kilometer	1/1000 km
Stiffness	Linear stiffness	kilonewton meter	kN/m
Tire Revolutions	Tire Data	revolution per kilometer	rev/km
Pressure	Tire	kilopascal	kPa
	Coolant		
	Lubricating oil		
	Fuel pump delivery		
	Engine compression		
	Manifold		
	Brake line (hydraulic)		
	Car heating & ventilation		
	Barometric pressure		
Luminous intensity	Bulbs	candela	cd
Accumulator Storage Rating	Battery	ampere hour	A-h

	U.S.A./Metric Comparison		
QUANTITY	USA	METRIC—SYMBOL	
Length	Inch-Foot-Mile	Meter	m
Weight (mass)	Ounce-Pound	Kilogram	Kg
Area	Square inch/Foot	Square Meter	m²
Volume-Dry	Cubic inch/Foot	Cubic Meter	m³
-Liquid	Ounce-Pint-Quart-Gallon	Liter	l
Velocity	Feet Per Second	Meter per Second	m/s
Road Speed	Miles Per Hour	Kilometer per Hour	km/h
Force	Pound-Force	Newton	N
Torque	Foot-Pounds	Newton meter	N-m
Power	Horsepower	Kilowatt	kW
Pressure	Pounds Per Square Inch	Kilopascal	kPa
Temperature	Degrees Fahrenheit	Degrees Kelvin	K
		and Celsius	°C

Decimal Equivalents

MILLI-METER	DECI-MAL	FRAC-TION	DRILL SIZE	MILLI-METER	DEC-IMAL	FRAC-TION	DRILL SIZE	MILLI-METER	DEC-IMAL	FRAC-TION	DRILL SIZE	MILLI-METER	DEC-IMAL	FRAC-TION	DRILL SIZE	MILLI-METER	DEC-IMAL	FRAC-TION
.1	.0039			1.75	.0689				.1570		22	6.8	.2677			10.72	.4210	27/64
.15	.0059				.0700		50	4.0	.1575			6.9	.2716			11.0	.4330	
.2	.0079			1.8	.0709				.1590		21		.2720		I	11.11	.4075	7/16
.25	.0098			1.85	.0728				.1610		20	7.0	.2756			11.5	.4528	
.3	.0118				.0730		49	4.1	.1614				.2770		J	11.51	.4531	29/64
	.0135		80	1.9	.0748			4.2	.1654		19	7.1	.2795			11.91	.4687	15/32
.35	.0138				.0760		48		.1660				.2811		K	12.0	.4724	
	.0145		79	1.95	.0767			4.25	.1673			7.14	.2812	9/32		12.30	.4843	31/64
.39	.0156	1/64		1.98	.0781	5/64		4.3	.1693			7.2	.2835			12.5	.4921	
.4	.0157				.0785		47		.1695		18	7.25	.2854			12.7	.5000	1/2
	.0160		78	2.0	.0787			4.37	.1719	11/64		7.3	.2874			13.0	.5118	
.45	.0177			2.05	.0807				.1730		17		.2900		L	13.10	.5156	33/64
	.0180		77		.0810		46	4.4	.1732			7.4	.2913			13.49	.5312	17/62
.5	.0197				.0820		45		.1770		16		.2950		M	13.5	.5315	
	.0200		76	2.1	.0827			4.5	.1771			7.5	.2953			13.89	.5469	35/64
	.0210		75	2.15	.0846				.1800		15	7.54	.2968	19/64		14.0	.5512	
.55	.0217				.0860		44	4.6	.1811			7.6	.2992			14.29	.5625	9/16
	.0225		74	2.2	.0866				.1820		14		.3020		N	14.5	.5709	
.6	.0236			2.25	.0885			4.7	.1850		13	7.7	.3031			14.68	.5781	37/64
	.0240		73		.0890		43	4.75	.1870			7.75	.3051			15.0	.5906	
	.0250		72	2.3	.0905			4.76	.1875	3/16		7.8	.3071			15.08	.5937	19/32
.65	.0256			2.35	.0925			4.8	.1890		12	7.9	.3110			15.48	.6094	39/64
	.0260		71		.0935		42		.1910		11	7.94	.3125	5/16		15.5	.6102	
.7	.0276			2.38	.0937	3/32			.1935		10	8.0	.3150			15.88	.6250	5/8
	.0280		70	2.4	.0945			4.9	.1979				.3160		O	16.0	.6299	
	.0292		69		.0960		41		.1960		9	8.1	.3189			16.27	.6406	41/64
.75	.0295			2.45	.0964			5.0	.1968			8.2	.3228			16.5	.6496	
	.0310		68		.0980		40		.1990		8		.3230		P	16.67	.6562	21/32
.79	.0312	1/32		2.5	.0984			5.1	.2008			8.25	.3248			17.0	.6693	
.8	.0315				.0995		39		.2010		7	8.3	.3268			17.06	.6719	43/64
	.0320		67		.1015		38	5.16	.2031	13/64		8.33	.3281	21/64		17.46	.6875	11/16
	.0330		66	2.6	.1024				.2040		6	8.4	.3307			17.5	.6890	
.85	.0335				.1040		37	5.2	.2047				.3320		Q	17.86	.7031	45/64
	.0350		65	2.7	.1063				.2055		5	8.5	.3346			18.0	.7087	
.9	.0354				.1065		36	5.25	.2067			8.6	.3386			18.26	.7187	23/32
	.0360		64	2.75	.1082			5.3	.2086				.3390		R	18.5	.7283	
	.0370		63	2.78	.1094	7/64			.2090		4	8.7	.3425			18.65	.7344	47/64
.95	.0374				.1100		35	5.4	.2126			8.73	.3437	11/32		19.0	.7480	
	.0380		62	2.8	.1102				.2130		3	8.75	.3445			19.05	.7500	3/4
	.0390		61		.1110		34	5.5	.2165			8.8	.3465			19.45	.7656	49/64
1.0	.0394				.1130		33	5.56	.2187	7/32			.3480		S	19.5	.7677	
	.0400		60	2.9	.1141			5.6	.2205			8.9	.3504			19.84	.7812	25/32
	.0410		59		.1160		32		.2210		2	9.0	.3543			20.0	.7874	
1.05	.0413			3.0	.1181			5.7	.2244				.3580		T	20.24	.7969	51/64
	.0420		58		.1200		31	5.75	.2263			9.1	.3583			20.5	.8071	
	.0430		57	3.1	.1220				.2280		1	9.13	.3594	23/64		20.64	.8125	13/16
1.1	.0433			3.18	.1250	1/8		5.8	.2283			9.2	.3622			21.0	.8268	
1.15	.0452			3.2	.1260			5.9	.2323			9.25	.3641			21.03	.8281	53/64
	.0465		56	3.25	.1279				.2340		A	9.3	.3661			21.43	.8437	27/32
1.19	.0469	3/64			.1285		30	5.95	.2344	15/64			.3680		U	21.5	.8465	
1.2	.0472			3.3	.1299			6.0	.2362			9.4	.3701			21.83	.8594	55/64
1.25	.0492			3.4	.1338				.2380		B	9.5	.3740			22.0	.8661	
1.3	.0512				.1360		29	6.1	.2401			9.53	.3750	3/8		22.23	.8750	7/8
	.0520		55	3.5	.1378				.2420		C		.3770		V	22.5	.8858	
1.35	.0531				.1405		28	6.2	.2441			9.6	.3780			22.62	.8906	57/64
	.0550		54	3.57	.1406	9/64		6.25	.2460		D	9.7	.3819			23.0	.9055	
1.4	.0551			3.6	.1417			6.3	.2480			9.75	.3838			23.02	.9062	29/32
1.45	.0570				.1440		27	6.35	.2500	1/4	E	9.8	.3858			23.42	.9219	59/64
1.5	.0591			3.7	.1457			6.4	.2520				.3860		W	23.5	.9252	
	.0595		53		.1470		26	6.5	.2559			9.9	.3898			23.81	.9375	15/16
1.55	.0610			3.75	.1476				.2570		F	9.92	.3906	25/64		24.0	.9449	
1.59	.0625	1/16			.1495		25	6.6	.2598			10.0	.3937			24.21	.9531	61/64
1.6	.0629			3.8	.1496				.2610		G		.3970		X	24.5	.9446	
	.0635		52		.1520		24	6.7	.2638				.4040		Y	24.61	.9687	31/32
1.65	.0649			3.9	.1535			6.75	.2657	17/64		10.32	.4062	13/32		25.0	.9843	
1.7	.0669				.1540		23	6.75	.2657				.4130		Z	25.03	.9844	63/64
	.0670		51	3.97	.1562	5/32			.2660		H	10.5	.4134			25.4	1.0000	1

Metric–English Conversion Table

Multiply	by	to get equivalent number of:	Multiply	by	to get equivalent number of:
Length			*Acceleration*		
Inch	25.4	millimeters (mm)	Foot/sec^2	0.304 8	meter/sec^2 (m/s^2)
Foot	0.304 8	meters(m)	Inch/sec^2	0.025 4	meter/sec^2
Yard	0.914 4	meters	*Torque*		
Mile	1.609	kilometers (km)	Pound-inch	0.112 98	newton-meters (N-m)
Area			Pound-foot	1.355 8	newton-meters
Inch^2	645.2	millimeters^2 (mm^2)	*Power*		
	6.45	centimeters^2 (cm^2)	Horsepower	0.746	kilowatts (kW)
Foot^2	0.092 9	meters^2 (m^2)	*Pressure or Stress*		
Yard^2	0.836 1	meters^2	Inches of water	0.249 1	kilopascals (kPa)
Volume			Pounds/sq. in.	6.895	kilopascals
Inch^3	16 387.	mm^3	*Energy or work*		
	16.387	cm^3	BTU	1 055	joules (J)
	0.016 4	liters (l)	Foot-pound	1.355 8	joules
Quart	0.946 4	liters	Kilowatt-hour	3 600 000.	joules (J = one W's)
Gallon	3.785 4	liters		or 3.6 × 10^6	
Yard^3	0.764 6	meters^3 (m^3)	*Light*		
Mass			Foot candle	1.076 4	lumens/meter^2 (lm/m^2)
Pound	0.453 6	kilograms (kg)	*Fuel Performance*		
Ton	907.18	kilograms (kg)	Miles/gal	0.425 1	kilometres/litre (km/l)
Ton	0.907	tonne (t)	Gal/mile	2.352 7	liters/kilometer (l/km)
Force			*Velocity*		
Kilogram	9.807	newtons (N)	Miles/hour	1.609 3	kilometers/hr. (km/h)
Ounce	0.278 0	newtons			
Pound	4.448	newtons			
Temperature					
Degree Fahrenheit	(†°F − 32) ÷ 1.8	degree Celsius (C)			

Glossary

AC Alternating current.

Active material The material used to coat the grids of the plates in a lead-acid battery. See *Lead peroxide* and *Sponge lead.*

Advance (spark) To change the ignition timing so that the spark occurs earlier.

Allen wrench A hex-shaped tool or bit that fits into a hex-shaped hole in the head of a bolt or screw.

Alternating current An electrical current that alternately changes polarity.

Alternator A device that converts mechanical energy to electrical energy in the form of alternating current.

Ammeter An instrument that measures current flow in amperes.

Ampere A unit of measurement of electrical current flow. With a pressure of 1 V, 1 A will flow in a circuit that has a resistance of 1 Ω.

Ampere hour capacity A measurement of a battery's ability to deliver a specified amount of current for 20 hours without cell voltage falling below 1.75 V.

Arbor press A piece of equipment used to apply pressure through leverage.

Arcing The spark formed when electricity jumps a gap. Arcing usually occurs when a circuit is broken.

Armature The rotating part in a motor or generator. The movable arm in a relay.

ATDC After top dead center.

Atom The smallest part of an element that retains all the characteristics of that element.

Available Voltage The maximum voltage produced by the ignition system. The voltage available at a load.

AWG American wire gauge.

Ball bearing An antifriction bearing that uses a series of steel balls held between inner and outer bearing races.

Ballast resistor A resistor used in a primary ignition circuit to stabilize voltage and current flow.

Battery An electrochemical device that converts electrical energy to chemical energy while charging, and converts chemical energy to electrical energy while discharging.

Bayonet base bulb A bulb with a metal base that has pins or lugs on its side that engage in locking slots in a socket.

BCI Battery Council International.

BDC Bottom dead center.

Bimetallic Made of two different metals.

Booster battery An auxiliary battery used to start the engine of a car that has a discharged battery.

Booster cables See *Jumper cables.*

Bound electrons Five or more electrons held tightly in the valence ring of an atom.

Breaker arm The movable part of a pair of ignition breaker points.

Breaker points A pair of contact points that are opened and closed by the action of a cam.

Brush An electrical conductor that contacts a commutator or a slip ring.

BTDC Before top dead center.

Cam angle See *Dwell.*

Capacitor See *Condenser.*

Car stands See *Jack stands.*

Carbon pile A variable resistance unit used to perform certain electrical tests.

Cell (battery) One unit or compartment of a battery.

Centrifugal advance A system that uses centrifugal force to advance spark timing as engine speed increases.

Centrifugal force The outward force from the center of a rotating object.

Charging rate The current flow, measured in amperes, from the generator or charger to the battery.

Circuit A connection of conductors that provides a complete path for the flow of current from a power source, through a load, and a return path to the power source.

Circuit breaker A circuit protection device that opens the circuit when current flow exceeds a predetermined amount.

Closed circuit A complete circuit.

Coil (ignition) A transformer that multiplies battery voltage to a voltage sufficient to push current across the gap of a spark plug.

Cold-cranking rating A measurement of the amount of current a battery can supply for 30 seconds at 0°F (−18°C) without the voltage of any cell dropping below 1.2 volts.

Color code Color markings used to identify wires in a circuit.

Commutator A series of metal bars or segments that are connected to the winding of an armature.

Condenser An electrical device that can absorb and store surges of current.

Conductor A material that has many free electrons, allowing the unrestricted flow of current.

Continuity A continuous, unbroken path for current flow. A closed or complete circuit.

Conventional theory (of current

flow) The theory that states that current flows from positive (+) to negative (−).

Core The iron center of an electromagnet that aligns and reinforces the magnetic lines of force.

Current The flow of electrons through a conductor.

D'Arsonval movement A movement used in meters. A current-carrying coil mounted so that it can move in a permanent magnetic field. Any change in the current flow in the coil causes a change in coil position.

DC Direct current.

Degree A unit used to measure angles. It is 1/360 of a circle. Usually abbreviated by the symbol ° placed behind a number.

Detonation The violent combustion of the fuel-air mixture in a cylinder.

Diagnosis The scientific process of determining the causes of problems.

Die (thread) A tool used to cut threads on a shaft.

Dimmer switch A two-way switch, usually SPDT, used to select the high beams or the low beams of the headlights.

Diode A solid state electronic device that allows current to flow in only one direction.

Direct current An electrical current that maintains a constant polarity.

Distributor (ignition) A rotating switching device that opens and closes the primary circuit and distributes secondary circuit voltage to the spark plugs.

Drive (starter) The movable gear and clutch assembly that enables the starter motor to turn the flywheel.

Dwell The amount of time, measured in degrees of distributor cam rotation, that the ignition points remain closed.

Dwell meter An instrument that measures dwell.

Electricity The directed movement of electrons.

Electrode A conductor used to form one side of an air gap.

Electrochemical action The chemical action that takes place between the plates and the electrolyte in a battery.

Electrolyte A mixture of sulfuric acid and water used in a lead-acid battery.

Electromagnet A nonpermanent magnet consisting of a coil of wire wrapped around a soft iron core. Magnetism is present only while current flows through the coil.

Electromotive force Electrical pressure; voltage.

Electron An atomic particle that has a negative charge.

Electron theory (of current flow) The theory that states that current flows from negative (−) to positive (+).

Electronic ignition system A system that uses solid state electronic components in the primary circuit to eliminate the need for breaker points.

Element Anything that is "pure" in that it is not combined with anything else. An element is made up of atoms.

Element (battery) A group of positive plates and a group of negative plates assembled with separators.

Energy The ability to do work.

Engine A device that converts heat energy into mechanical energy.

Exhaust stroke That stroke of a four-stroke cycle during which the burnt gases are forced from the cylinder.

Feeler gauge A thin metal strip or wire of known thickness used to measure the clearance between two parts.

Field The area of magnetic force that surrounds a magnet.

Filament The wire conductor in a bulb. The filament glows to produce light.

Firing order The sequence in which the cylinders of an engine fire.

Flasher A rapidly operating circuit breaker used to alternately make and break a circuit. Usually used in turn signal and emergency light circuits.

Flux A compound used to remove traces of surface oxidation and prevent additional oxidation during soldering.

Foot pound A unit of measurement for torque. In tightening a bolt or nut, 1 ft·lb is the torque obtained by a force of one pound applied to a wrench handle twelve inches long.

Force A pulling or pushing effort measured in pounds.

Four stroke cycle A term used to describe the operation of a particular type of internal combustion engine. The strokes indicate the action that occurs with each movement of the piston: (1) intake, (2) compression, (3) power, and (4) exhaust.

Frame The foundation of an automobile. The steel structure to which the body is attached.

Free electrons Three or fewer electrons held loosely in the valence ring of an atom.

Friction The resistance to motion between two objects in contact with each other.

Fuse An electrical safety device. A fuse will allow a limited amount of current to flow through it. When the current flow exceeds that limit, the fuse melts, or "blows," breaking the circuit.

Fuse block An assembly of fuse holders for various circuits.

Fusible link A length of fuse wire used in a circuit to protect the circuit from excess current.

Gap The air space between two electrodes or contacts.

Generator A device that converts mechanical energy to electrical energy.

Grid The meshlike framework of a plate in a lead-acid battery.

Ground A common return route in electrical circuits. The metal parts of a car are usually used as a ground to provide a return path to the negative (−) battery terminal.

Ground cable The cable connecting the battery to the engine or to the frame of a vehicle.

Growler A device used to test armatures for open and shorted windings.

Heat range (spark plug) The operating temperature range of a spark plug.

Heat sensitive switch A switch that is activated by temperature change.

Heat sink A heat-dissipating mounting for diodes and other components. A heat sink prevents those parts from damage caused by overheating.

High rate discharge test See *Load test*.

High tension wires See *Secondary wires*.

Hold-in winding The winding in a solenoid that creates the magnetic field that holds the plunger or core in position after it has been moved by the pull-in winding.

Holddown The clamp that holds a battery in its tray or mount.

Hydrometer An instrument used to measure the specific gravity of a liquid.

Idle speed The slowest specified engine operating speed.

Ignition system The system that boosts battery voltage and distributes it to each spark plug at the proper time.

Incandescent Glowing white hot. Light is produced by the incandescence of the filament in a bulb.

Induction The transfer of electricity by means of magnetism.

Infinity An ohmmeter reading that indicates an open circuit or an infinite resistance.

In-line engine An engine whose cylinders are arranged in a single row.

Insulation Any material used to prevent the flow of current or heat.

Insulator A material that has many bound electrons, restricting electron flow.

Intake stroke That stroke in a four-stroke cycle during which the fuel-air mixture enters the cylinder.

Ion An atom that has gained or lost an electron and is therefore unbalanced.

IVR (Instrument voltage regulator) A thermal device that provides a constant voltage for thermal gauges regardless of changes in system voltage.

Jack A device for raising a car.

Jack stands Pedestals used to support a car after it is raised from the floor by a jack or lift.

Jumper cables Cables used to start a car that has a discharged battery.

Jumper wires Tools used to temporarily reroute current by bridging or by-passing components for test purposes.

Lead-acid battery A battery that operates through the electrochemical action of lead, lead peroxide, and sulfuric acid.

Lead peroxide The active material used to form the positive plates of a lead-acid battery.

Lines of force The lines by which a magnetic field can be visualized.

Load Any device that converts electrical energy to another form of energy.

Load test A measurement of battery voltage taken while the battery is delivering a specified amount of current.

Lock washer A washer designed to prevent a bolt or a nut from loosening.

Lubricant Any material, usually liquid or semiliquid, that reduces friction when placed between two moving parts.

Magnetic field The area surrounding a magnet that is made up of lines of magnetic force.

Magnetic poles The points where magnetic lines of force enter and leave a magnet.

Magnetism The ability of a substance to attract iron.

Matter Anything that has weight and occupies space.

Mechanical advance See *Centrifugal advance.*

Millimeter A metric unit of measurement equal to 0.039370 inches. Usually found abbreviated as mm, as in 3 mm.

Motor A device that converts electrical energy to mechanical energy.

Neutral start (safety) switch A switch operated by the transmission selector that prevents the starter motor from operating unless the transmission is placed in Park or Neutral.

Neutron An atomic particle that is electrically neutral.

NIASE National Institute for Automotive Service Excellence.

Nucleus The center or central core of an atom. The nucleus contains the protons and neutrons of an atom.

Octane rating A measurement of the ability of a fuel to resist detonation.

Ohm A unit of measurement of resistance. A pressure of one volt is required to push a current of one ampere through a resistance of one ohm.

Ohmmeter An instrument that measures resistance in ohms.

Ohm's law Usually stated as $I = \dfrac{E}{R}$. A pressure of one volt is required to push a current of one ampere through a resistance of 1Ω.

One-way clutch See *Overrunning clutch.*

Open circuit A circuit in which a break prevents the flow of current.

Orbit The path followed by an electron around the nucleus of an atom.

Overrunning clutch A device that allows torque to be transmitted in only one direction.

Parallel circuit An electrical circuit that provides a separate path for current flow to and return from each of two or more loads.

Permanent magnet A piece of steel or alloy that acts as a magnet without the need for an electric current to create a magnetic field.

Petcock A drain valve.

Pinging The metallic knocking caused by detonation.

Plates (battery) The coated metallic grids in a battery. Positive plates are composed of lead peroxide. Negative plates are composed of sponge lead.

Point gap The distance between breaker points when they are held open by the highest part of a cam lobe.

Point resistance The resistance to current flow between breaker points.

Polarity Having poles, such as the north and south poles of a magnet, or the positive and negative terminals of a power source or of a load. Polarity determines the direction of a magnetic field or the direction of current flow.

Pole shoes The iron cores of the electromagnets that form the fields in a motor or DC generator.

Power stroke That stroke in a four-stroke cycle during which the fuel-air mixture burns.

Pre-ignition Ignition of the fuel-air mixture in a cylinder before the spark jumps the plug gap.

Pressure cap A radiator cap designed to hold in some of the pressure exerted by expanded coolant.

Pressure differential switch A hydraulically operated switch that is closed when a difference of pressure occurs in a dual braking system.

Pressure sender A variable resistor actuated by pressure change.

Pressure sensitive switch A switch actuated by a change in pressure.

Primary circuit The circuit in the

ignition system that creates the magnetic field in the coil.

Primary winding The winding in an ignition coil that uses battery current to create a magnetic field.

Primary wire Small gauge wire insulated to prevent leakage of primary or battery voltage. Wire used in the primary circuit of an ignition system.

Proton A positively charged atomic particle.

Puller A tool used to remove parts from a shaft or from a hole.

Pulley A wheel, usually with a V-shaped groove or grooves, that drives or is driven by a belt.

Pull-in winding The winding in a solenoid that pulls the plunger or core.

Reach (spark plug) The distance between the firing end of a spark plug and its seat.

Rectifier A device used to change AC to DC.

Regulator (alternator) A device used to control output voltage.

Relay An electromagnetic switch.

Required voltage The voltage necessary to force current across the gap of a spark plug.

Reserve capacity rating A battery rating based on the amount of time that a battery at 80°F (27°C) can deliver 25 A without the voltage of any cell dropping below 1.75 V.

Residual magnetism The magnetism remaining in a material after the current flow that produces the magnetism is stopped.

Resistance The ability of a conductor to restrict the flow of current.

Retard (spark) To change the ignition timing so that the spark occurs later.

Rheostat A variable resistor.

Ring gear (flywheel) The gear fitted around the circumference of a flywheel so that the starter motor drive gear can mesh with it and crank the engine.

Rotor (alternator) The rotating field

coil that creates a moving magnetic field.

Rotor (distributor) The rotating switch contact that distributes the high voltage produced in the coil to each of the spark plug wires.

rpm Revolutions per minute.

SAE Society of Automotive Engineers.

Schematic See *Wiring diagram*.

Sealed beam bulb A unitized bulb consisting of a lens, reflector, and a filament.

Sealed bearing A bearing that has been lubricated and sealed at the time of manufacture.

Secondary circuit The ignition system circuit that transmits the high voltage from the coil to the spark plugs.

Secondary wire Small gauge wire insulated to prevent the leakage of secondary or high voltage. Wire used in the secondary circuit of an ignition system.

Self-discharge Chemical action in a battery that causes it to slowly discharge.

Semiconductor A material whose atoms have four electrons in the valence ring. They are neither good conductors nor good insulators.

Separators Insulators in a battery that prevent the plates from contacting each other.

Series circuit An electrical circuit that provides only one path for the flow of current through two or more loads.

Series-parallel circuit A circuit that has some components in series with the power source and some components in parallel with each other and with the power source.

Short circuit A defect in an electrical circuit that allows current to return to the power source before passing through the load.

Shunt An electrical branch circuit or by-pass in parallel with another circuit.

Single wire system A method of using a single wire to conduct current to a load and using the metal frame and body of the car as the return conductor or ground.

Slip ring A part of an alternator rotor contacted by a brush.

Solenoid An electromagnet with a movable core.

Spark plug A device that provides a fixed air gap across which current jumps to provide a spark.

Spark timing See *Timing*.

Specific gravity The weight of a substance compared with the weight of an equal volume of water.

Specifications Measurements recommended by the manufacturer.

Splice To join together.

Stator (alternator) The stationary winding in which current is induced by a moving magnetic field.

Stroke The distance a piston travels as it moves from TDC to BDC.

Stud A headless bolt that is threaded on both ends.

Sulfated The condition of a battery when the composition of the plates has changed to lead sulfate.

Switch A control device used to open and close an electrical circuit.

Tach-dwell meter An instrument that measures both engine speed and ignition system dwell.

Tachometer An instrument that measures engine speed.

Tap (thread) A tool used to cut threads in a hole.

TDC Top dead center.

Temperature sender A variable resistor actuated by temperature change.

Temperature sensitive switch A switch actuated by a change in temperature.

Test lamp A bulb with wires attached to it so that it can be inserted in a circuit to confirm the presence of voltage.

Thermal switch See *Temperature operated switch*.

Thermistor A resistor whose resistance decreases as its temperature increases.

Timing (ignition) The timing of the firing of the spark plug in relation to the position of the piston.

Timing light A strobe light used to "stop" the motion of a crankshaft pulley or flywheel so that the alignment of the timing marks can be observed.

Timing marks Marks, usually on the crankshaft pulley and on the front of an engine, that indicate the position of the crankshaft.

Tolerance A permissible variation, usually stated as extremes of a specification.

Torque A force that tends to produce a twisting or turning motion.

Torque wrench A wrench or handle that indicates the amount of torque applied to a bolt or nut. A tool used to tighten bolts and nuts to a specific torque.

Transistor A semi-conductor device used in electronic circuits.

Vacuum A pressure less than atmospheric pressure.

Vacuum advance A system that utilizes engine vacuum to advance spark timing as engine load decreases.

Valence ring The outer ring of an atom.

Volt A unit of measurement of electrical pressure. A pressure of one volt is required to push one ampere of current through a resistance of one ohm.

Voltage The electrical pressure that causes current to flow in a circuit.

Voltage drop The loss of electrical pressure as it pushes current through resistance.

Voltmeter An instrument that measures electrical pressure in volts.

Watt A unit of measurement of electrical power. Volts times amperes equals watts.

Watts rating A battery rating obtained by multiplying the current flow from a battery by the battery voltage at 0°F (−18°C).

Wedge bulb A bulb that has no metal base.

Wire gauge Wire size numbers based on the cross section area of the conductor.

Wiring diagram A drawing of the components and conductors in a circuit.

Working device See *Load*.

Answer Key
with Text References

Chapter 1
1 D page 2
2 B page 3
3 B page 4
4 C page 5
5 D page 6
6 C page 6
7 C page 15–18
8 A page 17–18
9 B page 21
10 D page 21

Chapter 2
1 A page 27
2 D page 32
3 B page 33
4 A page 34
5 A page 35
6 C page 35
7 C page 38
8 D page 40
9 B page 42
10 B page 47–48

Chapter 3
1 A page 55
2 C page 55
3 A page 56
4 C page 57–59
5 C page 66
6 A page 69
7 D page 69
8 B page 71
9 A page 73
10 D page 74

Chapter 4
1 C page 78
2 D page 78
3 C page 79–80
4 B page 80
5 A page 83
6 D page 89
7 A page 93
8 B page 84–86
9 A page 86
10 C page 97–98

Chapter 5
1 D page 103
2 A page 104
3 D page 105
4 C page 105
5 B page 105
6 D page 107
7 C page 108
8 B page 109
9 A page 113
10 D page 112

Chapter 6
1 A page 125
2 A page 125
3 C page 135–136
4 C page 136
5 C page 140
6 C page 142
7 A page 146
8 D page 146
9 B page 151
10 C page 153

Chapter 7
1 A page 157
2 C page 157
3 B page 157
4 A page 160
5 D page 165
6 C page 169–173
7 D page 174
8 C page 174–176
9 B page 174
10 C page 179

Chapter 8
1 C page 189
2 A page 188–190
3 A page 192
4 C page 194
5 C page 197
6 B page 197
7 B page 200
8 B page 203
9 D page 203
10 B page 216

Chapter 9
1 C page 222
2 D page 222
3 A page 225
4 D page 225
5 A page 226–227
6 D page 227
7 B page 232–233
8 B page 238–239
9 D page 241
10 A page 249

Chapter 10
1 C page 262
2 D page 262
3 C page 262
4 C page 262
5 A page 263
6 C page 263–265
7 B page 268
8 A page 270
9 B page 272
10 D page 272

Chapter 11
1 D page 282
2 C page 282
3 C page 284–285
4 A page 285
5 A page 292
6 C page 293
7 A page 293–294
8 D page 293
9 D page 294–295
10 A page 294

Chapter 12
1 D page 315
2 C page 316
3 A page 322
4 A page 323
5 B page 323
6 D page 324
7 A page 325
8 A page 325
9 B page 329
10 B page 329

Chapter 13

1 A page 336
2 D page 338
3 C page 339
4 B page 342
5 C page 348
6 D page 348
7 C page 348
8 D page 350
9 D page 358
10 D page 359

Chapter 14

1 C page 366
2 A page 365
3 C page 365
4 B page 366–369
5 B page 371
6 B page 377
7 C page 386
8 B page 389
9 A page 389
10 C page 391

Chapter 15

1 D page 398
2 C page 398
3 C page 399
4 C page 402
5 C page 407
6 B page 413–417
7 B page 417
8 C page 420–422
9 B page 422
10 B page 422

Index